ROOF PLUMBING

Owen Smith

ROOF PLUMBING

PLUMBING SKILLS Series

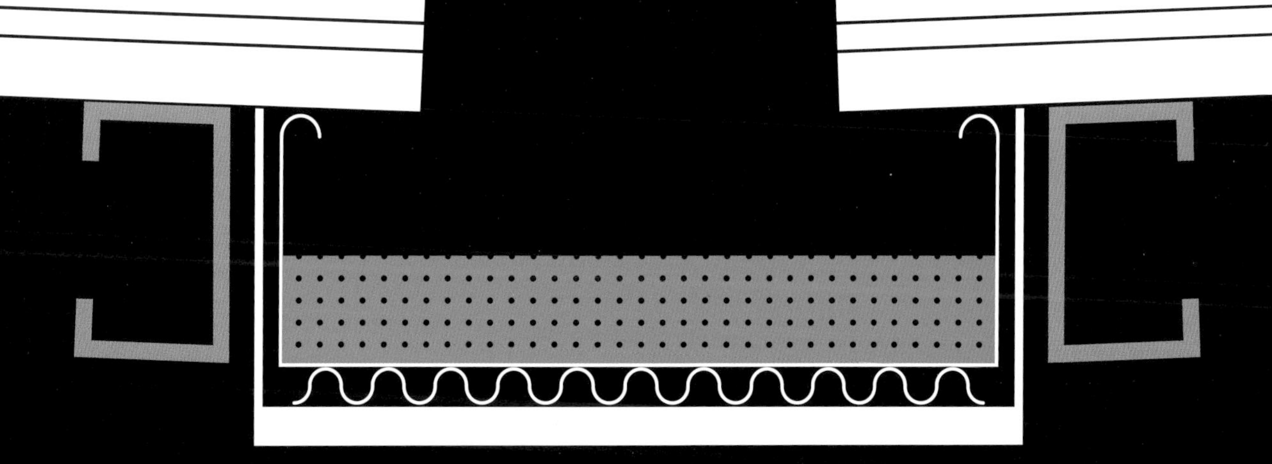

3e

Roof Plumbing
3rd Edition
Owen Smith

Head of content management: Dorothy Chiu
Content manager: Sandy Jayadev/Chee Ng
Content developer: Sarah Payne
Project editor: Raymond Williams
Editor: Julie Wicks
Proofreader: James Anderson
Cover design: Justin Lim / Linda Davidson
Text designer: Linda Davidson
Permissions/Photo researcher: Wendy Duncan
Indexer: Max McMaster
Typeset by KnowledgeWorks Global Ltd

second edition published by Cengage in 2018.

For product information and technology assistance,
in Australia call 1300 790 853;
in New Zealand call 0800 449 725

For permission to use material from this text or product, please email
aust.permissions@cengage.com

National Library of Australia Cataloguing-in-Publication Data
ISBN: 9780170439251
A catalogue record for this book is available from the National Library of Australia.

Cengage Learning Australia
Level 7, 80 Dorcas Street
South Melbourne, Victoria Australia 3205

Cengage Learning New Zealand
Unit 4B Rosedale Office Park
331 Rosedale Road, Albany, North Shore 0632, NZ

For learning solutions, visit cengage.com.au

Printed in China by 1010 Printing International Limited.
5 6 7 25 24

BRIEF CONTENTS

CONTENTS

Guide to the text

As you read this text you will find a number of features in every chapter to enhance your study of Plumbing and help you understand how the theory is applied in the real world.

CHAPTER OPENING FEATURES

The **Chapter overview** lists the topics that are covered in the chapter.

10 FABRICATE AND INSTALL ROOF DRAINAGE COMPONENTS

Roof drainage components are those parts of the roof system that collect and direct water from the roof to an approved point of discharge. Such components include gutters, rainwater heads, sumps and downpipes. Many drainage components require fabrication to suit the individual requirements of each job and you will need to develop the knowledge and hand skills necessary to create watertight joints and products to a high standard.

Before working through this section, it is absolutely important that you complete or revise each of the chapters in Part A Roof Plumbing Fundamentals.

Overview

In this chapter you will be introduced to many of the most common roof drainage components, gain an understanding of their purpose and be able to identify them by sight. You will then have the opportunity to work through step-by-step fabrication processes with advice and guidance from your teacher and supervisor.

FEATURES WITHIN CHAPTERS

Engage actively with the learning by completing the practical activities in the **NEW Learning task** boxes.

LEARNING TASK 1.1

Work with your teacher or supervisor to determine the following important requirements.

1 Are roof plumbing tradespeople required to be licensed in your State or Territory?
2 Where licensing is required, what government department administers this requirement?
3 Are there any mandatory training requirements for roof plumbing licenses in your State or Territory?

From experience boxes explain the responsibilities of employees, including skills they need to acquire and real-life challenges they may face at work to enhance employability skills on the job site.

FROM EXPERIENCE

On new projects or larger commercial installations, the plans and specifications will provide all the basic product selection and design information for the job. However, this is not the case for most re-roof and maintenance jobs. It is often the roof plumber alone who must advise the customer and select fit-for-purpose products for each situation. You must get in the habit of regularly researching all roofing products and materials.

FEATURES WITHIN CHAPTERS

The **Know your code** icons highlight where Plumbing Standards are addressed, to strengthen knowledge and hone research skills.

Caution boxes highlight important advice on safe work practices for plumbers by identifying safety issues and providing urgent safety reminders

If using a powered blower to remove swarf, ensure that other workers and the public are not exposed to any flying metal and other debris.

Green tip boxes highlight the applications of sustainable technology, materials or products relevant to plumbers and the plumbing industry.

GREEN TIP

In order to meet the energy efficiency provisions set out by the National Construction Code, designers and builders select very specific insulation solutions. These products will only work to standard if they are installed correctly. As such, the modern roof plumber must know as much about roofing insulation materials

How to boxes highlight a theoretical or practical task with step-by-step walkthroughs.

HOW TO

STEPS TO DETERMINE MARK-OUT ALLOWANCE

Step 1	Using a bevel, measure the angle that you wish to make. For example, this could be an existing downpipe angle or the gutter angles around a bay window.
Step 2	Transfer the bevel angle to a piece of paper, then sketch the object to the width required. Width in this instance is shown as 'x'.

Learn how to complete mathematical calculations with **Example** boxes that provide worked equations.

EXAMPLE 10.10

Calculation of gradient (ratio)

You have been asked to install a new box gutter on an old factory roof. You know that the roofing code requires that box gutters must have a minimum fall of 1:200.

Using an automatic builder's level, you determine that the fall in the existing gutter is 100 mm over 30 m, from the highest point to the lowest.

What is the existing grade of the gutter support? Are you able to lay your new box gutter on these existing sole boards?

$$G\,(\text{ratio}) = \frac{L(\text{m})}{F(\text{m})} = \frac{30}{0.1} = 300 = 1{:}300$$

A fall of 1:300 is flatter than the minimum permissible in the roofing code and therefore you will have to modify or replace the existing gutter support system in order to comply.

END-OF-CHAPTER FEATURES

At the end of each chapter you will find several tools to help you to review, practise and extend your knowledge of the key learning objectives.

Review your understanding of the key chapter topics with the **NEW Summary**.

Worksheets give you the opportunity to test your knowledge and consolidate your understanding of the chapter competencies.

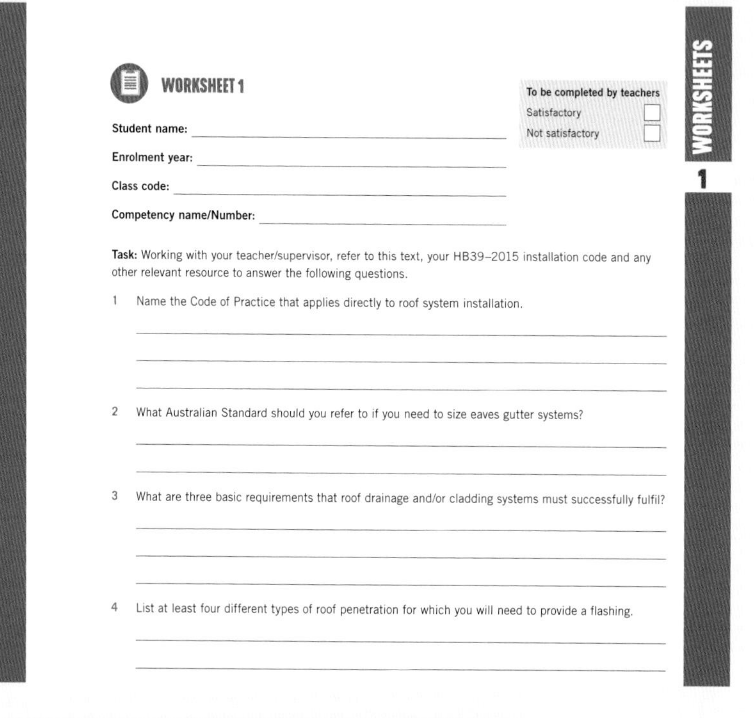

NEW

PART A 6

SUMMARY

Knowledge of fasteners is vital to your work as a roof plumber. In this chapter you have been introduced to:

- fastener drive types
- fastener point types
- corrosion coatings
- thread types
- how fasteners are classified and measured.

You also covered the requirement to use only neutral cure sealants on all rainwater drainage products.

WORKSHEETS 1

WORKSHEET 1

To be completed by teachers	
Satisfactory	☐
Not satisfactory	☐

Student name: ____________

Enrolment year: ____________

Class code: ____________

Competency name/Number: ____________

Task: Working with your teacher/supervisor, refer to this text, your HB39–2015 installation code and any other relevant resource to answer the following questions.

1 Name the Code of Practice that applies directly to roof system installation.

2 What Australian Standard should you refer to if you need to size eaves gutter systems?

3 What are three basic requirements that roof drainage and/or cladding systems must successfully fulfil?

4 List at least four different types of roof penetration for which you will need to provide a flashing.

Worksheet icons indicate in the text when a student should complete an end-of-chapter worksheet.

Guide to the online resources

FOR THE INSTRUCTOR

Cengage is pleased to provide you with a selection of resources that will help you to prepare your lectures and assessments, when you choose this textbook for your course.
Log in or request an account to access instructor resources at au.cengage.com/instructor/account for Australia or nz.cengage.com/instructor/account for New Zealand.

MINDTAP

Premium online teaching and learning tools are available on the *MindTap* platform - the personalised eLearning solution.

MindTap is a flexible and easy-to-use platform that helps build student confidence and gives you a clear picture of their progress. We partner with you to ease the transition to digital – we're with you every step of the way.

The *Cengage Mobile App* puts your course directly into students' hands with course materials available on their smartphone or tablet. Students can read on the go, complete practice quizzes or participate in interactive real-time activities. *MindTap* is full of innovative resources to support critical thinking, and help your students move from memorisation to mastery!

The *Series MindTap for Plumbing* is a premium purchasable eLearning tool. Contact your Cengage learning consultant to find out how *MindTap* can transform your course.

SOLUTIONS MANUAL

The **Solutions manual** provides detailed answers to every question in the text.

MAPPING GRID

The **Mapping grid** is a simple grid that shows how the content of this book relates to the units of competency needed to complete the Certificate III in Plumbing.

WEBLINKS

References from the text are provided as easy-to-access **weblinks** and can be used in presentations or for further reading for students.

INSTRUCTORS' CHOICE RESOURCE PACK

These resources save you time and are a convenient way to add more depth to your classes, covering additional content and with an exclusive selection of engaging features aligned with the text. The Instructor Resource Pack is included for institutional adoptions of this text when certain conditions are met. The pack is available to purchase for course-level adoptions of the text or as a standalone resource. Contact your Cengage learning consultant for more information.

COGNERO® TEST BANK

A **bank of questions** has been developed in conjunction with the text for creating quizzes, tests and exams for your students. Create multiple test versions in an instant and deliver tests from your LMS, your classroom, or wherever you want using Cognero. Cognero test generator is a flexible online system that allows you to import, edit, and manipulate content from the text's test bank or elsewhere, including your own favourite test questions.

POWERPOINT™ PRESENTATIONS

Cengage **PowerPoint lecture slides** are a convenient way to add more depth to your lectures, covering additional content and offering an exclusive selection of engaging features aligned with the textbook, including teaching notes with mapping, activities, and tables, photos, and artwork.

ONLINE CHAPTER

This **online-only chapter** provides additional coverage of the Plumbing qualification.

ARTWORK FROM THE TEXT

Add the **digital files** of graphs, tables, pictures and flow charts into your learning management system, use them in student handouts, or copy them into your lecture presentations.

FOR THE STUDENT

MINDTAP

MindTap is the next-level online learning tool that helps you get better grades!

MindTap gives you the resources you need to study – all in one place and available when you need them. In the *MindTap Reader*, you can make notes, highlight text and even find a definition directly from the page.

If your instructor has chosen MindTap for your subject this semester, log in to MindTap to:

- Get better grades
- Save time and get organised
- Connect with your instructor and peers
- Study when and where you want, online and mobile
- Complete assessment tasks as set by your instructor

When your instructor creates a course using *MindTap*, they will let you know your course link so you can access the content. Please purchase *MindTap* only when directed by your instructor. Course length is set by your instructor.

PREFACE

As a roof plumber, you are about to commence your training and skills development in one of the key sectors of the construction industry. In fact, there is little that is built in Australia that does not need the expertise of a roofing professional in one way or another. If you care about the quality of your work, can learn to solve problems on the run and like the challenge of a fast-paced and ever-changing occupation, then you will find that your skills will be in high demand.

Roof plumbers come from various employment backgrounds, depending on the regulatory requirements and past employment traditions in each state or territory. You may be employed as a plumbing apprentice or be learning roofing as part of your broader plumbing trade, and your employer may undertake new roof installations or perhaps concentrate on maintenance and repairs. Alternatively, you might be employed as a trainee or apprentice in a specialist roofing company tendering for major projects. Whether it is the roofing of a commercial warehouse, cladding of an industrial production plant or a single residential re-roof, roof plumbers are found in all areas of the construction sector.

Australia's roofing industry is diverse. However, there is one aspect of this work that should unite all successful roof plumbers and can be summed up in one phrase:

Performance and quality!

Consider your responsibilities – a leaking roof or overflowing gutter can cause thousands of dollars' worth of damage to the building's structure, foundations, and internal fittings and furniture; a roof must remain watertight and not blow off in high winds; your rainwater goods must drain the water efficiently to an approved point of discharge; and the wall cladding and window flashing must be impervious to wind-driven rain.

To achieve these goals, you must adhere to all relevant codes, Australian Standards and accepted trade practice, developing along the way the knowledge and high-level hand skills that will enable your work to not only perform well but look good too. Cutting corners is not an option in the roofing trade.

About this text

Roof plumbing is very much an outdoor, hands-on trade. You cannot become a good roof plumber by simply reading a textbook. You must have good instruction on the job and from your teachers and, perhaps most importantly, have the personal drive and dedication to work towards improving your skills. However, this text can guide you as a reference in the classroom or on the job to assist you in this goal.

The intention of this text is to provide an easy-to-read, instructional guide to basic roof plumbing installation. It does not seek to replicate the codes and standards that apply to roof plumbing, but it will call on these where relevant to interpret and explain a process or requirement. The text also makes use of numerous cross-section diagrams, perspective drawings and photographs that provide a visual link to the ideas contained within the chapter itself.

How is the text structured?

As an apprentice or trainee, you will be working towards a plumbing qualification at the Certificate III level. A Certificate III qualification is made up from numerous units of competency that are designed around actual job tasks. When both the practical and theoretical sections of all units have been successfully completed, you will have earned the full Certificate III qualification.

The text is structured with four main features:

1. **Roofing fundamentals**. These are many fundamental principles and knowledge requirements that apply to almost every unit of competency. These include having knowledge of roofing

structures, corrosion and electrolysis, expansion and contraction, basic safety and capillary action. Rather than repeat such items for each unit, these subject areas have been isolated as sub-sections within Part A 'Roof plumbing fundamentals'. As you work through the text, refer to this section as and when you need.

2. **Units of competency.** Chapters 10 to 18 are titled the same as each unit of competency within the Certificate III qualification. The textbook will isolate key aspects of each unit with a view to assisting in your understanding of the requirements of each job task. The chapters have been sequenced as close to the actual job progression as possible, but are not dependent on each other. This allows your teacher to alter your learning and assessment progression as needed. Of course, remember that roof plumbing is a 'hands-on' trade and there are therefore many things you will have to learn over time that cannot be included in a textbook. Nevertheless, the basic principles covered here will give you a great start!
3. **Worksheets.** As you read and complete each section of the text, you will come across a direction asking you to turn to a worksheet at the end of the chapter. These worksheets consist of questions or tasks that enable you to consolidate your understanding of what you have just worked through. Worksheets can also be used for homework and revision purposes prior to your assessments. Of course, this is at the discretion of your teacher.

What is the scope of this text?

In the context of what is generally defined as roof plumbing in Australia, the scope of this text covers the installation of roof sheeting, cladding and water drainage systems such as flashings, gutters and downpipes, and connection of these systems to an approved point of discharge.

Generally, this description is based on the use of various metallic roofing products, but it also includes non-metallic complementary components such as translucent sheeting, polymer flashings and plastic rainwater goods.

This text does not cover the installation of masonry/terracotta tile products. Roof tiling is classed as a separate trade area to general roof plumbing, though a roof plumber is still required to install flashings and rainwater drainage to such roofs. There is also no reference here to the low-pitch membrane roofing often used for flat-roof domestic and commercial high-rise buildings.

It is anticipated that, as a roof plumbing apprentice or trainee, you have already commenced many of the general plumbing units that make up a full qualification. Therefore, your proficiency in trade-level literacy, numeracy and standard plumbing terminology is assumed. Of course, roof plumbing-specific calculations and terminology are explained and defined as required. However, always ask your teacher or supervisor if something is not clear!

Changes in this 3rd edition

In this edition of *Basic Plumbing Services Skills: Roof Plumbing*, teachers and supervisors should be aware of the following core changes:

- Upgrade of content and references to complement the following industry reference documents:
  - SA HB39:2015 'Installation Code for Metal Roof and Wall Cladding'
  - AS/NZS 3500.3:2021 'Stormwater Drainage'
  - AS 3959:2018 'Construction of buildings in bushfire prone areas'.
- Numerous changes to images and drawings.
- Additional content relating to the selection of materials and components required to comply with construction plans that must comply with a Bushfire Attack Level (BAL) rating.

The author has made every effort to ensure that this text accurately reflects current roof plumbing trade practice and regulatory compliance. However, Codes and Standards may change at any time during the life of this edition, and as such teachers and supervisors must ensure students are made aware of any differences to the contents of this book.

Owen Smith

ABOUT THE AUTHOR

Owen Smith is a fully licensed plumber and gas fitter, a former plumbing contractor and since 2001 has been both a teacher and departmental manager in the TAFE sector. He holds a Bachelor degree in Adult and Vocational Education, was a foundation member of the National Plumbing and Services Training Advisory Group (NPSTAG) and is currently teaching full time in the area of Hydraulic Services Design with TasTAFE.

Owen is also the author of another Cengage text, *Basic Plumbing Services Skills: Gas Services*.

ACKNOWLEDGEMENTS

I would like to offer my particular thanks to BlueScope Steel Limited, BlueScope Lysaght® and Buildex®, who generously provided access to their installation guides, product information and sales brochures for use in the development of this text. Such assistance was invaluable and demonstrates each company's clear commitment to training and the future of the roofing industry.

Thanks and appreciation must also go to Heath Bussell of the Queensland Building and Construction Commission and Martin Richardson from TasTAFE, for the numerous new and high quality photographs included in this edition.

Lastly, a note of thanks to my father, the late Bill Smith, a proud plumber who, having started work in Wollongong, NSW in 1947, was 'on the tools' both there and in Queensland for the following 50 years. It was he who taught me from an early age the fundamental principles that will always apply in the trade, regardless of inevitable changes in materials and techniques. I am forever in his debt.

LYSAGHT®, ZINCALUME®, COLORBOND®, BlueScope and ® colour names and ® product names are registered trademarks and ™ colour names are trademarks of BlueScope Steel Limited.

The LYSAGHT® range of products is exclusively made by BlueScope Steel Limited trading as BlueScope Lysaght®.

From the publisher

The author and Cengage Learning would like to thank the following reviewers for their incisive and helpful feedback:

- Ben Freer – Central Regional TAFE
- Greg Mazey – TasTAFE
- Darren Schiavello – Victoria University

UNIT CONVERSION TABLE

TABLE 1 Length units

Millimetres	Metres	Inches	Feet	Yards
mm	*m*	*in*	*ft*	*yd*
1	0.001	0.03937	0.003281	0.001094
1000	1	39.37008	3.28084	1.093613
25.4	0.0254	1	0.083333	0.027778
304.8	0.3048	12	1	0.333333
914.4	0.9144	36	3	1

TABLE 2 Area units

Millimetre square	Metre square	Inch square	Yard square
mm^2	m^2	in^2	yd^2
1	0.000001	0.00155	0.000001
1000000	1	1550.003	1.19599
645.16	0.000645	1	0.000772
836127	0.836127	1296	1

TABLE 3 Volume units

Metre cube	Litre	Inch cube	Foot cube
m^3	*L*	in^3	ft^3
1	1000	61024	35
0.001	1	61	0.035
0.000016	0.016387	1	0.000579
0.028317	28.31685	1728	1

TABLE 4 Mass units

Grams	Kilograms	Pounds	Ounces
g	*kg*	*lb*	*oz*
1	0.001	0.002205	0.035273
1000	1	2.204586	35.27337
453.6	0.4536	1	16
28	0.02835	0.0625	1

TABLE 5 Volumetric liquid flow units

Litre/second	Litre/minute	Metre cube/hour	Foot cube/minute	Foot cube/hour
L/sec	*L/min*	m^3/hr	ft^3/min	ft^3/hr
1	60	3.6	2.119093	127.1197
0.016666	1	0.06	0.035317	2.118577
0.277778	16.6667	1	0.588637	35.31102
0.4719	28.31513	1.69884	1	60
0.007867	0.472015	0.02832	0.01667	1
0.06309	3.785551	0.227124	0.133694	8.019983

TABLE 6 High pressure units

Bar *bar*	Pound/square inch *psi*	Kilopascal *kPa*	Megapascal *mPa*	Kilogram force/ centimetre square *kgf/cm²*	Millimetre of mercury mm *Hg*	Atmospheres *atm*
1	14.50326	100	0.1	1.01968	750.0188	0.987167
0.06895	1	6.895	0.006895	0.070307	51.71379	0.068065
0.01	0.1450	1	0.001	0.01020	7.5002	0.00987
10	145.03	1000	1	10.197	7500.2	9.8717
0.9807	14.22335	98.07	0.09807	1	735.5434	0.968115
0.001333	0.019337	0.13333	0.000133	0.00136	1	0.001316
1.013	14.69181	101.3	0.1013	1.032936	759.769	1

TABLE 7 Temperature conversion formulas

Degree Celsius (°C)	(°F – 32) × 0.56
Degree Fahrenheit (°F)	(°C × 1.8) + 32

TABLE 8 Low pressure units

Metre of water	Foot of water	Centimetre of mercury	Inches of mercury	Inches of water	Pascal
mH₂O	*ftH₂O*	*cmHg*	*inHg*	*inH₂O*	*Pa*
1	3.280696	7.356339	2.896043	39.36572	9806
0.304813	1	2.242311	0.882753	11.9992	2989
0.135937	0.445969	1	0.39368	5.351265	1333
0.345299	1.13282	2.540135	1	13.59293	3386
0.025403	0.083339	0.186872	0.073568	1	249.1
0.000102	0.000335	0.00075	0.000295	0.004014	1

LIST OF TABLES AND FIGURES

List of tables

List of figures

COLOUR PALETTE FOR TECHNICAL DRAWINGS

Colour name	Colour	Material
Light Chrome Yellow		Cut end of sawn timber
Chrome Yellow		Timber (rough sawn), Timber stud
Cadmium Orange		Granite, Natural stones
Yellow Ochre		Fill sand, Brass, Particle board, Highly moisture resistant particle board (Particle board HMR), Timber boards
Burnt Sienna		Timber – Dressed All Round (DAR), Plywood
Vermilion Red		Copper pipe
Indian Red		Silicone sealant
Light Red		Brickwork
Cadmium Red		Roof tiles
Crimson Lake		Wall and floor tiles
Very Light Mauve		Plaster, Closed cell foam
Mauve		Marble, Fibrous plasters
Very Light Violet Cake		Fibreglass
Violet Cake		Plastic
Cerulean Blue		Insulation
Cobalt Blue		Glass, Water, Liquids
Paynes Grey		Hard plaster, Plaster board
Prussian Blue		Metal, Steel, Galvanised iron, Lead flashing
Lime Green		Fibrous cement sheets
Terra Verte		Cement render, Mortar
Olive Green		Concrete block
Emerald Green		Terrazo and artificial stones
Hookers Green Light		Grass
Hookers Green Deep		Concrete
Raw Umber		Fill
Sepia		Earth
Vandyke Brown		Rock, Cut stone and masonry, Hardboard
Very Light Raw Umber		Medium Density Fibreboard (MDF), Veneered MDF
Very Light Van Dyke Brown		Timber mouldings
Light Shaded Grey		Aluminium
Neutral Tint		Bituminous products, Chrome plate, Alcore
Shaded Grey		Tungsten, Tool steel, High-speed steel
Black		Polyurethane, Rubber, Carpet
White		PVC pipe, Electrical wire, Vapour barrier, Waterproof membrane

PART A

ROOF PLUMBING FUNDAMENTALS

Part A of this text will assist you in developing a stronger understanding of the fundamental knowledge and basic principles that you will require to effectively undertake roofing installations. Work through this section in partnership with your teacher, supervisor and peers in preparation for installation practice.

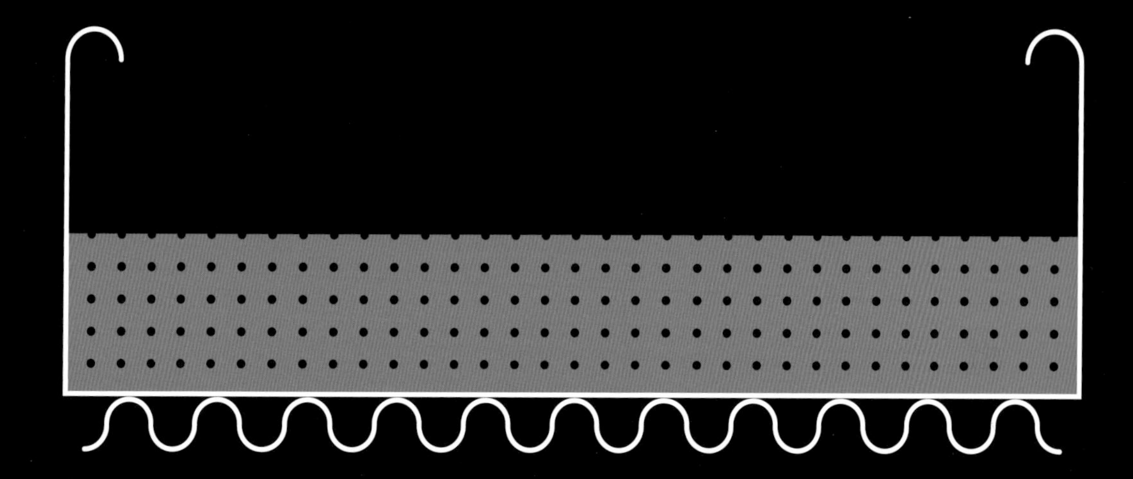

1 OVERVIEW OF ROOF PLUMBING

This chapter will provide you a broad overview of the industry and what a roof plumber actually does. You may find that there is more to roof plumbing than you had at first thought, and it is certainly one field of the plumbing trade in which the visual quality of your work is often on display for all to see and judge. Your reputation in the trade and community depends upon roofing work that demonstrates the following attributes:

- Your work should be completed to the highest standard of quality and professional competence. Quality roof work is based upon the development of high-level hand skills, extensive trade knowledge and a personal commitment to maintaining professional standards. Proficiency can take time to achieve, but it should always be your goal as a roofing professional.
- Your work should always comply with the relevant Codes of Practice and Australian Standards. Even in areas where licences are not required for roof plumbers, all building work must by law still comply with the National Construction Code and Plumbing Code of Australia. From the outset, you should determine to use these codes and standards as the primary guide when completing your installations. Together with manufacturer's technical literature, adherence to such codes and standards will ensure that your roofing systems perform according to design requirements.

LEARNING TASK 1.1

Work with your teacher or supervisor to determine the following important requirements.

1. Are roof plumbing tradespeople required to be licensed in your state or territory?
2. Where licensing is required, what government department administers this requirement?
3. Are there any mandatory training requirements for roof plumbing licences in your state or territory?

Codes and standards

General plumbing is a licensed trade in all areas of Australia, and roofing specialists are also licensed in a number of jurisdictions. Plumbers play a key role in maintaining high levels of public health and safety through the provision of safe drinking water, effective effluent disposal and the control of stormwater. Without a strongly regulated plumbing sector working in accordance with codes and standards, diseases such as typhoid, cholera, malaria and dysentery would develop very quickly.

The control, discharge and storage of stormwater is vital to the high standards of health we enjoy in Australia, and as a roof plumber you will play a major role in ensuring that this continues. To do so, you will be required to refer to and carry out your work in reference to certain *Codes of Practice* and *Australian Standards* (see Figure 1.1). These documents should be regarded as essential tools of trade, as they prescribe the minimum acceptable standards of installation required to ensure satisfactory system performance.

FROM EXPERIENCE

A key attribute that you must develop at the start of your roof plumbing career is to develop the consistent habit of referring to codes, standards and manufacturers' recommendations for every aspect of your work.

FIGURE 1.1 Codes and standards are essential tools

Installation according to codes and standards

Prior to working through this part of the text, it is appropriate that you gain a further understanding of the relationship between codes and standards and how they impact upon your installation practice.

Each state and territory has different regulatory requirements that cover roof plumbing. Therefore, the description below can only be regarded as indicative and you must discuss this further with your teacher and/or supervisor for specific requirements in your area.

The National Construction Code (NCC)

The NCC is a document that provides descriptions of what are known as Performance Requirements for all aspects of building and plumbing. These describe how any system is meant to perform. It then details options as to how industry can meet these requirements in their installations.

Most states and territories, through a range of different legal mechanisms, have largely adopted the NCC as the key document to which all industry activity must refer. The NCC is made up of three primary documents that cover commercial and multistorey construction, residential construction and plumbing.

FIGURE 1.2 Basic relationship between different parts of the NCC

The performance requirements within each of these volumes can only be satisfied by using one of the following options:

a Performance solution (formerly known as an 'Alternative solution')
b Deemed-to-satisfy solution (DTS)
c A combination of a and b.

Volumes 1–3 of the NCC also contain some specific installation descriptions known as Acceptable Construction Practices.

Deemed-to-satisfy solutions (DTS)

For most plumbing and roofing work it is DTS solutions that are the most straightforward for plumbers to apply in their daily work. In most cases, having detailed the performance requirements of an installation, the NCC will then nominate an Australian Standard to be used as a means of satisfying those same performance requirements. This is known as the DTS solution. As long as a plumber follows the Standard, then the job is 'deemed-to-satisfy' and no other verification is required.

In respect to general roof plumbing and depending on where you work in Australia, you may find that the DTS solution to enable compliant installation and design of roof drainage systems could be one or both of the following as called up within both Volume II and III of the NCC:

- NCC Volume Two – Building Code of Australia
  - Part 3.5.1 Sheet Roofing (largely reflects content within HB39:2015)
  - Part 3.5.3 Gutter and Downpipes (focused mainly on roof drainage system sizing)
- NCC Volume Three – Plumbing Code of Australia
  - AS/NZS 3500.3 Stormwater Drainage

Your teacher and/or supervisor will be able to assist in answering the following two questions:

LEARNING TASK 1.2

1 Does your state/territory require roof and gutter installations to be carried out in accordance with any part of the NCC?
2 Does your state/territory specify any particular DTS solution/s from the NCC to satisfy roof stormwater performance requirements in both residential and commercial construction?

Due to the variability among states and territories, there is no universal answer as to which installation and design standards should apply to roof plumbing. As such, this text will continue to refer to the following most commonly used primary documents for roof and gutter installation requirements:

- SA HB39:2015 'Installation code for metal roof and wall cladding'.

HB39:2015 SECTION 1.1 'SCOPE'

Furthermore, if you are to learn how to size eaves and box gutter systems later on in your training you will also need to refer to the following, wherever applicable to your region or circumstances:

- AS/NZS 3500.3 'Stormwater drainage'
- Plus any relevant statutory state or territory requirements.

AS/NZS 3500.3 'STORMWATER DRAINAGE'

The scope of the roof plumber's work

Roof plumbing is a key part of the plumbing sector and a significant trade within the broader construction industry. As a whole, the scope of a roof plumber's work includes the fabrication, installation, replacement, repair, alteration, maintenance, testing and commissioning of any roof covering or roof flashing and any part of a roof drainage system involved in the collection or disposal of stormwater, which connects to ground level.

More specifically, in this part of your training as a general plumber or roofing specialist, your primary focus will be on the installation, repair and maintenance of *metallic roof drainage systems*. However, the modern Australian roof plumber can be involved in a very wide range of work beyond that of traditional roof installations. Your training and scope of work also extends to include the use of non-metallic ancillary components, metal flashings for non-metallic roofs, and the metallic cladding/flashing of exterior and interior walls and ceilings.

Almost every structure that is built must have an effective roof and cladding system and, as such, must fulfil some basic requirements:

- The roof/cladding system must protect the occupants from rain, wind, heat and cold.
- The roof/cladding system must protect the building and its contents from water and wind damage.
- The roof system must collect and drain water away to an approved point of discharge and/or storage.

To understand more clearly how a roof and cladding system can achieve these requirements, we can break the system down into its main component parts:

- gutters and downpipes
- flashings
- roof sheets and decking
- the flashing of roof penetrations
- cladding
- rainwater storage.

The following brief descriptions are to assist you in gaining a better understanding of the scope of work that a roof plumber may undertake. Of course, this list is only a guide, and the components will vary according to the type of work you are involved with. As you read through each section, refer back to your HB39 installation code so that you are able to locate the various parts of a roof drainage system (see Figure 1.3).

- Eaves gutter
- Box gutter
- Flashing
- Rainhead
- Ridge capping
- Soaker flashing
- Sump
- Downpipe

Use your HB39:2015 to confirm the definition of these and other common roof plumbing terms. Refer to Figure 1.3 'Common Terms' and Section 1.3 'Terms and Definitions'.

FIGURE 1.3 Common roof terms shown in the HB39 installation code

Gutter and downpipes

The water that runs off the roof is collected in a gutter. Types include eaves gutters (also known as 'spouting' in some areas), concealed gutters, valley gutters and box gutters. A gutter may be external to the building (eaves gutter) or it may be within the building line itself (valley and box gutters). The gutter must be installed with a fall, as this ensures that the collected water is directed towards a downpipe outlet, rainhead or sump (see Figure 1.4).

FIGURE 1.4 An eaves gutter and downpipe

If a gutter or downpipe were to become blocked, water must be prevented from flowing back into the building. To ensure this does not happen, specific overflow measures and devices are also incorporated in the roof drainage system.

The downpipes carry the water from the roof to an approved point of discharge or storage. Gutters and downpipes ensure that roof water does not erode building foundations or collect in pools that may harbour disease. While in certain tropical areas of Australia that experience extremely high downpours during the wet season it is not uncommon for buildings to be built without gutters and downpipes, in all other regions gutters and downpipes *must* be fitted.

The gutters, downpipes and overflow measures/devices form a complete roof drainage system that must be sized, designed, installed and maintained to dispose of stormwater flows caused by extreme rainfall events.

As a roof plumber, to install the gutter and downpipes you will complete the following steps:

1. On a new house you will normally install the metal fascia system to the ends of the roof trusses. This metal fascia will support the eaves lining and the gutter itself. Some houses may have a timber fascia installed by the builder. You will fix gutter brackets to this timber fascia using screws or nails. In some northern areas, spike brackets may be used.
2. Depending upon the fascia/gutter system used, you will install gutter brackets, ensuring that the gutter will fall towards the downpipes. Again depending upon the system, this may be done before or at the same time as the gutter is installed.
3. You will then fabricate gutter angles (mitres) and stop-ends where necessary and install the gutter.
4. Any valley and box gutters must be installed before the roof sheets are installed.
5. Box gutter systems will require the installation of rainwater heads and/or sumps and associated overflow solutions.
6. You will usually fabricate and install downpipes after the exterior wall finish is completed. The downpipe must be securely clipped to the wall.

Flashings

There are many sections of any roof that the roof sheets cannot cover. Flashings are used to form a watertight seal at the intersections of roof sheets to walls, ridgelines, penetrations and valleys (see Figure 1.5).

To install the flashings, you will complete the following steps:

1. Before laying the sheets, you will install the valley gutter if required and any other tray, barge roll or flashing that must go under the roof covering.
2. Having laid the sheets, you will install ridge capping, barge capping, apron flashing, parapet capping, chimney flashing and step flashing as required.

Roof sheets and decking

The roof sheets and/or decking of a building form the largest and often most visible part of the roof system (see Figure 1.6). The roof must be installed to ensure that there is a watertight seal over the building, and it

FIGURE 1.5 A one-piece soaker/apron flashing topped with a parapet capping

FIGURE 1.6 Roof sheets drain the water to the gutters

must be laid with a fall to ensure that water is drained away effectively. A wide choice of materials and sheet profiles are available to suit almost any requirement.

To install roof sheets and decking, you will complete the following steps:

1 You will normally lay some form of thermal insulation first and, where necessary, do so over a protective wire mesh that is designed to prevent workers from falling through the frame.
2 You will then measure, cut and fit the sheeting, fixing it to the roof structure with screws or special concealed fastenings.
3 Some jobs may require the fitting of translucent polycarbonate or fibreglass sheeting. These sheets come in matching profiles and allow light into the building. Particular installation methods apply to translucent sheets.

Flashing of roof penetrations

Roof penetrations include vent pipes, flues, chimneys, skylights, roof ventilators and ducting (see Figure 1.7). Some roof penetrations are flashed during the installation of the roof covering, some are completed

FIGURE 1.7 A gas flue penetration, soaker flashing and synthetic collar

after the roof covering is laid and some are flashed on existing roofs where alterations or new fixtures and appliances are installed. These penetrations must be carefully flashed, not only to keep water from leaking into the building but also to ensure that the penetration does not obstruct the efficient flow of water down the roof.

To flash penetrations, you will use a range of materials compatible with other roofing components that may include synthetic rubber, sheet metal, zinc and lead to ensure a watertight seal that does not impede the flow of water down the roof sheet.

Cladding

Roof plumbers are increasingly being called on for their expertise in the installation of metal roofing and cladding products on exterior and interior walls (see Figure 1.8), soffits and ceilings. Metallic profiled sheeting and cladding products make an ideal lightweight building material, and designers often specify these products for buildings, particularly in locations or circumstances where traditional construction methods and materials would be too expensive or heavy to use.

Depending upon requirements, you may need to fabricate and/or install window and door flashings, foot moulds and sheets, and finish all exterior corner and return flashings. Sheets may be fitted horizontally, diagonally or vertically, depending upon the designer's

FIGURE 1.8 Profiled sheet-metal cladding used on exterior walls

requirements. Particular care must be taken around windows to ensure that the flashing prevents the ingress of wind-driven rain.

Water storage

In country areas, water tanks have been a critical consideration for decades. Now, as weather patterns become more unpredictable and Australia's population continues to grow, the storage of water in city areas has become extremely important in conserving our valuable water supplies.

FROM EXPERIENCE

Be aware that in most areas the use of onsite water tanks is subject to local authority approval and guidelines.

Water from the roof can be drained via the downpipe into a storage tank with the overflow connected to an approved point of discharge (see Figure 1.9). To install a water tank, you must be able to size the tank to suit the owner's needs and ensure that it is constructed of a suitable material for the job. If you are also licensed to undertake water supply plumbing, you might then connect the tank to an approved point of use.

Source: © BlueScope STEEL®

FIGURE 1.9 A steel water storage tank

FROM EXPERIENCE

As you progress with your training you will come across many more technical terms and industry jargon. Remember that terms and definitions may vary across Australia, so if you are unsure, ask your teacher or supervisor.

LEARNING TASK 1.3

1 Name three types of gutters.
2 Identify three types of roof penetrations that might be encountered.
3 Which direction can exterior wall cladding be fitted?

SUMMARY

In this chapter you were introduced to the requirements for and importance of undertaking all aspects of roof plumbing in accordance with relevant Codes of Practice and Australian Standards. You were also introduced to a basic overview of what type of work is carried out by roof plumbers, including descriptions of the following:

- Gutters and downpipes
- Flashings
- Roof sheets
- Penetration flashing
- Wall cladding
- Water storage

COMPLETE WORKSHEET 1

WORKSHEET 1

To be completed by teachers	
Satisfactory	☐
Not satisfactory	☐

Student name: ____________________

Enrolment year: ____________________

Class code: ____________________

Competency name/Number: ____________________

Task: Working with your teacher/supervisor, refer to this text, your HB39:2015 installation code and any other relevant resource to answer the following questions.

1 Name the Code of Practice that applies directly to roof system installation.

2 What Australian Standard should you refer to if you need to size eaves gutter systems?

3 What are three basic requirements that roof drainage and/or cladding systems must successfully fulfil?

4 List at least four different types of roof penetration for which you will need to provide a flashing.

5 What is the purpose of roof flashings?

6 Referring to your HB39:2015, find the definitions for the following terms:

a pressure flashing

b eaves gutter

WORKSHEETS 1

c barge capping

____

____

7 Using your HB39, glossary and any resources your teacher may provide, draw a line to match the following terms with the definitions shown in this table.

Term
Barge
Pop
Rainhead
Spouting
Penetration
Valley gutter
Mitre
Ridge capping
Step flashing
Bead
Fascia

Definition
An external gutter receptacle used to connect downpipes to roof gutters and also provide an external overflow point.
In some areas, another name used for an eaves gutter.
A gutter at the intersection of two sloping roofs.
Formed metal designed to weatherproof the junction at the apex of opposing roof slopes.
A flashing at the gable end of a roof that is fixed parallel to the roof slope.
A projection through the roof (e.g. a vent pipe, flue or skylight).
Shaped pieces of material used for flashing where a sloping roof abuts a vertical wall or chimney.
The formed curve or fold at the front edge of an eaves gutter.
A board or metal profile section fixed along the line of the eaves and the end of trusses to which the eaves gutter may be secured.
More commonly known as a gutter angle.
A gutter outlet point, also known as a 'pop', 'downpipe outlet' or 'drop'.

TYPES OF ROOF STRUCTURE

2

This chapter will address clarification of roof plumbing terminology, introduce basic roof types (low pitch, skillion and lean-to, gable, hip roof, composite, curved and saw tooth) and roof type variations (Dutch gable, gambrel, mansard, jerkin head, butterfly and hyperbolic paraboloid).

Overview

Roof types and styles are determined at the design stage of a project and are specified for each individual structure to suit certain requirements that will vary from place to place. These include:

- local climatic conditions
- the intended purpose of the building
- aesthetic requirements (aesthetic design is all about the proportion and visual appeal of a building)
- the relationship of the roof to the building design and structure
- cost.

Before you proceed to examine each of the many roof styles used, you will need to get a grasp of how the terms 'pitch', 'slope' and 'fall' are used in roof plumbing.

FROM EXPERIENCE

There is a wide range of roof types to be found in domestic, commercial and industrial applications. It is important that you are able to understand the differences between each and can identify them as required.

Expressing and measuring 'pitch', 'slope' and 'fall'

As with many terms used in plumbing, variations will occur around the country and even between tradespeople. Essentially the terms 'pitch', 'slope' and 'fall' can all be used to describe the same thing, and variations will occur even within the codes and standards.

Generally, however, 'slope' and 'pitch' are normally applied to the roof itself, while 'fall' is usually used in relation to the gutter and downpipe. For the sake of clarity, this will be the application used in this text.

Roof pitch

The pitch of a roof can be expressed and measured either in degrees or as a ratio.

Degrees

The angle of the roof in degrees is measured between the roofline at its intersection with the horizontal plane formed by the eaves. Always express the angle from the horizontal. For example, the minimum recommended pitch to lay corrugated steel sheeting is 5°. Figure 2.1 shows how the roofline rises at 5° from the horizontal.

Ratio

The pitch may be expressed in the form of a ratio as *rise:run* in metres (rise is to run). Using the example of corrugated steel roof sheets once again, you can see that the minimum recommended pitch of 5° is now expressed as a 1:12 ratio. Figure 2.2 shows how the roofline rises at a 1:12 pitch.

Gutter fall

Gutter fall is almost always expressed as a ratio. For example, you might encounter a length of gutter that has a 1:500 fall. This is a ratio that is read as *fall:run* (fall is to run). Such a gutter has a fall of 1 mm for every 500 mm of length. Therefore, a 10-m length of eaves gutter would have a minimum of 20-mm fall.

How you choose to express pitch and fall will ultimately be up to your personal preference, your teacher's advice and local custom. However, for consistency in this text, roof pitch will be expressed in degrees and gutter fall as a ratio.

Why do I need to know the pitch of a roof?

There are a number of reasons why you need to know the pitch of a particular roof before commencing work. Firstly, roof sheets are designed and selected to match different minimum roof pitch angles. You need to know that the selected roof sheet profile is suitable for the pitch of the roof. As in our previous example, standard corrugated steel sheets cannot be used on a roof pitch of less than 5°. If the roof pitch was only 3°, you would need to choose another profile that is suitable for the job.

Scaffolding and fall arrest requirements will vary according to roof pitch. For example, a roof of 30°

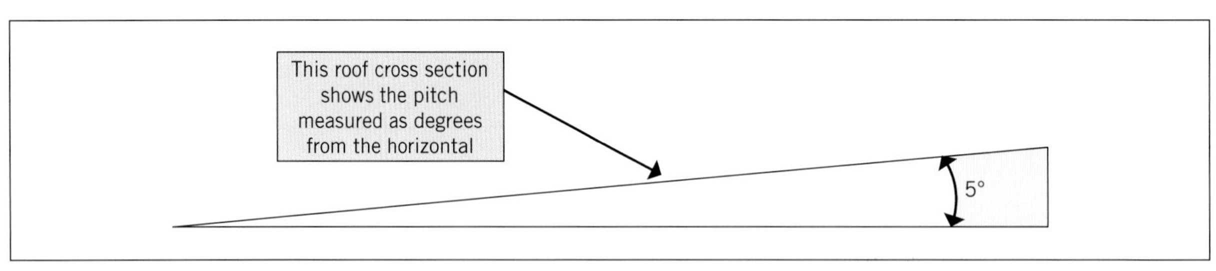

FIGURE 2.1 Roof pitch expressed in degrees

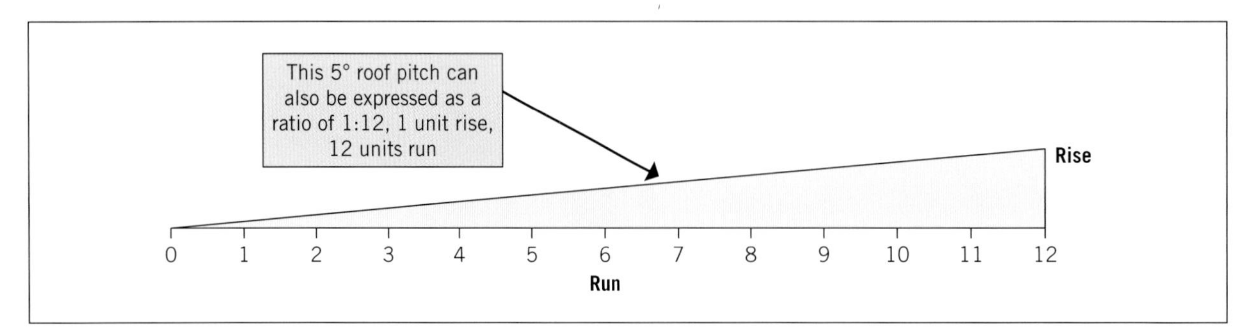

FIGURE 2.2 Roof pitch expressed as a ratio

or more will require a different form of perimeter scaffolding from a roof of lower incline. You will need to be able to recognise when this is the case.

The process used to determine the correct size of roof drainage components also requires that you know the pitch of a roof. Roof pitch can be determined from the approved plans for new work, or by using a pitch gauge on existing roofs (see Figure 2.3).

FIGURE 2.3 A pitch gauge reading of 6.6 degrees means this roof pitch is suitable for a corrugated sheet profile (min. 5°)

LEARNING TASK 2.1

1 A roof's pitch can be expressed in two ways. What are they?
2 Which tool can be used to identify existing roof pitch?

Roof types

There are seven main roof types used in general construction. The other roof structures that can be found are considered to be a combination of or variation on these seven basic types. The following list includes a selection of styles that can be found in many places around Australia.

Basic roof types

Low-pitch roof

A low-pitch roof is one that has a fall of less than 10° and normally drains in one plane towards a box or eaves gutter. In some instances, a fall as low as 1° may be achieved using suitable steel profile sheets in continuous lengths (see Figure 2.4).

These roofs are often incorrectly referred to as 'flat' roofs. However, a roof must have some fall so that water drains efficiently towards a point of collection and discharge, and therefore the term 'low-pitch' is a more accurate description.

Another form of low-pitch roof may be found on top of multi-storey or concrete-roofed commercial

FIGURE 2.4 Example of a low-pitch roof laid at approximately 1°

buildings, where a trafficable roof is covered with a synthetic membrane or flexible bituminous sheeting and sealed with bitumen and slip-resistant gravel. These roofs also fall gradually towards a point of discharge. The laying of this type of membrane is a specialist area, and falls outside the scope of this text.

Skillion and lean-to roof

Skillion and lean-to roofs are very similar in that they are both pitched in one plane from the highest point to the gutter. For this reason, their names are often used interchangeably. However, the accepted differences are as follows:

- A *lean-to roof* is always joined to another building (see Figure 2.5). Where one building abuts the wall of another higher building, the roof of the lower structure is usually covered in sheeting that falls in one plane towards the gutter. While the accepted practice is to pitch a lean-to at an angle of more than 10°, it is not uncommon to find them laid at a much flatter angle.

FIGURE 2.5 A lean-to roof abutting the wall of a higher part of the main building

- A *skillion roof* is a form of simple roof found on a structure that stands alone and normally has a pitch of more than 10°, with the sheets falling in one plane towards the gutter (see Figure 2.6).

FIGURE 2.6 Skillion roofs on a carport and main building

Gable roof

A gable roof is laid in two opposite planes between the ridgeline and the eaves on either side of the building (see Figure 2.7). The walls at each end of the building continue up towards the ridge itself. The pitch may vary, from the very steep roofs often found on alpine buildings and churches to low pitch structures that are covered in one continuous sheet curved over the ridge.

FIGURE 2.7 A simple gable roof

A raked gable is a roof that extends past the gable end wall to create shelter over a deck or veranda (see Figure 2.8).

FIGURE 2.8 A raked gable roof

Hip roof

The simple hip roof is laid in all four planes between the ridgeline and the eaves of each of the four walls of the building (see Figure 2.9). The hip itself is the intersection between two of the planes.

FIGURE 2.9 Example of a hip roof

Composite roof

In its simplest form, a composite roof is regarded as being the combination of a hip roof and a gable roof in one structure (see Figure 2.10). This can be within the one plane in a rectangular fashion, but more commonly includes one or more right-angle changes in direction. The intersection of the roof sections between two parts of the building is known as the 'valley'. Composite roofs may also incorporate other structural variations including multiple hips, Dutch gables, low-pitch and curved sections.

FIGURE 2.10 Example of a composite roof

Curved roof

Curved roofs were once restricted to short-sheet veranda applications and temporary military structures to house soldiers and stores. However, with the advent of advanced roll-forming technology and the use of continuous-length sheeting, designers now have the freedom to use curved roofs in multiple applications that were previously not possible or economical. Sheet profiles may be pre-curved (see Figure 2.11) or, in certain applications, straight lengths may be 'sprung' over convex or concave roof structures (see Figure 2.12).

Saw-tooth roof

The saw-tooth roof, with its unmistakable, characteristic shape (see Figure 2.13), is mainly found in commercial and industrial applications. The vertical sections of the

FIGURE 2.11 A pre-curved roof found on a public building

FIGURE 2.12 Example of a domestic spring-curved roof

FIGURE 2.13 A saw-tooth roof in an industrial area

roof are normally fitted with large clerestory windows that provide natural light (and sometimes ventilation) to what would otherwise be the darkened interior of the building. Water is collected at the bottom of each roof slope into box gutters and drained away.

Roof style variations

As is common in the construction industry, the names of many structural sections or building characteristics may vary considerably from region to region and between different countries. For example, what we may know as a Dutch gable, gambrel or mansard roof in Australia is often known by a quite different name in Europe or the United States. For clarity, the most commonly known terms found in Australia are used below. However, be aware that you will encounter differences between regions.

Dutch gable (variation of a hip roof)

In Australia, the term Dutch gable refers to a hip roof where the ridge has been extended past the normal intersection with the hips to form a small, triangular gable shape (see Figure 2.14). This is largely done for aesthetic reasons and makes a simple hip roof look more appealing.

FIGURE 2.14 Example of a Dutch gable roof section

Gambrel (variation of a gable roof)

Familiar to many people as the classic American barn-style roof, the gambrel roof has two different pitches on either side of the ridge (see Figure 2.15). From the eaves, the first section of the roof rises steeply to around two-thirds of the height of the roof, before changing to a lower pitch before the ridge. The gambrel roof is common in the United States and Europe, where it traditionally provided a simple method of maximising attic space without the use of additional masonry or wall material.

FIGURE 2.15 A gambrel roof

Mansard roof (variation of a hip roof)

The mansard roof is often confused with the gambrel roof, as they both feature steep sides and a double pitch. However, the key difference is that the gambrel roof is pitched on only two sides, with full-height gables at each end, while the mansard has sloped roof planes *on each side of the building* (see Figure 2.16). The upper roof slope is often difficult to see from the

FIGURE 2.16 A mansard roof used on a colonial-era building

ground, but it is generally finished in a very low-pitch hip style. The mansard profile was a popular roof construction method during the 1800s.

Jerkin head (variation of a gable roof)

The jerkin head roof varies from the gable roof in that the upper third of the gable end is replaced by a small hip section (see Figure 2.17). In this way, the jerkin head roof is halfway between a hip and a gable roof.

FIGURE 2.17 Example of a jerkin head roof

Butterfly (variation of a skillion roof)

The butterfly roof is characterised by two skillion-style roofs falling towards each other and meeting above a box gutter that drains the water away. Early forms of butterfly roofs were commonly used for petrol stations before the advent of large vaulted frames over the service areas. Both roofs could be supported via central columns, leaving the rest of the ground area free of obstructions. Today the term has broadened to include any roof construction where the roof area falls towards the middle box gutter (see Figure 2.18).

FIGURE 2.18 Example of a domestic butterfly roof

Hyperbolic paraboloid (variation of a curved roof)

The geometric nature of a hyperbolic paraboloid roof allows for a double-curved roof to be constructed using straight lengths of framing material and sheeting. This means that although the roof looks curved, the roof plumber lays straight sheets, making installation fairly simple. As you can see from Figure 2.19, the roof features a convex curve around one axis while a concave curve follows the opposite axis, creating a saddle-type shape.

Hyperbolic paraboloid-style roofs are rarely found in domestic applications, but are normally used to provide a visually striking roofline in public and commercial buildings.

FIGURE 2.19 Geometric shape of a hyperbolic paraboloid roof

LEARNING TASK 2.2

1 How many basic roof types are identified?
2 A skillion roof normally has a pitch greater than:
3 How many planes does a gable roof have?

SUMMARY

In this chapter we have clarified some very basic forms of terminology related to:

- fall
- pitch
- slope.

Using the correct terminology is an important part of learning the trade. This extends also to your knowledge of roof types. Describing a roof type is a fundamental trade skill and knowing the basics will enable you to better understand manufacturer product data and converse within the industry with confidence.

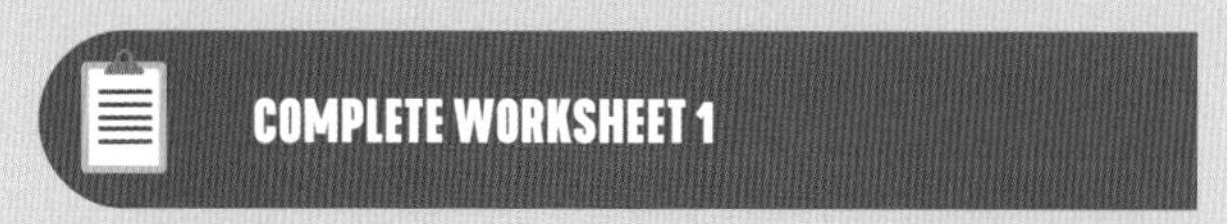

WORKSHEET 1

To be completed by teachers	
Satisfactory	☐
Not satisfactory	☐

Student name: ______________________

Enrolment year: ______________________

Class code: ______________________

Competency name/Number: ______________________

Task: Working with your teacher/supervisor, refer to this text and any other relevant information to answer the following questions.

1 List two or more reasons why it is important for a roof plumber to know the pitch of any particular roof.

____

____

____

2 What are five factors that are considered by designers when planning the design of a roof drainage system?

____

____

____

____

3 Where a roof pitch is specified in degrees, is this measured from the horizontal or the vertical plane?

____

____

4 What would be the common industry name given to a roof that had a pitch of 2°?

____

____

5 Define what is regarded as being the simplest form of composite roof.

____

____

____

6 Describe the difference between a simple gable roof and a hip roof.

____

____

____

7 What is the key difference between a gambrel and mansard roof?

____

____

8 Think about the building where you live. What roof type does it have?

____

____

9 Match each of the following roof profiles with the correct name:

Name	Profile
Hip roof	
Skillion roof	
Gable roof	
Saw-tooth roof	
Composite roof	
Low-pitch roof	

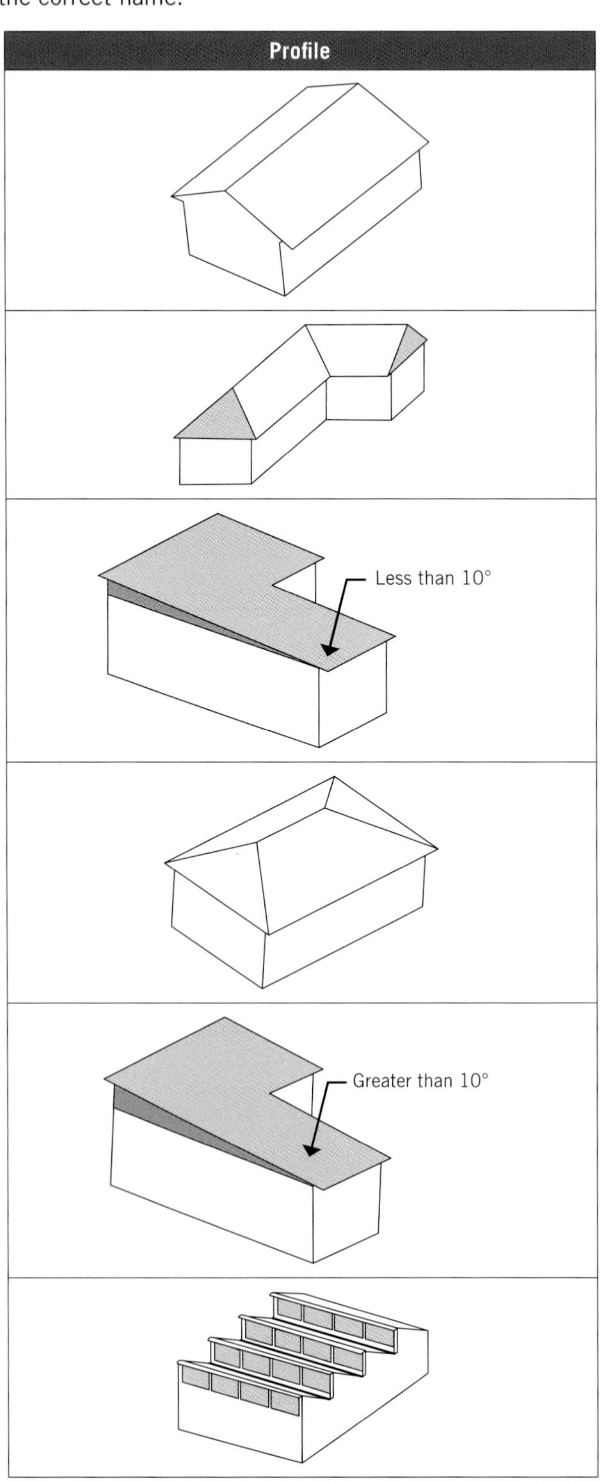

BASIC ROOF SAFETY

3

By its very nature, roof work is hazardous. Working at heights leaves little margin for error, and workers who disregard the basic safety requirements endanger their own lives as well as those of their co-workers.

Falling from a roof is not just something that happens to other people. If you become complacent and fail to apply appropriate safety procedures on the job, you could end up being seriously injured or killed. Statistics indicate that the cause of most deaths and serious injuries in the Australian construction industry is falls from height.

Therefore, if you are going to work in the roofing industry you need to get serious about safety!

Overview

As you have already covered occupational health and safety as part of your earlier training, this chapter will revise the basic safety requirements as well as introducing some considerations specific to roof work. It includes:

- regulatory requirements
- applying safe work procedures
- basic height access
- skin protection
- hearing and eye protection.

Note: The hazards associated with asbestos exposure in the construction industry are especially important for you as a roof plumber. Therefore, the subject of asbestos is covered in more detail in Chapter 4 'Asbestos' of this text.

Fall arrest and perimeter edge protection are assessed under the unit 'Work safely on roofs' within the plumbing training package. This unit has been dealt with in the Cengage Learning publication *Construction Skills* by Glenn P. Costin and therefore will not be covered in this book.

Regulatory requirements

The *Work Health and Safety Act 2012* is the nationally harmonised legislation that is intended to provide workers with the same standard of health and safety protection regardless of where they live. At the time of publication, two states had not adopted this legislation and were still operating under separate laws. Therefore, it is important that you confirm which authority administers WHS legislation in your state/territory and confirm the Acts and Regulations that apply.

What is an Act?

An Act is a written and approved law that has been passed through state or federal parliament. In respect to workplace health and safety, an Act sets out the broad guidelines and responsibilities to which all employers and workers must adhere (see Example 3.1).

EXAMPLE 3.1

The WHS legislation that applies to all workers and employers in Victoria is the *Occupational Health and Safety Act 2004* (the year denotes when the Act was passed by parliament).

What are Regulations?

Regulations are typically used to provide more detail and context in the application of the broader Act that they sit beneath. Such detail often includes specific references to documents or situations that may frequently change. This allows the law to be interpreted without having to go to the trouble of changing the Act (see Example 3.2).

EXAMPLE 3.2

In Victoria the activities of all employers and workers are regulated by the Occupational Health and Safety Regulations 2007.

What is a Code of Practice?

A Code of Practice is a practical guide for achieving the requirements and standards stipulated in the Act and Regulations. It tells you *how* things are to be done. An approved Code of Practice is a statutory document that must be followed unless an alternative course of action provides the same or better outcome (see Example 3.3). While a Code of Practice may not carry direct legal standing, it is still an admissible document in court proceedings where failure to observe its guidelines may be considered an offence under the Act or Regulations.

EXAMPLE 3.3

In NSW, the two specific Codes of Practice that apply to roof plumbers working at heights are:

- NSW Code of Practice: Safe Work on Roofs, Part 1, 'Commercial and industrial buildings'
- NSW Code of Practice: Safe Work on Roofs, Part 2, 'Residential buildings'.

What are Australian Standards?

Australian Standards are detailed technical documents that set the minimum design and performance standards for equipment and systems (see Example 3.4).

EXAMPLE 3.4

Eaves and box gutters are sized and installed according to the AS/NZS 3500.3 'Stormwater Drainage'.

AS/NZS 3500.3 'STORMWATER DRAINAGE'

In some instances, Australian Standards are combined with New Zealand Standards and are abbreviated AS/NZS, as seen in Examples 3.4 and 3.5.

EXAMPLE 3.5

If you purchase a fall arrest harness that is manufactured to comply with AS/NZS 1891.1, you can be confident that the design has been evaluated and certified to meet minimum industry performance requirements.

AS/NZS 1891.1 'FALL ARREST HARNESS'

Standards are written by key stakeholders and relevant technical experts and are regularly updated to reflect new requirements. Always refer to the latest version! While standards are not generally legal documents in their own right, they are often referred to, or 'called up', within Regulations and therefore must be adhered to.

To complete the worksheet at the end of this chapter successfully and go on to attain competence in most of your roofing units, you will need to have access to the Codes of Practice applicable in your area.

The websites shown in Table 3.1 have been included to assist you in finding out the Regulations for working at heights that apply in your region.

PART A
3

TABLE 3.1 Sourcing regulations for working at heights

State/ territory	Regulating body	Website
ACT	WorkSafe ACT	www.worksafe.act.gov.au
NSW	SafeWork NSW	www.safework.nsw.gov.au
NT	NT WorkSafe	www.worksafe.nt.gov.au
QLD	Workplace Health and Safety Qld	www.worksafe.qld.gov.au
SA	SafeWork SA	www.safework.sa.gov.au
TAS	Worksafe Tasmania	www.worksafe.tas.gov.au
VIC	WorkSafe Victoria	www.worksafe.vic.gov.au
WA	WorkSafe WA	www.commerce.wa.gov.au/WorkSafe/

Investigate which Acts, Regulations and Codes of Practice apply to your work area and insert them in the box below for future reference.

EXAMPLE 3.6

Your state/territory ....

Act of parliament ....

Regulations ....

Code/s of Practice ....

Applying safe work procedures

The only way of ensuring that safe work procedures are implemented is through the use of a consistent and methodical system of evaluation and recording on standardised documents. Using an established system of evaluation minimises errors and omissions and also provides a formal mechanism of review to ensure that work practices are revised and upgraded whenever necessary.

The steps involved in worksite risk assessment can be summed up under the easily recalled acronym S-A-F-E as shown in Figure 3.1.

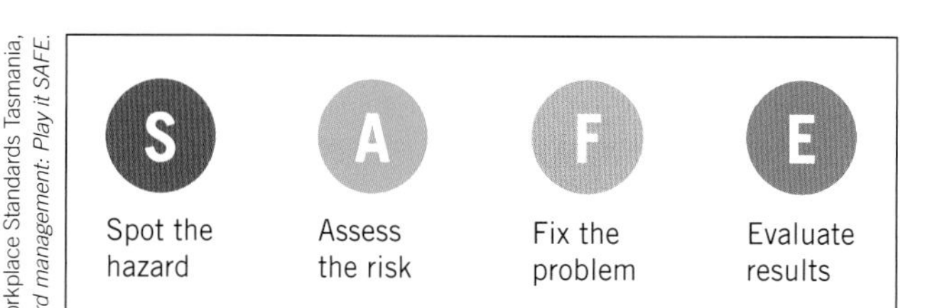

Source: WorkCover Tasmania and Workplace Standards Tasmania, *Hazard management: Play it SAFE.*

FIGURE 3.1 The four steps of worksite risk assessment

These simple headings form the centre of your worksite hazard control process, and as they are easy to remember, you will be able to apply the steps as and when required during your daily work.

Referring to Figure 3.2, you can see that the 'SAFE' process has been incorporated into a more detailed flowchart that sets out the steps of workplace hazard control.

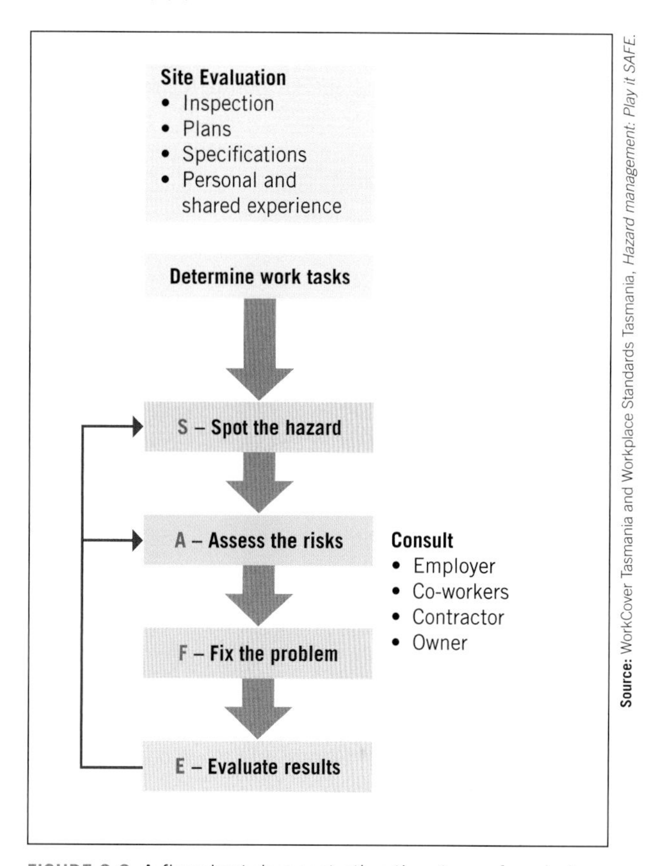

Source: WorkCover Tasmania and Workplace Standards Tasmania, *Hazard management: Play it SAFE.*

FIGURE 3.2 A flowchart demonstrating the steps of workplace hazard control

Site evaluation

Every roof job is different in many ways. It may be a new domestic or commercial construction, a roof replacement or a repair of flashings and penetrations. Before you can proceed with any job planning, risk assessment and hazard control, you need to evaluate the site conditions and job requirements. While some preliminary planning can be completed from building plans, an effective evaluation cannot be completed until you have actually walked over the site.

General site evaluation considerations include, but are not limited to, the following:

- general site access for worker and delivery vehicles
- the site's slope and stability
- clearances to site boundaries

- the expected weather and seasonal characteristics over the expected duration of the job
  - wind
  - temperature
  - dew
  - reflective glare
- the location of services, including power and phone lines
- the height and size of the structure
- the structural integrity of older buildings and the location of sound anchor points for fall arrest systems
- the characteristics of the existing roof membrane:
  - the pitch of the roof
  - the brittleness of sheets/tiles
  - the presence of asbestos
- the impact of other trade contractor activities on roof workers:
  - noise
  - dust
  - proximity of the work
- suitable arrangements for waste material storage and disposal.

With safety requirements as your focus, the plans and site visit enable you to identify the basic worksite factors that are likely to have an impact upon your job scheduling, material delivery locations, choice of height access method, fall protection systems and other general hazards.

Having conducted the basic evaluation of the site and recorded the outcome, you now need to determine what actually needs to be done and what hazards are created by the combination of the work tasks and the site environment.

Determine the work tasks

On each individual site, different work tasks may create entirely different hazards. For example, a job may require the installation of skylights into a 30° sloping roof in addition to low-level downpipe installation around the building. A single site may have significantly different hazard and risk control requirements for each individual task.

You need to ask yourself what the job is all about and record each aspect of it in sequence. Once you have determined how the worksite's characteristics combine with the actual job task, you can then identify the hazards associated with each step of the job itself.

Consultation

You will notice from the flowchart in Figure 3.2 that each step in the hazard control process depends on consistent consultation with all job site stakeholders. Such consultation is required under the relevant WHS Acts, and it must be a formalised and documented process between employer, employees, safety representatives, unions and contractors.

Spot the hazard

Hazards are identified as a combination of worksite conditions and work task requirements. These add up to potential hazards that need to be identified and documented, and decisions must then be made on how to control each hazard.

Each state and territory authority recommends certain ways of documenting hazards and risks. Working with your teacher/supervisor, access your state or territory WHS authority and search for relevant local documents that cover hazard and risk identification and assessment.

Assess the risk

On any job you will find a number of hazards, each of which has to be assessed individually. You should look at each hazard and ask the following questions:

- What is the potential *impact* of the hazard?
- What is the risk of death, serious injury, long-term illness and so on?
- How *likely* is the hazard to cause harm?
- How *often* does the hazard occur (frequently, rarely)?

Number each risk in order of the urgency to resolve it, with 1 being the most urgent. You should then deal with each of these risks in order.

Fix the problem

Having identified the hazards and assessed the risk, you need to apply some form of control to fix the

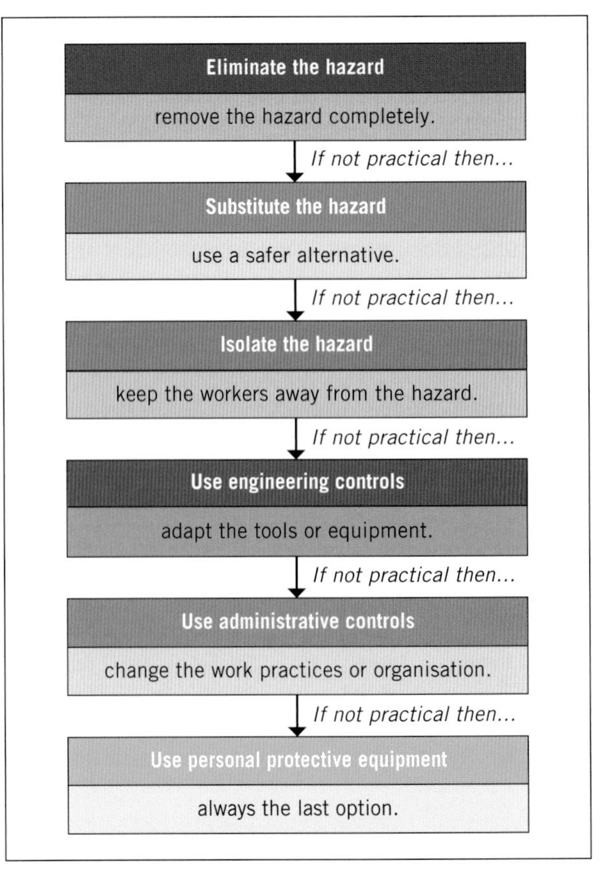

FIGURE 3.3 Control measures

problem. Construction sites can be very complex environments and you need to apply a methodical rigour to the hazard control process. This is best done using a system called the *hierarchy of control measures*. In this system, hazard control measures are ranked in order from 1 to 6, the aim being to implement the highest ranked option possible. Where this is not possible, you then move to the next-ranked measure.

In certain circumstances, more than one measure may be warranted. Work through each hazard from the most urgent, and look for a solution, starting at the highest ranked control. Each control measure is listed in Figure 3.3 in order from highest ranking to lowest.

Evaluate the results

The effective management of hazards and risks requires the consistent evaluation of safety solutions. This is done through consultation with all levels of an organisation so that any deficiencies in the process can be identified early and rectified before something goes wrong. The evaluation process should be documented to enable appropriate follow-up and review.

Using a risk assessment worksheet

Figure 3.4 shows an example of how a risk assessment worksheet can be used. It is based on a simple scenario in which the owner of a workshop requires skylights to be installed in order to provide internal natural lighting. An evaluation of each task or step is conducted so that risks and hazards are revealed. Once identified, the hierarchy of controls is applied to reduce the risk of each individual task.

Workplace location:	*Davo's Tyres and Brakes*
Assessor:	*Nick Smyth*
Job task:	*New workshop skylights*
Date:	*12.11.2021*

SPOT THE HAZARD		ASSESS THE RISK	FIX THE PROBLEM			EVALUATE RESULTS
Identify the work task or activity.	What are the hazards with each activity or the steps in each activity?	Is the risk associated with the hazard low, moderate, significant or high?	If the risk is deemed unacceptable for the task, what will be done to reduce or remove the risk?	By whom?	By when	Go through the first 3x steps again to ensure risk levels are now at an acceptable level Revised risk level
Roof access	*Roof height is 6 m. Fall hazard from ladder access.*	*High*	*Fit top ladder bracket to access point. Ensure base of ladder is secured.*	*Jonesy*	*14.12.21*	*Low*
Getting polycarb sheets and wire on roof.	*Sheets and wire are too large and heavy to manhandle up scaff or ladder.*	*High*	*Use small crane to lift sheets and wire onto roof.*	*Peter to arrange crane.*	*14.12.21*	*Low*
Lift old sheets where new skylights are going.	*Fall hazard through hole.*	*High*	*Anchor points to be installed on ridge adjacent to each skylight position.*	*Peter to site each anchor. Jonesy to install.*	*30.12.21*	*Low*
etc.	etc.	etc.	etc.	etc.	etc.	etc.

Source: Adapted from WorkCover Tasmania and Workplace Standards Tasmania, *Hazard management: Play it SAFE.*

FIGURE 3.4 An example of a risk assessment worksheet

LEARNING TASK 3.1

In the following table and using numbers from 1 to 6, identify each of the steps in the hierarchy of control measures.

Order of application	Risk control measure
	Isolate the hazard
	Use engineering controls
	Substitute the hazard
	Use personal protective equipment
	Eliminate the hazard
	Use administrative controls

Your teacher or supervisor should be able to provide you with some other examples or layouts commonly used in your particular region.

Roof hazard considerations

The following section details some considerations, equipment and systems that relate to common roof plumbing hazards.

Basic height access

Height access methods involve the use of various forms of equipment, including:

- ladders (stepladders or extension ladders)
- scaffolding
- elevated working platforms (scissor lifts or boom lifts/cherry pickers).

The use of scaffolding and elevated working platforms is outside the scope of this chapter, but you must ensure that you are trained to a competent level before using such equipment. Be aware that you may also require industrial licensing.

Ladders

Ladders should be regarded as a means of access only and not as equipment that you would actually work from, other than for short-duration, light-duty tasks. Where they are used for commercial purposes, all ladders must be of industrial strength, rated to at least a 120-kg capacity and meet the design requirements of AS/NZS 1892 'Portable Ladders' (see Figure 3.5). A compliant ladder should quote this Standard on an attached product sticker or literature. Lightweight, 'domestic-style' stepladders and extension ladders are not acceptable on the modern construction site.

FIGURE 3.5 An industrial-grade stepladder

The following points should be considered when using stepladders:

- Always check for defects before use.
- Use only industrial-grade products.
- Ensure that the stepladder is fully extended during use, with the locks/latches engaged.
- Do not stand on the top two rungs.
- Use proprietary brackets and attachments to hold/support tools while in use.
- If you can't reach what you are after, move the ladder. Do not over-reach.

When using extension ladders:

- Always check for defects before use.
- Use only industrial-grade products.
- Pitch the ladder at a 1:4 incline (1 m out for every 4 m up; see Figure 3.6).
- The ladder must *always* be secured at the bottom.
- The base of the ladder must sit on a firm and even surface (see Figure 3.7).
- Secure the top of the ladder as soon as possible and as often as required. If it is in regular use from one position, consider using a temporary/permanent ladder bracket.
- Extend the top of the ladder to 1 m above the landing.

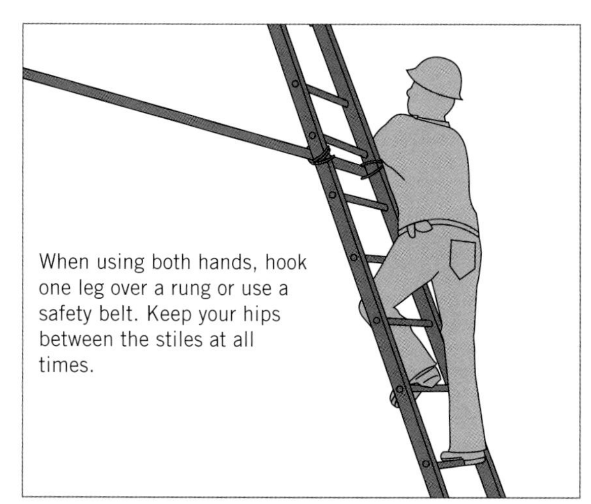

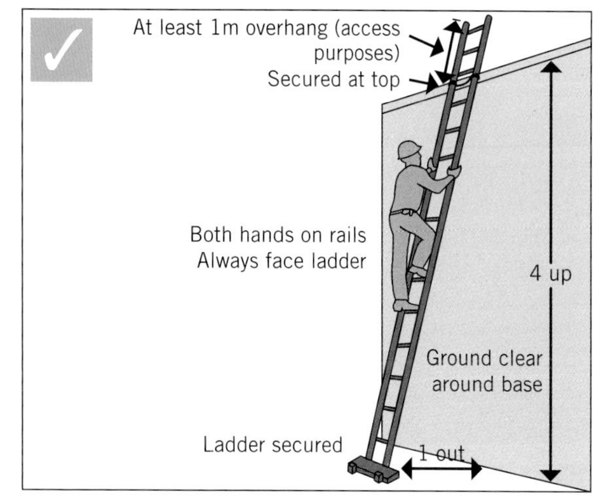

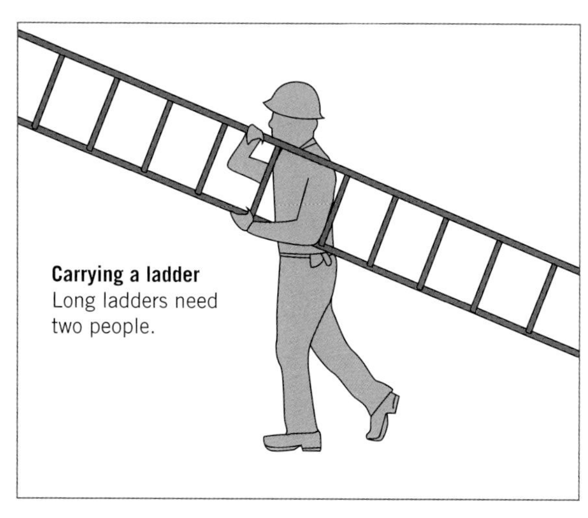

Stepladders
A bucket hook or tool tray makes work easier. Stay below the top two steps

Source: Safe Work Australia, 'Preventing falls in housing construction code of practice', p. 25.

FIGURE 3.6 Choose the right ladder for the job and use it correctly.

FIGURE 3.7 A ladder should be placed on an even surface.

- If you can't reach what you are after, move the ladder. Do not over-reach.
- Do not use metal ladders or ladders with wire reinforcing near any electrical hazards. Use approved fibreglass products instead.
- Ensure that there are three points of contact – either two hands and one foot or two feet and one hand.

On commercial and public buildings, it is now becoming more common to see permanently fixed galvanised ladder brackets in use (see Figure 3.8). A ladder bracket is a safe and efficient way of securing the top of the ladder. Such brackets are available in a number of configurations and provide a secure point of ladder placement for regular access. A range of temporary brackets is also available and should be part of any roof plumber's kit.

Scaffolding

Apart from quick and minor repairs, work should not be carried out from ladders. For safety reasons, you are required to use some form of scaffolding or approved work platform when undertaking roof work.

FIGURE 3.8 A permanently fixed bracket designed to secure the top of a ladder

Scaffolding is a temporary structure that is used to provide a means of safe access and a secure working platform during the repair or construction of a building. On some sites the builder may already have the scaffolding in place for the use of all relevant sub-contractors. On others, the roof plumber may need to determine what scaffolding system to specify and order.

FIGURE 3.9 Frame scaffolding with internal stairs and toe boards

Roof plumbers generally use one or more of the following scaffolding systems:

- modular
- mobile
- edge protection guardrails.

Modular scaffolds

A modular scaffold is made up of a range of prefabricated components with a proprietary integrated fixing system on each part. Such systems are generally made from steel, though aluminium is also available where conditions may dictate the use of a lightweight material.

Scaffolding provides a safer means of access and a working platform. In Figure 3.9 you can see how a modular scaffold has been erected to enable access for the installation of new roof ventilators.

It is important to note that where a modular scaffold is to be erected at heights above 4 m, the scaffold installer must possess some form of accreditation or industrial licensing to do so.

Mobile scaffolds

Mobile scaffold systems (see Figure 3.10) are made from aluminium or fibreglass and are manufactured as individual components that enable the erection of a mobile working platform on castors. For a roof plumber, a mobile system may be used as a means of

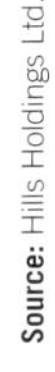

FIGURE 3.10 A mobile scaffold system

access to a roof, to fix downpipes or perhaps to repair a section of guttering.

Edge protection guardrails

Edge protection guardrails are a special component-based system designed for attachment to the roof perimeter (see Figure 3.11), allowing workers to move freely around the roof. The rails may be secured to the end of rafters, or some systems are designed to attach to the actual roof sheet rib profile.

Source: Rhino Rail

FIGURE 3.11 Edge protection guardrails on a roof

Edge protection systems enable roof workers to move freely around the protected roof area without the need to wear personal fall arrest equipment.

You will learn more about edge protection guardrails and personal fall protection systems at another point in your training. This subject is not covered in detail within this text.

Elevated work platforms

The two common forms of elevated work platforms that roof plumbers will regularly use are scissor lift mobile platforms and boom lift platforms.

Scissor lift platforms

A scissor lift is normally specified where it can be used on a flat and even surface around or within a building, although rough terrain versions are also available. These lifts are powered by electricity or combustion engine (see Figure 3.12) and can be controlled and manoeuvred by the operator from the platform. A scissor lift is particularly useful where access to a roof is required, or where the task to be carried out is directly above the machine.

FIGURE 3.12 A scissor lift mobile working platform providing access on a commercial roof job

Boom lift platforms

Boom lift platforms are also available in both electric and combustion engine-powered versions. They offer the advantage of being able to extend both vertically and horizontally from the base of the machine. The working platform is positioned at the end of the boom where the operator is also able to control the machine's movement (see Figure 3.13). Boom lifts are work platforms only and not designed for access purposes.

FIGURE 3.13 Roof plumbers being trained in boom lift operation

You will undertake training in the use of elevated work platforms as part of your roof plumbing training, and therefore it is not covered in further detail within this text.

Skin protection

Roof plumbing is almost always an outdoors job – a situation that most roof plumbers enjoy and thrive in – so it is unavoidable that you will be exposed daily to the sun as part of your work. We all need a moderate dose of sun to keep healthy, as this encourages the production of vitamin D, but if you allow your exposure to be too intense or for too long a period, you run a high risk of developing skin cancers. Of these, melanoma is the most deadly.

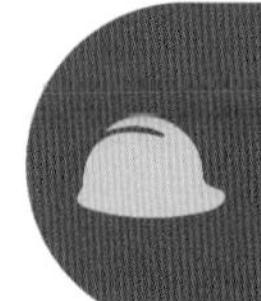

Australia has the highest incidence of the skin cancer melanoma in the world. Melanoma can strike at any time in your life and is the most common cancer in people aged between 15 and 44. Roof plumbers are especially at risk!

While regular exposure may protect your skin from the obvious effects of a light burn, damage to the skin and the possible DNA mutation that leads to skin cancers can still occur. As part of your daily routine and in combination with protective clothing and a broad-brimmed hat, apply minimum SPF 30+ sunscreen 20 minutes before any sun exposure and reapply it at recommended intervals during the day.

Roof glare

When you are up on a roof you are not only exposed to direct solar radiation; you may also be subject to reflected glare, particularly from light colours and unpainted ZINCALUME® steel sheets. Failure to protect yourself in such a situation will lead to severe sunburn and eye damage.

In simple terms, in order to protect yourself you need to:

- cover up
- wear a hat
- wear UV or polarised eye protection
- wear minimum SPF 30+ sunscreen.

To get the latest information on how to look after yourself in the sun, refer to the Cancer Council Australia website at www.cancer.org.au. The Cancer Council regularly updates its advice through its position statements. Search the site to find the latest information.

Note that sunscreens containing semi-conducting metal oxides such as titanium dioxide (TiO_2) and zinc oxide (ZnO) may accelerate the degradation of organic materials, including roof product paint finishes. It is recommended to avoid the contact of such sunscreens with painted finishes.

Hearing protection

Roof plumbing often involves the use of tools and equipment that produce dangerous noise levels. Angle grinders, power saws and drills can all reach sound levels that will cause damage to your hearing. Industrial deafness is normally an affliction that accumulates over years, though some workers may be affected sooner.

Roof plumbers are exposed almost every day to the sound of Tek guns and impact drivers. The sound levels of such tools and repeated use can lead to industrial deafness and/or tinnitus. Wear hearing protection!

Some workers may develop a condition called tinnitus after only one severe event, often leaving them with a lifetime of continuous, loud ringing sounds in the ear. To avoid the risk of hearing damage, wear task-specific hearing protection that complies with the Australian Standards as and when required.

Eye protection

When you carry out a task that may involve the risk of flying debris or particles, ensure that you wear a pair of Australian Standard safety glasses to protect your eyes. In particular, be careful of the very fine and sharp steel offcuts that can fly off when cutting sheet metal. Such particles can embed themselves in your eye much faster than you are able to blink.

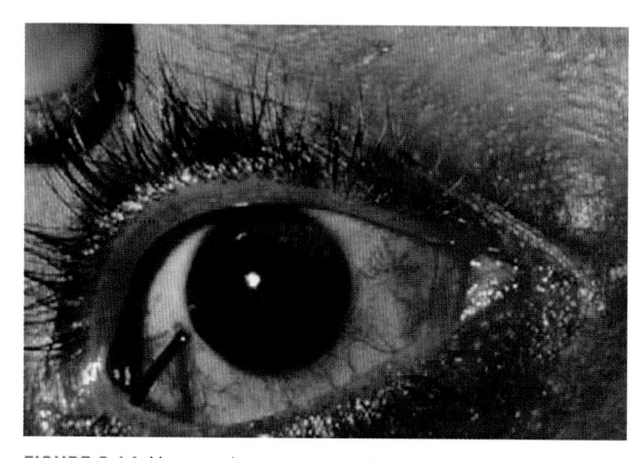

FIGURE 3.14 You rarely get a second chance with eye injuries

Electrical hazards: look up and look around!

Many tradespeople have suffered electrical shocks when moving materials and equipment such as ladders around a job site. A roof plumber is particularly at risk when manoeuvring long lengths of gutter or sheets into position or erecting an extension ladder. Always be aware of your surroundings and identify any electrical hazards before beginning work.

LEARNING TASK 3.2

1. Name three forms of equipment to access height.
2. Identify the weight load capacity of portable ladders according to AS/ NZS 1892.
3. Name two materials mobile scaffolding is constructed from.

SUMMARY

In this chapter you have covered some of the safety issues relevant to roof plumbing. Roof plumbers must abide by strict requirements relating to:

- safe work procedures
- height access and working at heights.

Furthermore, you also have an individual responsibility to comply with the use of personal protective equipment, clothing and sunscreen to ensure you remain safe and healthy on the job.

COMPLETE WORKSHEET 1

WORKSHEET 1

To be completed by teachers

Satisfactory ☐

Not satisfactory ☐

Student name: ______________________

Enrolment year: ______________________

Class code: ______________________

Competency name/Number: ______________________

Task: Working with your teacher/supervisor, refer to this text and relevant state or territory information to answer the following questions

1 In regard to workplace risk assessment, what does 'SAFE' stand for?

____

____

____

2 When carrying out a site evaluation, what are four considerations about the weather that you need to take into account?

____

____

____

3 At what incline ratio should you place a ladder before using it?

____

____

____

4 Are standard domestic stepladders and extension ladders permitted for use on domestic roofing jobs?

____

____

____

5 Would a roof worker be permitted to alter the modular scaffolding around a house frame where the height is 6 m?

____

____

____

6 How far should a ladder extend past the edge of the structure being climbed?

---

7 What are three forms of scaffolding commonly used by roof plumbers?

---

8 Name two forms of mobile elevated work platforms.

---

9 What are four things you can do to protect yourself from severe sunburn?

---

10 The glare from unpainted and light-coloured roof sheets can cause damage to your eyes. What form of eye protection does the Cancer Council recommend in its eye protection position statement?

---

11 When moving lengths of gutter or extension ladders around the job site, what is one of the most important things that you need to look out for?

---

12 True or False: The first hazard control measure you must apply on any job is the selection and use of personal protective equipment.

4

ASBESTOS

The intent of this chapter is simply to provide a basic level of awareness about the hazardous nature of asbestos and how it was used historically in the building and plumbing industry. It is not in any sense a guide on how to work with asbestos or undertake its removal. State and territory authorities must be consulted in all instances relating to asbestos. If at any stage in your work activities you suspect that asbestos may be present, you must immediately stop work and inform your employer.

Overview

Asbestos is a hazardous substance that was for many years used in a wide range of products, including construction and roofing materials. General plumbers and, in particular, roof plumbers undertaking renovations, repairs and re-roofing jobs can be considered at risk of asbestos exposure if appropriate precautions are not observed.

Most asbestos roofs do not display the warning signs shown in Figure 4.1, so you need to learn how to identify asbestos-containing materials. Learning about asbestos and how to recognise its presence in the workplace is the only way of protecting yourself and others from its effects.

FIGURE 4.1 Warning signs on an asbestos roof

The inhalation of asbestos dust can lead to the development of:

- malignant lung cancer – uncontrolled malignant cell growth within the lung
- asbestosis – a severe inflammatory disease of the lung tissue that will often progress to lung cancer or mesothelioma
- mesothelioma – a particularly deadly form of cancer most commonly formed in the lining of the lungs which can take 20–40 years to develop. Once a very rare disease, Australia now has the highest incidence of mesothelioma in the world. There is currently no cure.

Asbestos and the illnesses that it can cause cannot be ignored or dismissed as just something that happens to someone else. Many of the older tradespeople you know may be able to tell you of individuals and families affected by asbestos exposure; indeed, the author of this text has lost two direct family members to mesothelioma. Many tradespeople will also have themselves handled asbestos many times in their earlier working life.

Before proceeding with this section, please be aware that asbestos is a *proven* Class 1 human carcinogen. This means that exposure can lead to the development of malignant cancers. It is important to note at the outset that there are no safe levels of exposure to asbestos. While exposure to asbestos does not necessarily mean that you will develop cancer in later life, you should not take the chance. It is not worth the risk.

What is asbestos?

Asbestos is a naturally occurring mineral and has been known to exist by a number of civilisations for thousands of years. However, it was only in the 19th century that it began to be used as an increasingly important resource for the mass production of a wide variety of goods around the world. Asbestos is primarily known for its very effective resistance to heat and chemical breakdown. This led to its widespread use as insulation in many domestic, commercial and industrial applications.

The fibrous structure of asbestos also made it useful as a form of reinforcement in many building products, where it was often mixed with cement and plaster. Mineral asbestos in its unrefined form was generally known under the following three classifications:

- white (chrysotile) asbestos – the most commonly encountered form of asbestos, this is a soft flexible mineral fibre often woven into threads and cloths, which was once used in products such as brake linings, gaskets, sheet cement and fireproof rope.
- brown (amosite) asbestos – this is less widely used, but is still to be found in insulation and sheet cement products
- blue (crocidolite) asbestos – regarded as the most dangerous asbestos product, this was mostly used in cement sheet products until the 1970s.

Products that contain asbestos are often known more collectively by the abbreviation 'ACM' (asbestos-containing materials). Over the last 100 years, at least 3000 products have been identified as using asbestos in their manufacture. The following list of ACM is just a small sample of where asbestos was once used and may still be encountered:

- sheet cement products used in roofing, cladding, ceiling and wall lining
- concrete roof tiles
- pipe lagging
- boiler and water heater insulation
- electrical cable insulation
- domestic stove-top heat diffusers
- water supply and drainage pipes
- plaster cornices
- render
- concrete plant pots
- fire doors
- automotive brake and clutch linings
- paint, bituminous treatments and roof membranes
- vinyl floor tiles and carpet underlay.

Asbestos use in Australia

In Australia, ACM began to be phased out in the 1980s, and they were banned from building products in 1989. However, the stocks of asbestos-containing building products took many years to be depleted, and there were many instances of installation occurring into the late 1990s. This means that the age of any particular installation should not be used solely to gauge its potential for exposure. Asbestos was finally banned in all forms of production or import in 2003, but even since then it has on a number of occasions been found to be illegally imported from overseas so vigilance is always required.

While ACM in construction can be found in every state and territory in Australia, they were particularly popular in those regions where substantial housing growth took place in the post-Second World War period. As a versatile and relatively low-cost product, many thousands of houses in new suburbs of the time were built almost entirely from ACM, including roofs, gutters, exterior and interior walls, ceilings and eaves linings.

Bonded and friable asbestos

ACM are also classified in two separate groups, based upon how the fibres were integrated within each product. Different risk control, working and removal requirements apply to each classification.

It is important to note that, in accordance with the Work Health and Safety Regulations, removal of friable and bonded ACM from the workplace may only be carried out by a licensed removalist. Very specific requirements also relate to the removal of limited quantities of bonded ACM by unlicensed people but this must be seen as subject to state and territory advice and no reference to such work will be made here. At this stage in your training as a roof plumber, you will not be required or permitted to undertake such work.

Bonded (firmly bound) asbestos

Many products were manufactured with asbestos fibres that were mixed and firmly bound within the material itself. A good example is the use of asbestos within cement sheets (fibro). If in good condition, painted and left undisturbed, bonded ACM pose a low hazard.

Examples of bonded ACM include:

- roof sheets and shingles
- exterior wall cladding
- fencing
- eaves lining sheets.

Shutterstock.com/Radovan1

FIGURE 4.2 Bonded ACM in the form of asbestos cement roof sheets

Friable (loosely bound) asbestos

Some products were manufactured with asbestos fibres that were only loosely bound within the parent material. Others were made from 100% free asbestos. Friable asbestos products are defined as being any ACM that, when dry, can be crumbled, pulverised or reduced to powder by hand pressure. Importantly, this may include ACM that were originally bonded but which, through weathering or other degradation, have been reduced to this state.

These friable ACM pose a much greater hazard than bonded ACM. Examples include:

- lagging around hot water pipes
- insulation within water heaters, stoves and domestic space heaters
- some ceiling insulation
- brake and clutch linings
- degraded roof sheets may fall into this category.

Shutterstock.com/Timothy L Black

FIGURE 4.3 Old pipe lagging in ceilings and under floors may contain friable asbestos

LEARNING TASK 4.1

1. Asbestos minerals had unique qualities that made them very valuable to the building industry. What were these qualities?
2. What does the acronym ACM stand for?
3. State the two manufactured forms of asbestos in building products.

Asbestos use in roof plumbing

As a roof plumber, you are likely to encounter asbestos when an existing building requires repair, a new roof or new cladding. You must develop the habit of always assessing each new job to identify the presence of asbestos in any form. While asbestos products were often painted, they could also be left bare and still remain largely watertight and corrosion proof. Originally a light grey in colour, the surface of unpainted asbestos exposed to the weather will often darken into a mottled grey/black colour. The following are examples of ACM that you may find in your daily work.

Roof sheeting, shingles and flashings

The most popular form of asbestos roof sheeting both in commercial and domestic use was a corrugated product known as 'Super Six'. The name came from the measurement of 6 inches (150 mm) between each crest of the corrugations. The sheets came in a range of lengths of between 1500 mm and 3000 mm. However, 1800–2400-mm lengths were most common; these were screwed down by hand to the roof battens (see Figure 4.4). An 8-inch (200-mm) version that was normally used in fencing or cladding can also sometimes be found on roofs.

Less common was a 3-inch (75-mm) profile similar to standard corrugated steel products of the time (see Figure 4.5).

Asbestos shingles were also produced and used on higher pitched roofs. Used in both a diamond and a square pattern, these were meant to provide a building with a look similar to that of traditional slate roofs (see Figure 4.6).

Ridge capping, parapet capping, ventilators and apron flashings were also made to complement the asbestos roof sheeting and shingles.

Gutters and downpipes

Being corrosion proof, asbestos-cement gutter and downpipe systems were particularly popular in seafront areas. Gutters were usually produced in 1800-mm lengths, held up with external galvanised brackets that had joints sealed with bitumen and bolted together (see Figure 4.7).

FIGURE 4.4 Examples of both painted and unpainted Super Six roofing

FIGURE 4.5 Example of the 3-inch asbestos roofing sheet profile

FIGURE 4.6 Example of asbestos roof shingles

A range of spigot/socket downpipe lengths and angles was available to carry the water to an approved point of discharge.

FIGURE 4.7 Example of asbestos gutters and downpipes

Bituminous felt membranes

While the products available today are asbestos free, early bituminous felt membranes containing asbestos were used in low-pitch roofing applications for many years. These included small domestic skillion awnings and bay windows, as well as the more common use on commercial roofs with large surface areas (see Figure 4.8)

Wall sheeting, eaves lining, weatherboards and faux-brick sheeting

Whether you are re-cladding a building or simply fixing downpipe astragals or clips to a wall, be aware that certain wall cladding and eaves lining products contain asbestos (see Figure 4.9). Disturbing such material during removal or screwing fixings through the surface can create hazardous dust. Where regulations permit, there are specific procedures for undertaking such work safely. However, in no circumstances should you proceed with work once you suspect asbestos is present. Inform your employer immediately.

FIGURE 4.8 Example of bituminous membrane roofing

FIGURE 4.9 Super Six wall cladding, gutter and downpipe

Legal requirements

There are national laws and regulations that cover all aspects of working with, removing and disposal of ACM. Through these laws, states and territories are able to 'call up' national Codes of Practice that provide specific detail on the safe management of asbestos issues. Codes of Practice are more easily updated than legislation, and where they are developed nationally they can provide greater consistency in application.

The two key Codes of Practice that you should be aware of are:

- How to Safely Remove Asbestos
- How to Manage and Control Asbestos in the Workplace

These are available for free download from the Safe Work Australia website, at https://www.safeworkaustralia.gov.au/law-and-regulation/model-whs-laws

However, you need to realise that each state and territory may apply different variations, interpretations of and requirements for asbestos hazard mitigation, and therefore you should become familiar with the laws of your own region. For instance, although the 'Code of Practice for the Safe Removal of Asbestos' states that angle grinders, sanders, power saws and high-pressure water sprays are prohibited from use with ACM, a state or territory authority may also list additional prohibited tools, or permit the qualified use of battery-operated tools. Do not assume anything. Check it out!

LEARNING TASK 4.2

1 What ACM is commonly found in the construction industry in your area?
2 Based upon local state or territory requirements, what can and can't you do in relation to the presence of ACM on your jobsite? What actions do you take?

Dealing with ACM

So what do you do if you find or suspect that ACM may be on the job?

If you are not sure of the material you are about to drill into, or you find an ACM flue/vent penetration through the roof, or you simply suspect that a product contains asbestos, immediately stop work and consult your supervisor or site safety representative. Do not start work again until you can be satisfied that appropriate measures have been put in place to remove or control the hazard.

If you suspect you have encountered an asbestos product STOP WORK IMMEDIATELY. Inform your employer and/or supervisor.

It is impractical to provide a detailed examination of the requirements of each state and territory within this text. The following website homepage will provide the current link to the responsible authority in each state or territory: http://www.safeworkaustralia.gov.au.

After you access the relevant link, your instructor and/or supervisor will assist you in determining what requirements apply in your area. Use the general information in this text and the website information for your jurisdiction to help you answer the questions in Worksheet 1.

FROM EXPERIENCE

Unlike previous generations of workers, you are fortunate that through your training as a roof plumber you now have the opportunity to learn about the dangers related to asbestos and thereby avoid future health issues. Always be on the alert for the presence of ACM!

SUMMARY

In this chapter you have been introduced to the following:

- What asbestos is
- The extreme risk asbestos poses
- Where and in what products asbestos was used
- What to do if you encounter asbestos

COMPLETE WORKSHEET 1

WORKSHEET 1

To be completed by teachers
Satisfactory ☐
Not satisfactory ☐

Student name: ______________________

Enrolment year: ______________________

Class code: ______________________

Competency name/Number: ______________________

Task: Working with your teacher/supervisor, refer to this text and relevant state or territory information to answer the following questions.

1 What does the acronym 'ACM' stand for?

____

2 List at least six products that were manufactured with asbestos.

____

____

____

____

____

3 Which of the three mineral forms of asbestos listed in this chapter is considered to be the most dangerous?

____

____

____

4 Referring to local requirements and Appendix G – Recommended Safe Working Practices, from the Code of Practice 'How to Manage and Control Asbestos in the Workplace', describe in a series of dot points the procedure to safely drill through an ACM sheet to affix a downpipe astragal in your state/territory.

____

____

____

____

____

____

____

5 What licence classifications do asbestos removalists require in your area?

6 Provide two examples each of bonded and friable ACM.

7 What is mesothelioma?

8 Who is required to maintain an accurate asbestos register for non-domestic premises?

9 Referring to Appendix G – Recommended Safe Working Practices, from the Code of Practice 'How to Manage and Control Asbestos in the Workplace' and your state/territory regulations, what procedure is recommended for cleaning leaves or debris from ACM roofs/gutters?

10 List at least four tools that are prohibited from use with ACM.

11 Referring to Appendix G – Recommended Safe Working Practices, from the Code of Practice 'How to Manage and Control Asbestos in the Workplace', and where permitted by local requirements, what type of respirator must be used for hand drilling an asbestos sheet?

MATERIAL TYPES

5

In this chapter you will be introduced to the basic characteristics of the following materials:

- galvanised steel
- ZINCALUME® steel
- COLORBOND® steel
- copper
- lead
- stainless steel
- zinc alloy
- aluminium
- plastics.

Sections also include suggested websites where you can find more detail about each product and its uses.

Overview

Before the introduction of mass-produced, profiled metal roofing materials in Australia during the latter half of the 19th century, roof coverings were generally selected from whichever materials were most easily sourced from the local environment. Such materials included bark sheets peeled from certain trees, thatch and sawn timber 'shingles' and 'shakes' individually split by hand before being nailed in an overlapping pattern on the roof frame (see Figure 5.1). It was even common to see flattened kerosene tins used as roof and wall cladding!

FIGURE 5.1 A colonial-era house roofed with hand-split timber shakes

As manufactured materials became easier to source, homes and public buildings were also clad in imported slate and clay tiles, lead and copper sheet. While such materials are still used in certain applications, it was profile metal roofing in the form of the almost iconic corrugated sheet that increasingly became the most common form of roof cladding material used in Australia.

Today there is a huge range of materials suitable for roofing and rainwater goods. On large jobs and most new domestic work, the materials will be predetermined in the specifications of the approved plans. However, in many other instances, including repairs and replacement work, the roof plumber will have a direct influence on the selection of the type and profile of materials, and for this reason you should develop a broad working knowledge of what is available and suitable for particular applications.

Considerations for choosing roof materials

There are many requirements to consider when choosing roof and cladding material. Because each job is different, your selection of material and the reasons for doing so will also change depending on the site, region and building design.

The following factors should be considered when choosing roof and cladding materials:

- cost
- appearance
- compliance with codes and standards
- thermal efficiency
- bushfire resistance
- resistance to high winds
- durability and corrosion resistance
- fixing system
- ease and speed of installation.

New terms

When discussing roofing materials, there are some terms that are often used to describe particular characteristics of a product. A selection of these terms is found in Table 5.1; your teacher may suggest some others that will be useful at this stage.

LEARNING TASK 5.1

1. Which environmental factors should be considered when choosing roof and cladding materials?
2. Which roofing material slowly develops a greenish patina when exposed to weather conditions?

Materials

The most commonly used Australian roof plumbing materials are reviewed below.

TABLE 5.1 General characteristics of roof plumbing materials

Term	Definition	Example
Alloy	A combination of two or more metals, usually to create desirable characteristics in the new matrix.	Steel is an alloy of iron, carbon and other metals in small amounts. Brass is an alloy of copper and zinc.
Ductile	The ability of a metal to be stretched under tensile stress.	Copper is a ductile metal.
Durable	In relation to roof plumbing materials, durability describes a product's ability to withstand corrosion and the effects of the weather and environment.	Aluminium is a durable product that is ideal for use in marine environments.
Malleable	The ability of a metal to be compressed or deformed under compressive stress.	Lead is a very malleable metal that can be worked to form the shape of an underlying material.
Rust	A common term describing iron oxide, formed by the reaction between iron, moisture and oxygen.	The term 'rust' is applied to the corrosion of iron and steel.
Patina	A tarnish that forms on the surface of metals such as bronze, copper and zinc.	Copper that is left to weather outside will slowly develop a greenish patina that prevents any further degradation.
Corrosion	The breakdown of a metal due to chemical reactions with its surroundings.	Iron will rust (corrode) in the presence of air and water.
Aesthetic	In its simplest definition, aesthetic means something that has the quality of being beautiful.	An architect may attempt to design a roof to look aesthetically pleasing. Materials are often chosen for their aesthetic qualities as much as for their performance attributes.

Galvanised steel

Galvanised iron and steel was the most commonly used metallic roofing product in Australia for well over 100 years. The product is made by passing the sheet steel into a bath of molten zinc, which then adheres strongly to the base metal.

The key points to remember about galvanised steel are that it:

- is easily identified by its characteristic large, crystalline surface coating (see Figure 5.2; note the difference in grain structure compared to ZINCALUME® steel)
- has a surface that becomes a dull, mottled-grey colour once weathered (see Figure 5.3)
- is available in selected roof sheet profiles, rainwater goods and flat sheet products
- is increasingly the product of choice on heritage buildings to maintain their original appearance
- may be sealed using solder or most modern sealant products
- has a shorter life than most modern alternatives (subject to environmental conditions)
- is recyclable.

FIGURE 5.2 Galvanised steel compared to ZINCALUME® steel

FIGURE 5.3 Galvanised steel weathered to a mottled grey appearance

ZINCALUME® steel

ZINCALUME® steel became commonly available during the 1980s and quickly replaced galvanised steel as the material of choice for most sheet and rainwater goods (see the LYSAGHT CUSTOM ORB® corrugated sheet in Figure 5.4). The sheet steel is coated with an alloy of 55% aluminium, 43.5% zinc and 1.5% silicone, which provides a high level of protection to the base metal. In fact, ZINCALUME® AZ150 steel has a life span that is four times longer than that of galvanised steel. The next generation AM125 ZINCALUME® is now even more resistant to corrosion due to the inclusion of magnesium to 'activate' the aluminium within the coating.

FIGURE 5.4 Example of a ZINCALUME® steel LYSAGHT CUSTOM ORB® corrugated sheet

The key points to remember about ZINCALUME® steel are that it:

- is easily identified by its characteristic small, silvery, crystalline surface coating
- has a very long service life under normal environmental conditions and correct installation
- has a scuff-resistant surface coating that retains its lustre for a long period
- is available in a wide range of products
- cannot be soldered, but is sealed with neutral-cure silicone
- is incompatible with lead
- is recyclable.

You can find more information on ZINCALUME® steel roofing products at the following BlueScope website: http://www.steel.com.au/products/coated-steel/zincalume-steel

COLORBOND® steel

COLORBOND® steel is the registered trade name used for the Australian-developed pre-painted steel roofing, cladding and rainwater goods product. First introduced in the 1970s, COLORBOND® steel is now available in a wide variety of colours and applications.

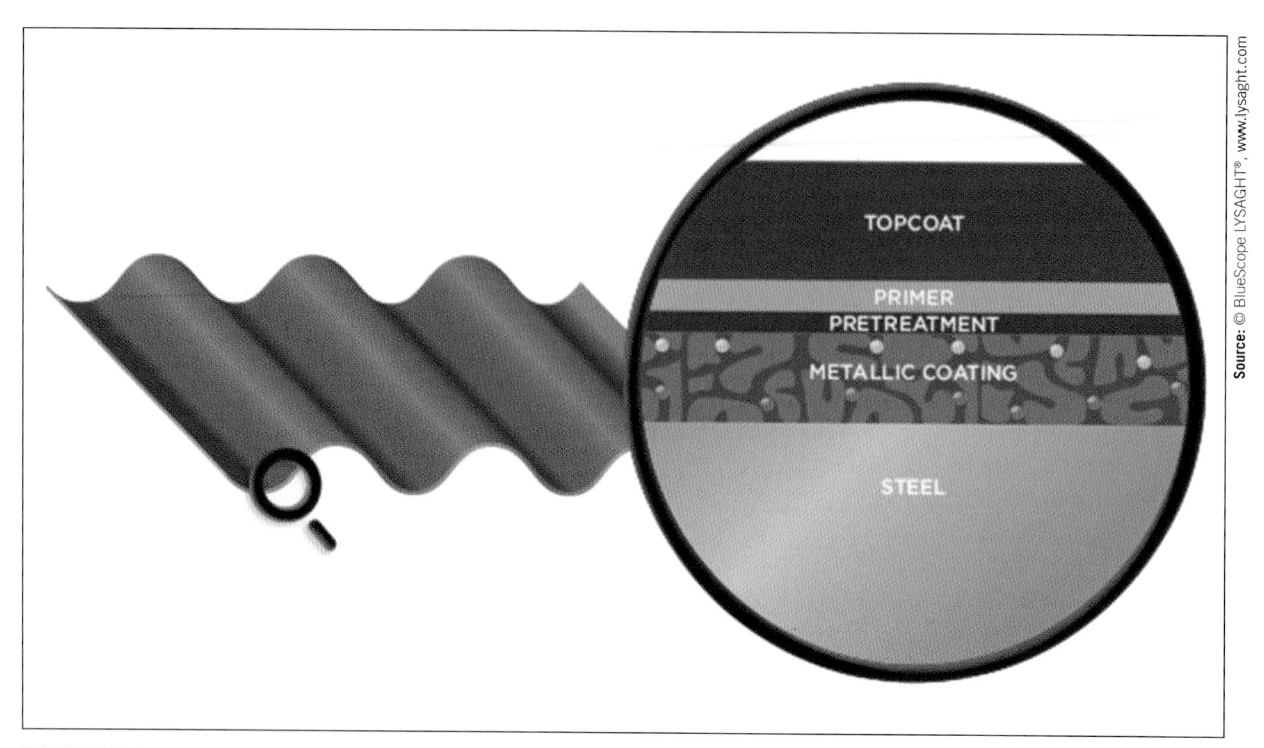

FIGURE 5.5 The different layers that make up COLORBOND® steel products

As shown in Figure 5.5, COLORBOND® steel roofing products have a base metal made from ZINCALUME® steel and are then painted in primers and a baked-on exterior coat.

You will notice that because COLORBOND® steel is based upon ZINCALUME® steel, any scratch in the painted surface will not cause immediate corrosion because the ZINCALUME® steel will sacrifice itself to protect the underlying steel. In addition to the standard COLORBOND® steel range, other products are available that should be specified in locations where aggressive corrosion is more likely, such as coastal and industrial areas. Figure 5.6 provides an indicative selection guide for each option.

The key points to remember about COLORBOND® steel are that it:

- is available in a wide range of classic and contemporary colours
- has a very long service life due to the use of AM125 ZINCALUME® steel base metal and high-quality painted coatings
- is available in a wide range of products, including stainless steel, to suit all environments
- is recyclable.

You can find more information on COLORBOND® steel roofing products at the following website: http://www.colorbond.com

Copper

Copper is available in flat sheet, roofing profiles and rainwater goods (see Figure 5.7). Copper may be used in all applications, but it is most commonly found on commercial and institutional buildings. A range of products is also available for domestic applications.

While copper is an expensive option, it does not rust and, if installed correctly, can last for hundreds of years.

The key points to remember about copper are that it:

- has a long life span
- can be used almost anywhere subject to lower drainage material compatibility
- is relatively expensive to purchase and install, though this is offset by its minimal maintenance requirements and long life
- has a surface finish that weathers to a soft brown/green patina
- is a premium product often specified for high-end customers and institutions (e.g. government buildings, churches, schools)
- can be soldered or sealed with silicone
- is recyclable.

You can find more information on copper roofing and cladding products at the following website: http://www.copperform.com

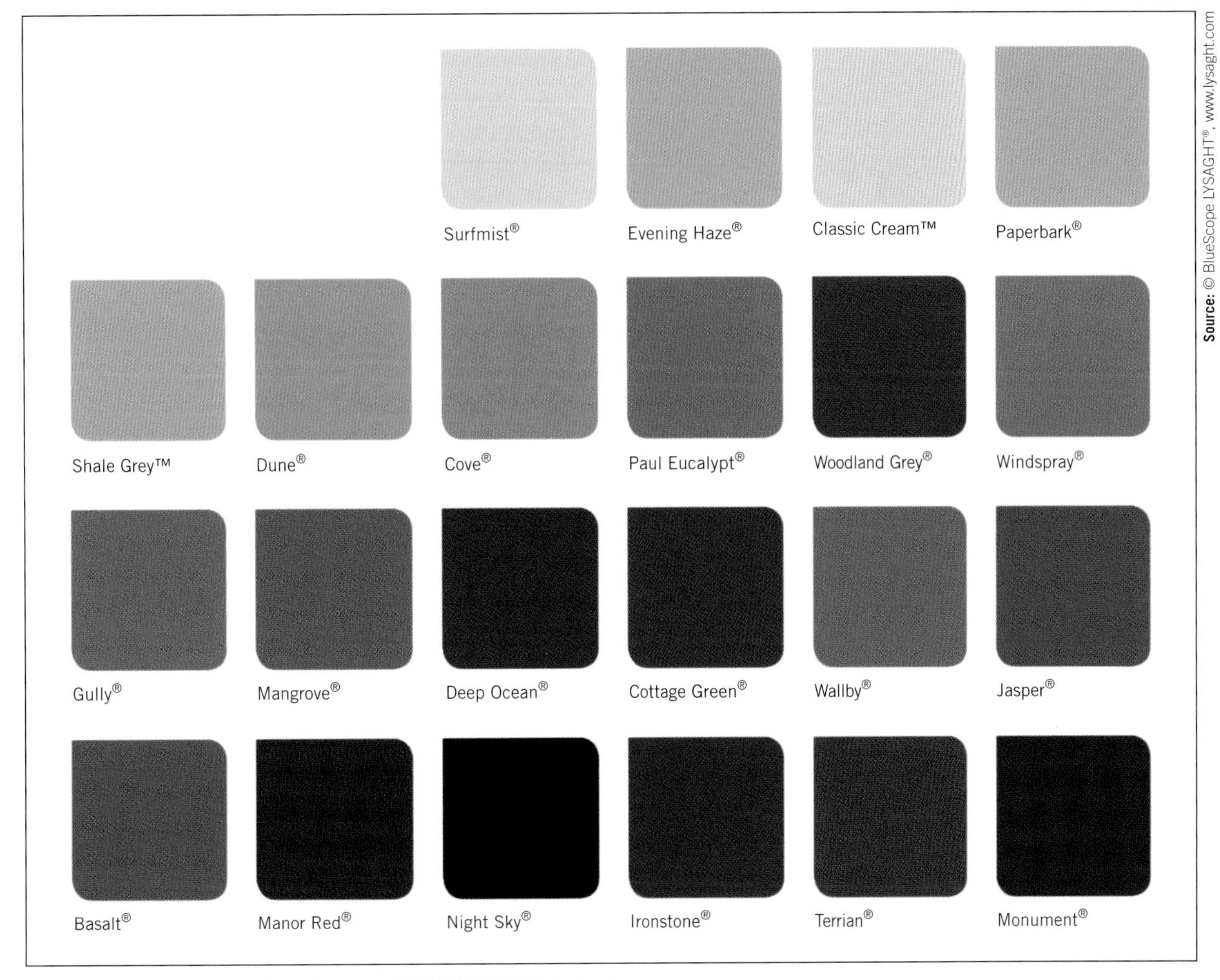

FIGURE 5.6 A small selection of current COLORBOND® steel colours

FIGURE 5.7 Copper standing-seam sheeting, gutter and downpipe

Lead

Lead is a heavy, soft, malleable metal that has been used extensively in the plumbing industry for many years. Lead pipes were used to supply water over 2000 years ago in ancient Rome, and as recently as the early 1970s plumbers would still use lead pipe to make the final connections to waste and soil fixtures. In fact, the very word 'plumber' is derived from the ancient Latin word for lead: *plumbum*.

Today, lead is still used in the modern roofing trade in the form of strip flashing (see **Figure 5.8**), penetration flashings (see **Figure 5.9**) and parapet cappings. Lead is highly resistant to corrosion, and for this reason some institutional buildings may still use lead sheet for roof coverings, gutters and cladding.

In accordance with the NCC Volume Two, lead products cannot be used where the roof water is collected for drinking water.

FIGURE 5.8 An acrylic coated lead slate flashing

The use of uncoated lead for plumbing purposes has declined dramatically in recent years because lead, as a heavy metal, is highly poisonous if ingested or inhaled. Lead poisoning causes extensive damage to the nervous system and can cause brain and blood disorders.

FIGURE 5.9 A weathered lead soaker flashing around an asbestos vent pipe

FROM EXPERIENCE

Never assume that because a product has been approved for use in the past that this is still the case. Check product compliance with the relevant Australian Standard on a regular basis. Labels, brochures and website should clearly state current compliance.

The key points to remember about lead are that it:

- has a long life span
- is highly resistant to corrosion
- is very malleable, and can be worked to take the shape of any surface
- is available in coated products
- is still used extensively on heritage buildings and as penetration and strip flashings for modern tiled roofs (subject to roof water use)
- incompatible with ZINCALUME®
- has a surface that, when uncoated, weathers to a mottled-grey colour
- is recyclable.

You can find more information on lead roofing products at the following website: http://www.cagroup.com.au

Stainless steel

Uncoated steel will rust if exposed to oxygen and moisture. This layer of iron oxide (rust) is an 'active' coating that causes more rust to develop and degrade the base metal. Some other materials will instead develop a 'passive' oxide layer. Once formed, this passive corrosion layer will actually protect the base metal from further degradation. Such metals include copper, bronze, lead and stainless steel.

Unlike bare steel, stainless steel contains a minimum of 11 % chromium. The chromium forms a passive film of chromium oxide over the surface of the steel. While quite thin and invisible to the naked eye, this layer will immediately protect the steel from any damage to the surface and prevent any corrosion of the metal's molecular structure. While some grades of stainless steel do actually stain in certain circumstances, the material will normally retain its original lustrous appearance.

Stainless steel is therefore often specified for projects where its lustre and aesthetic appeal are important (see Figure 5.10). It is also used in severe marine or industrial environments where other products would soon corrode. Rainwater goods can be manufactured from stainless steel sheet to suit any application. It also forms the base metal of COLORBOND® stainless steel painted products.

FIGURE 5.10 A stainless steel roof on a public building

The key points to remember about stainless steel are that it:

- has a long life span
- is highly resistant to corrosion
- is available as coloured products
- can be left with a 'mill' finish or can be highly polished
- is relatively expensive, but can be used for all roofing purposes with almost no maintenance
- is recyclable.

You can find more information on stainless steel roofing products at the following website: https://steelselect.com.au/materials/supadura-stainless-steel

Zinc alloy

Not only is zinc used to coat other products such as galvanised steel and ZINCALUME® steel, but it is also used on its own as a sheet product to produce roof coverings and rainwater goods.

Zinc is a soft, malleable metal that forms a passive, protective patina when exposed to the atmosphere. This provides zinc with a long lifespan, even in severe environments.

Zinc is available as a 99.95% pure alloy (with titanium and copper) for use in roof sheeting, rainwater goods and flashing products (see Figure 5.11).

FIGURE 5.11 A zinc alloy soaker flashing and roll

The key points to remember about zinc are that it is:

- a soft, malleable material that is ideal for flashings
- non-rusting, but will form a protective dull patina
- available in a range of products for use in all environments
- compatible with ZINCALUME® steel and COLORBOND® steel
- easily soldered
- able to be used on any roof to collect rainwater for drinking
- recyclable.

You can find more information on zinc roofing products at the following website: https://euroclad.com.au/zinc-roofing/

Aluminium

Aluminium is a soft, lightweight, ductile and malleable metal that is often used in the roofing industry where resistance to corrosive marine and industrial environments is required (see Figure 5.12). As it is lightweight, it is also a product of choice for certain structural requirements where steel would be too heavy. Aluminium is available in a range of roof covering profiles and rainwater goods, and in a range of colours or a standard mill finish.

Source: © BlueScope LYSAGHT®, www.lysaght.com

FIGURE 5.12 A bulk sulphur storage warehouse in WA clad entirely with aluminium

Bare aluminium has excellent corrosion resistance due to a passive protective layer of aluminium oxide that forms over its surface. Some aluminium gutter installers manufacture continuous-length gutters onsite using portable roll-forming machines mounted in the back of trucks.

The key points to remember about aluminium are that it is:

- lightweight
- durable and requires only low maintenance
- available in a range of colours
- available in rainwater goods and flashings
- recyclable.

You can find more information on aluminium roofing products at the following website: http://www.permalite.com.au/products/roofing

Plastics

A wide range of plastic roofing products exists for use in roof plumbing. Materials include:

- polycarbonate (see Figure 5.13)

FIGURE 5.13 A polycarbonate roof sheet providing natural light for a shed

- PVC
- PVC/acrylic
- polyester
- fibreglass.

Each has its own properties, and you will need to research these materials before specifying on a particular job. Plastic roof coverings are often specified where natural light is required beneath a roof structure. Translucent sheets are available to match many popular roof profiles.

Complete gutter systems are available using PVC and, if installed according to manufacturers' specifications, provide a long-lasting, low-maintenance alternative to other products.

The key points to remember about plastics are that they:

- are lightweight
- are available in a wide range of products
- are corrosion free
- will become brittle over time (most materials)
- may be recyclable (some products).

You can find more information on plastic roofing products at the following website: https://www.palram.com/au/product/suntuf-polycarbonate-corrugated-sheets/

This chapter has given you a basic introduction to roofing and cladding materials. Such products must be selected with reference to their environment and the building requirements. For every job you must also get into the habit of referring to the manufacturer's technical specifications in order to confirm a material's suitability for the application and its compatibility with other materials.

LEARNING TASK 5.2

1. Read the SDS for traditional lead to understand why lead is a hazardous material.
2. Stainless steel is coated with which oxide to prevent rusting?

SUMMARY

In this chapter you have been introduced to the material types utilised for roof sheeting, flashing and drainage products in Australia. Knowing the characteristics of your roofing products and maintaining this knowledge is a core requirement of any roof plumber.

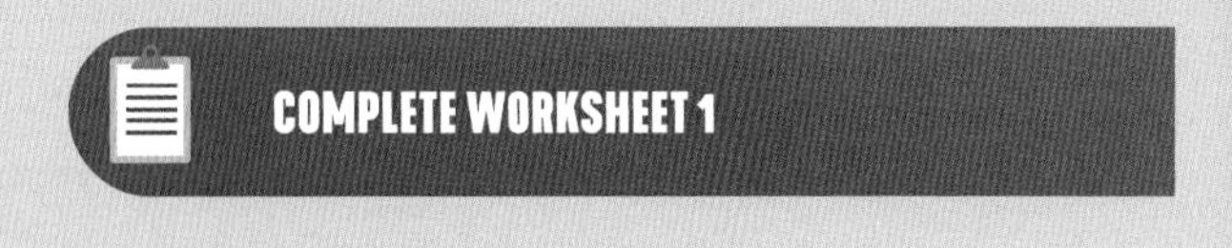

WORKSHEET 1

To be completed by teachers	
Satisfactory	☐
Not satisfactory	☐

WORKSHEETS 5

Student name: ______________________

Enrolment year: ______________________

Class code: ______________________

Competency name/Number: ______________________

Task: Working with your teacher/supervisor, refer to this text and relevant state or territory information to answer the following questions.

1 List at least six factors that you would need to consider when choosing a suitable roof material.

2 Define the following terms.

i ductile

ii malleable

iii patina

iv alloy

3 You have been asked to install a rainwater tank to hold drinking water on a rural property. To ensure that the roof catchment area is safe for the collection of drinking water, which metal would you have to look for on the existing roof and why is it unsuitable?

4 When choosing a product to replace the material mentioned in Question 3, how would you confirm that the new product is suitable for use on a drinking water catchment area?

5 Which metal is used as the protective layer on galvanised steel?

6 Imagine that you have been asked to replace the roof cladding on this boat-shed jetty, but must still retain the look and colour of the original. What would influence you in your choice of new roofing material and what would you choose as a solution?

7 What is the name given to the oxide coating that forms on weathered copper?

8 List at least six factors that you should consider when choosing a roof and cladding material.

9 What are the four substances that are combined to create the protective coating used on AM125 ZINCALUME® steel?

10 List at least four types of plastic roof sheeting.

11 Is a bare lead flashing compatible with a new ZINCALUME® steel roof?

FASTENERS AND SEALANTS

6

In this section you will be introduced to the key fasteners and sealants used in the roofing industry. This chapter will provide you with a basic understanding of the range of products available and their applications and characteristics. The actual selection processes used will be covered later, in Part B of this book.

Overview

A huge range of ancillary roofing products is available today. A roof plumber needs to consider the right product choice every working day, as this is vital for the performance and compliance of every installation.

FROM EXPERIENCE

A compliant roof installation is more than a collection of individual roofing products; it is a complete *system* in which the performance of each product is dependent on the suitability and installation compliance of other components used on the job.

Definitions

There is a considerable variation in terminology across the different states/territories or between regions. The words 'fastener' and 'fixing' are often used interchangeably, and this can be confusing for the learner. Therefore, for the purposes of clarity, the following definitions will apply in this text:

- Fasteners are the products used to actually secure or fasten another roofing product to the building structure. These products include:
  - nails
  - screws
  - rivets
  - bolts
  - nuts and washers
  - chemical, nylon and metallic anchors.
- Fixings are generally regarded as some form of bracket, clip or strap used to secure a roof sheet or rainwater product to the building structure. The fixing is itself secured using some form of fastener, as described above (see Figure 6.1).

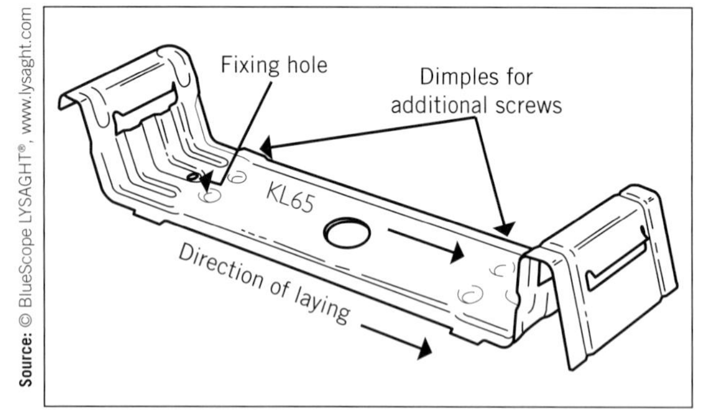

FIGURE 6.1 Diagram showing how roof sheets are secured to the roof using concealed *fixing clips* which are in turn secured by appropriate *fasteners*

Such terms may be different in your region, so discuss with your teacher and supervisor.

Fasteners

The range and quality of fasteners now available to the roof plumber is huge, with innovations and new products appearing on a regular basis. So that you are able to select the correct fasteners for every application, you need first to develop an understanding of the basic characteristics of fasteners and the terminology used to describe them.

In this section you will review the most commonly encountered fasteners used in roof plumbing. These are:

- nails
- screws
- rivets.

Nails

For many years, nails were the principal fastener used for general steel profile roofing. Figure 6.2 provides an example of roofing nails that were once used to secure roof sheets.

FIGURE 6.2 Spring-head and lead-head roofing nails were once used to secure roof sheets

Roof sheets were generally secured with lead-head or galvanised spring-head nails. While initially creating a reasonably effective seal, these nails would often split the underlying timber batten lengthways and, despite the later introduction of a twisted shank, over time they would tend to work loose. With the broader introduction of self-drilling screws in the 1970s, the use of roofing nails quickly declined. However, although roofing nails are no longer used, you may still encounter them during roof repair or replacement work.

Where specified by the manufacturer, it is still permissible to use galvanised clouts and twist-shank nails to secure the fixing clips of some concealed fixed decking profiles to timber battens (see Figure 6.3). Another example of where nails are used is to secure valley gutters. The HB39 states that it is prohibited to penetrate any valley gutter with a screw or nail. Instead, the Code specifies that galvanised nails are to be driven into the valley board and then bent over the edge of a valley gutter. This secures it in position while not penetrating the gutter itself. However, in almost all other instances, screws are now preferred due to their greater holding power and ease of use.

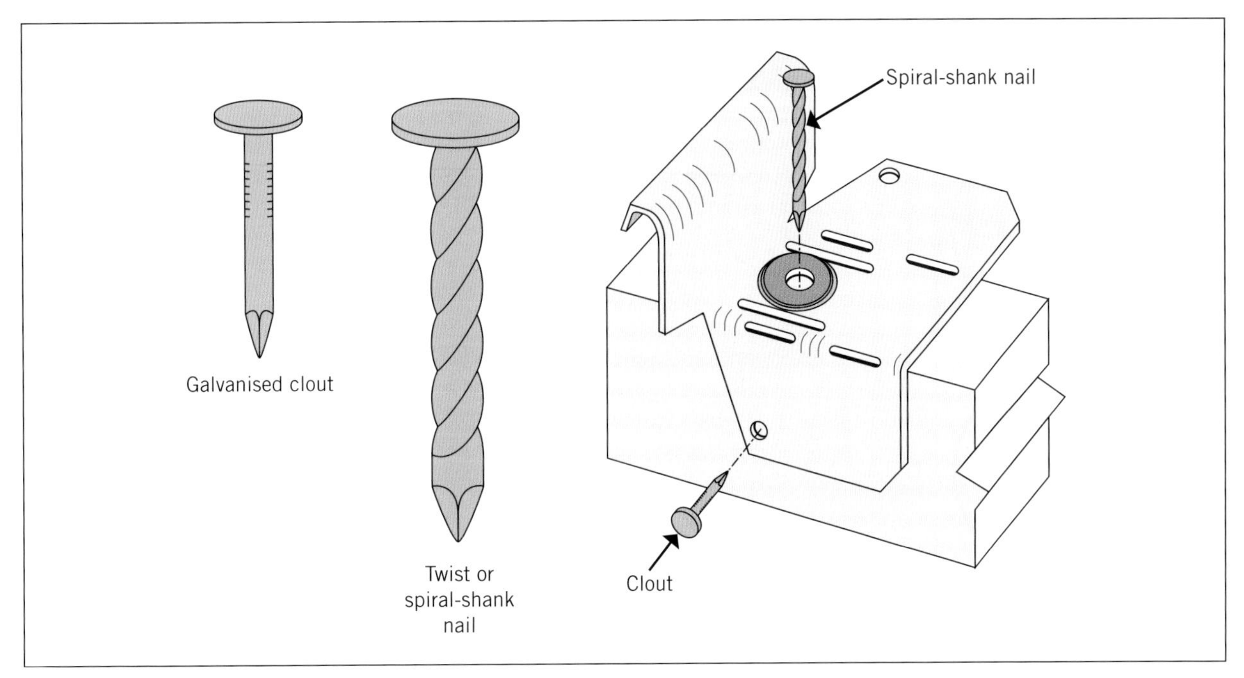

FIGURE 6.3 Diagram showing the use of galvanised clouts and twist-shank nails (subject to manufacturer specification)

Screws

So that you are able to effectively research and specify the correct type of screw for each individual job, it is important that you are able to identify and describe each part or characteristic of a particular product.

Screws can be described according to their *design*. These include:

- types of drive
- head type
- point type
- corrosion coating.

Screws are also described according to their *dimensions*:

- gauge (or outside diameter)
- TPI – threads per inch (or thread pitch in metric use)
- length (mm).

Some of these characteristics are presented as an overview in Figure 6.4, with more detailed explanation throughout this section.

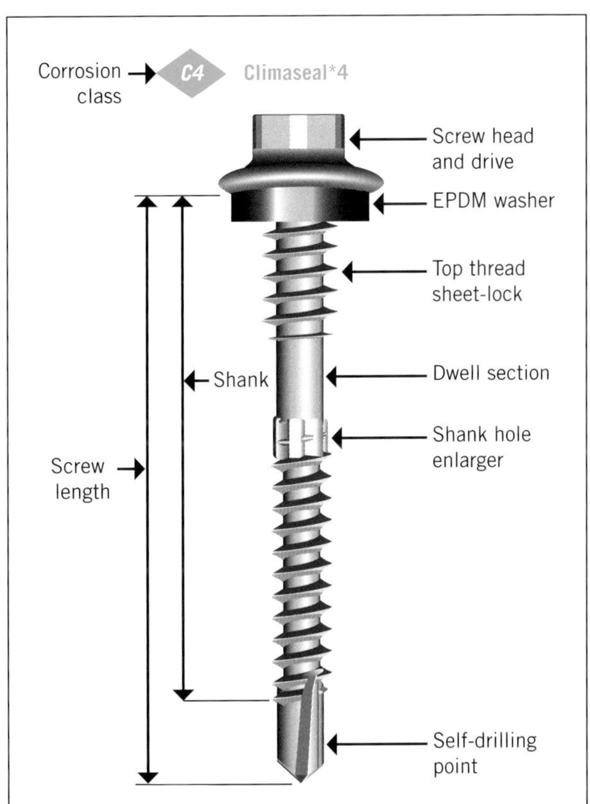

FIGURE 6.4 Characteristics of a typical roofing screw

Source: Courtesy of Buildex®.

Description of screws according to design

Knowing the particular features and design of a fastener allows you to select solutions to meet individual job requirements.

Types of drive

The screw 'drive' is the particular design and method used to turn the screw into place. You need to be able to identify different drives and choose the correct driver bits to suit. Some of the most commonly encountered screw drive systems can be seen in Table 6.1.

TABLE 6.1 Common screw drive designs found in roof plumbing

Drive	Description
Slotted	The single slotted screw features a simple slot across the head of the screw into which a flat-blade screwdriver is placed. Slotted screws were once commonly used in roof plumbing, particularly with asbestos sheets. After pre-drilling with a separate drill, screws were simply driven into place by hand with a screwdriver, or an attachment placed within a hand-driven brace drill. A single slot screw is not suitable for machine driving, and therefore these are rarely seen on construction sites nowadays.
Phillips	The basic Phillips drive is commonly encountered in the roofing industry. It is purposely shaped to cause the driver to 'cam out' when the screw is stalled, in order to prevent damage to the head. There are a number of variations on this form of screw drive, but the most common types are designated according to a number system of increasing size: N°000, N°00, N°0, N°1, N°2, N°3 and N°4. Of these, *the Phillips N°2* is most commonly found in general plumbing. Driver bits will be normally marked as *PH2*.
Pozidriv®	Identified by its characteristic four cross marks adjacent to the driver slots, the Pozidriv® is a later improvement on the original Phillips drive. With straight-sided drive slots, the Pozidriv® lessens the tendency of the driver to 'cam out' and allows greater torque to be applied. While used extensively in Europe, the Pozidriv® is less commonly found in Australia. The Pozidriv® screw should be used only with a matching driver, normally marked as *PZ*. While a standard Phillips driver can be used, it fits poorly and will often result in damage to the screw head.
Internal hexagon 	Roof plumbers will normally encounter the internal hexagon drive when using bugle head batten screws. Such screws are driven with a 5 mm internal hex driver.
Hexagon	Of all the screws used in roof plumbing, it is the external hexagon drive that is the most common. The six-sided shape allows high-torque power drivers to be used, and is generally manufactured with an integral flange washer that increases the bearing surface area. Drivers are available in three main sizes: 1/4", 5/16" and 5/8". Driver bits will normally include an integral magnet to hold the screw in place better.

Source: Courtesy of Buildex®.

Head type

Screws can also be described according to the style of head used. Each shape has particular characteristics that enable different materials to be secured. The correct screw head designs must be selected to suit the job materials and applications, as using the incorrect screw head shape may mean that your material is poorly secured or may leak. A small selection of screw head types is shown in Table 6.2.

Point type

Modern roof product installation relies on a wide range of self-drilling screws to be used on a very wide selection of structural materials. Therefore, to ensure that your installation proceeds efficiently and is also compliant, you need to select the correct point type for the job (see Table 6.3).

TABLE 6.2 Screw head designs

	Head	Description	Driver type
	Hexagon head with integrated washer face	The integral washer face applies better loading on the material surface and can accommodate a watertight EPDM sealing washer.	Hexagon drive socket: ¼", 5/16" or 3/8"
	Wafer head	Often used by roof plumbers to fasten clips and concealed sheet fixings. The wafer head prevents dimpling of the roof sheet when walked on.	Phillips N°2 cross recess bit
	Countersunk head	Designed to countersink into a timber surface and leave a flush or recessed finish. You would not use such a head to affix sheet metal.	Phillips N°2 cross recess bit
	Bugle head batten	The ribbed bulge shape embeds into timber to provide a flush surface. Used to fasten timber battens to rafters in cyclonic and high-wind areas.	5 mm internal hexagon bit
	Special pan head	Low profile, self-sealing screw used to fix mini-corrugated sheeting without a washer or deforming the sheet.	Phillips N°2 cross recess bit
	Headlok® security head	Where sheeting or components must be vandal- and tamper-proof, unique security-type screws may be specified.	Buildex® security head socket
	Designer head	A smooth profile screw that can be specified where the aesthetic appearance of the job has a high importance.	Hexagon drive bit

Source: Based on information from Buildex®.

TABLE 6.3 Different point designs for different materials

Drill point	Type	Use
	Type 17	A self-drilling screw for fastening into timber.
	Teks®	A self-drilling screw for fastening into standard-thickness steel purlins.
	Teks® Series 500	An extended point for self-drilling into thick section steel.
	Zips® point	A multi-purpose point for screwing into soft and hardwood timber, lightweight metal battens and purlins.

Source: Based on information from Buildex®.

EXAMPLE 6.1

A standard Teks® screw will often jam, spin or even break if used for drilling through thick section steel. Teks® Series 500 with an extended point would be a more suitable choice.

Corrosion coating

All screws used for the installation of roofing products must be protected from corrosion. The standard that specifies the minimum performance requirements for fasteners is AS 3566. Fasteners can be selected based on four corrosion class categories to suit the environment (see Figure 6.5).

AS 3566 'SELF DRILLING SCREWS'

Class 3 and Class 4 screws are coated with high-density sacrificial coatings that protect the screw from corrosion for many years. In some cases, modern impact drivers may damage this coating so always check the fastener manufacturer's installation guide before using. While coastal areas are nominally regarded as Class 4 zones, screw selection should always take into account local topography, wind direction and levels of industrial fallout. This means that you may need to use a higher-class screw even if you live further inland. Figure 6.6 provides an example of Australian corrosion zones.

Description of screws according to dimension

Job specifications and manufacturer installation guides will provide very specific details relating to fastener dimensions that must be adhered to.

Class 1
Only used internally where corrosion resistance is a minor concern.
Example: zinc/yellow plasterboard screws.

Class 2
Internal use where condensation is likely to occur.
Example: electroplated zinc/yellow screws.

Class 3
Intended for mild mid-industrial and marine applications.
Example: all external roofing applications outside of Class 4 zone.

Class 4
External marine and moderately severe corrosive environments within 1 km of marine surf.
Example: all external roofing applications within Class 4 zone.

FIGURE 6.5 Corrosion class coatings for fasteners

Gauge

Screw 'gauge' is a traditional measurement method that describes the screw outside thread diameter as a single, arbitrary number (not related to actual diameter).
At the time of writing, Australia is in the process of shifting screw diameter descriptions to the full metric format, but this transition will take some time and therefore you must be familiar with the traditional system. In Table 6.4 you can see the most common screw gauge sizes in use compared with nominal metric sizing.

TABLE 6.4 Screw gauge and nominal diameter

Screw gauge	Nominal diameter	
6g	M3.5	
8g	M4	
10g	M5	
12g	M5.5	
14g	M6.5	
15g	M7	

Source: Courtesy of Buildex®.

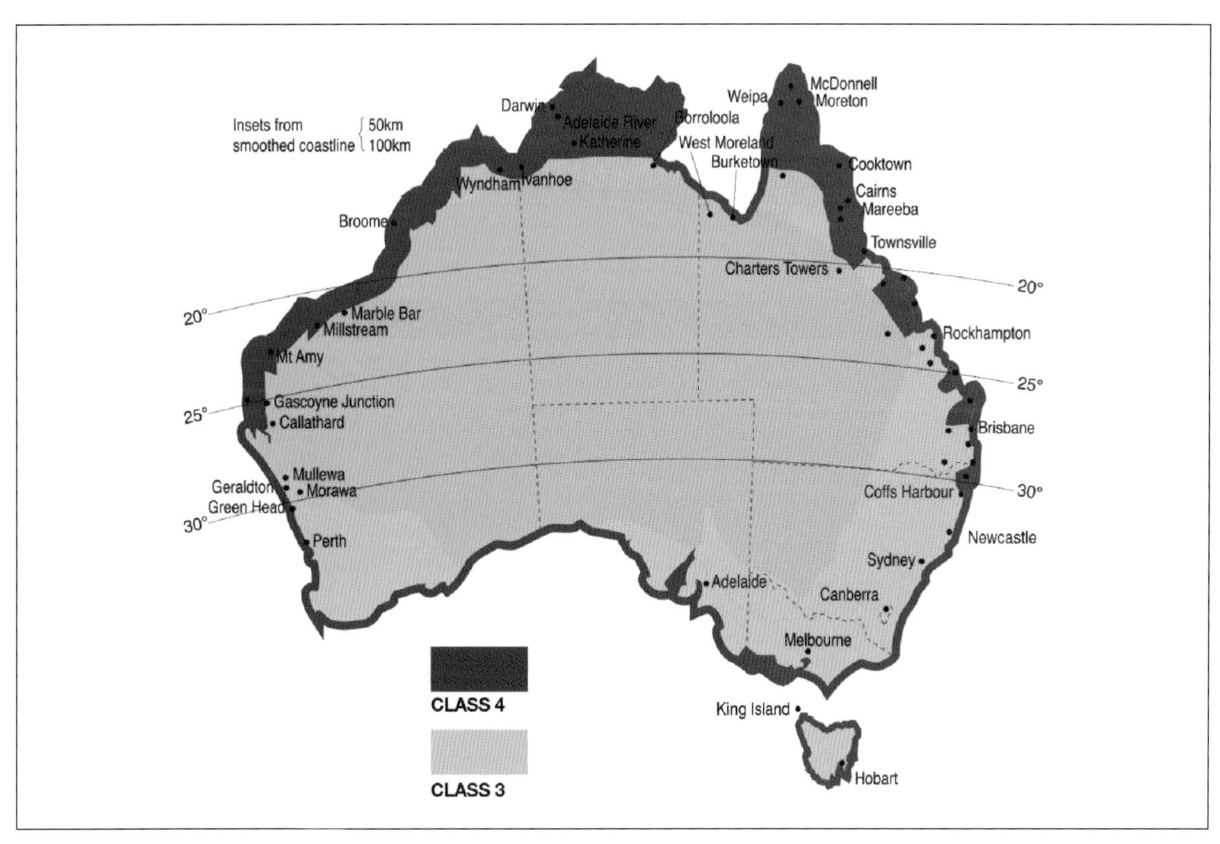

Source: Courtesy of Buildex®.

FIGURE 6.6 A simple guide to different corrosion zones around Australia

Thread types and TPI

A thread is a helical, machined ridge around the outside of the screw shank that converts rotational force into lineal movement. In other words, the thread leads the screw into the material as it is rotated. Once the screw is in position, the frictional surface area and pitch of the thread also prevent the linear forces (such as wind or mass) from causing the screw to turn out. Table 6.5 lists the applications of thread types and TPI (threads per inch).

Threads are often described as being 'coarse' or 'fine'. These terms relate to the 'pitch' of the thread and also to the number of 'threads per inch' (TPI). *Pitch* describes the distance from the crest of each thread (see Figure 6.7). The term 'pitch' will eventually replace TPI when describing threads.

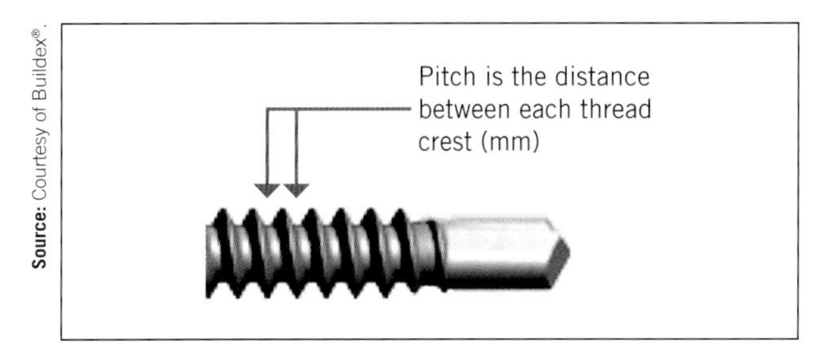

Source: Courtesy of Buildex®.

FIGURE 6.7 The pitch of screw threads

Coarse threads have a larger pitch and fewer threads per inch (up to 16 TPI) and as such have a larger thread form in relation to screw diameter.

Fine threads have a small pitch and more threads per inch (over 16 TPI), and as such have a smaller thread form in relation to screw diameter.

EXAMPLE 6.2

A Type 17 self-drilling screw designed to use on timber battens has a coarse thread of 12g–11 TPI.

EXAMPLE 6.3

For securing cladding, clips or fittings to a thick-walled, structural steel frame (3–12 mm) you could use a Series 500 Teks® with a fine thread of 12g–24 TPI.

Why do I need to know about threads?

As a roof plumber, you must gain an understanding of fastener threads in order to deal with the many structural materials to which you need to secure roofing products. Using the wrong fastener for particular materials could be disastrous.

TABLE 6.5 Applications of different thread types and TPI

TPI (threads per inch)			
Thread types and relationship to TPI		Application	Examples
	Coarse threads – up to 16 TPI	Normally used with timber and thinner metal applications (metal 1–5 mm thick).	10g–24 TPI, 12g–24 TPI
	Fine threads – over 16 TPI	Only used in thicker steel applications (metal 2–12 mm thick)	12g– 24 TPI, 14g–20 TPI
	Buttress thread	This type of thread has the best holding power in thin metal. The common Teks® screw uses this thread.	15g–15 TPI, 14g–20 TPI
	Taptite thread	This form of thread has superior locking and resistance to loosening in both steel and plastics. They are screwed into pre-drilled holes and tap their own thread as they go.	14g–20 TPI
	Twin start thread	The unique shape of the twin start thread causes the screw to move forward at twice the normal speed during use.	7g–16 TPI, 8g–10 TPI

Source: Based on information from Buildex®.

EXAMPLE 6.4

If you were undertaking a re-roof of an older, timber-framed dwelling and purchased metal batten Teks® (15g–15 TPI) instead of the correct Type 17 (12g–11 TPI), you might end up being held responsible when all the roof sheets tear out of the timber battens during the next high-wind period.

Knowing about thread characteristics is vital to your work as a roof plumber!

Length

When you come to order screws from your supplier, you must quote the length required in millimetres. The type of screw head determines how screws must be measured and specified. Screw length is measured in two different ways depending on whether it is pan head or countersunk. The difference can be seen in Figure 6.8.

- Screws with a flat bearing face are measured from the *underside* of the head to the point end.
- Screws with a countersunk or embedded head design must be measured from the *top* of the screw head through to the point end.

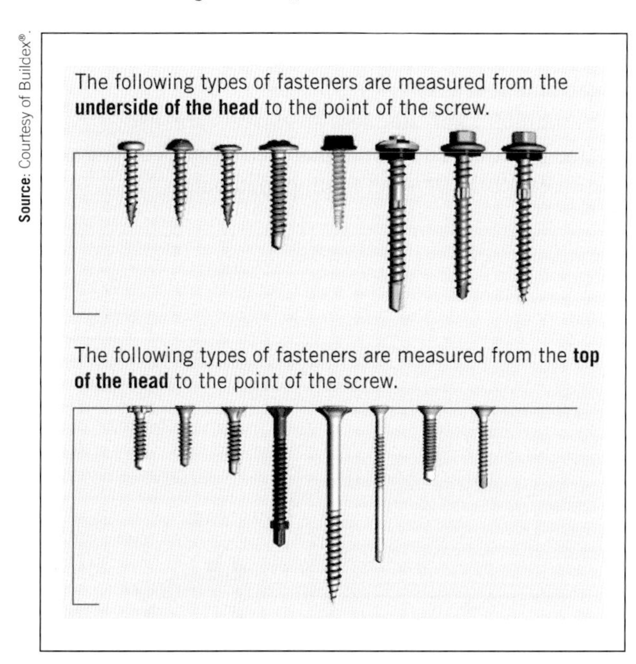

Source: Courtesy of Buildex®.

FIGURE 6.8 Measuring and specifying the length of screws

Standard screw identification format

For consistency, screws are identified and specified using a standard number code and format. The use of a standard description means that whether you are reading approved plan specifications, looking at product brochures or ordering screws from a supplier, everyone is familiar with the format and inaccuracies are reduced.

The number code layout is typically as seen in Figure 6.9.

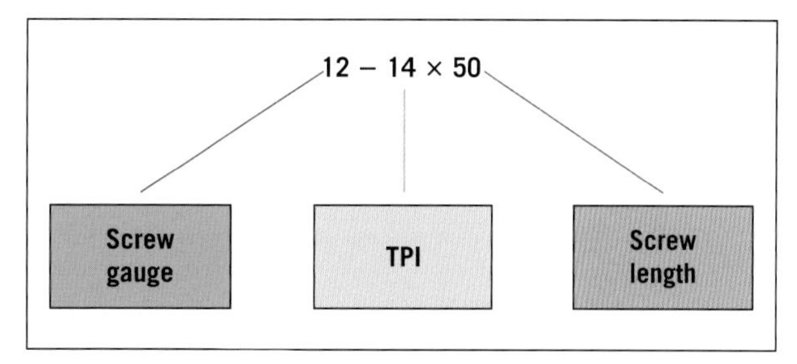

FIGURE 6.9 Standard screw identification format

When selecting or ordering your screw fasteners from your supplier, you can check all of these details from the carton or packet descriptions to ensure that they meet the job requirements (see Figure 6.10).

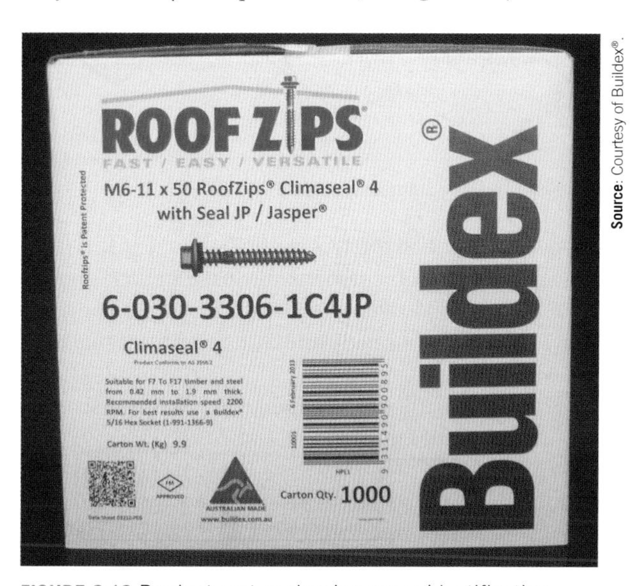

Source: Courtesy of Buildex®.

FIGURE 6.10 Product carton showing screw identification

Rivets

Early rivet types used in the roofing industry required access to *both* sides of the sheet material. Generally known as tinman's rivets, they were inserted into the pre-drilled hole then drawn and crowned (mushroomed) off with a riveting tool, hammer and backing bar (see Figure 6.11). While you will occasionally encounter old installations with short-length gutters joined in this fashion, this style of jointing is now obsolete in the modern roofing industry, and it is mentioned here only for the purpose of comparison with the modern 'pop' or 'blind rivet' (the term 'pop' rivet is actually a commercial trade name).

The introduction of blind rivets enabled sheet metal to be joined together from one side only. The far side of the sheet was the 'blind' side, and therefore these fasteners became known as blind rivets. In recent years the term blind rivet has begun to be used erroneously in reference to what are actually 'closed-end' (sealed) rivets. Basically, all rivets with a detaching mandrel are blind rivets, but some have a sealed end instead

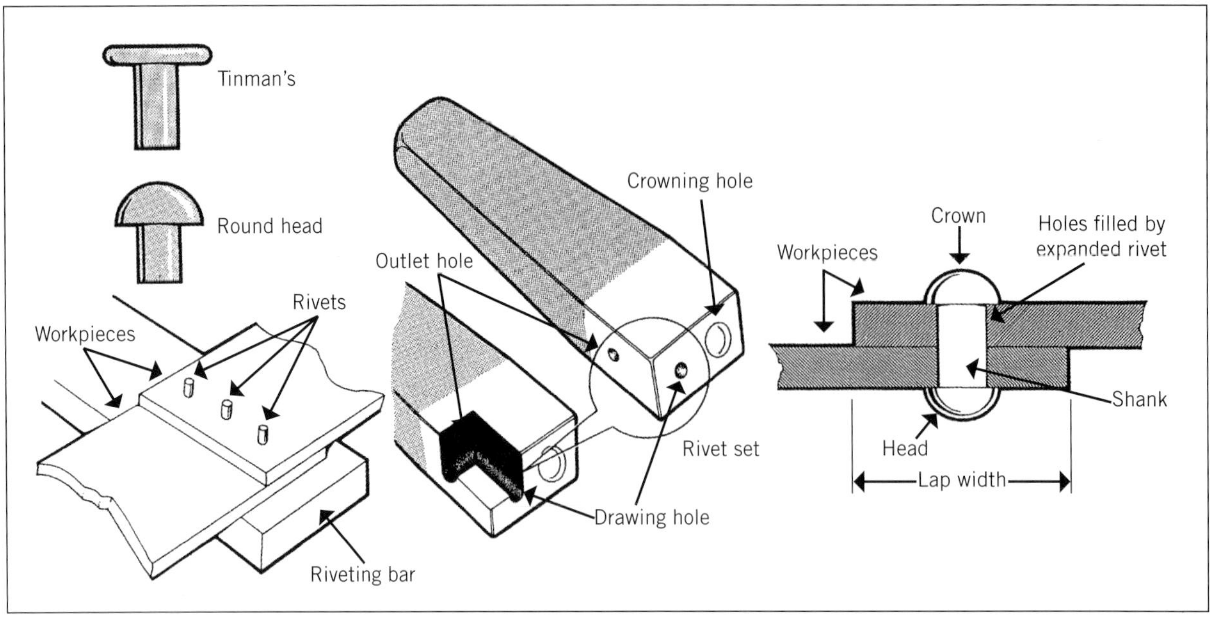

Source: Based on Basic Training Manual 12-1 *Roof Plumbing, Introduction and Downpipes*, Australian Government Publishing Service, 1981.

FIGURE 6.11 Early tinman's rivets

of open tubular body. Check with your supervisor and teacher to confirm the terminology used in your area.

A typical blind rivet consists of a compressible rivet body, a rivet head or body and a snap-off steel mandrel (see Figure 6.12).

Blind rivet bodies are available in a range of sizes and material types, including:

- steel
- aluminium
- stainless steel
- copper.

Like all other fasteners, it is vital that you select the rivet body and matching mandrel material to suit the product you are working with in order to avoid any sort of galvanic reactions taking place.

Aluminium and stainless steel rivets are also available in over 20 colours matched to suit current and past COLORBOND® steel roofing materials.

Source: Based on Basic Training Manual 12-1 *Roof Plumbing, Introduction and Downpipes*, Australian Government Publishing Service, 1981.

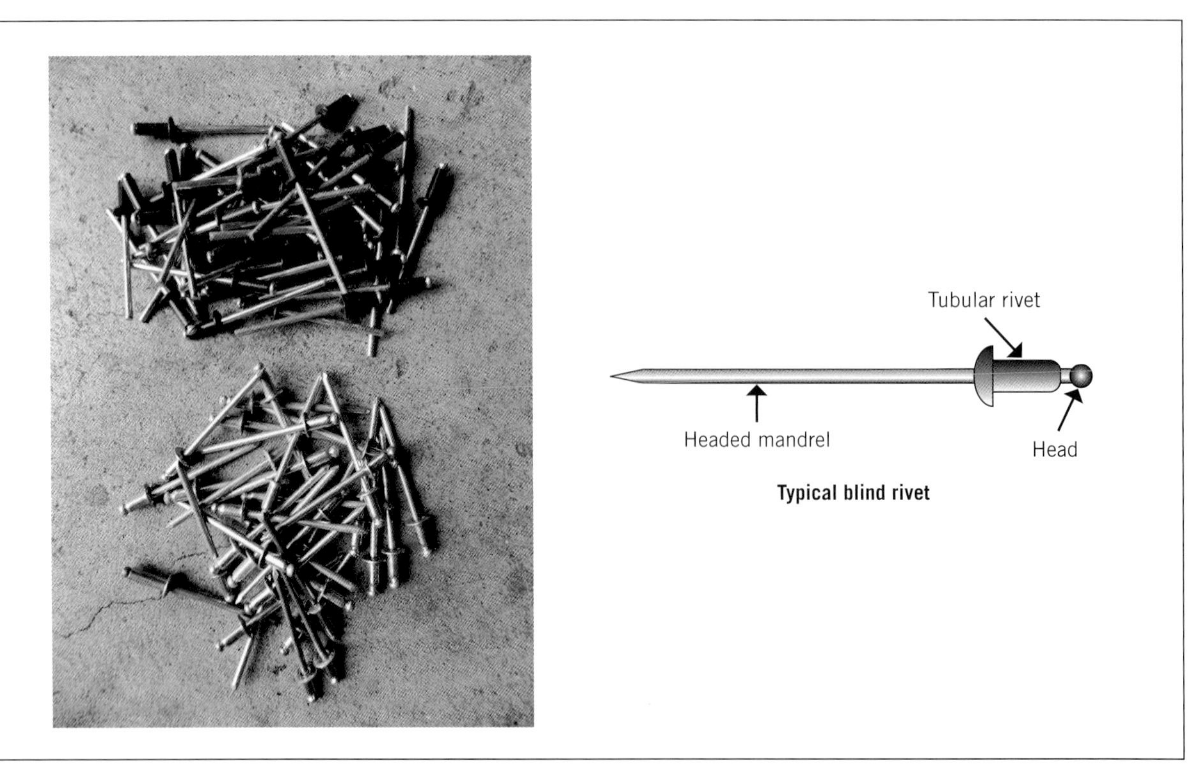

FIGURE 6.12 (a) Typical blind rivet design (b) Plain and painted blind rivets

PART A

6

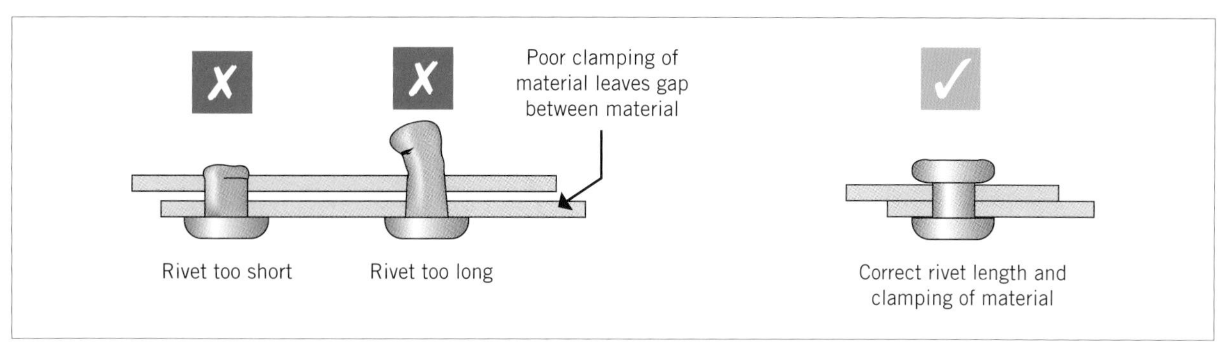

FIGURE 6.13 Ensuring that rivets are not too short or too long

EXAMPLE 6.5

- If you are working with COLORBOND® stainless steel, you need to use rivets with a stainless steel body and a stainless steel mandrel.
- Installation of standard ZINCALUME® steel or COLORBOND® steel requires the use of aluminium rivets.

Rivet lengths should be chosen so that the rivet protrudes the correct distance through the material, ensuring that when the rivet body is compressed, the material is tightly clamped together (see Figure 6.13). If the rivet is too short or too long, a poor joint will result.

The rule of thumb when choosing the correct sized rivet is to allow the body of the rivet to protrude through the material approximately one and a half times the diameter of the rivet body. This is fine in most cases, but it is always better to refer to the manufacturer's specification charts and determine which material thickness a rivet is suitable for.

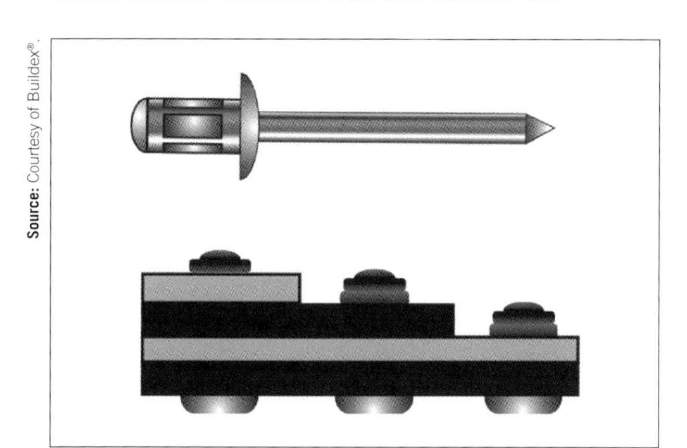

Source: Courtesy of Buildex®.

FIGURE 6.14 The wide grip range of the Buildex® multigrip rivet

LEARNING TASK 6.1

1 Identify the four important characteristics when describing the design of screws.
2 Identify the three important characteristics when describing the dimensions of screws.
3 What does the acronym TPI stand for?
4 Why is it important to select the correct rivet length?

Fastener technology is always advancing and rivets are now available that are capable of securing material of multiple thicknesses within a specified range. These enable you to use one rivet for multiple tasks around the site (see Figure 6.14). You need to refer to manufacturers' websites regularly in order to keep up to date with the latest innovations.

Sealants for roof plumbing

The most commonly used and most appropriate sealing product in the roofing industry is neutral cure silicone rubber. It is important to emphasise the term 'neutral cure' sealant. A huge range of sealants is available, but many contain aggressively corrosive by-products that will damage the surface of roofing materials during the curing phase. Such sealants can often be identified by a very strong vinegar or ammonia-type smell.

Before using a silicone sealant, check the product information on the tube or carton and ensure that it is labelled 'neutral cure' (see Figure 6.15) or sometimes 'acetic-acid free'.

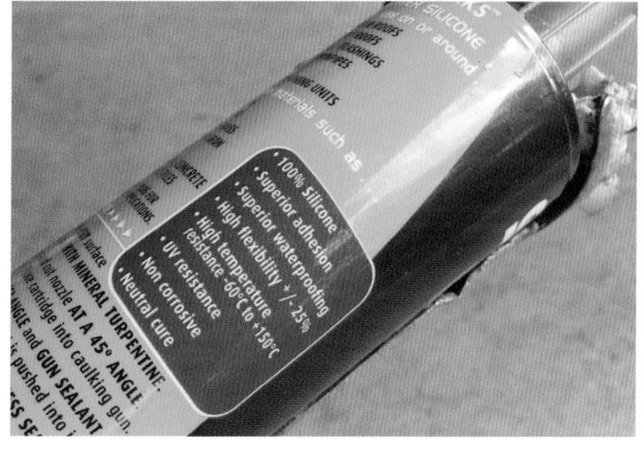

FIGURE 6.15 Sealant product information showing it is neutral cure

Silicone sealant has a number of advantageous characteristics for roof plumbing use, including that it:

- is easy to use and apply with the use of simple caulking guns.
- allows a reasonable 'working time' before curing.

- can be purchased from almost all hardware stores and comes in a range of colours to suit pre-painted steel installations
- is water resistant, UV resistant and non-corrosive
- is resistant to extremes of heat and cold
- can be used as a gap-filler where required
- is very flexible
- has a long service life (of at least 20–30 years).

There are some limitations with some silicone sealant products, and these include the following:

- Standard sealants cannot be painted over, so you must limit the amount of 'squeeze-out' left on the job.
- Silicone sealants can be difficult and messy to clean up where excessive amounts have been used.
- The sealant must be applied to dry joints that are free from production oils. An appropriate solvent must be used to remove any oil (mineral turpentine is often recommended).
- Neutral cure sealants are not generally designed or classed as adhesives. You must always ensure that the joint to be sealed is appropriately fixed with some form of mechanical fastener.

LEARNING TASK 6.2

1 Identify two advantageous characteristics that silicone sealant has when exposed to environmental conditions commonly found on a roof.

COMPLETE WORKSHEET 1

Acknowledgement

The author would like to thank Buildex® for its generous permission to access the technical information drawn upon in this chapter. Readers are encouraged to learn more about fasteners from the following website: buildex.com.au

SUMMARY

Knowledge of fasteners is vital to your work as a roof plumber. In this chapter you have been introduced to:

- fastener drive types
- fastener point types
- corrosion coatings
- thread types
- how fasteners are classified and measured.

You also covered the requirement to use only neutral cure sealants on all rainwater drainage products.

WORKSHEET 1

To be completed by teachers	
Satisfactory	☐
Not satisfactory	☐

Student name: ______________________________

Enrolment year: ______________________________

Class code: ______________________________

Competency name/Number: ______________________________

Task: Working with your teacher/supervisor, refer to this text and relevant state or territory information to answer the following questions.

1 List at least six different types of fastener used in roof plumbing.

____

____

____

2 What is one application where the use of galvanised nails may still be appropriate?

____

____

____

3 How would you tell the difference between a Pozidriv® and Phillips N°2 driver?

____

____

____

4 What are the three main sizes of hexagon driver used in the roofing industry?

____

____

____

5 Which of the following four screw point types are suitable for securing into timber? Circle the correct answer.

Zips® Point   Teks®   Type 17   Teks® Series 500

6 In the table below, draw a line to match the correct name with the image.

Screw feature	Name
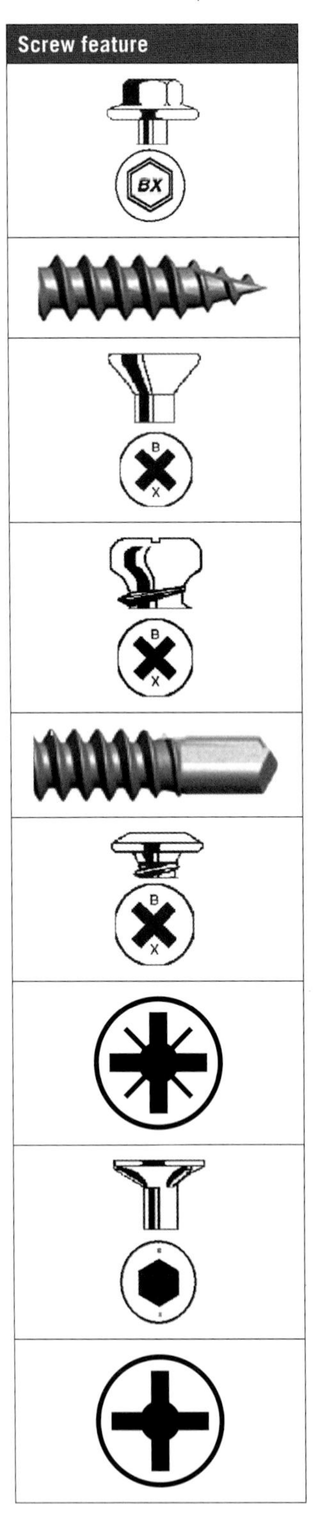	Teks® point
	Special pan head
	Bugle head batten
	Type 17 point
	Pozidriv®
	Wafer head
	Phillips
	Hexagon head
	Countersunk head

Source: Courtesy of Buildex®.

7 If you were planning to replace the roof of a house that was situated approximately 400 m from the beachfront, what corrosion coating would you specify for your roof screws?

____

____

____

8 What class of screw corrosion coating should be specified in your own area?

9 You purchase some cartons of Teks® for your next job. On the side of the box the following information is listed:

i 12 – 14 × 80

ii Class 3

What do these references mean?

10 When purchasing silicone sealant that is suitable for roof and gutter work, what is the key characteristic you must look for to confirm that it is appropriate?

11 Would it matter if you used steel rivets on a ZINCALUME® steel flashing? Explain your answer.

12 Prior to installing a new flashing on a roof, you notice that the flashing material is covered in a light, oily substance. What should you do before proceeding to apply silicone?

13 What size and drive type would you typically find on a bugle-head batten screw used for securing timber battens to the rafter?

CAPILLARY ACTION

When it all comes to down to it, the primary task of a roof plumber is to keep the weather out. You therefore need to develop an eye for anticipating where wind and water can get into a building, and one very significant phenomenon that a roof plumber must be aware of is capillary action.

Overview

Capillary action and the effects of water surface tension have a significant influence on your work as a roof plumber. Failure to install roof components in a manner that will counter the effects of capillary action will lead to leaks into the building. In this chapter you will review how capillary action is caused and what you must do to stop it.

How capillary action works

In most cases, the roof plumber is concerned about the water that flows *down* the roof. In some instances, however, water can actually flow *upwards* against gravity and leak into a building. This physical phenomenon is known as capillary action.

If the gap between two surfaces is small enough, the combination of water surface tension and the forces of molecular adhesion between the water and the close physical surfaces act to lift the liquid upwards. Look carefully at Figure 7.1. Two sheets of glass are held in a container filled with water. As the two pieces of glass are brought together, water is drawn *upwards* through capillary action.

The same thing can happen between two sheets of metal. Unless some form of capillary break is built into a flashing or roof sheet, water can be drawn upwards and leak into the building.

You have now seen how water rises between two sheets of glass. Now look at a real-life example in Figure 7.2. The incorrectly lapped sheets on this roof were in direct contact causing water to be drawn upwards and leak directly into the roof space.

Source: Heath Bussell

FIGURE 7.2 Poor sheet installation causing capillary action leakage into a building

In this experiment, two sheets of glass (representing sheet-metal flat surfaces) have been placed in a container filled with blue-coloured water. As the two flat surfaces are gradually brought together, you can see how the water rises due to capillary action. Though not shown here, the water eventually reached a height of 150 mm!

Capillary action revealed

Surfaces 5 mm apart

Surfaces 3 mm apart

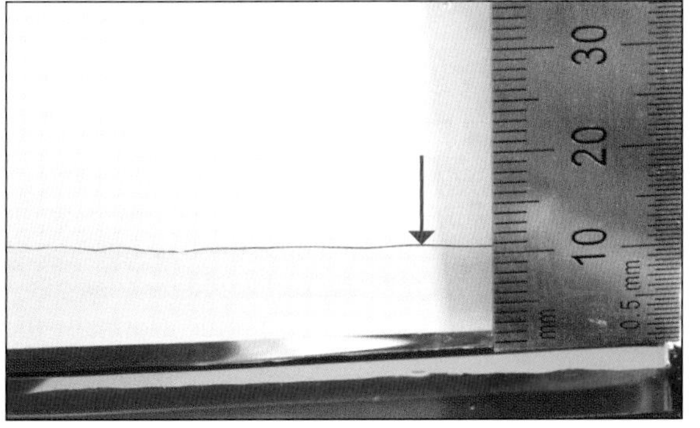

Surfaces 1 mm apart

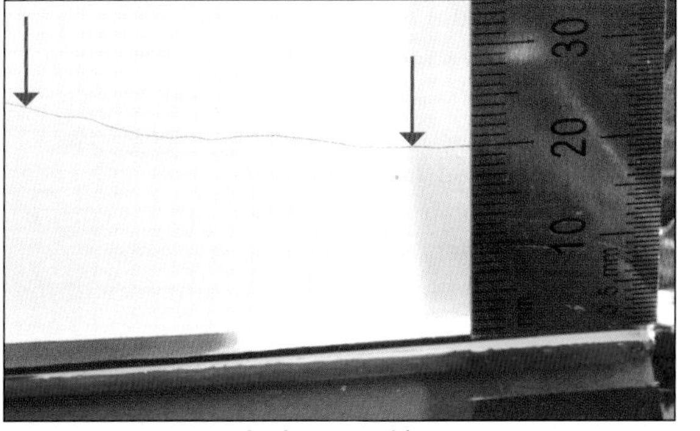

Surfaces touching

FIGURE 7.1 Water rising against gravity between two sheets of glass

Preventing capillary action

Wherever two pieces of flat material (normally metal) are in contact, it is likely that capillary action may occur. To counteract this, you need to provide a 'break' between the two surfaces. An anti-capillary break can be built into materials by providing an air gap between the two sheets, most commonly by creating a fold in the overhanging sheet.

Figure 7.3 shows the difference between a hanging flashing with no capillary break and one with a break included. The cross-section on the left shows how water droplets that are flowing down the hanging flashing are drawn into the tight space between the two sheets (the gap between the metal has been increased for clarity in the diagram). Capillary action then draws the water upwards where it may leak behind the apron flashing.

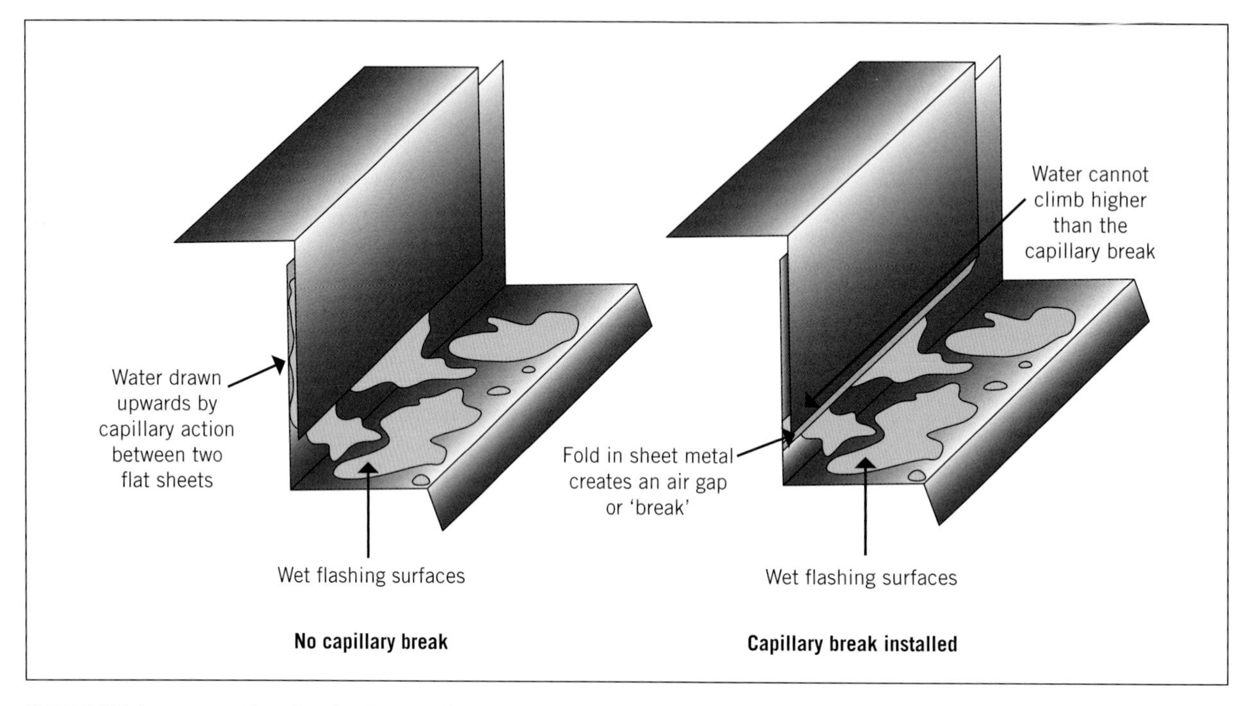

FIGURE 7.3 A cross-section showing how capillary action can be controlled

Look now at the cross-section on the right. A slight fold has been placed at the bottom of the top hanging flashing. The water is captured by the contact between both sheets of metal but the wider gap created by the capillary break prevents the water from rising any further. Any hanging, parapet or over-flashing of any kind must have some form of break folded into the metal.

Capillary action in roof sheet laps

Capillary action could also happen where roof sheets overlap each other, except that manufacturers have designed capillary breaks into the shape of the sheet lap profile (see Figure 7.4). This capillary break runs the full length of one side of the sheet and each sheet has a specific over-lap and under-lap. Always ensure that the sheet is oriented the correct way and fastened carefully to maintain the performance of the capillary break. As long as sheets are fitted to the correct roof pitch in accordance with manufacturers' specifications and screwed correctly, they will remain watertight.

FROM EXPERIENCE

Careless use of torque-adjusted impact drivers will lead to crushing of the over-lap and failure of the capillary break. For this reason, purpose designed 'Tek-guns' with a depth-override adjustment should be the preferred roofing screw power tool.

Take care with penetrations!

Where a penetration passes through the lap of two sheets, special attention needs to be given to how the upstream rib is terminated before the flashing. You can see in Figure 7.4 how the capillary break is built into the design of the roof sheet profile. Normally the

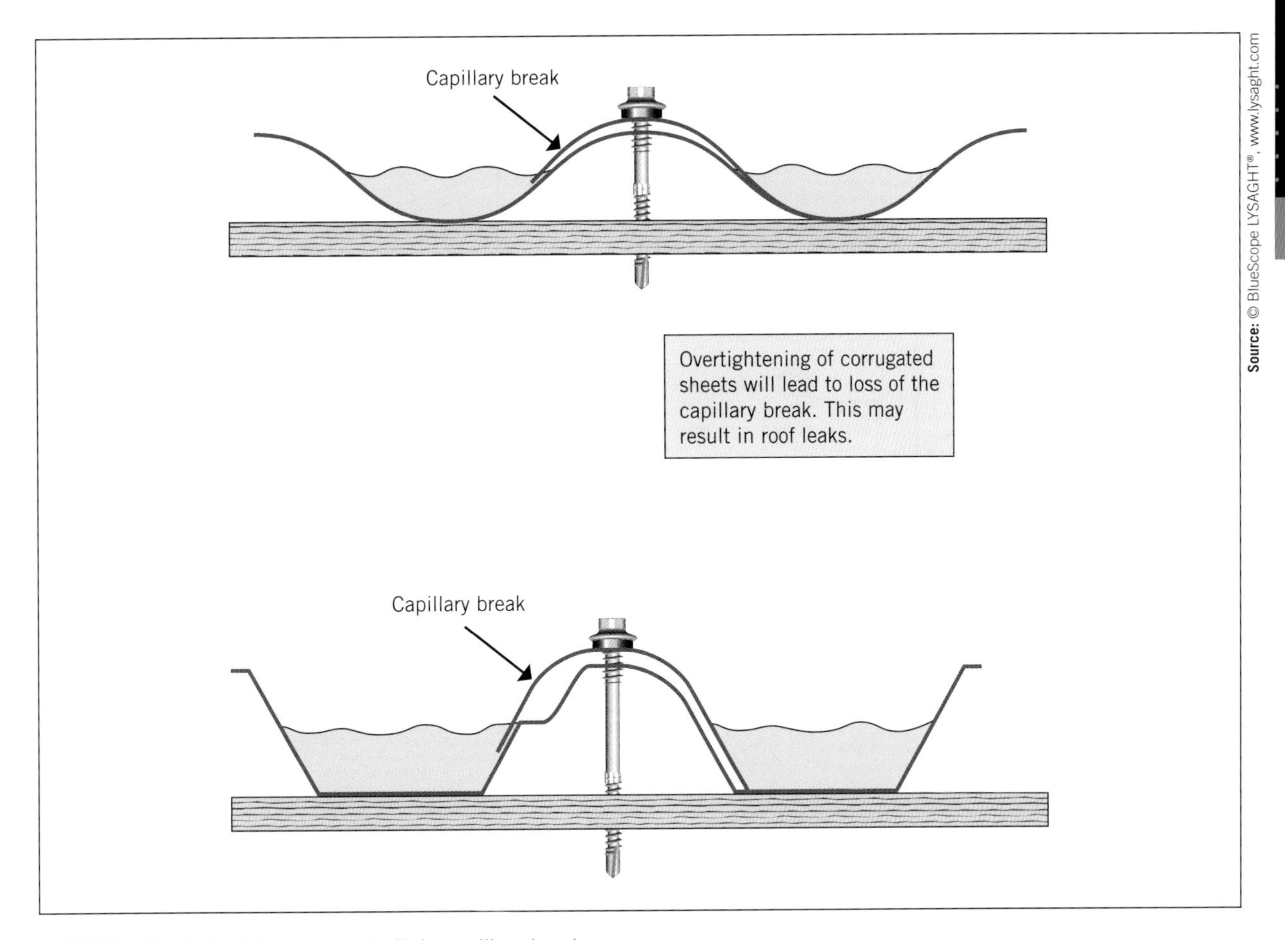

FIGURE 7.4 Roof sheet laps have a built-in capillary break

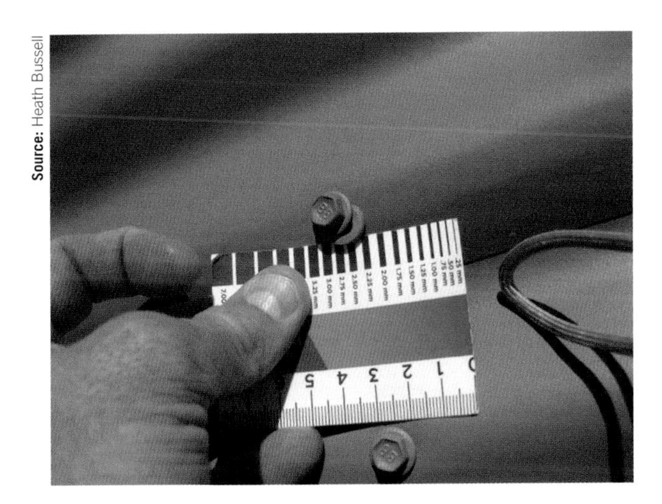

FIGURE 7.5 An over-screwed lap will compromise the sheet anti-capillary gap

small amount of water retained by the lap is drained gradually towards the gutter. However, when you cut a penetration through a roof sheet lap, you must ensure that a drain is cut into the rib *immediately upstream of the flashing*. Failure to do this will result in water leaking underneath the flashing and directly into the building! As shown in Figure 7.6, the most effective way to get around this problem is to drain the capillary break by cutting a small V-slot in the over-sheet immediately upstream of the flashing.

At times you will find that someone has used silicone to try and seal the lap from the penetration all the way up to the highest point of the sheet. *This does not work!* This method is a poor work practice that often results in a greater volume of water being held in the capillary break, causing it to fill and leak into the building.

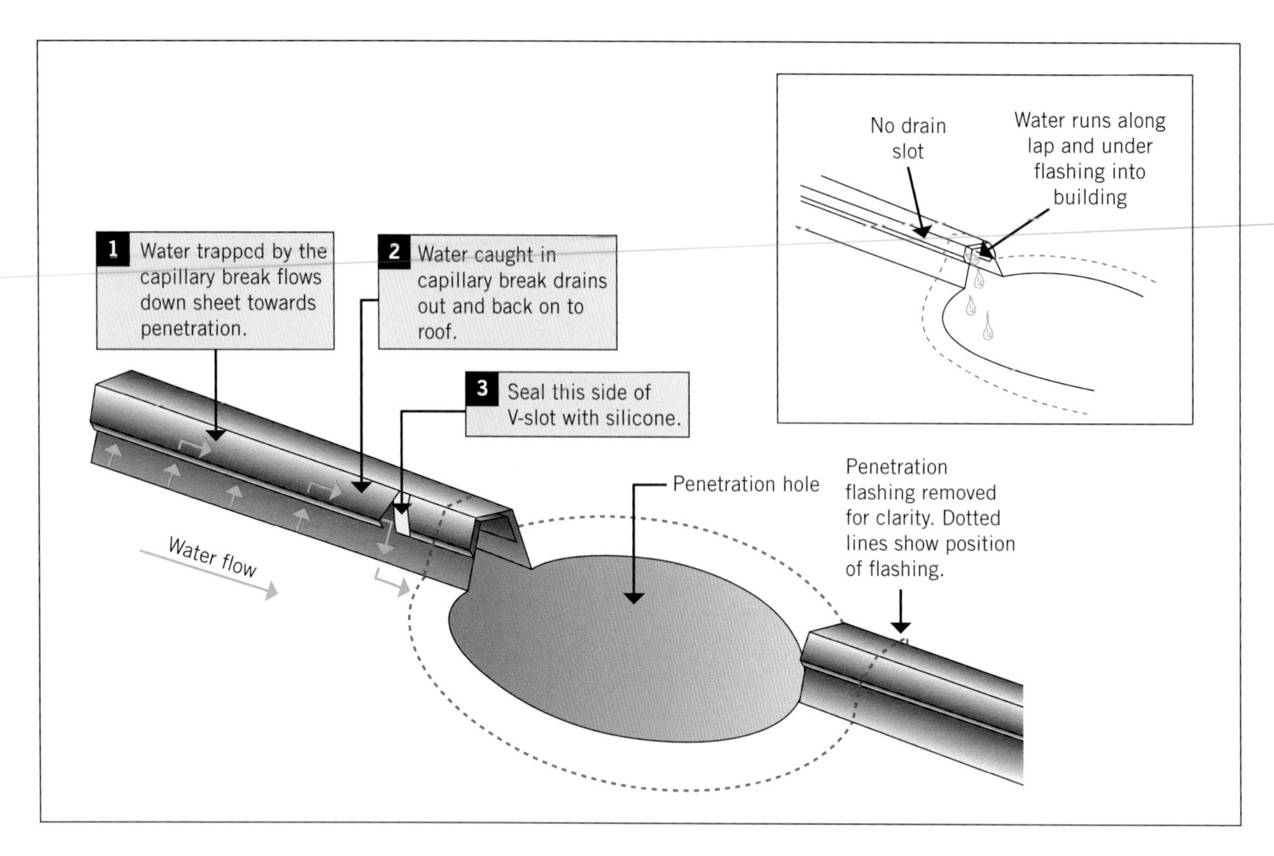

FIGURE 7.6 Cutting a small V-slot just above the flashing to drain the capillary break

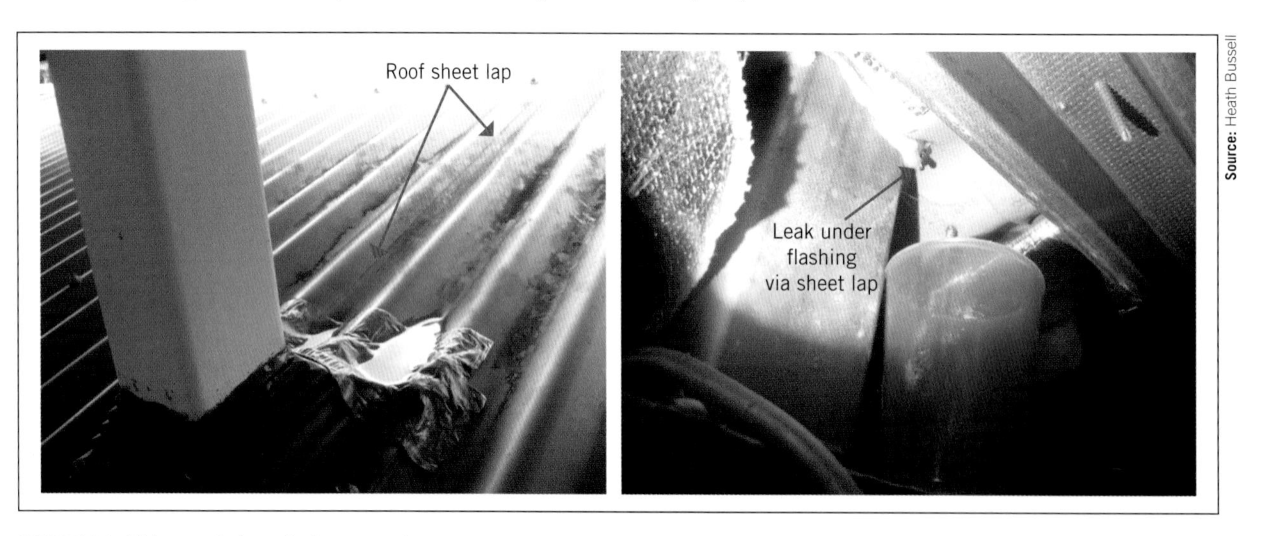

FIGURE 7.7 This poorly installed penetration passed through a sheet lap. No V-slot drain was cut into the overlap sheet and investigation revealed significant leakage under the polymer flashing.

Shape of capillary breaks

The actual shape of a capillary break sheet-metal fold is normally determined by location and cosmetic and design considerations. If the flashing is out of sight, then a simple fold will be sufficient to do the job. Where seen from the ground, capillary folds may need to be bolder to blend in with cladding and roof design. Some examples can be seen in Figure 7.8.

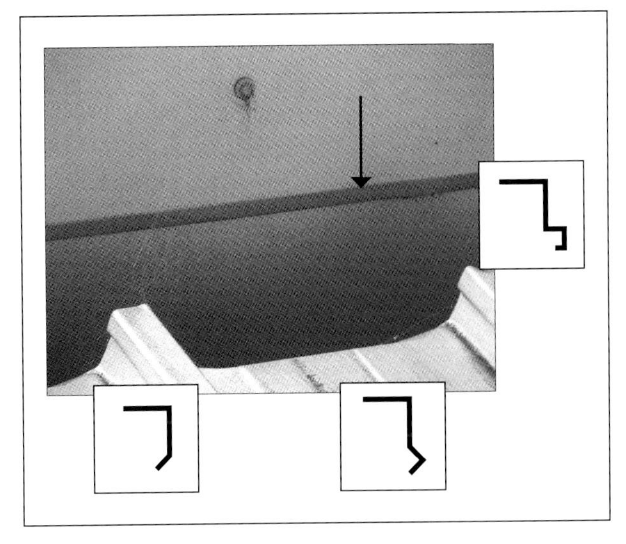

FIGURE 7.8 Examples of capillary folds

Sheet end turn-down and water surface tension

Sheets that are laid at less than approximately 3° are particularly susceptible to the action of water surface tension drawing water back on the underside of the sheet. Figure 7.9 shows how the water surface tension and molecular adhesion allow the water to creep back on low-pitch roof sheeting. This may lead to leaks in the building and/or premature corrosion of the sheet end section.

To prevent this from happening, you must turn down all of the sheet pans, allowing the water to drain more effectively into the gutter. Special folders can be fabricated or purchased to turn down roof sheet pans (see Figure 7.10).

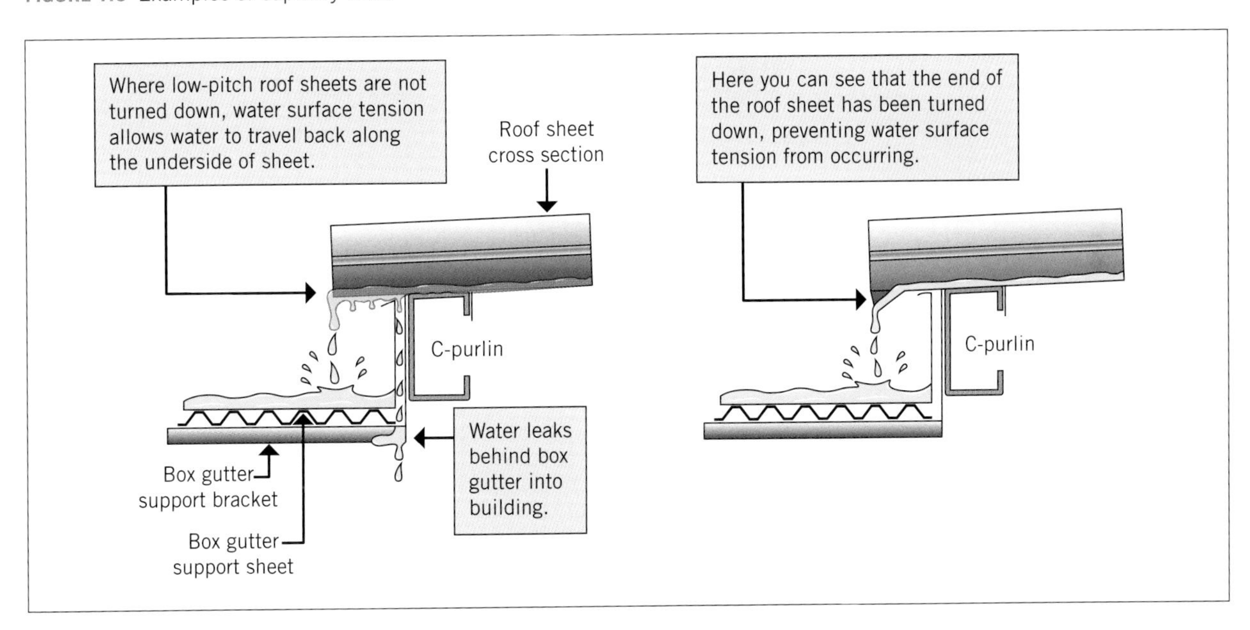

FIGURE 7.9 Turning down low-pitch roof sheets to prevent water surface tension leakage

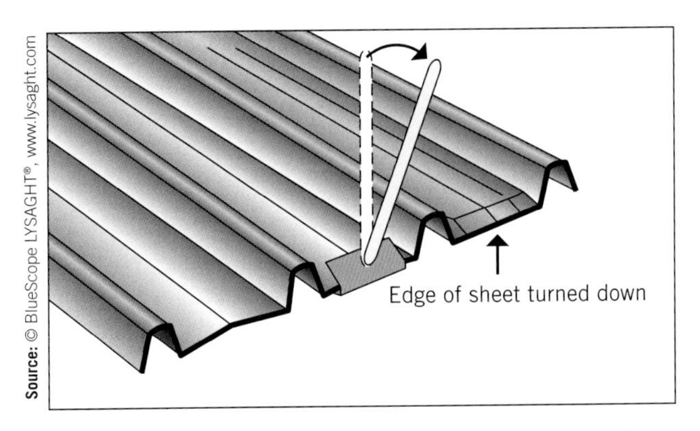

Source: © BlueScope LYSAGHT®, www.lysaght.com

FIGURE 7.10 Special folders are used to turn down roof sheet pans

LEARNING TASK 7.1

1. What is the main reason for capillary action when installing roof sheeting?
2. At what angle is low-pitch roof sheeting susceptible to the actions of water surface tension?
3. What installation technique prevents water running back under the end of sheets?

SUMMARY

Capillary breaks are frequently required on all roofing jobs. A good roof plumber is always on the lookout for any part of a roof design or installation that may be subject to capillary action or leakage from the action of water surface tension, and will provide for some form of break wherever necessary. You will find more specific references to capillary breaks as you work through further chapters in this text.

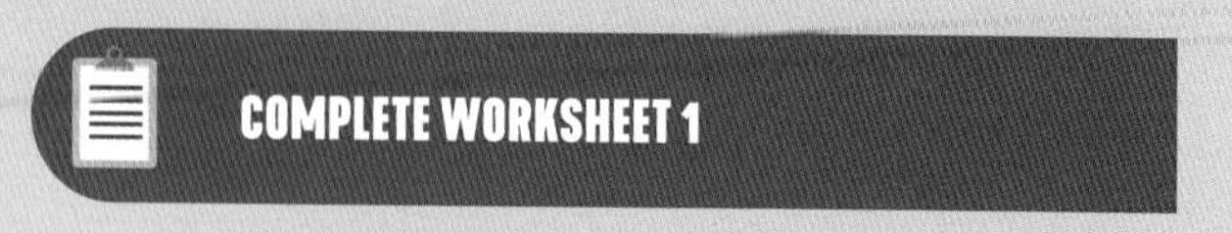

WORKSHEET 1

To be completed by teachers

Satisfactory ☐

Not satisfactory ☐

Student name: ____________________

Enrolment year: ____________________

Class code: ____________________

Competency name/Number: ____________________

Task: Working with your teacher/supervisor, refer to this text and relevant state or territory information to answer the following questions.

1 In your own words, describe what capillary action is.

____

____

____

2 Name at least two methods of controlling capillary action in relation to roof plumbing.

____

____

____

3 If you do not set the depth adjustment mechanism or torque setting on a screw gun correctly, it is possible to over-tighten screws when laying sheets. What is likely to happen if you over-tighten the screw securing the lap join on a corrugated roof sheet?

____

____

____

4 With an arrow, identify where the capillary break is on the following drawings.

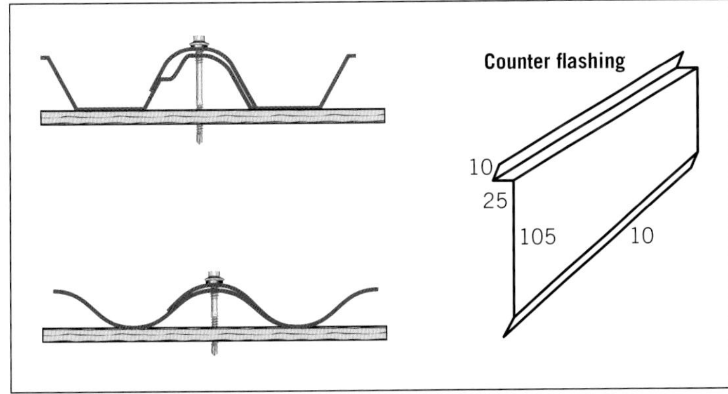

5 You have cut a penetration through a roof to fit a new gas appliance flue. The penetration intersects the lap of the high-profile roof sheeting. Do you cut in a V-slot above or below the penetration flashing?

____

____

____

6 Why must you turn down the ends of some roof sheet profiles into the gutter?

____

____

____

CORROSION

8

Metallic roof coverings need to be durable. This means that a product should be able to withstand damage and degradation from the sun, cold, wind and rain. It is also important that it should have a degree of corrosion resistance.

Overview

In relation to roof plumbing, corrosion is the electrochemical oxidation of a metal through contact with an electrolyte (water) and oxygen. You are probably familiar with the term 'rust', a common term describing the corrosion of steel.

Products such as copper and stainless steel are resistant to corrosion in themselves. Other materials such as steel require a protective coating of zinc and other alloys to provide the longevity required of a roofing product.

A roof plumber must be mindful of corrosion potential at all times. In this section you will gain a better understanding of corrosion, how it impacts on your selection of roofing materials, and the manner in which these materials need to be installed in relation to each other. The following subjects will be discussed:

- sacrificial protection and the galvanic series
- material compatibility
- inert catchment and drip-spot corrosion
- cut-edge and scratch protection
- ponding
- crevice corrosion
- swarf
- timber and corrosion
- patina.

Sacrificial protection

Sacrificial protection describes the chemical phenomenon where one metal will 'sacrifice' itself to protect another metal from corrosion. Metals have different physical and chemical characteristics. One characteristic of metals is the degree to which each of them corrodes. If two dissimilar metals are placed together in the presence of water (the electrolyte), an electrolytic circuit is created and one of the metals will start corroding while the other is relatively protected or untouched. This is commonly known as electrolytic corrosion.

In fact, all metals can be ranked in descending order according to this level of electrolytic activity. This ranking is known as the galvanic series. Sacrificial protection depends on the position of a metal in the galvanic series. Less-active metals at the bottom of the list are classed as noble, and those near the top are less noble. Figure 8.1 shows a slightly modified version of the galvanic series, as shown in your HB39.

If two of these metals are placed together in the presence of an electrolyte, the higher ranked, more active metal will sacrifice itself and protect the lower ranked metal. This characteristic of metals allows manufacturers to coat products that would normally corrode away in the weather with another metal that will protect it.

EXAMPLE 8.1

Bare sheet steel will rust if exposed to water and air. However, if the steel is coated in a layer of zinc it will be protected. This is because the zinc is a more active metal than steel and will sacrifice itself against the effects of corrosive solutions or scratches and cuts in the steel (see Figure 8.1). While the steel remains coated in zinc it will be protected. However, through sacrificing itself, the zinc coating gradually becomes thinner until it reaches a point at which there is no longer enough zinc left to protect the underlying steel.

Size relationship

Any corrosion that occurs also depends on the relative difference in size between each dissimilar metal. If two pieces of dissimilar metal were in contact with each other, the more active piece would corrode first before the more noble piece. However, if the more active piece were considerably larger than the less active one, corrosion would tend to be much slower and perhaps even negligible. You can gain a broader appreciation of the implications of contact between metals in your HB39.

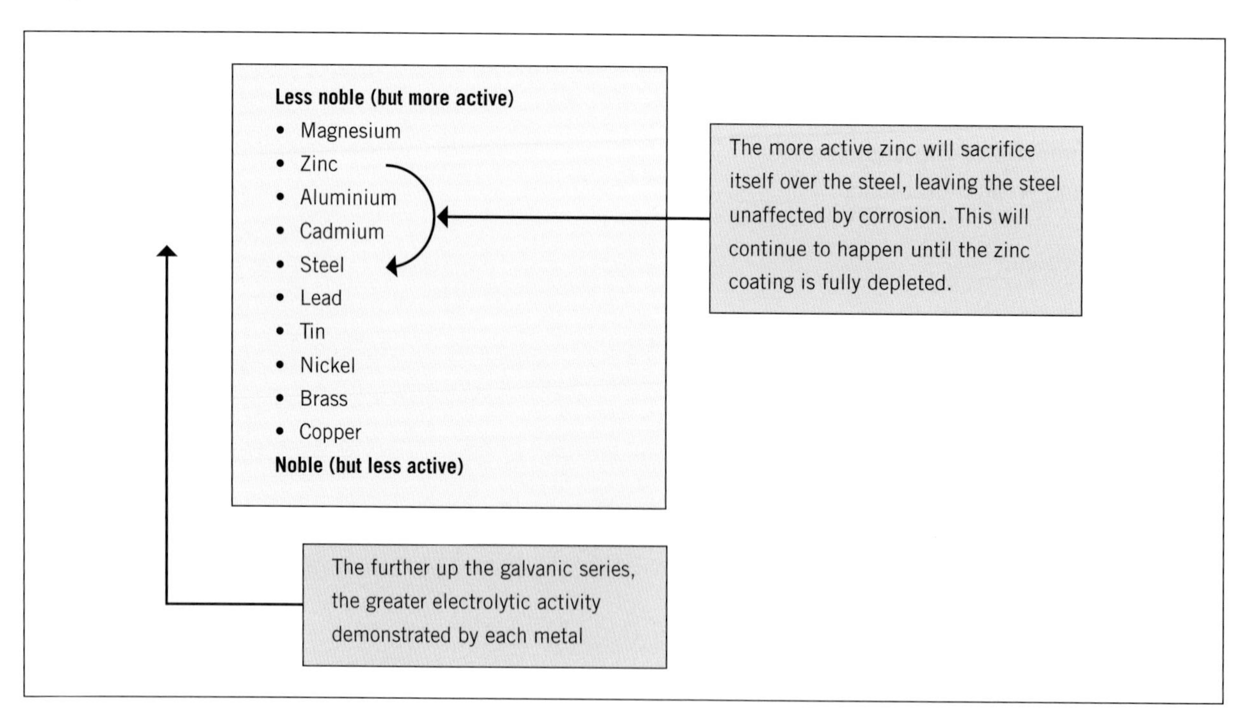

FIGURE 8.1 A modified version of the galvanic series

Material compatibility

The galvanic series demonstrates how certain metals are more active than others. From this you can see how some materials can work together quite well while others are not compatible with each other. A roof plumber must be very conscious of material compatibility at all times.

Compatibility extends to not only materials in direct contact, but also the order in which they are placed in relation to the flow of water.

The following points and examples may help you to understand this further:

- Two roofing products made of the same material will be naturally compatible with one another. This means that neither will cause corrosion of the other and you can use them in any way you please.
- However, when dissimilar materials are used, you must ensure that the more active material is always higher than the less active metal in relation to water flow.
- If a more noble material is placed above a less noble product, it is likely that corrosion will occur.

EXAMPLE 8.2

If you install a ZINCALUME® steel roof above ZINCALUME® steel or COLORBOND® steel gutters and downpipes, neither system will suffer corrosion from being in contact with the other (see Figure 8.2).

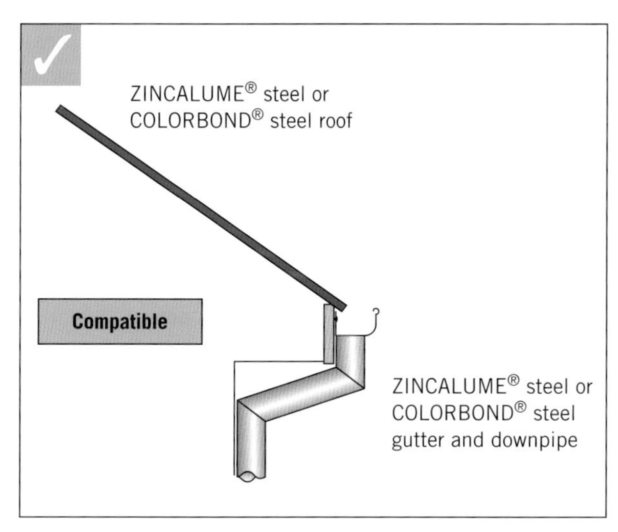

FIGURE 8.2 Compatible materials

EXAMPLE 8.3

If you install new ZINCALUME® steel or COLORBOND® steel gutter *under* an existing galvanised (zinc-coated) roof, no corrosion will occur because the lower material is more noble than the higher. Therefore, the water that runs off the roof will not aid in the corrosion of the lower material. In other words, a zinc-coated roof will not affect the more noble metals beneath it (see Figure 8.3).

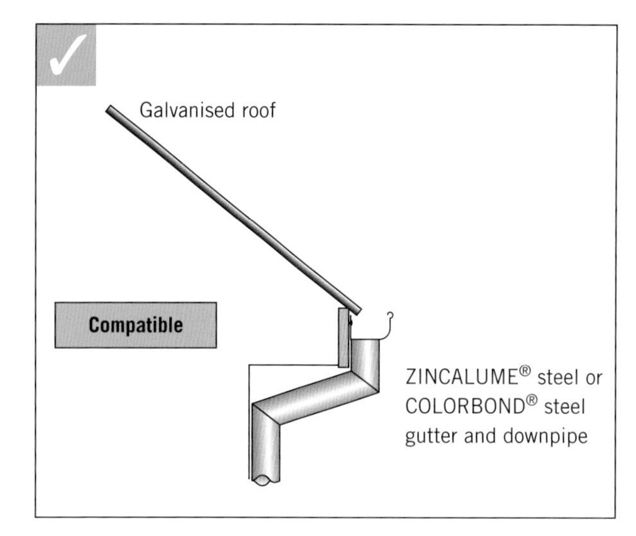

FIGURE 8.3 Cross-section of a zinc-coated (galvanised) roof with ZINCALUME® steel or COLORBOND® steel gutter and downpipe

EXAMPLE 8.4

If you install a ZINCALUME® steel or COLORBOND® steel roof over existing galvanised gutters and downpipes, the lower materials will suffer accelerated corrosion (see Figure 8.4).

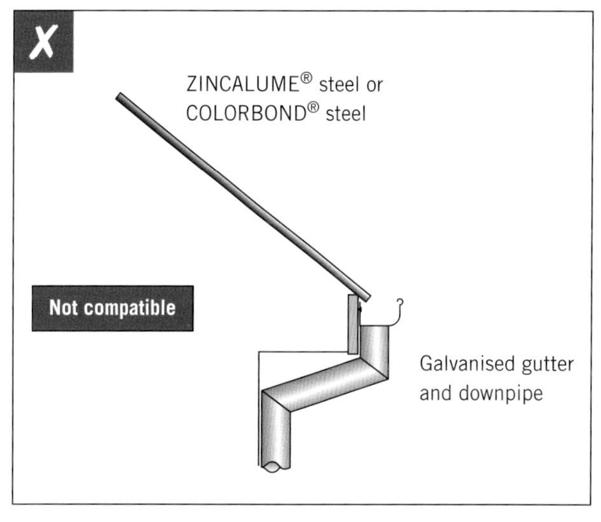

FIGURE 8.4 Cross-section of a zinc-coated gutter beneath a more noble metal

It is not necessary that you remember the entire galvanic series by rote. Whenever you are planning for a job and are not entirely sure which material is compatible with another, you need only refer to your HB39 (see Figure 8.5).

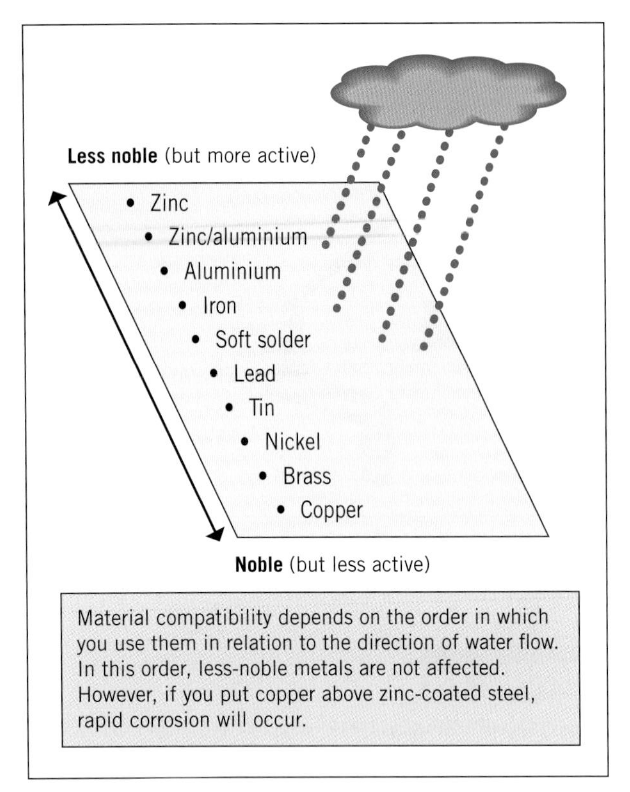

FIGURE 8.5 Checking the position of a metal in the galvanic series to ensure sacrificial protection of materials

FROM EXPERIENCE

Selecting, installing and maintaining roof drainage materials with corrosion prevention in mind is not just good trade practice. A roof plumber is actually required to adhere to the primary corrosion guidelines set out within the AS/NZS 3500.3 'Stormwater drainage' and the details set out in both the HB39:2015 and relevant manufacturer's specifications. Always refer to your Standards!

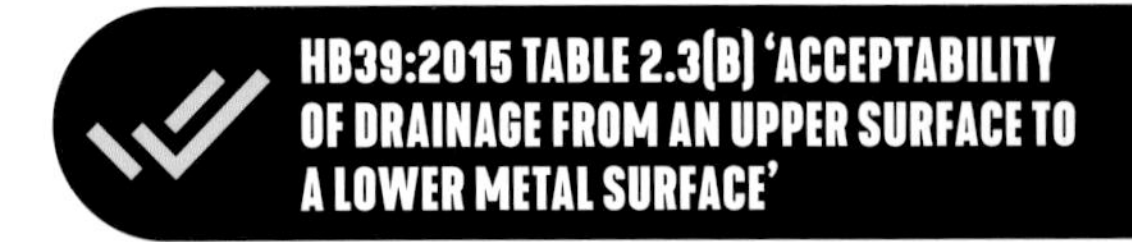

Inert catchment and drip-spot corrosion

The quality of rainwater that falls onto our roofs, the reactivity of certain metals and the way in which the water and materials interact have a major effect on the durability of your roofing system. To understand this more fully you need to have a working knowledge of some basic chemistry.

The pH scale

The acidity or basicity of any liquid solutions can be expressed using the term pH. The pH scale runs from 0 to 14: solutions less than 7 are acidic and solutions greater than 7 are alkaline. A liquid solution with a pH of 7 is known as neutral. Pure water is generally regarded as having a neutral pH.

At each extreme of the scale are solutions that are extremely acidic or alkaline. For example, if you refer to Figure 8.6 you can see that the battery acid you would find in a car has a pH of 0, while simple soapy water may have a pH of 12. Remember: neutral is pH 7, so a pH of 0 is extremely corrosive!

Unpolluted rainwater has a pH of approximately 6 and therefore it is slightly corrosive. In certain areas, raindrops will tend to accumulate airborne pollutants and chlorides as they fall from the sky. This is particularly so near cities and coastlines. These pollutants make the rainwater more acidic and therefore potentially more corrosive. In some areas where heavy industrial fallout is common, the rain may become so polluted that it is classed as 'acid rain' and has a very detrimental effect on metallic surfaces. From Figure 8.6 you can see that acid rain has a pH of 4.

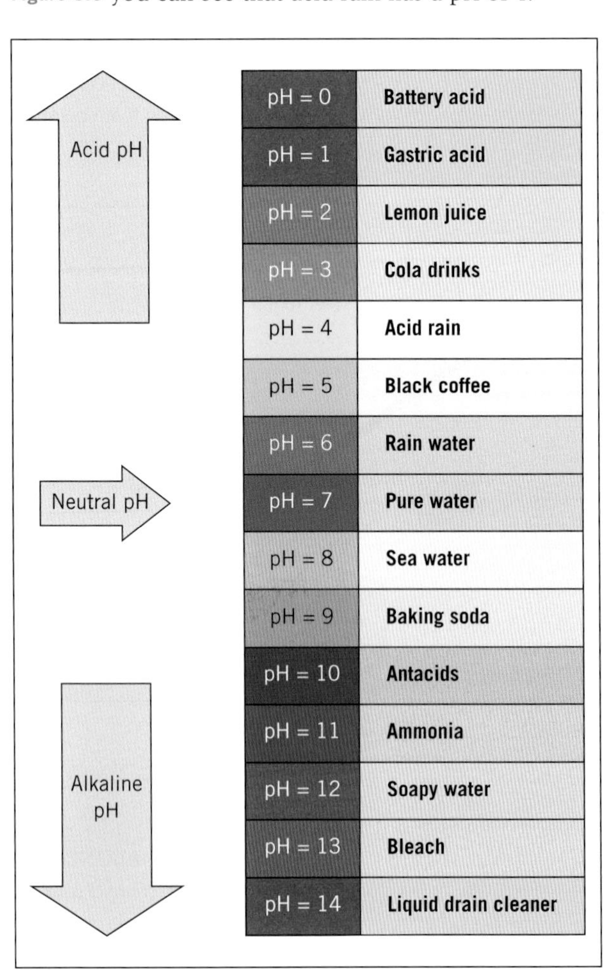

FIGURE 8.6 A simplified pH scale demonstrating the difference between acid and alkaline substances

A large roof or catchment made from materials such as galvanised steel and unglazed cement tiles acts to neutralise the acidity of rainwater to some degree. As the slightly acidic rainfall flows down the roof, it absorbs metal salts and minerals. These salts act to lessen the acidity of the rainwater solution, and by the time the water reaches an area of concentration, such as the gutter, the water has become far less reactive. Less acidic water allows the zinc coating within the gutter to sacrifice and protect the base metal steel.

Rainwater and inert catchment materials

Unlike galvanised steel and unglazed cement tiles, a catchment area (roof) made from materials such as glazed tiles, glass (solar panels), ZINCALUME® steel, COLORBOND® steel or plastic is classed as inert, and as such is known as an 'inert catchment'. Such materials do not give off the necessary salts and minerals that would lessen the acidity of the water. This means the water that flows across these materials continues on to the gutter unchanged. As a result, the water is able to concentrate in the gutter in a relatively acidic state. If the gutter is made from galvanised steel, it will suffer accelerated corrosion from this acidic water (see Figure 8.7).

This form of corrosion is often identifiable as spots of rust beneath an overhanging roof sheet, accessory or other metallic object and is called 'drip-spot corrosion'. It can also be identified as rusted flow patterns across the sheet or flashing. Such corrosion is evidence of the incorrect placement of materials in relation to the galvanic series. The most common example is rusty drip spots and flow patterns in a galvanised gutter that are caused by water run-off from a ZINCALUME® steel or glazed tile roof covering (see Figure 8.8).

Figure 8.9 shows an example of drip-spot corrosion caused by water dripping off bare copper pipe of a solar hot water installation onto a ZINCALUME® roof.

Source: Heath Bussell

Source: Heath Bussell

FIGURE 8.7 Painted roof sheets draining into a zinc coated gutter is an example of inert catchment

FIGURE 8.8 Inert catchment corrosion in a valley gutter beneath glazed terracotta tiles

Source: Heath Bussell

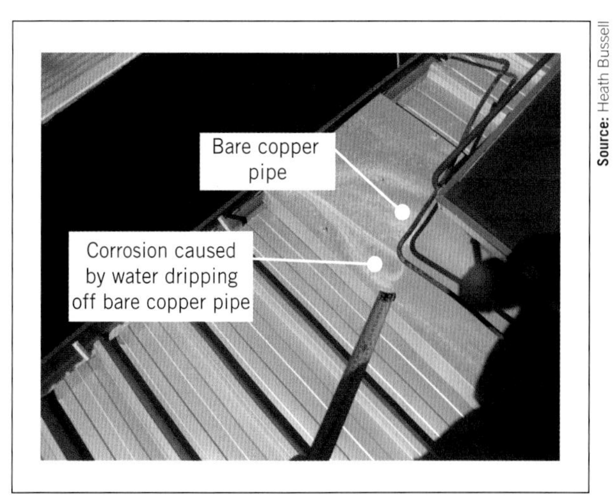

FIGURE 8.9 Drip-spot corrosion caused by metallic incompatibility

ZINCALUME® steel is rarely affected by drip spot corrosion, but copper is a very noble metal and should never be left uncovered on roofs.

Figure 8.10 shows how an incompatible soaker flashing has been damaged through inert catchment corrosion. Note how the corrosion pattern matches the flow of water from the upstream inert catchment, across the soaker to the downstream side.

FIGURE 8.10 Water flowing from an inert catchment has corroded this soaker.

All roof plumbers need to be mindful of the potential for inert catchment corrosion as your material selection may be the difference between a job that lasts for decades or one that is damaged by corrosion in just months.

LEARNING TASK 8.1

1 What is the neutral point of the pH scale?
2 Which two roofing materials can have a slightly neutralising effect on rainwater?

Cut edge and scratch protection

With all this talk of corrosion, you may be wondering what happens when you cut a sheet with a pair of snips. Won't the freshly cut edge start to rust?

Fortunately, this is not the case. Snips, electric shears and nibblers make a clean cut that does not damage the zinc or zinc/aluminium coating next to the cut. Because the thickness of the cut edge is relatively small in relation to the large sheet of coated steel, the zinc coating automatically begins to sacrifice itself over the cut edge, protecting it from any corrosion. The same applies to minor scratches in the surface of galvanised sheet, ZINCALUME® steel or COLORBOND® steel.

HB39:2015 SECTION 2.7 'CUT EDGE PROTECTION'

HOW TO

Certain cutting processes can damage the corrosion coating. Never use abrasive carborundum discs to cut metallic roof products! These discs cause the zinc coating to burn back from the cut edge and as a result, the zinc cannot sacrifice itself over the cut. Corrosion will always occur. (See Figure 8.11.)

Roofing products should only ever be cut with snips or electric shears/nibblers.

FIGURE 8.11 Use of carborundum cutting discs will cause corrosion around the cut edge.

Pencils

A quick note about pencils: avoid using graphite pencils on bare ZINCALUME® steel or galvanised steel. The graphite within the core of the pencil sets up an electrolytic 'cell' that will corrode the zinc coating wherever it is used. This means that the mark-outs and notations you make on the material will be indelibly etched into the surface. Use a fine marker pen or scriber instead.

LEARNING TASK 8.2

1 Which type of cutting disc should not be used to cut metallic roof products?
2 Which marking tool will leave an indelibly etched mark into the surface of bare ZINCALUME® steel or galvanised steel?

Ponding

Ponding occurs when a section of roof or gutter retains water for extended periods. Such water will cause premature deterioration of the material. Ponding can be caused by the following problems:

- a gutter laid below the minimum grade
- decking sheets laid below the manufacturer's minimum recommended pitch
- poorly installed soaker flashings
- excessive use of silicone sealant around downpipe pops, sumps and other gutter laps, preventing normal flow of water
- too great a span between the purlins on low-pitch roofs
- damage to the rib profile of low-pitch roofing sheets
- a build-up of dirt, leaves or other debris that prevents the normal flow of water.

Figure 8.12 shows what can happen when ponding is allowed to occur. The ZINCALUME® steel box gutter shown was only six years old when photographed, but it had quickly deteriorated. In this instance it is likely that this section of the gutter support system was installed below grade and was not rectified when the replacement box gutter was installed. Remember: there is little point in replacing gutters without checking and rectifying any problems with the fall!

FIGURE 8.12 Water ponding in a box gutter

Low-pitch roofs are especially prone to ponding if due care is not exercised in their installation, or if the distance between purlins exceeds the manufacturer's recommendations. (See Figure 8.13.)

FIGURE 8.13 This badly installed lean-to roof will quickly corrode if not soon rectified

Compared to higher pitched roofs, low-pitch sheets do not receive the same degree of 'rainwash' that would normally flush away salts, contaminants and dirt. Any minor change in fall within the length of the sheet can result in ponding and increase the risk of premature deterioration.

Crevice corrosion

The AS/NZS 3500.3 requires that all roofs be designed and installed to ensure that complete drainage and drying occurs. Any roof sections, corners and crevices that do not readily dry out or that are contaminated by accumulated solids or coatings may suffer localised corrosion.

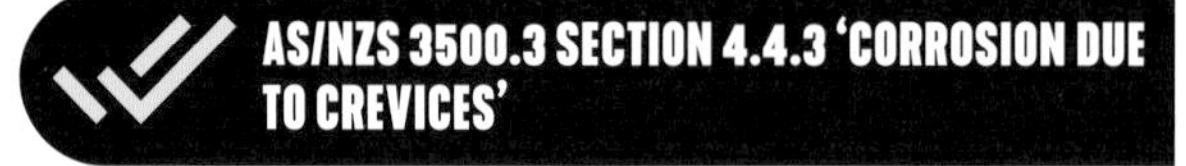

Swarf

Prevent corrosion by cleaning up! Many otherwise good roof and gutter installations have been ruined because the job was not cleaned down after each day's work.

Working on any roof requires the drilling of holes and the cutting and trimming of metal. This work produces large amounts of metal swarf, offcuts and general fabrication debris. Swarf is the name given to the small particles of metal that are left behind from drilling and cutting processes. These small swarf particles, pop rivet shanks, screws and trimmings rust very quickly. If left to lie on the roof or in the gutter, this debris will cause discoloration and eventual corrosion of the metal upon which it lays. Figure 8.14 shows an example of the rust marks that can be left by drill swarf. The roof in this photograph was only months old.

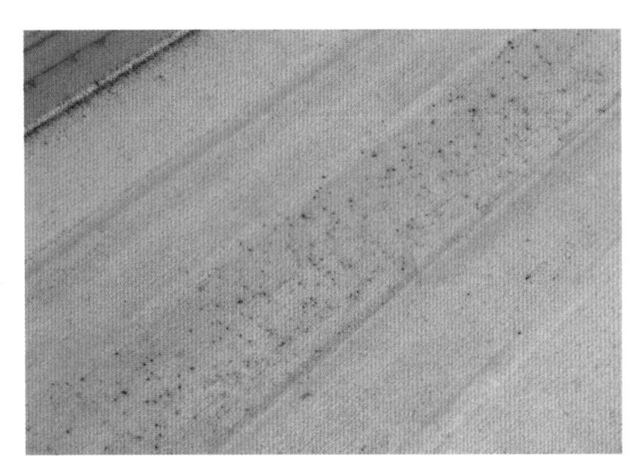

FIGURE 8.14 Permanent rust marks on a ZINCALUME® sheet caused by drill swarf

Always look for and remove the following debris from your job:

- drill swarf
- nibbler swarf
- pop rivet shanks (a common cause of premature corrosion)
- screws
- nails and sarking tabs
- all offcuts
- lengths/offcuts of metal and timber fabrication materials.

It is not acceptable to wait until the end of the job before cleaning up. Swarf will start rusting once exposed to the first rain or overnight dew and will leave rust marks on roof sheeting and gutters. Furthermore, swarf and small offcuts will embed themselves in your footwear and cause scratching to both painted and unpainted metal surfaces. Wherever practicable, clean up once each section of the work is completed and at least every day (see Figure 8.16). Use a soft brush to sweep the swarf off the roof and out of gutters.

Source: Heath Bussell

FIGURE 8.15 COLORBOND® sheets are also affected by swarf corrosion

FIGURE 8.16 A rusted pop rivet shank on a ZINCALUME® steel roof

If using a powered blower to remove swarf, ensure that other workers and the public are not exposed to any flying metal and other debris.

Other debris

Sometimes tradespeople may leave debris on a newly completed roof *after* you have left the job. Always point out to the builder or owner the importance of having the roof cleaned of all contaminants, such as:

- lime mortar and cement waste
- render mix droppings
- offcuts of copper wire and pipe near appliances
- timber offcuts.

During maintenance inspections or quotes on other sites, you may encounter site-specific issues that will lead to premature corrosion if not corrected. These may include:

- a build-up of animal and bird droppings
- an accumulation of leaves on the roof and in gutters
- the long-term storage of materials on roof
- deposits of creosote concentrations from wood-burning heater flues.

LEARNING TASK 8.3

1 What are two roofing processes that might cause swarf?
2 What are two additional site-specific issues that might cause premature corrosion?

Timber and metal corrosion

Premature corrosion can occur if certain timbers are placed above and/or in contact with underlying roof sheets. This is particularly so with chemically treated timber, green hardwood and western red cedar. Such timber is often used by following trades to construct walkways, bases for air-conditioning systems, wall cladding and decks.

CCA is an abbreviation for copper/chrome/arsenate, a preservative treatment used to stop the decomposition of pine. It is identified by the green appearance of the timber product. Run-off from such timber contains soluble salts and copper leachate and has been shown to accelerate corrosion through electrolysis.

Western red cedar (WRC) is a timber product often used for the weatherboard cladding of homes. However, caution should be exercised where the timber is in direct contact with galvanised roofing products or water run-off is allowed to flow onto a lower roof or gutter. As it ages, WRC will be subject to tannic acid bleed, which can cause corrosion on galvanised surfaces. The same also applies to many other 'green' (not kiln-dried) timbers.

Patina

Metals such as copper, bronze and brass do not rust like steel or iron. However, the surface of these metals does oxidise through a natural weathering process, creating a greenish surface coating consisting of a mixture of chlorides, carbonates and sulfides known as patina (also known as verdigris). For example, the copper and bronze sheeting on the clock tower in Figure 8.17 has developed a patina over many years.

Once the oxidation process has developed such a coating, the patina then acts to protect the base metal from further corrosion. The longevity that patina provides to copper makes it a premium roofing product, often specified for high-end private, public and institutional buildings where long-term maintenance issues are to be avoided due to high costs and access problems.

FIGURE 8.17 Patina on the copper and bronze covering of a clock tower

SUMMARY

In this chapter you have been introduced to some of the corrosion problems that a roof plumber will encounter on a daily basis. A good tradesperson always aims for a quality installation, and ensuring that all possible causes of corrosion are eliminated is a fundamental requirement of a professional job.

Having worked through this chapter, you may find it useful to access some more in-depth information on corrosion. You will find a range of corrosion-specific technical bulletins on the following website: **http://steelproducts.bluescopesteel.com.au/home/technical-library**

COMPLETE WORKSHEET 1

WORKSHEET 1

To be completed by teachers	
Satisfactory	☐
Not satisfactory	☐

Student name: ______________________

Enrolment year: ______________________

Class code: ______________________

Competency name/Number: ______________________

Task: Working with your teacher/supervisor, refer to this text, your HB39:2015, AS/NZS 3500.3 and relevant technical bulletins to answer the following questions.

1 What is meant by the term 'electrolysis' in roof plumbing?

____

____

____

2 Place the following list of metals in the correct order according to the galvanic series, with the most noble metals at the bottom:

- steel
- copper
- brass
- zinc
- nickel
- lead
- magnesium
- aluminium

3 Which metal classification is more active in the galvanic series: noble or less noble?

____

____

____

4 If two dissimilar metals are placed directly in contact with each other with no electrolyte present, will they corrode?

____

____

____

5 True or false: The life of any metallic-coated steel depends upon the thickness of the steel itself. Provide a reference from your HB39 to back up your answer.

____

____

____

6 A customer asks you to install a new galvanised steel water tank to collect water from the roof of the house. You inspect the job and notice that the roof is made from glazed concrete tiles and the gutter is made from factory pre-painted steel. What should you do and why? Provide a reference from your HB39 to back up your answer.

7 What is wrong with using a graphite pencil on bare ZINCALUME® steel when marking out your gutter angles?

8 You are asked to roof a new veranda with polycarbonate translucent sheeting. The veranda roof drains into an unpainted zinc/aluminium-coated steel gutter. Is this satisfactory? Provide a reference from your HB39 to back up your answer.

9 You are sent to a job on a warehouse to cut out a section of existing ZINCALUME® steel decking and fit a new skylight. Your workmate grabs a power saw fitted with a carborundum cutting disc. Is this the right tool for the job? Explain your answer and back it up with a reference to your HB39.

10 What are at least four causes of ponding that lead to corrosion?

11 Of the following list of materials, which are classed as an inert catchment? Circle your answers.

- zinc-coated steel
- factory pre-painted steel
- polycarbonate sheet
- glazed concrete tiles
- zinc/aluminium sheet
- unglazed concrete tiles
- copper sheet

WORKSHEETS 8

12 After a day's work, what are at least four types of workplace contaminants and debris that you need to clean from the roof and gutters?

13 What is the name given to the greenish colour found on weathered copper rainwater products?

14 Would it be acceptable for a plumber to run an uncoated copper gas line across ZINCALUME® steel decking to a water heating system? Explain your answer.

15 Areas on a roof that are shielded from sunlight, contaminated with foreign materials or otherwise prevented from completely drying out may suffer what kind of corrosion? Support your answer with a reference from your AS/NZS 3500.3.

9 ROOF PLUMBING TOOLS

This chapter provides an introduction to many of the tools commonly used by roof plumbers. While you would already be familiar with many of these, you may also encounter some new tools further into your training and on the job.

Overview

So far in your plumbing training you will have been introduced to a range of plumbing hand, power and workshop tools. In this section you will have the opportunity to review and perhaps revise the tools specific to roof plumbing that you may need to use or be aware of. Personal preference is a big part of tool selection but most importantly they must be fit for purpose and enable you to carry out your work safely and efficiently.

Tools used in roof plumbing

There is an almost endless supply of great tools available to the roof plumber. Unfortunately, it is not possible to include all of them here, but as you are starting to put your tool kit together it is worth considering some of the tools listed in Table 9.1, which gives a brief description and photograph of each. Your teacher should be able to provide examples of and instruction for these tools and many others.

TABLE 9.1 A small selection of basic roof plumbing tools

Tool name	Comments	Image
Spirit level	For greater accuracy, longer spirit levels are often useful in roof plumbing. In addition to a standard 600-mm level, you will need a 1200-mm level and perhaps a line level.	
Plumb-bob	A plumb-bob is a heavy metal weight at the end of a string that enables you to mark a point either exactly below or above another position. For example, you can suspend a plumb-bob above a stormwater drain connection and project the exact downpipe outlet centre on the gutter above it.	
Bevel gauge	A bevel enables you to copy another angle and transfer that angle to a new pattern. A roofer can't be without one when matching metal downpipe angles or setting out bay-window gutter angles.	
Square and/or tri-square	A square is used all the time. It is vital for quality work. A number of options are available, including adjustable squares that also measure angles.	

Source: Imagery courtesy Milwaukee Tool

>>

Tool name	Comments	Image
Compass dividers	Metalwork compass dividers enable you to scribe neat circles for downpipe and overflow outlets.	
Pitch gauge	If you need to check the pitch of an existing roof to ensure that your profile choice is compliant, the easiest way to do so is to use a pitch gauge. Use it on top of a level or straight-edge for a quick and accurate answer in degrees.	
Scribers, pencils and marking pens	Accuracy is essential if you want a good job. Ensure that your marking equipment is sharp and able to be seen clearly. Remember: do not use graphite pencils on zinc-coated steel as the marks will cause corrosion! Fine marker pens are a better option.	
Chalk-lines	Coloured chalk-lines enable you to 'spring' a line over considerable distance and can be ideal for marking cut lines or screw lines. Remember that red chalk is indelible, so it is better to use blue or white.	

Source: Imagery courtesy Milwaukee Tool

Source: Imagery courtesy Milwaukee Tool

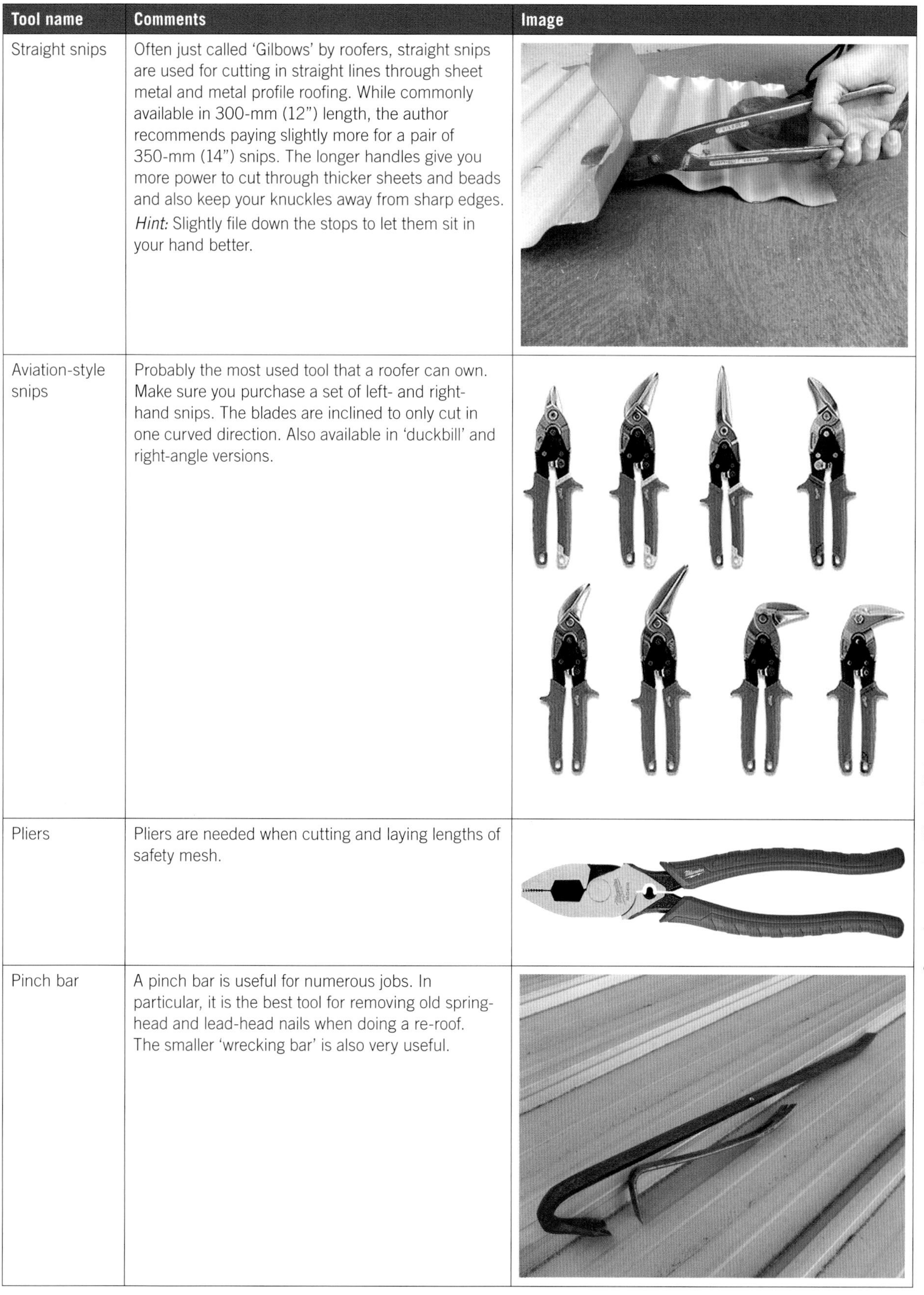

Tool name	Comments	Image
Straight snips	Often just called 'Gilbows' by roofers, straight snips are used for cutting in straight lines through sheet metal and metal profile roofing. While commonly available in 300-mm (12") length, the author recommends paying slightly more for a pair of 350-mm (14") snips. The longer handles give you more power to cut through thicker sheets and beads and also keep your knuckles away from sharp edges. *Hint:* Slightly file down the stops to let them sit in your hand better.	
Aviation-style snips	Probably the most used tool that a roofer can own. Make sure you purchase a set of left- and right-hand snips. The blades are inclined to only cut in one curved direction. Also available in 'duckbill' and right-angle versions.	**Source:** Imagery courtesy Milwaukee Tool
Pliers	Pliers are needed when cutting and laying lengths of safety mesh.	**Source:** Imagery courtesy Milwaukee Tool
Pinch bar	A pinch bar is useful for numerous jobs. In particular, it is the best tool for removing old spring-head and lead-head nails when doing a re-roof. The smaller 'wrecking bar' is also very useful.	

>>

Tool name	Comments	Image
Pop riveter	Used every day! A good quality riveter will last for years. Swap heads around for different-sized rivets.	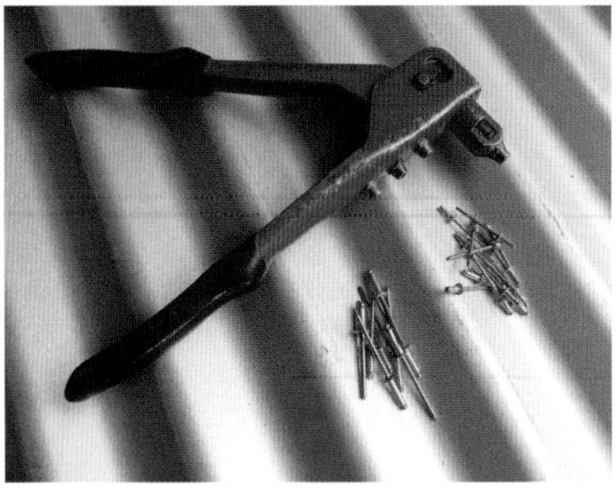
Multigrips	Multigrips are an all-purpose tool for numerous applications. Roofers often use them to turn up the ends of corrugated sheets. Must have.	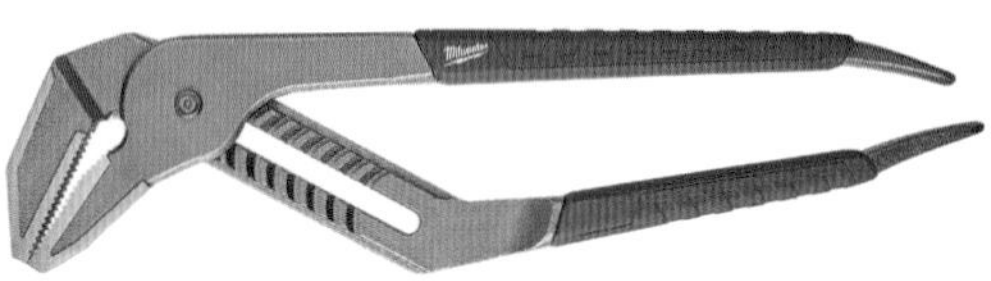
Self-locking C-clamps	Ideal to give you that extra set of hands while fabricating rainwater goods. Self-locking C-clamps hold your work in place while you drill, seal and insert pop rivets.	
Hand folder	While fabricating flashings on the roof you will often need to fold a lap. This makes the job quick and neat.	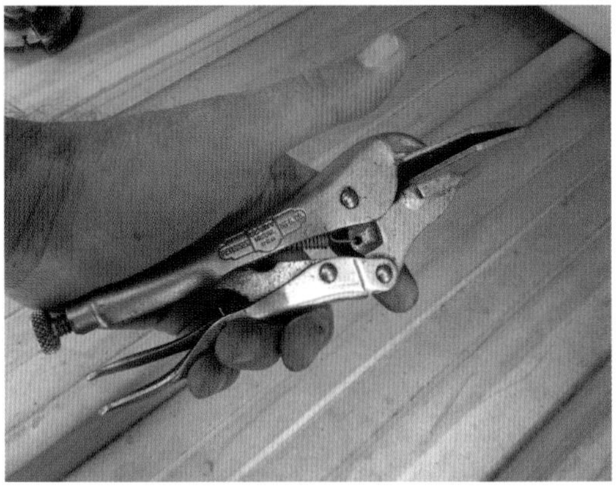

Source: Imagery courtesy Milwaukee Tool

>>

>>

Tool name	Comments	Image
Roof sheet folders	Wide pan roof sheet profiles will in most cases require the pans to be turned down into the gutter and upper ends turned up under the ridge or apron flashing. These can be purchased or fabricated to suit. © BlueScope LYSAGHT®, www.lysaght.com	
Hacksaw	A good hacksaw with quality bi-metal, 32-tooth/inch blades makes easy work of cutting guttering and downpipes.	
Crimper	Ideal for maintenance work when you need to make a spigot end out of round downpipe.	
Raking chisel (plugging chisel)	Some flashings need to be inserted into brick courses. The raking chisel enables you to hammer out the mortar to fit the flashing.	

Source: Imagery courtesy Milwaukee Tool

>>

Tool name	Comments	Image
Old wood chisel	Ideal for starting a penetration quickly before neatly cutting out with snips. Just don't use the boss's best wood chisel!	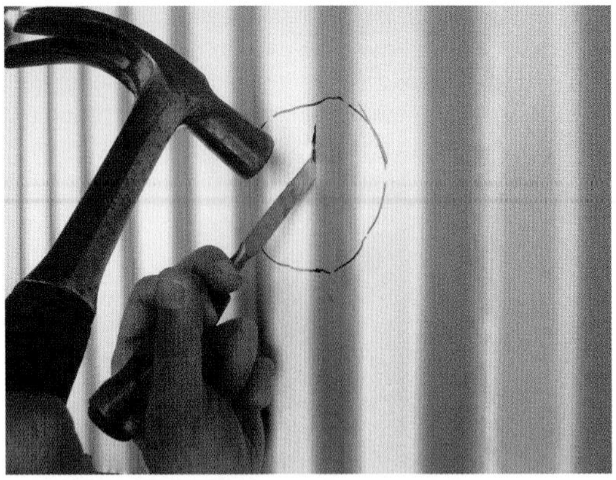
Sealant gun	Used for roofing silicone, other sealants and adhesives. You will need this every day.	
Bossing mallet and lead dressers	Not used that often now, but if you are installing lead flashing and collars, these are the only tools that work the lead correctly. Also work well with soft zinc flashing.	
Cordless drill	Indispensable! There is a wide range of brands. Roof plumbing demands powerful gear so look for drills and drivers with adequate voltage and amp/hour ratings.	

Source: Imagery courtesy Milwaukee Tool

>>

Tool name	Comments	Image
Screw gun	Screwing down a roof requires a screw gun ('Tek gun') with constant torque application to avoid damage to screw head coatings and a depth override mechanism that prevents over-tightening or under-tightening of roof screws. Once the depth override locator is set it is impossible to overdrive screws.	
Impact driver	The flexibility and range of attachments makes an impact driver indispensable for a wide range of fastening tasks. Check with the fastener manufacturer before using on coated screws.	
Assorted drivers	The variety of fasteners used in roof plumbing demands that you have a wide assortment of drivers on hand to suit any drive head.	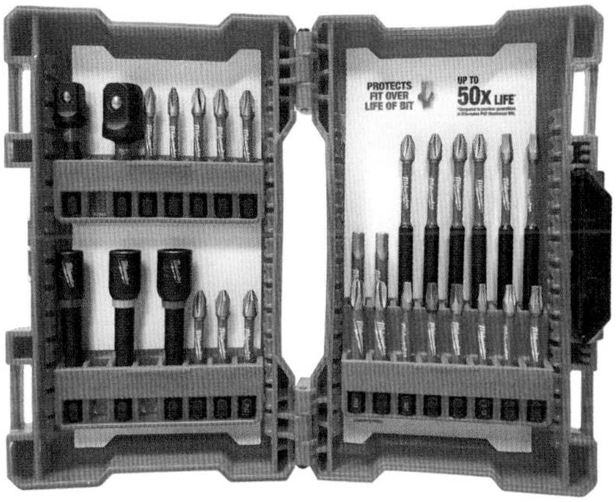
Rotary hammer drill	A rotary hammer drill is required when drilling into any concrete or brickwork. You will often need to do this when installing flashings on walls and parapets.	

Source: Imagery courtesy Milwaukee Tool

Source: Imagery courtesy Milwaukee Tool

>>

>>

Tool name	Comments	Image
Shears	Cordless shears enable you to cut long roof sheet lengths with ease. Some models are also suitable for work across the profile as well.	
Nibbler	Instead of blades, a nibbler cuts with a reciprocating notched pin that enables you to cut with and against the run of profile sheeting. Nibblers are available as dedicated tools or as attachments for power drills.	
Circular saw	A circular saw is useful where timber battens need replacement or alteration, particularly on re-roof jobs.	
Diamond blade for angle grinder and/or power saw	For quick removal of mortar from brick courses or cutting in a raked flashing line in brickwork, you need a diamond blade or masonry disc on a power saw and/or angle grinder.	

Source: Imagery courtesy Milwaukee Tool

Imagery courtesy Milwaukee Tool

>>

>>

Tool name	Comments	Image
Utility belt	To be safe and efficient, a roofer must have tools at hand. A good utility belt is just the shot, but don't overload it. Only carry your most-used tools and fasteners.	
Earth leakage device	Whenever you need to use 240V power tools remember to use an earth leakage device that will shut off the power supply when a fault occurs.	

LEARNING TASK 9.1

1 Which hand tool enables you to copy and transfer different angles?
2 Which hand tool can be used to identify the pitch of an existing roof?
3 Pliers have rubber-covered handles to provide grip, and for what other important safety reason?

Tool care

To achieve safe, high-standard roof installations, it is important that you always take care of your tools and equipment. Tools that are blunt, damaged or inefficient are not only frustrating to use but may also be dangerous.

FROM EXPERIENCE

Your tools are your living! Treat all your equipment with care and ensure that all maintenance is carried out promptly. Ensure all gear is kept clean and locked away safely at night. Safety and profitability go hand in hand with the right attitude to your tools

The section below is a guide to what you may need to consider when maintaining your tools both before and after the job.

Cordless power tools

To maintain cordless power tools, the following points should be considered:

- Once a battery is flat, always put it in the charger as soon as possible. Forgetting to charge batteries leads to significant lost productivity and profitability. Ensure you have matching spare batteries ready to go.
- When you pack up for the day, always ensure that the replacement battery is charged.
- Match the tool to the job. Take care to ensure that you do not overload the tool during use, as this will shorten its working life. If the task is too much for the tool, use a more powerful equivalent and/or review your approach.
- Examine cordless tools for:
  - cracks in the tool casing
  - loose battery clips
  - damaged chargers
  - missing handles and guards
  - worn and missing components
  - faulty switches and buttons
  - worn drill chucks

- an unusually short battery life
- missing replacement batteries.
- Report all problems and faults to your supervisor as soon as possible!

240V power tools

For safety and convenience, a roof plumber will normally use cordless power tools wherever possible. 240V power leads can present a trip hazard, and if carelessly used are prone to being cut by the edges of sheets and flashings. However, where 240V power tools must be used for a specific job, you need to consider the following:

- check that each tool is tagged and the test date is valid (see **Figure 9.1**)
- never allow power tools to become wet
- ensure that all power tool consumables, such as diamond blades, cutting blades and drill bits, are serviceable, and sharpen or replace them as appropriate
- where relevant, ensure that clutch mechanisms operate as required
- examine the tool for:
  - cracks in the tool casing
  - damaged leads
  - exposed wires
  - a damaged plug
  - missing handles and guards
  - worn and missing components
  - faulty switches and buttons

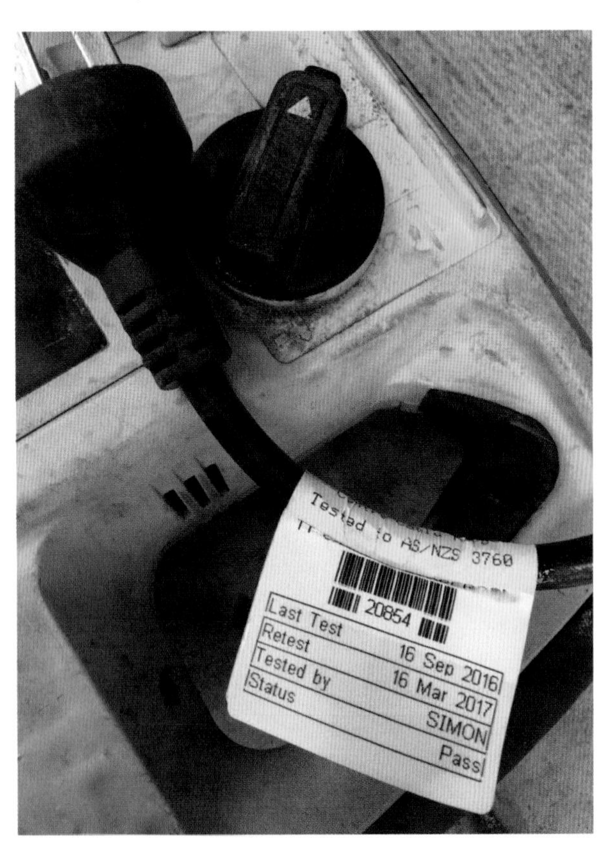

FIGURE 9.1 The re-test date on the tag of a 240V power tool

- report all problems and faults to your supervisor as soon as possible!

Whenever you need to use a 240V power tool, ensure you plug it into a fixed or portable power outlet fitted with an 'earth leakage device'. Such a device will shut off the power in the event of tool or lead malfunction.

Hand tools

It is impossible to list the maintenance considerations for all hand tools. However, review the following as a general guide:

- Check all tools for broken, missing or loose components.
- Ensure that all blades are sharp and cut accurately.
- Check snips cut cleanly and do not 'fold' the edge of the sheet as you cut. If this happens, gently tighten the central clevis pins to bring the blades together.
- Where appropriate, lightly lubricate blades, threads and other working parts to ensure smooth operation.
- Remove burred edges from prick punches and raking/cold chisels.
- Ensure that hacksaw blades are sharp and that spares are available.
- Report all problems and faults to your supervisor as soon as possible!

FROM EXPERIENCE

With all faults or service requirements for tools and equipment, always ensure that the appropriate people are informed of the problem or that the correct processes are followed. Never simply return the tool to its storage place in an unserviceable condition with the attitude that it is a problem for someone else to fix.

LEARNING TASK 9.2

1 Tool care is a part of your company's quality assurance policy. How many principles of tool care can you identify in this section that your company may have informed you of?

SUMMARY

This chapter has included a small selection of the key tools used by a roof plumber. As you move through your training and work in the field take note of what tools are being used by experienced tradespeople. Listen to the opinion of those who look after their tools, use them safely and for the correct purpose as you will generally find their advice is worth taking into account when purchasing your own kit.

WORKSHEET 1

To be completed by teachers	
Satisfactory	☐
Not satisfactory	☐

Student name: ______________________________

Enrolment year: ______________________________

Class code: ______________________________

Competency name/Number: ______________________________

Task: Working with your teacher/supervisor, refer to the text to answer the following questions.

1 You may already have many of the tools described in this chapter. List at least four other tools or items of equipment that you own or use on the job.

____

____

____

____

2 List at least six checks and/or tasks that you would need to conduct for the correct care of a circular saw.

____

____

____

____

____

____

3 During a job you notice that your left-hand snips are not cutting correctly and the sheet metal folds up and gets caught between the blades. The snips are not that old and still look sharp. What do you think is wrong, and how might this be remedied?

____

____

____

4 Which cordless power tool would you use to cut a circle penetration hole through a profiled roof sheet?

____

____

____

____

5 Before using a 240V power tool you check the tag on the lead and notice that the required re-test date has now passed. What should you do?

i proceed to use the tool, intending to tell the boss before knock-off.

ii cut the test tag off and get to work.

iii do not use the tool and report to your supervisor as soon as possible. Inform others in your team that the test date is invalid and not to use the tool.

____

____

____

6 What is the name of the device that all 240V power tools need to be plugged into?

____

____

____

PART B

ROOFING PRACTICE

Part B of this text has a focus on actual installation practice. Each chapter allows you to explore the different aspects of roof work installation in close reference to the Code of Practice and relevant Australian Standards. With interpretation and guidance, you will be able to apply these requirements to your roof installation practice in a methodical, compliant and safe manner.

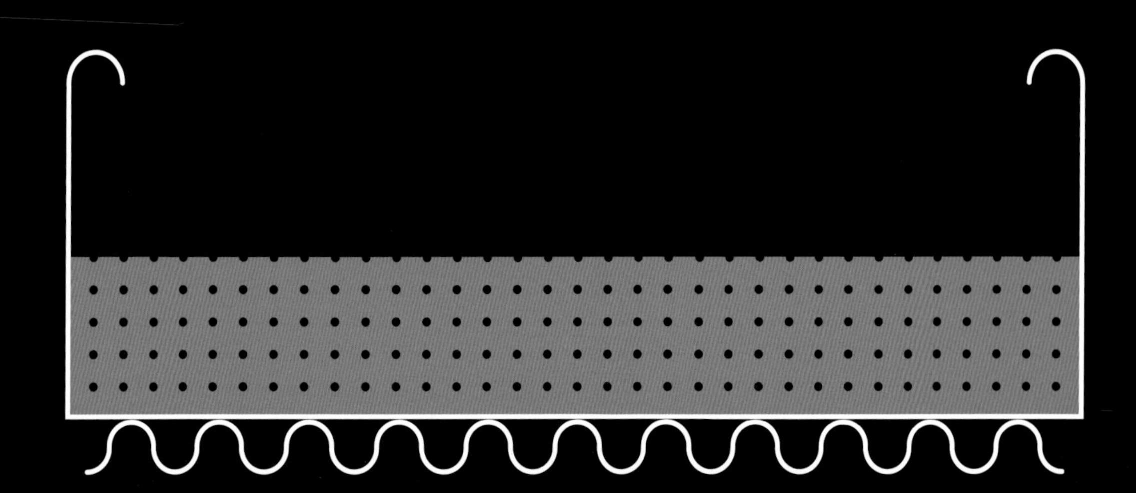

10 FABRICATE AND INSTALL ROOF DRAINAGE COMPONENTS

Roof drainage components are those parts of the roof system that collect and direct water from the roof to an approved point of discharge. Such components include gutters, rainwater heads, sumps and downpipes. Many drainage components require fabrication to suit the individual requirements of each job and you will need to develop the knowledge and hand skills necessary to create watertight joints and products to a high standard.

Before working through this section, it is absolutely important that you complete or revise each of the chapters in Part A Roof Plumbing Fundamentals.

Overview

In this chapter you will be introduced to many of the most common roof drainage components, gain an understanding of their purpose and be able to identify them by sight. You will then have the opportunity to work through step-by-step fabrication processes with advice and guidance from your teacher and supervisor.

Identification of roof components

In this section you will be introduced to the most commonly encountered roof drainage components found in both residential and commercial applications.

Eaves gutter

An eaves gutter is generally installed external to the building line at the end of some form of eaves overhang. Less commonly, some eaves gutters are installed internally and concealed behind some form of fascia. Roof water flows into the eaves gutter and is then drained towards the downpipes.

You will see many different eaves gutter shapes. These shapes are called 'profiles' and are represented and named in relation to the cross-sectional shape of the gutter. Gutter profile shapes and sizes have developed over time to suit both functional and aesthetic purposes.

Not all gutter profiles are found in all areas. Some profiles are unsuitable for high rainfall areas, while in other parts of the country certain profiles are preferred simply because they were traditionally used in that area over many years.

As a roof plumber, you will regularly be responsible for the selection of gutter profiles that not only have a minimum effective cross-sectional area that provides effective drainage, but also have a shape to suit the style and age of the building. For example, installing a modern ZINCALUME® steel slotted LYSAGHT SHEERLINE® gutter on a colonial-era sandstone cottage may well be a functional solution, but the profile would be a poor choice from an aesthetic and heritage perspective. A galvanised LYSAGHT OGEE® Gutter or 'quad' profile may be a better choice.

Some examples of gutter profiles are shown in Figure 10.1.

Downpipes

If water was permitted simply to fall from the roof or gutters directly to the ground, unacceptable erosion of the building footings could result. Water could also pool and provide an environment that is suitable for the development of disease, parasites and mosquitoes.

To avoid such problems, downpipes are connected at low points in a gutter drainage system with the purpose of controlling and directing the water to an approved point of discharge. Such a discharge point may include:

- a rainwater storage tank
- an underground absorption system
- a municipal stormwater drainage network (e.g. underground drains or street kerbing).

Like eaves gutters, downpipes also come in various materials and profiles that can be matched to different gutter types. For example, where quad gutter is installed it is more usual to install round downpipes rather than square or rectangular ones.

Some examples of basic downpipes are shown in Figure 10.2.

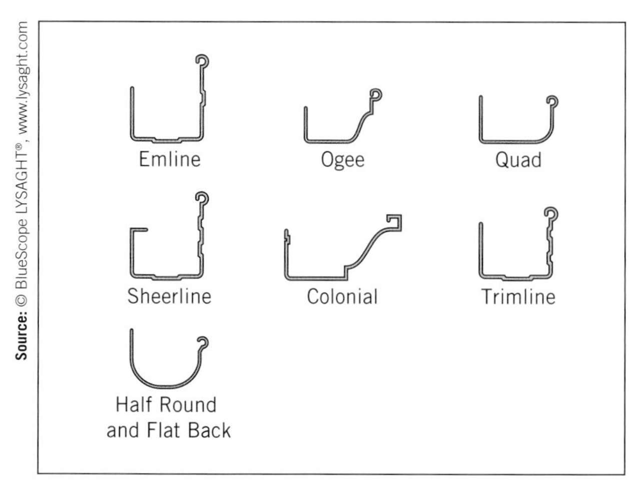

Source: © BlueScope LYSAGHT®, www.lysaght.com

FIGURE 10.1 Examples of eaves gutters and manufacturer profiles

FIGURE 10.2 Examples of commercial, domestic and heritage downpipes

Valley gutter

A valley gutter is an inclined channel used to drain water away from the intersection of two sloping roof planes and direct it to eaves or box gutters. Valley gutters are normally encountered on residential hip roof buildings and are generally supported on flat timber valley boards or metal purlin sections.

Valley gutters are available in a range of profiles to suit tile and metal roof sheeting and must be sized to handle different rainfall intensities. Some examples of valley gutter profiles are shown in Figure 10.3.

Box gutter

A box gutter is a rectangular-shaped graded channel located within the building line that drains water to sumps and rainwater heads.

Box gutters are almost exclusively custom fabricated to suit particular building requirements or applications.

An example of a box gutter is shown in Figure 10.4.

Rainwater heads (rainheads)

A rainwater head (rainhead) is a fabricated component connected to the top of a downpipe that is installed external to a building at the end of a box gutter run. Rainwater heads may also be used on eaves gutter installations. A rainwater head increases the head of water at the entry to the downpipe, increasing its drainage capacity. Rainwater heads also provide a point of overflow to atmosphere outside of the building line.

As rainwater heads are visible components that are mounted on the outside of the building, they are often designed in a style that will aesthetically enhance and match the age and application of the building itself. Colonial- and Federation-era buildings often featured ornate cast iron or copper rainwater heads. Modern designs are now sized to suit higher rainfall intensities and normally incorporate a more prominent overflow outlet. Be aware that there is only one design of rainwater head that is classed as a Deemed-to-Satisfy solution in the Standard. Your teacher or supervisor will explain local requirements where other designs are proposed. Some rainwater head designs are shown in Figure 10.5.

Valley gutter—three break
0.4 × 400 mm
0.55 × 400 mm

Tile valley gutter
0.35 × 465 × 2400 mm

Structural valley
145°
145 145

Source: © BlueScope LYSAGHT®, www.lysaght.com

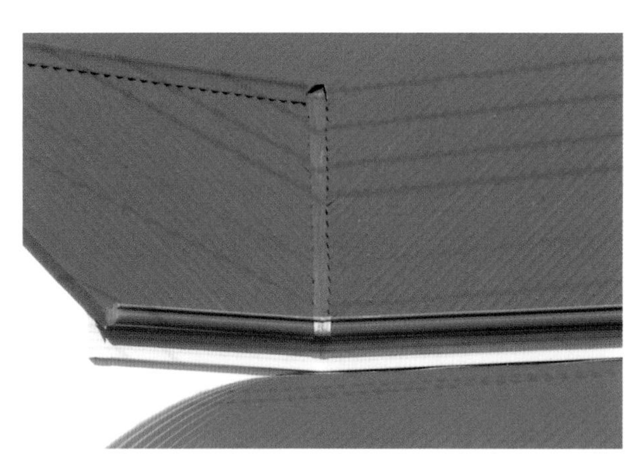

FIGURE 10.3 Examples of valley gutters

FIGURE 10.4 A box gutter draining a low-pitch residential roof

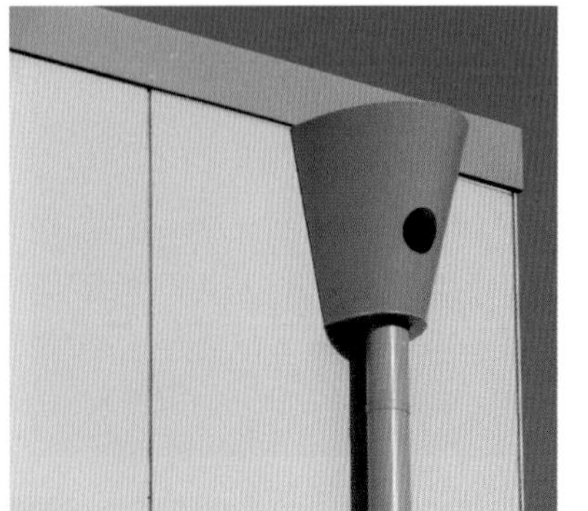

FIGURE 10.5 Examples of modern and heritage rainwater heads

Rainwater sumps

The width of a box gutter is normally many times greater than the cross-sectional area of a downpipe. If the downpipe were connected directly to the base of a box gutter, considerable turbulence and a back-up of water would occur around the outlet. The box gutter would then be at risk of overflowing into the building.

To avoid this problem, a sump is installed across the full sole (base) of a box gutter. This creates a concentrated water collection point that increases the efficiency of water flow into the downpipe.

In the event of a downpipe blockage, water overflowing into a building can potentially cause very significant damage. Therefore, as a sump is fitted to a box gutter system within the building line, overflows of the correct dimension must be fitted at the side or within the sump itself.

An example of a rainwater sump is shown in Figure 10.6.

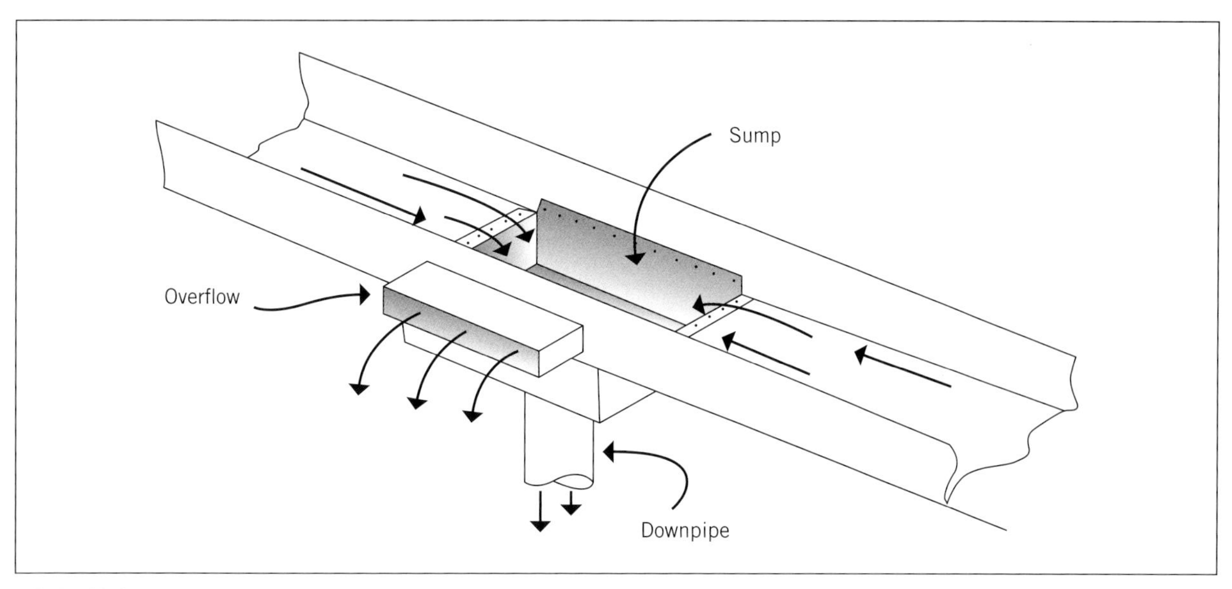

FIGURE 10.6 Example of a box gutter rainwater sump and overflow

Overflows

Overflows are used to prevent water overflowing the gutter and into the building due to component blockage or during unusually high rainfall intensity or duration. Overflows must be sized and positioned in accordance with the requirements of the relevant codes and standards. All box gutter sumps and rainwater heads require some form of overflow that must discharge to atmosphere. An example of an overflow is shown in Figure 10.7.

FIGURE 10.7 The outlet from a box gutter sump overflow

Sizing

It is important to understand at this point that where state and territory regulations require reference to the NCC, all the rainwater drainage systems and components just discussed must be sized in accordance with the AS/NZS 3500.3 'Stormwater drainage'. You cannot simply select any random off-the-shelf item or just copy what other roof plumbers do. Installation must follow the requirements of any approved plans; or in the case of roof and gutter replacement work, a licensed roof plumber (subject to the local licensing system) must calculate requirements to ensure the new installation meets current Australian Standards.

AS/NZS 3500.3 CLAUSE 1.2 'APPLICATION'

LEARNING TASK 10.1

1 What is the name given to the rainwater device located on top of a downpipe at the end of a box gutter?
2 All box gutters must have overflows. Where must overflows discharge to?

COMPLETE WORKSHEET 1

'Take-off', selection and quantity calculation of materials

Before commencing the job, you need to undertake a materials quantity calculation from the approved plans. You will need to account for all consumables in order to schedule their purchase and delivery at the appropriate stage during the work. (More detail on delivery sequencing will be given in Chapter 11 'Receive roofing materials'.)

GREEN TIP

Following sustainable building practices through the minimisation of waste and recycling are important aspects of quantity calculation. This relies upon accurate estimation of project requirements.

In this section, you will be introduced to some basic roof drainage quantity calculations that are common to many jobs.

Product selection

Building designers and licensed roofing contractors are responsible for sizing each component and section of a roof drainage system so that it complies with the Performance Requirements of the NCC and the nominated DTS Standard AS/NZS 3500.3 'Stormwater drainage'. These details are outlined in the approved plans and specifications list. This requirement also applies to domestic re-roofing jobs, even where formal plans and specifications are not developed.

During your preparation for the job, you will need to identify all of the relevant details and then order your materials to match. While the specifications might include basic size requirements, details relating to the complete system or manufacturer accessories are often not included. These are important items of information that you will need to source yourself.

Fortunately, almost everything you need to find is available through manufacturers' product and technical data, which can be found on their websites. The information provided includes:

- product dimensions
- drainage capacity rating
- features (e.g. slotted or unslotted gutter)
- coating and finish
- length restrictions
- support systems
- state/territory availability
- matching accessories and flashings.

These sources enable you to complete your preparation for the job knowing that the selection will comply with the specifications. Figure 10.8 shows an example of manufacturer's product selection guides.

Calculations and waste minimisation

One of the ways to ensure that a job is profitable is to minimise waste. Sustainable work practices that reduce environmental impact also dictate that waste be kept to a minimum. Therefore, when calculating the amount of gutter required you must allow enough extra material for laps and angles, but at the same time reduce the amount of costly offcuts.

In the days when gutter lengths were available in only 6-m and 8-m lengths, plumbers would calculate a total quantity, add a percentage for waste and then divide this amount by the 6-m or 8-m length to arrive at the total number of lengths required. Unfortunately, this would often result in what today is considered an unacceptable quantity of offcuts.

For standard residential installations in most areas today, it is more usual to order pre-cut sections that are specific to the design of the building. For instance, if you need a section that is 4.5 m long, this is what you order (plus lap allowance), not a 6-m length, which was once the only option. In fact, many suppliers can offer the services of an experienced estimator, who will carry out these calculations for you and have delivered to your jobsite all materials sequentially numbered to a matching schedule.

In addition, builders and property owners are keen to keep joins in gutters to an absolute minimum for aesthetic reasons, so installing lengths corner-to-corner wherever possible is considered best practice.

The only restrictions you need to consider in ordering continuous lengths are:

- the maximum transportable lengths in your area
- the efficiency and safety of handling extra-long lengths relative to the number of personnel you have available on the job. For example, handling a 10–12-m length of gutter by yourself is possible, but it could be unsafe and the gutter will likely end up twisting and buckling in the middle as you struggle with it on the scaffold. If you are by yourself, order sizes that you can handle.

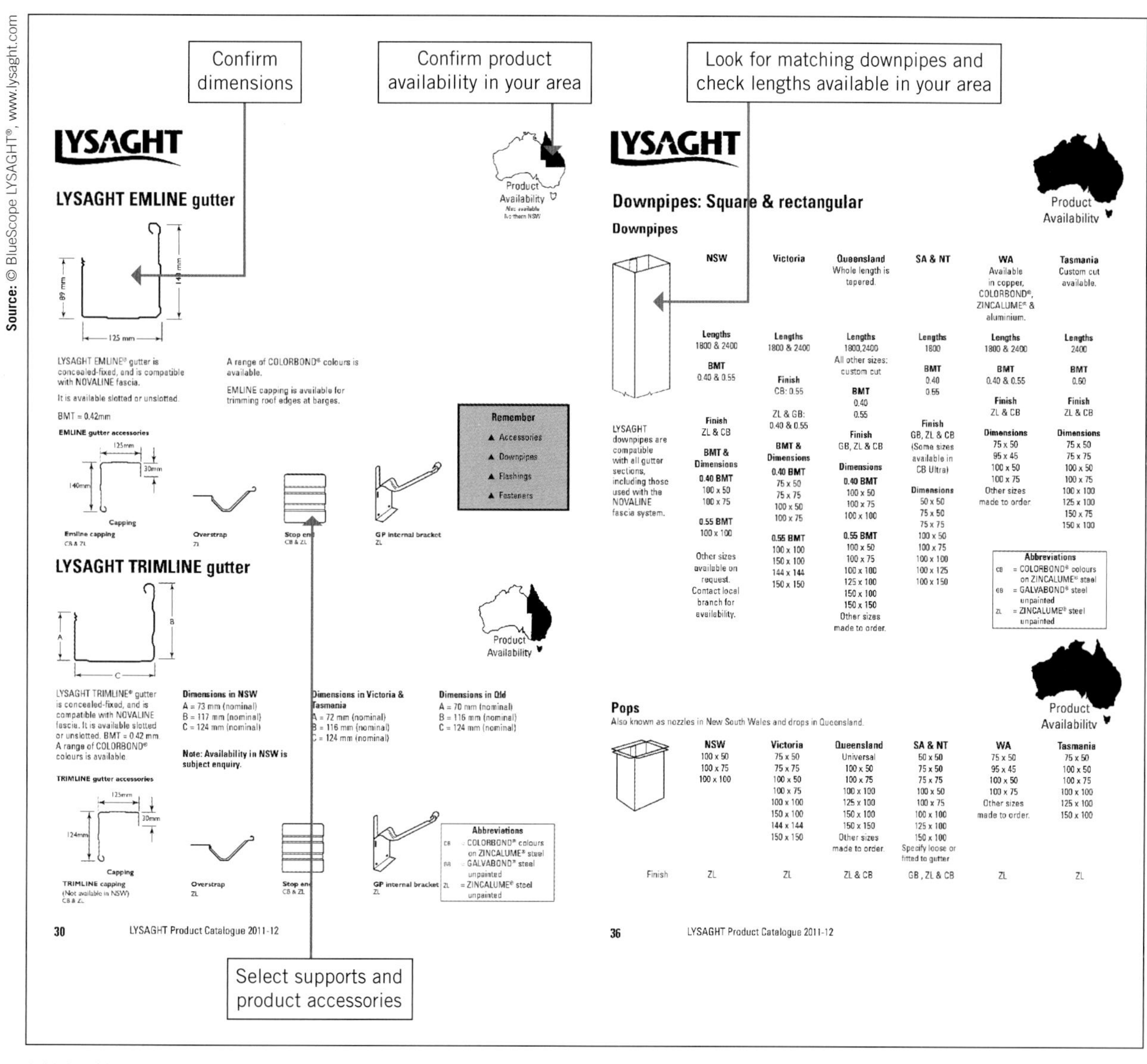

FIGURE 10.8 Example of product selection guides

Note: The catalogue from which these pages were taken has now been moved online and so they are pictured here as an example only.

FROM EXPERIENCE

Caution: While it is now often possible simply to order the lengths you need, be careful that you are not too precise. While it is important to minimise waste, you also need to be practical and still make some allowance for slight deviations from plan and minor measurement and installation errors.

Allowances for external angles

For all external angles, you need to include an allowance of at least 2 × gutter width for each angle. To this you should add a small amount extra to cover errors in mark-out or minor measurement inaccuracies.

EXAMPLE 10.1

If your gutter has a width of 150 mm, you could allow approximately 400 mm extra for each external angle, or 200 mm extra on each length.

What about internal angles?

If you fabricate an internal angle and fold it around, there is actually no loss of length so an allowance is not necessary. However, if you are installing continuous lengths from corner to corner you would need to allow at least the 25-mm lap necessary at the back of the gutter. In practice, you would normally round off to the nearest 50–100 mm for internal corner laps and slip joints.

Review the plan in Example 10.2. This basic outline shows the plan view of a small, low-set hip-and-gable roofed dwelling. A comparison is made between gutter length calculation based upon working in continuous lengths and ordering off-the-shelf 6-m lengths. The following examples go on to detail the materials and accessories required for such a job.

Calculate gutter system ancillary components and materials

The additional components needed for the gutter drainage system include:

- gutter brackets (timber fascia)
- gutter bracket fasteners
- prefabricated stop-ends
- downpipe outlets
- rivets.

EXAMPLE 10.2

1 Calculate gutter lengths
- Measurements indicate fascia dimensions.
- Allow 200 mm for each side of external angles.
- Gutters terminate at the gable end with a prefabricated stop-end.

Process (see Figure 10.9):

a Number each section.

b Measure each section and add any allowance.

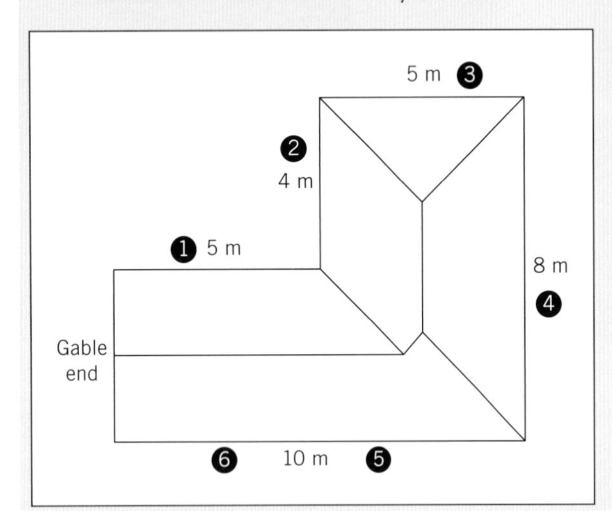

FIGURE 10.9 Calculating gutter lengths

c Calculate the total measurements if required to work out the cost per metre.

d Record the results for your order.

2 Ordering continuous lengths, corner to corner

Section measurements: Section 1: 1 × 5.05 m (allowance for internal angle lap)

Section 2: 1 × 4.2 m (allowance of 200 mm at one end)

Section 3: 1 × 5.4 m (allowance of 200 mm at each end)

Section 4: 1 × 8.4 m (allowance of 200 mm at each end)

Section 5: 1 × 5.2 m (allowance of 200 mm at one end)

Section 6: 1 × 5.05 m (allowance of up to 50 mm for lap)

Total length: 5.05 + 4.2 + 5.4 + 8.4 + 5.2 + 5.05 = 33.3 m

This represents approx. 375 mm waste and only one slip joint.

3 Comparison: ordering by 6 m lengths

Plan length: 5 + 4 + 5 + 8 + 10 = 32 m

Plan length + 5% allowance: 32 + 5% = 33.6 m

Number of 6-m lengths: 33.6 ÷ 6 = 5.6, or 6 full lengths

This represents a total of 36 m of purchased material, approximately 2.4 m waste and 6 slip joints.

EXAMPLE 10.3

Calculate gutter system components and materials (see Figure 10.10)

- Timber fascia.
- Brackets spaced at 1.2-m intervals with 3 screws per bracket.
- Allow average 10 rivets per angle, stop-end and downpipe outlet.

Number of gutter brackets:

Number of brackets = Total metres ÷ 1.2 m

$$= \frac{(5+4+5+8+10)}{1.2}$$

$$= \frac{32}{1.2}$$

$$= 26.6$$

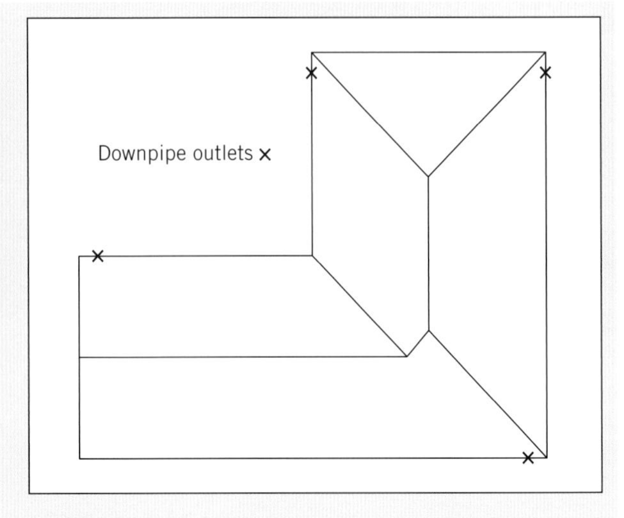

FIGURE 10.10 Calculating components and materials for a gutter system

>>

>>

Round up and add 4 brackets for additional support where required:
- Number of brackets = 31

Number of gutter bracket fasteners:

Number of fasteners = Total brackets × 3 fasteners per bracket
= 93 fasteners

(Round up to nearest packet size.)

- Number of stop-ends for gable end = 2
- Number of downpipe outlets on plan = 4

Number of rivets = Total angles + stop-ends + outlets × 10 rivets
= 6 angles + 2 stop-ends + 4 outlets × 10 rivets
= 12 items × 10
= 120 rivets (3.2 mm grip × 3.2 diameter coloured)

(Round up to nearest packet size.)

EXAMPLE 10.4

Calculate downpipe lengths and supports (see Figure 10.11)
- Assume a 3.2 m outlet height to NGS all round.
- The available downpipe lengths are 1.8 m and 2.4 m.
- Allow 2 brackets per downpipe.
- Allow an average of 10 rivets per downpipe.

Downpipe lengths:

Lengths per downpipe = 1 × 2.4 m and 1 × 1.8 m
Total for 4 downpipes = 4 × 2.4 m and 4 × 1.8 m

Number of rivets:

Number per downpipe = 10
= 10 × 4 downpipes
= 40 rivets (3.2 mm grip × 3.2 diameter coloured)

Number of brackets:

Number per downpipe = 2
= 2 × 4 downpipes
= 8 brackets

Number of fasteners:

Number per bracket
= 2 × 25 mm × 5 mm nylon anchors

Total for 4 downpipes
= 16 × 25 mm × 5 mm nylon anchors

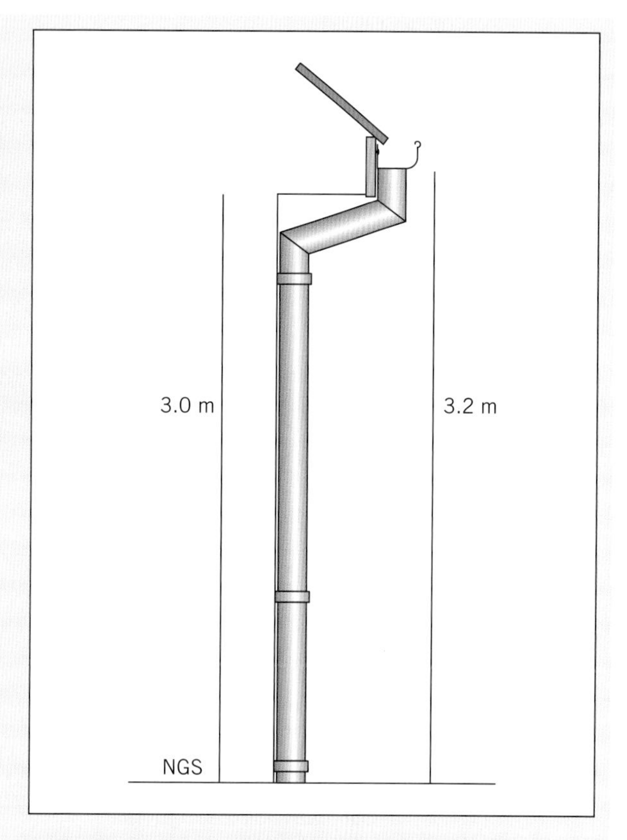

FIGURE 10.11 Calculating downpipe lengths and supports

Calculate downpipe lengths and supports

To calculate the quantity of downpipe lengths you need to review the approved plan elevation views and determine the following:
- the width of the eave overhangs
- the distance from gutter outlet to natural ground surface (NGS) level.

Calculate valley length

As the valley is installed on an angle, its length cannot be determined from just looking at a plan view of the building. However, from the plan view and elevations, you are able to determine the following:
- rise – this is the vertical distance from the bottom of the valley to the top, and is seen in the elevation view
- run – this is the horizontal distance from the bottom of the valley to the top. It can be determined adequately by simply measuring it with a scale rule off the plan view. (Scaling off a plan is not normally recommended, but it will suffice here because of the allowance added.)

You can then use Pythagoras' theorem to find the length.

Summary

Having worked out the material requirements for each section, you need to create a summary list that will be used for costing and/or ordering. Don't forget consumables such as drill bits, hacksaw blades and drivers – they all have to be paid for. A simple table is all that is needed, such as in the example given in Figure 10.13.

EXAMPLE 10.5

Calculate valley length (see Figure 10.12)

- The elevation view shows a 'rise' of 2 m.
- The plan view shows a 'run' of 2.9 m.
- The valley gutter is to be one continuous length.

Pythagoras' theorem:

$$H^2 = rise^2 + run^2$$

Where

H = incline or length of valley

Valley length^2 = (2 × 2) + (2.9 × 2.9)

= 4 + 8.41

= 12.41

Valley length = √ 12.41

= 3.52 m

To allow for gutter turn-down and ridge turn-up add an extra 200 mm

= 3.7 m length

Number of galvanised nails = 1 × 500 mm on either side

= 16 nails

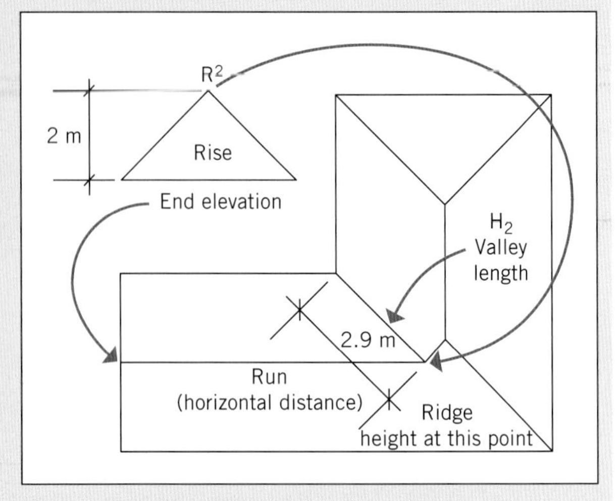

FIGURE 10.12 Calculating the length of a valley

Joe Bloggs' fishing shack			
Item	**Description**	**Sub Q'ty**	**Total Q'ty**
Gutter	*Lysaght Ogee – Colorbond 'Manor Red' – All angles fabricated onsite*	*S1/ 1 x 5.05 m S2/ 1 x 4.2 m S3/ 1 x 5.4 m S4/ 1 x 8.4 m S5/ 1 x 5.2 m S6/ 1 x 5.05 m*	*33.3 m*
Brackets	*Internal GP*		*31*
Bracket fasteners	*30 mm x 10 g wafer head screws*	*93*	*1 pkt of 100*
Stop ends	*Lysaght Ogee – Colorbond 'Manor Red'*		*2*
Valley	*Standard 190 x 190 profile*	*1*	*3.7 m*
Valley nails	*Galvanised 75 mm x 3.75 mm nails*	*16*	*1 pkt of 50*
D/pipe outlets	*75 mm round – Colorbond 'Manor Red'*		*4*
D/pipe	*75 mm round – Colorbond 'Manor Red'*		*4 x 1.8 m 4 x 2.4 m*
D/pipe brackets	*75 mm round – Colorbond 'Manor Red'*		*8*
Fasteners	*25 mm x 5 mm nylon anchors*	*16*	*1 pkt of 50*
Rivets	*3.2 mm grip x 3.2 mm diam. – Colorbond 'Manor Red'*	*Angles/stop ends/ d/pipe outlets 120 D/pipes 40 160*	*2 pkt of 100*
Silicone sealant	*– Colorbond 'Manor Red'*		*2 x tubes*
Hacksaw blades	*32 TPI bi-metal*		*2*
Drill bits	*1/8" double ended panel drills*		*6*
Phillips driver	*100 mm Phillips driver*		*1*

FIGURE 10.13 Example of a materials list showing all consumables

LEARNING TASK 10.2

1 To calculate the length of downpipe, which two pieces of information will you need?
2 Which theorem was used to calculate the length of valley?

Basic fabrication skills for roof drainage components

The fabrication of roof drainage components is an area within the plumbing sector where high-level hand skills, judgement and attention to detail are still required. Each component that you work on adds to the quality of the whole installation and must not only perform well, but look good too. Badly-made roof components 'attract the eye'. Regardless of how well you install the rest of the job, one poorly made gutter angle, downpipe or flashing will detract from the whole project. Roof plumbing is a 'finishing trade', and your work standard will always be on display.

Coupled with good instruction from your teachers and supervisor, the ability to produce high-quality work requires repetition, patience and personal determination to improve your output. This section introduces you to some basic fabrication skills.

Standard of silicone sealed joints

Silicone sealant is the most commonly used form of joint sealing in general roof plumbing. Used correctly, silicone sealant joints will remain watertight for the life of the plumbing product. However, the sealant's performance depends on its correct application. It is not unusual to find leaking joints in gutter installations where silicone was applied only to the top of the joint *after* fabrication. This is poor trade practice and such joints are subject to failure over time due to repeated expansion and contraction of the metal. For example, the lap of the flashing shown in Figure 10.14 was never sealed correctly, with silicone only applied after installation. Leakage over many years caused substantial rotting of the underlying timber framework.

Referring to your HB39:2015, follow the steps in the following 'How to' box to see how a quality silicone sealed joint is fabricated.

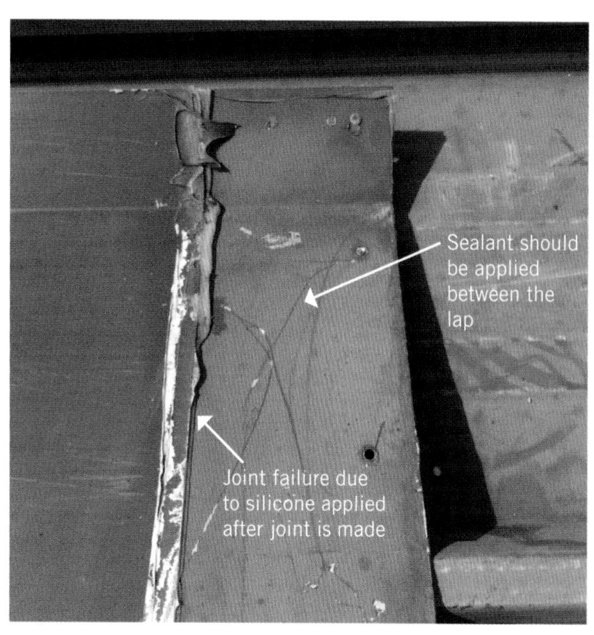

FIGURE 10.14 Leaks will always occur unless silicone is applied between each lap.

HOW TO

BASIC JOINTING PROCEDURE USING SILICONE SEALANT AND FASTENERS

Step 1	a Remove any protective plastic wrap from the product. b Clean the material of all oils and contaminants. c Align the work with a minimum 25-mm lap and pre-drill holes at approximately 40 mm centres.	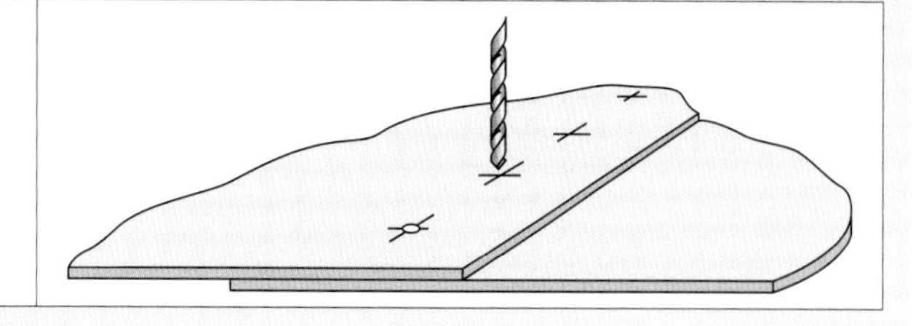

>>

>>

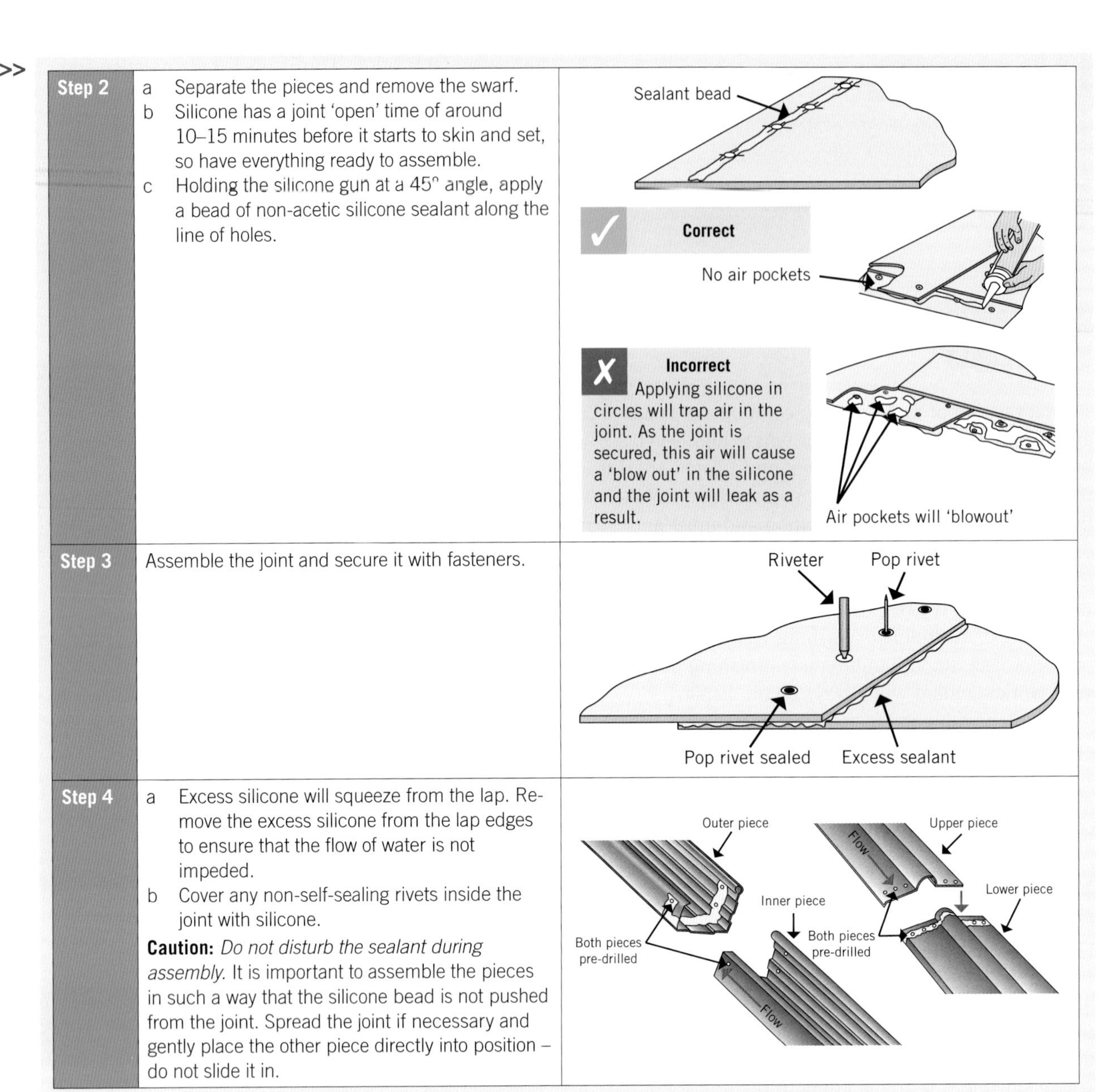

Step	Instructions
Step 2	a Separate the pieces and remove the swarf. b Silicone has a joint 'open' time of around 10–15 minutes before it starts to skin and set, so have everything ready to assemble. c Holding the silicone gun at a 45° angle, apply a bead of non-acetic silicone sealant along the line of holes.
Step 3	Assemble the joint and secure it with fasteners.
Step 4	a Excess silicone will squeeze from the lap. Remove the excess silicone from the lap edges to ensure that the flow of water is not impeded. b Cover any non-self-sealing rivets inside the joint with silicone. **Caution:** *Do not disturb the sealant during assembly.* It is important to assemble the pieces in such a way that the silicone bead is not pushed from the joint. Spread the joint if necessary and gently place the other piece directly into position – do not slide it in.

Source: Based on Basic Training Manual 12–1 *Roof Plumbing, Introduction and Downpipes*, Australian Government Publishing Service, 1981.

During your fabrication layout and assembly, always ensure that you align the laps of each piece so that they lie in the direction of the water flow (see Figure 10.15). Particularly where items have a low pitch or fall, any impediment to the flow of water can cause ponding, a build-up of silt and premature corrosion. Double check your laps and also ensure that any excess sealant around laps and fasteners is removed.

Angles

To fabricate an angle in a section of gutter or downpipe, you need to remove a certain amount of material from the original straight length to create a

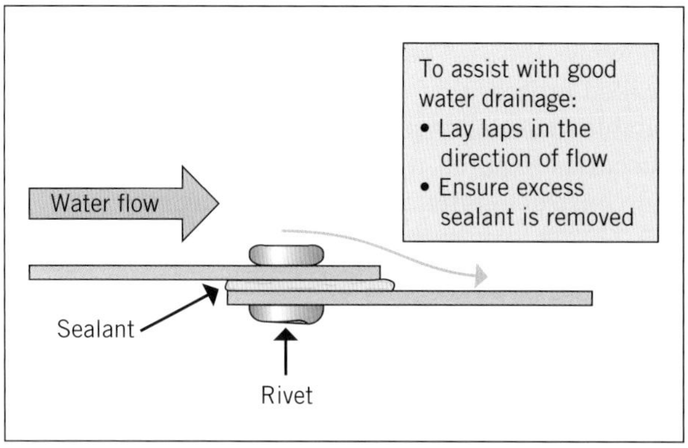

FIGURE 10.15 Assemble laps in the direction of water flow.

change in direction and a new angle. Knowing how to do this is a core skill for any roof plumber. In order to do so, there are two principles that you need to understand:

1 To create a new angle from a straight length, *the same amount of material must be removed from either side of centre or the 'vertex'* (excluding an allowance for laps). The shapes in Figure 10.16 show how this works. If necessary, trace and cut out these shapes to understand this concept better.

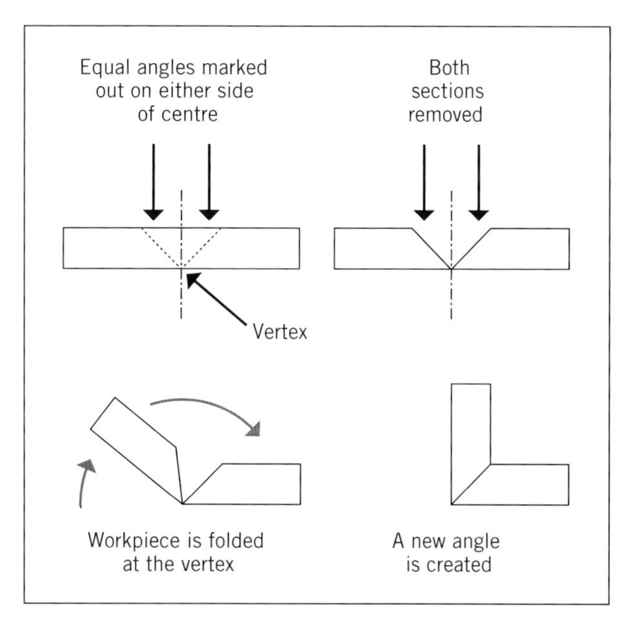

FIGURE 10.16 Creating angles from straight lengths of material

Angles are measured in degrees. The most commonly used angles when constructing gutter angles are the 90° angle or 'right-angle' and the 45° angle. The diagram in Figure 10.17 shows how angles can be measured.

2 A roofing angle is measured from and made up of *two angles of equal dimension projected from the vertex*.

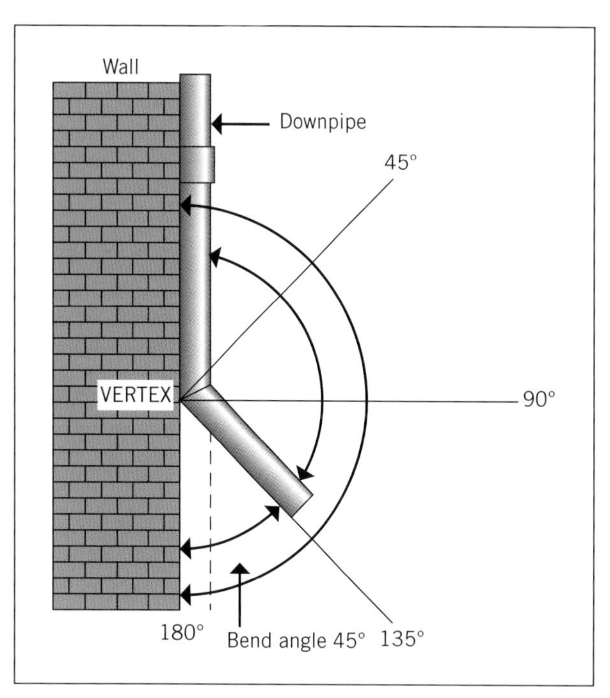

FIGURE 10.17 Measuring angles in degrees

EXAMPLE 10.6

A 90° angle is made up of 2 × 45° angles
(45° + 45° = 90°)
A 60° angle is made up of 2 × 30° angles
(30° + 30° = 60°)
A 135° angle is made up of 2 × 67.5° angles
(67.5° + 67.5° = 135°)

Therefore, to fabricate an angle in gutter or downpipes you need to do the following:

1 Determine at what location you need to swing the angle (this point is the vertex).
2 Measure the angle required (e.g. 90°, 45°).
3 Mark the vertex on the material.
4 Mark out and remove an equal amount of material on each side of the vertex (allowing extra for laps on the upstream piece).
5 Join the halves together to create the angle.

Determine mark-out allowance where the angle is unknown

The 'mark-out allowance' of any angle is the amount of material that needs to be removed to create the angle.

If all building frames were perfectly built to dimension, you could fabricate angles with protractors and other measuring instruments to match the required angle perfectly. However, this is not usually the case and, particularly with the installation of gutter/downpipes on older buildings, you may need to customise each angle to accommodate 'out-of-square' or warped fascia and walls. For example, sometimes you may need to transfer a bevel angle from older existing downpipes to your new material to ensure that the fall on the new downpipe matches the old.

On the job, you are rarely equipped with specialised drawing equipment. Fortunately, there is a reasonably easy method of determining any mark-out allowance for a given situation without necessarily needing to know the exact degrees required. This method, which is shown in the following 'How to' box, is a simple and easily recalled process that can be marked out on any suitable material or surface. Refer back to this process as required when learning how to mark out and construct gutter and downpipe angles.

HOW TO

STEPS TO DETERMINE MARK-OUT ALLOWANCE

Step 1	Using a bevel, measure the angle that you wish to make. For example, this could be an existing downpipe angle or the gutter angles around a bay window.	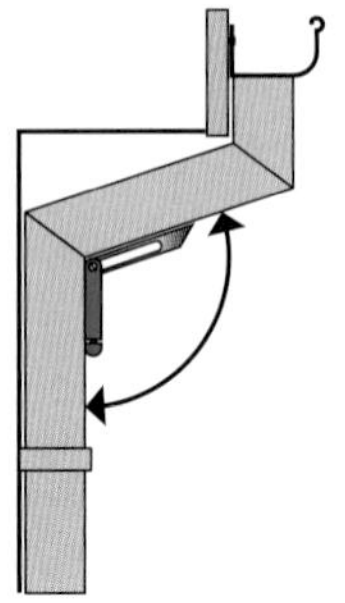
Step 2	Transfer the bevel angle to a piece of paper, then sketch the object to the width required. Width in this instance is shown as 'x'.	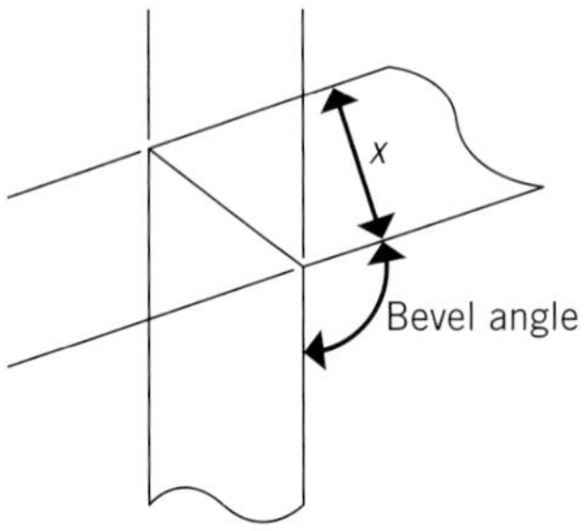 x Bevel angle
Step 3	From the inside 'throat' of the angle, project 2 × 90° lines to the back of the angle. These are marked as line 1 and 2 in red.	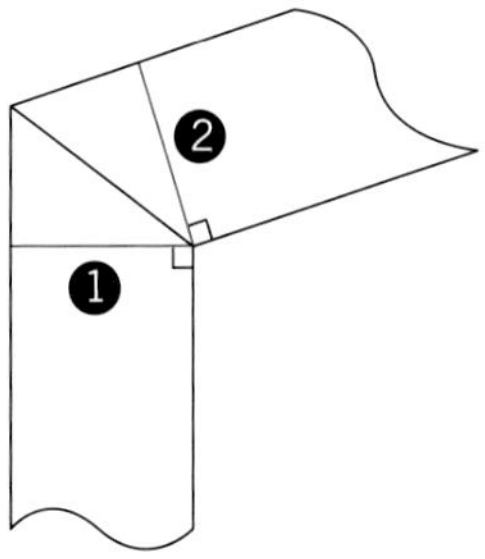 1 2
Step 4	Measure distance 'a' from each line to the vertex of the angle. This is the mark-out allowance (laps not included).	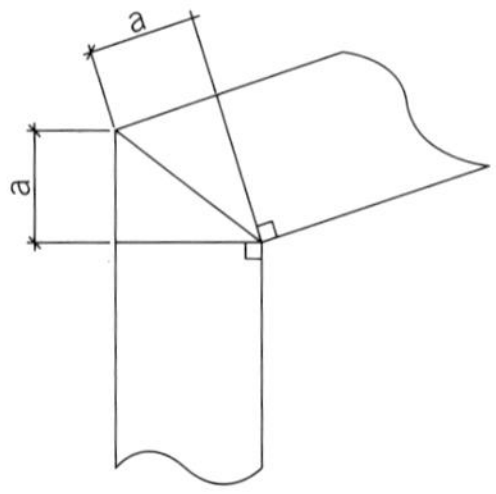 a a
Step 5	Transfer distance 'a' to your work piece.	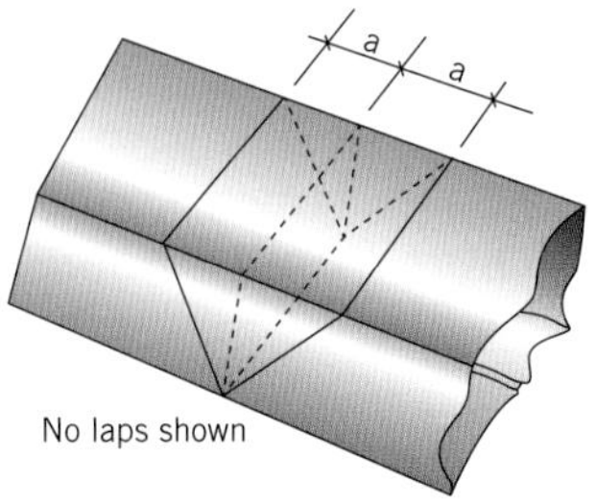 a a No laps shown

LEARNING TASK 10.3

1 To assist with good drainage, do the laps need to be laid with or against the flow of water?
2 When creating a new angle from a straight length, the same amount of material must be removed from either side of centre line, which is also known as what?

Work standard requirements

There is no single 'right way' to mark out and fabricate rainwater goods. While the basic principles remain the same for all angles, there are many variations on how the angle is cut, lapped and fastened. Your teachers and supervisors will show you various methods and eventually you may adapt and develop your own process and product.

In this text you will see a sample of mark-out methods that may or may not be the ones used in your area or place of employment. However, while methods of fabrication vary considerably, the standard and result of your work should comply with the following:

- The joints must not show gaps.
- The laps must lay in the direction of water flow.
- The laps must be fully sealed.
- The joints must be appropriately fastened.
- The product angles and measurements must match the building requirements.

Marking out rainwater goods

Getting your mark-out completed in a neat and accurate manner is the first step in the successful fabrication of rainwater goods. Double-check all your measurements before cutting with care.

FROM EXPERIENCE

The following adage is something that you should remember in all the trade work you undertake:
Measure twice and cut once!

Preparation

The mark-out process will generally require the following items of equipment:

- rule and tape measure
- square – fixed and/or adjustable
- bevel
- fine marker pen and/or scriber (no graphite pencils)
- relevant personal safety equipment.

As you work through the following sections on mark-out allowances, refer to Figure 10.18 as required to confirm the various terms used.

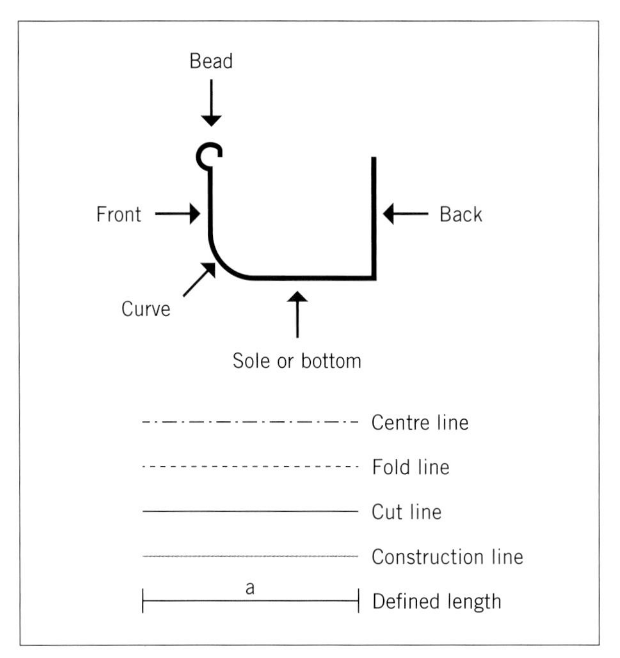

FIGURE 10.18 Component references and line identification

Marking out eaves gutter angles

The following mark-out examples are suggested solutions for the fabrication of two of the more commonly used gutter profiles in Australia: the various square-profile eaves gutters and quad eaves gutters. All other gutter types and relevant mark-outs are simply variations on this procedure.

Modern versus traditional fabrication processes

Before starting the mark-out, it is worth noting some of the differences that exist between the fabrication methods used in 'new work' and the more traditional fabrication approach.

Particularly in relation to residential dwellings, the majority of new eaves gutter installations today are coupled with metal fascia systems, and gutter is delivered pre-cut in continuous lengths to suit each section of the fascia. This means that for any length of gutter up to approximately 15 m (subject to transport limitations), there are no joins between each angle and change of direction, thus providing a clean, unbroken gutter line. As a result, angles are fabricated at the end of each gutter length, then assembled, sealed and fastened while in position. This is often known as 'corner-to-corner' assembly.

However, before the advent of continuous lengths, gutter was supplied off the shelf in standardised measurements. For many early installations, the maximum length was 1.8 m and each section was jointed with tinman's rivets and solder. Gutter angles were normally fully fabricated as a separate component on the ground. Wet-weather days were often spent in workshops making angles ready for the next job.

The mark-out methods shown here, for quad gutter in particular, are largely based upon the traditional approach, as this suits the shorter gutter lengths used in training and also because it includes the addition of corner gussets and wider laps that are often dispensed with in new installations.

Traditional fabrication methods are still important where repair angles need to be robust enough to be pushed in under stubborn ridge caps or joined to older profile gutter. The same methods also lend more authenticity to heritage installations and demonstrate that the work was done by a tradesperson and not merely bought off the shelf.

Where your focus is generally new work, simply apply a similar approach to each end of the separate gutter lengths for corner-to-corner assembly and follow your normal company protocols.

Square-profile gutter angles

Square gutter profiles are probably the most common type installed on residential and small commercial work today. There are many manufacturer variations to the actual profile line and the range comes with slotted overflows, unslotted, pre-painted and bare metal.

Gaining proficiency in square profile fabrication is advisable before moving onto other profiles such as quad and LYSAGHT OGEE® gutters. In this section, you will find generic mark-out procedures that can be varied as required to suit slightly different brands. Your teacher will most likely advise you of any variation to these examples. The basic angles shown include:

- 90° external (fold-around method)
- 90° external (separate lengths)
- 90° internal
- return stop-end.

Each of the following 'How to' boxes details a step or series of steps to assist you in marking out your work. The numbers in brackets in each step correlate with the numbers in the photographs or diagrams. *Where photographs are shown, please note that sealant has been omitted for better image clarity.*

Ensure that you are satisfied with each step before moving on to the next. Once completed, have your mark-out checked before cutting.

HOW TO

MARK OUT A 90° EXTERNAL SQUARE-PROFILE GUTTER ANGLE (FOLD-AROUND METHOD)

Step	Instructions
Step 1	a Measure and mark the corner of the fascia on the rear bottom edge of the gutter *(1)*. b Add the width of the gutter *(a)* to locate and mark the centreline *(2)*. c Add the width of the gutter *(a)* on the other side of the centreline *(3)*.
Step 2	a Project the centreline across the bottom, back and up to the bead *(1)*. b Using a 45° square, project angle lines from the back of the gutter to the bottom front *(2)*. c Turn the gutter over and, using a 45° square, mark 2 × 45° lines from the front face of the bead, across and to the back of the bead *(3)*.
Step 3	a Confirm the direction of the water flow *(1)*. Mark 25 mm laps on the upstream section *(2)*. b With a marker pen, cross-hatch the material to be removed *(3)*. (In this diagram, the shaded areas show which section is to be removed.) c The top bead gusset can be marked out from 50 × 50 mm scrap. Mark 4 holes for pre-drilling *(4)*.

HOW TO

MARK OUT A 90° EXTERNAL SQUARE-PROFILE GUTTER ANGLE (CORNER-TO-CORNER METHOD)

This mark-out will assume that you are measuring *back from the ends of each gutter length* at the fascia corner. The process is essentially the same when determining the corner position from the other end of the gutter, the only difference being that the waste material is at the angle end.		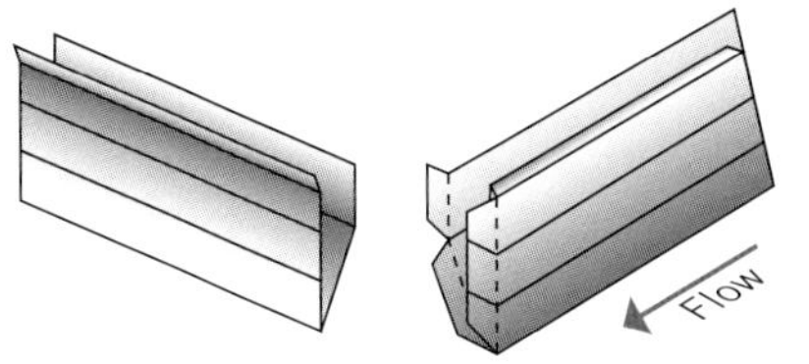
Step 1	a On the upstream length, measure back the distance from the end of the gutter *(a)* + a 25-mm lap to determine and mark the fascia corner position *(1)*. b On the downstream length, measure back the distance from the end of the gutter *(a)* to determine and mark the fascia corner position *(2)*.	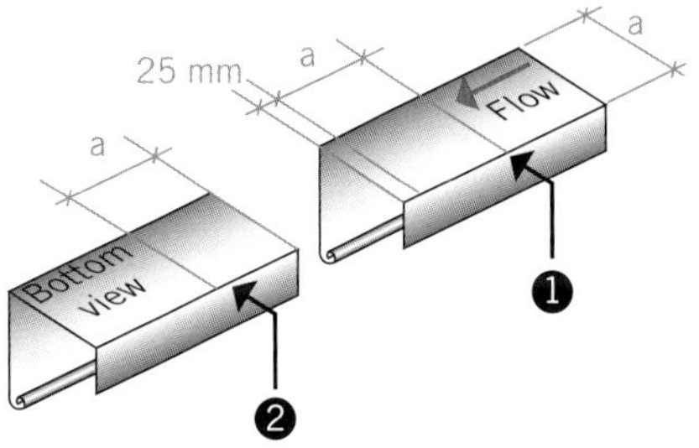
Step 2	a On the upstream length, project a centreline across the bottom and up to the bead *(1)*. b Using a 45° square, project the angle line from the back of the gutter to the bottom front *(2)*.	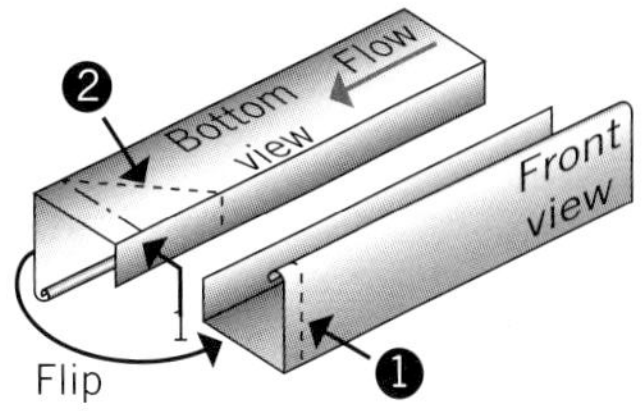
Step 3	a On the upstream piece, mark 25 mm laps on the upstream bottom, back *(1)* and front *(2)*. b Turn the gutter over and, using a 45° square, mark a 45° line from the front face of the bead, across and to the back of the bead *(3)*. c With a marker pen, cross-hatch the material to be removed *(4)*. (In this diagram, the shaded areas show which section is to be removed.)	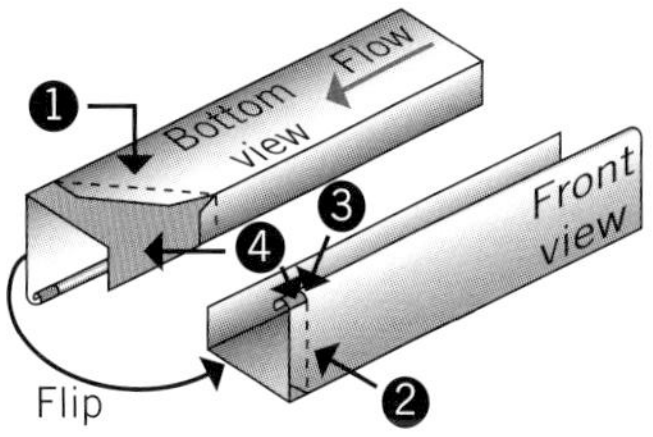
Step 4	a On the downstream piece, project angle lines from the back of the gutter to the bottom front *(1)* using a 45° square. b Turn the gutter over and, using a 45° square, mark a 45° line from the front face of the bead, across and to the back of the bead *(2)*. c With a marker pen, cross-hatch the material to be removed *(3)*. (In this diagram, the shaded areas show which section is to be removed.)	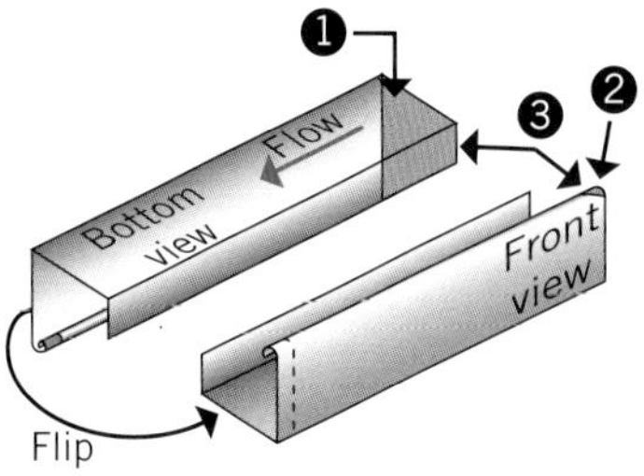

HOW TO

MARK OUT A 90° INTERNAL SQUARE-PROFILE GUTTER ANGLE

This image is of a completed 90° internal square profile angle after following these steps.

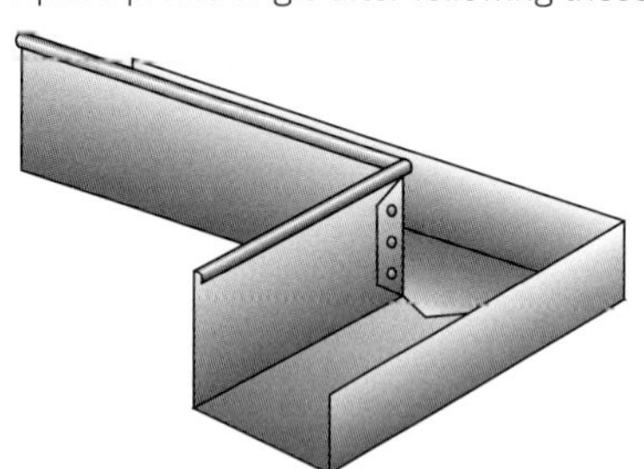

Step 1

a Measure and mark the internal corner of the fascia on the rear-bottom edge of the gutter *(1)*. This is also the vertex of the angle.
b Project this line across the bottom, front and up to the bead *(2)*.
c Add the width of gutter *(a)* upstream of the centreline *(3)*.
d Add the width of gutter *(a)* downstream of the centreline *(4)*.

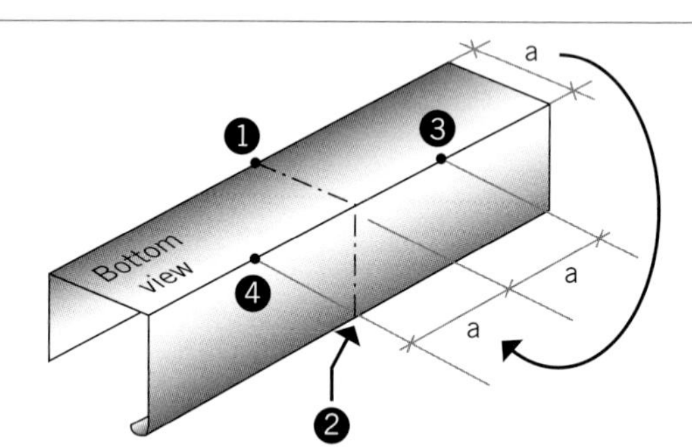

Step 2

a Using a 45° square, project angle lines from the bottom-front of the gutter to the vertex *(1)*.
b Project the lines up the front to the bead *(2)*.
c Confirm the direction of water flow then mark 25-mm laps on the bottom and front of the upstream section *(3)*.

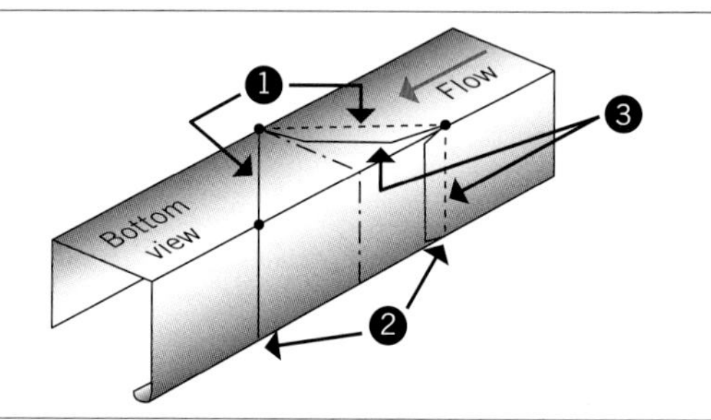

Step 3

a Turn the gutter over and, using a 45° square, mark 2 × 45° lines from the front face of the bead, across and to the back of the bead *(1)*.
b With a marker pen, cross-hatch the material to be removed *(2)*. (In this diagram, the shaded areas show which section is to be removed.)
c Note that the back of gutter at vertex point is folded around, not cut *(3)*.

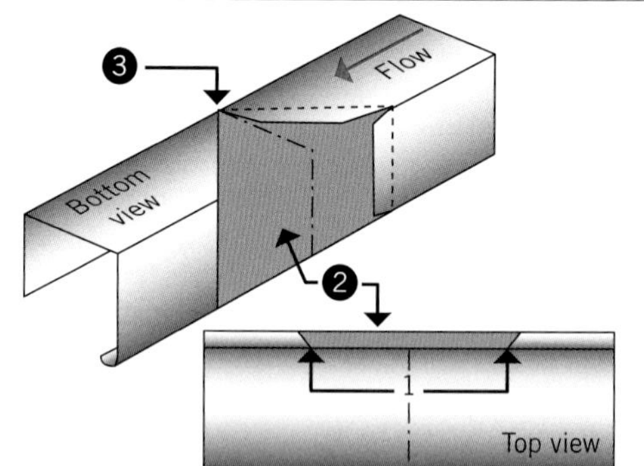

HOW TO

MARK OUT A RETURN STOP-END FOR A SQUARE-PROFILE GUTTER (FOLD-AROUND METHOD)

This image is of a completed return stop-end for a square profile gutter after following these steps.		
Step 1	a From the end of the length, measure back the same distance as the width of the gutter *(a)*. This is point *(1)*. b Project a line from point *(1)* across the bottom, back and up the front to the top of the bead *(2)*.	a a Bottom view Front view Flip 1 2
Step 2	a Add a 25-mm lap at the front bottom of the gutter *(1)*. b Add a 25-mm lap at the back of the gutter *(2)*. c Turn the gutter over and, using a 45° square, mark 2 × 45° lines from the front face of the bead, across and to the back of the bead *(3)*.	Bottom view Front view Flip 1 2 3
Step 3	a With marker pen, cross-hatch the material to be removed *(1)*. (In this diagram, the shaded areas show which section is to be removed.) b The top bead gusset can be marked out from 50 × 50 mm scrap. Mark 4 holes for pre-drilling *(2)*.	Bottom view Front view Flip 1 2 3

Quad-profile gutter angles

The quad gutter profile is a very traditional shape and has been in use for many decades. It is still popular today in both the domestic and commercial roofing sectors and is available in a very wide range of sizes, shapes and coatings.

A well-made quad angle depends on an accurate mark-out, with particular emphasis on the front curve. Getting the curve right is the key, and your teacher should be able to demonstrate a number of different methods to do this. In the following section, you will find generic mark-out procedures that can be varied as required to suit slightly different brands and your personal preference. Your teacher will most likely advise you of any variation to these examples.

The basic angles shown include:

- 90° external (including three variations for curve lap)
- 90° internal
- return stop-end.

Where photographs are shown, please note that the author has omitted the sealant for better image clarity.

Each 'How to' box details a step or series of steps to assist you in marking out your work. Ensure that you are satisfied with each step before moving on to the next. Once completed, have your mark-out checked before cutting.

HOW TO

MARK OUT A 90° EXTERNAL QUAD-PROFILE GUTTER ANGLE (INCLUDES THREE OPTIONS FOR CURVE LAP)

This image is of a completed external quad-profile gutter after following these steps.

Step 1

a Measure and mark the corner of the fascia on the rear-bottom edge of the gutter *(1)*.
b Add the width of the gutter *(a)* to locate and mark the centreline *(2)*.
c Add the width of gutter *(a)* on the other side of the centreline *(3)*.
d Project the centreline across the bottom and up the front to the bead *(4)*.

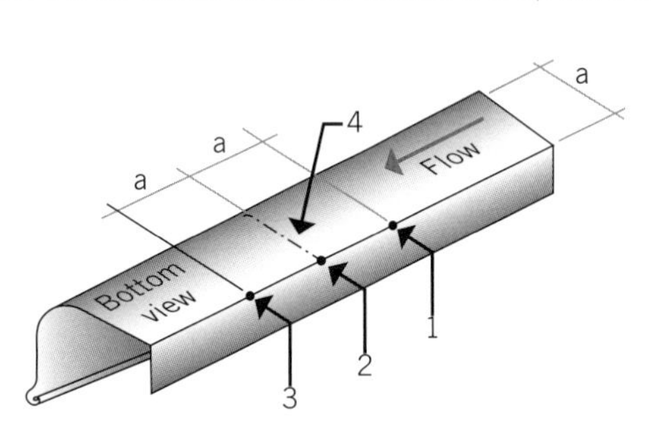

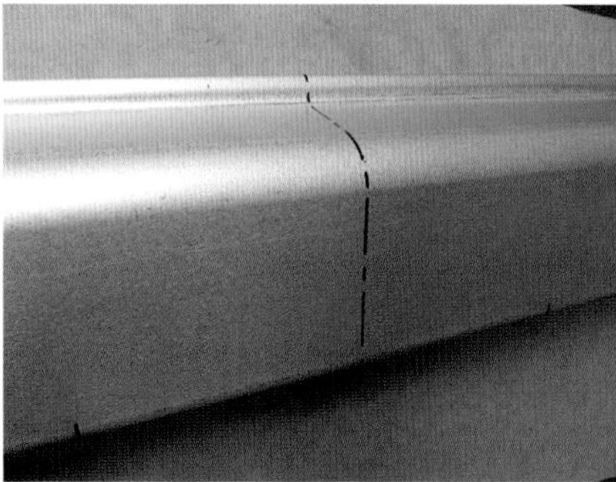

Step 2

a Using a 45° square, project angle lines from the back of the gutter to the front leading edge of the bead.
b With the square in place, position your view directly over the centreline on the front curve *(1)*.
c Without moving your head, position the very tip of the marker pen to follow the line of the square from the front edge of the bead and up the curve *(2)*.
d Do the same on the other side *(3)* and both sides should look the same.

This method takes some time to master but it is ultimately quicker and more accurate than others. Your teacher may show you another method using a pre-bent strip of hoop iron.

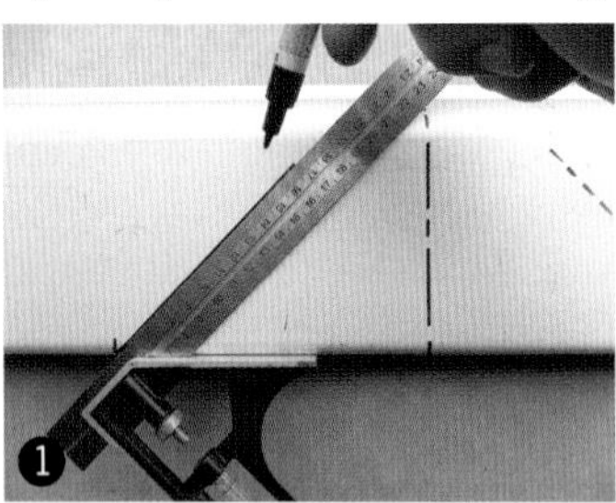

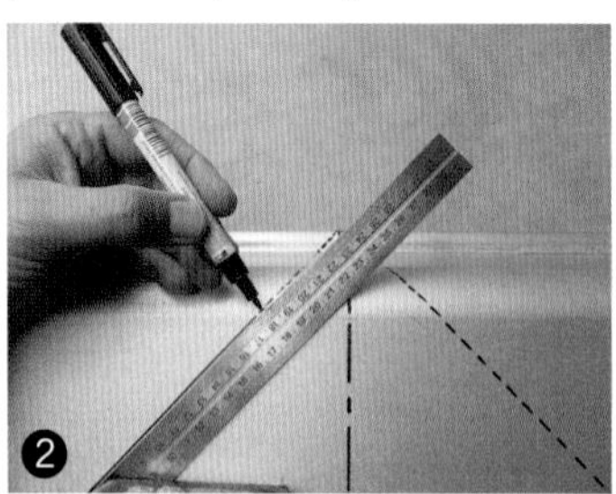

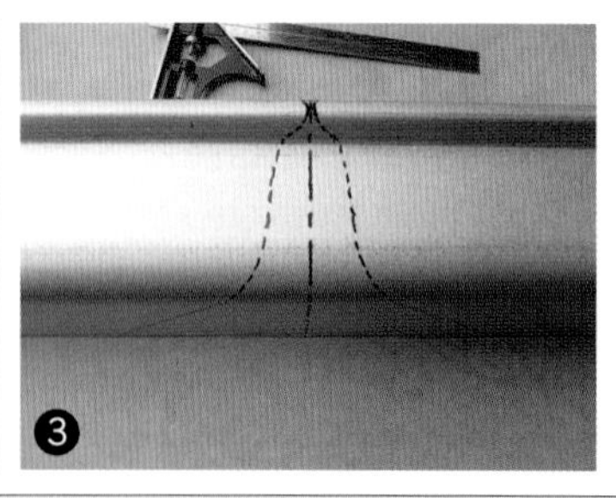

>>

>>

Step 3

a Use a 90° square to project lines down the back of the gutter *(1).*
b Turn the gutter over and, using a 45° square positioned on the centreline, mark 2 × 45° lines from the front face of the bead, across and to the back of the bead *(2).*
c This is what the bead should look like *(3).*

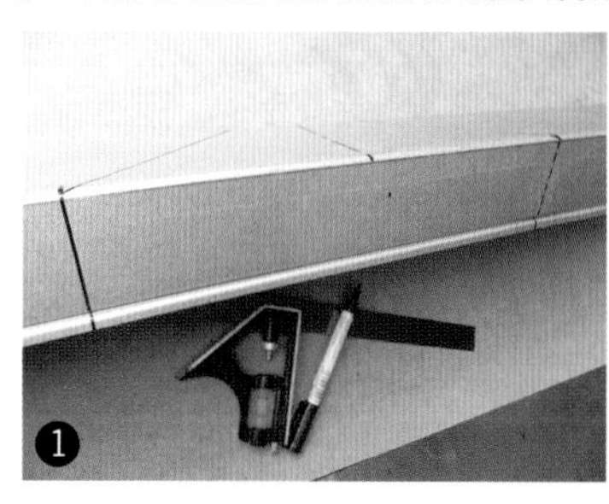

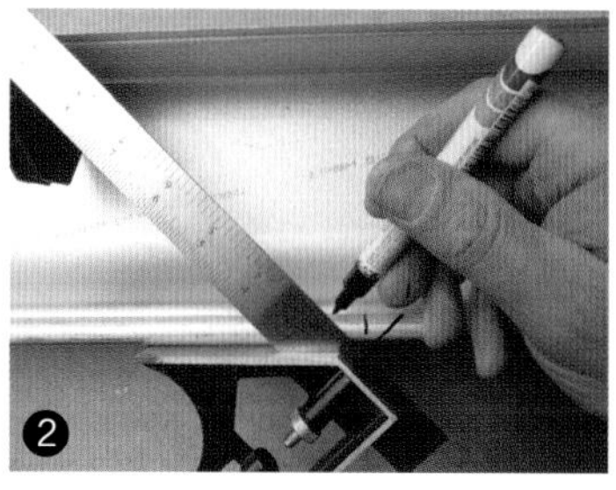

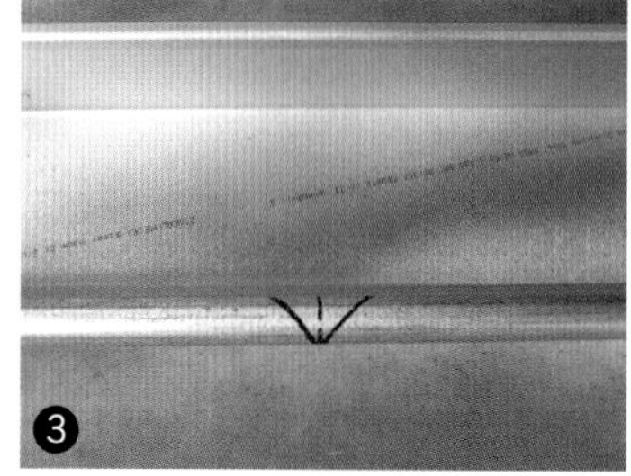

Step 4

a Confirm the direction of the water flow, then mark 25-mm laps on the upstream section *(1).*
b Keep the bottom lap away from the front curve *(2).*

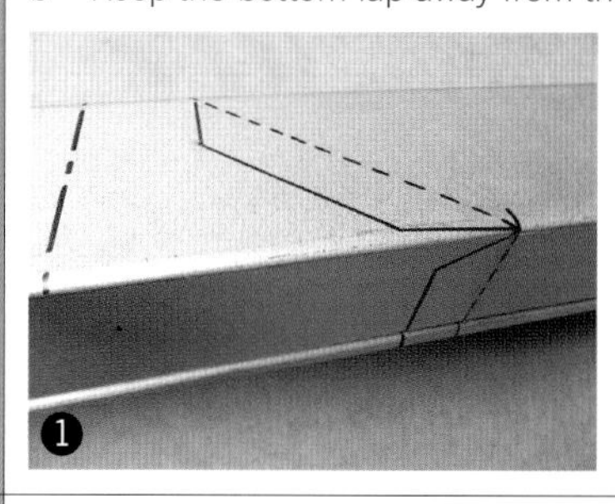

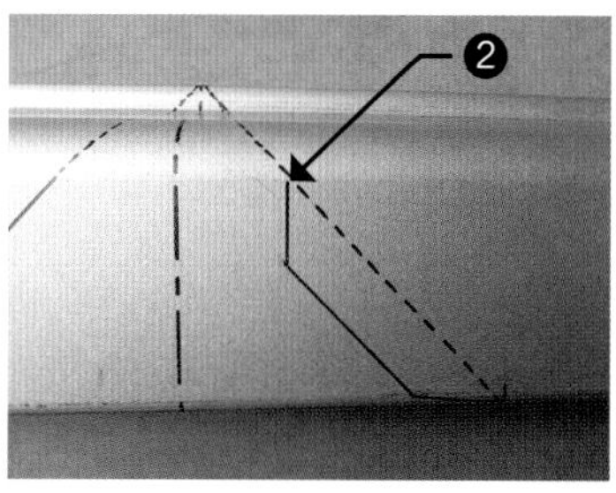

Step 5

Choose one of the following mark-outs for the front curve.

Option description	Mark-out	Example 
Option 1: Exposed lap and lug This method is ideal for unpainted gutter and provides a strong joint, with the lap dressed around the upstream piece. The lug can be reduced in length to take only one rivet if desired.	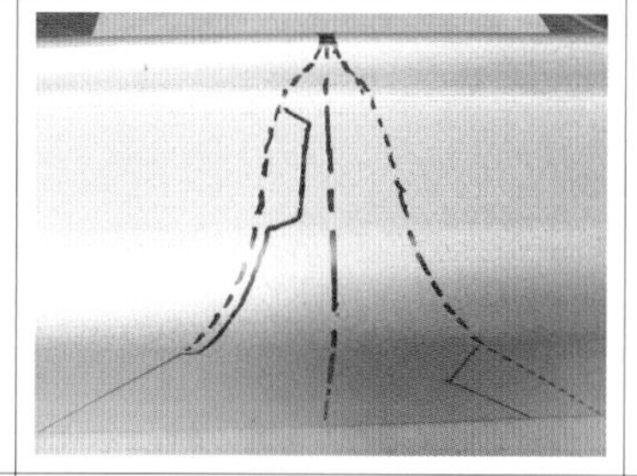	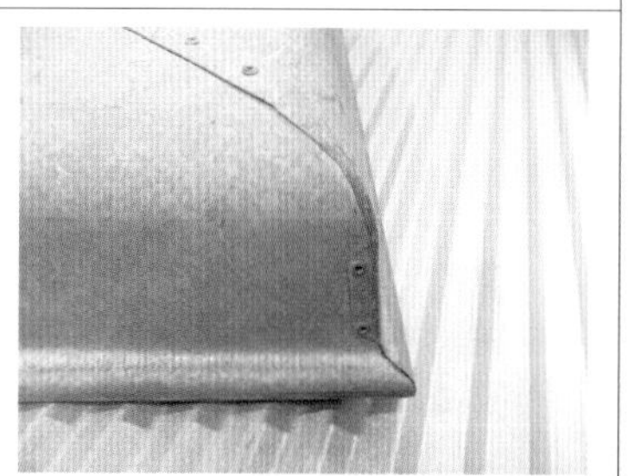
Option 2: Hidden lug The lug on the upstream piece is folded first so that it sits *inside* the downstream piece and is riveted. The downstream lap is then dressed *over* the joint, creating a very strong and clean-looking angle.	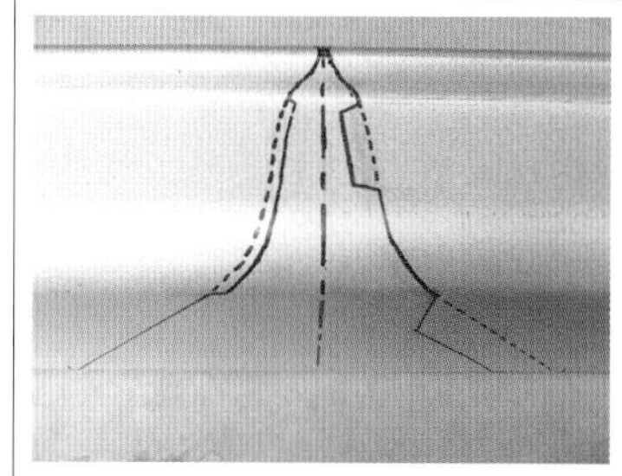	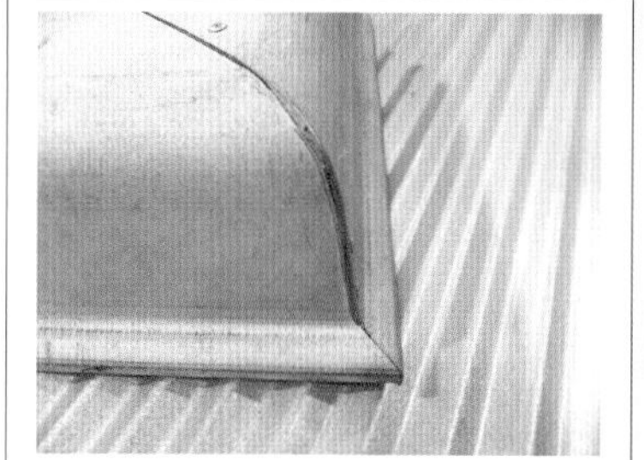
Option 3: Pre-painted gutter hidden lap and lug Ideal for pre-painted gutter, as *both the lap and lug are inside the downstream piece* and there is no dressing required that would mark the paint. This method takes considerable practice to get right and should not normally be attempted until your mark-outs are consistently accurate.	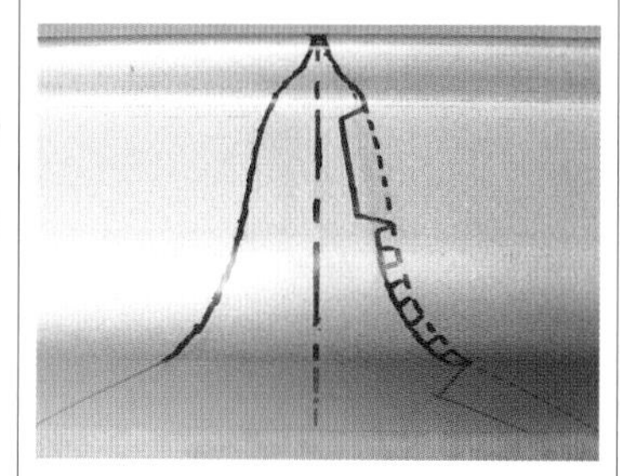	

>>

>>

Step 6

a With a marker pen, cross-hatch the material to be removed *(1)*. (In this diagram, the shaded areas show which section is to be removed.)

b The top bead gusset can be marked out from 50 × 50 mm scrap. Mark 4 holes for pre-drilling *(2)*.

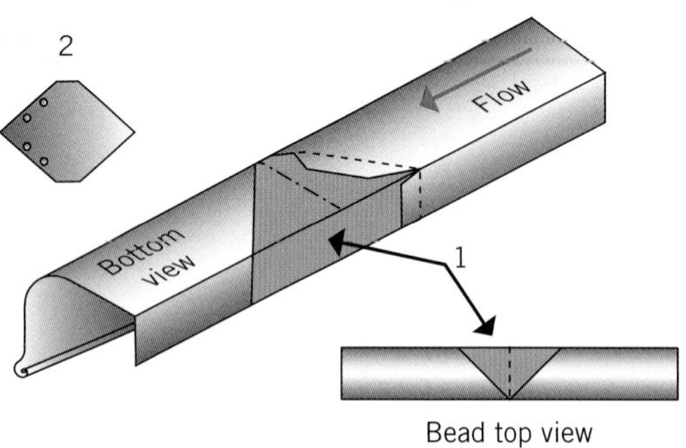

HOW TO

MARK OUT A 90° INTERNAL QUAD-PROFILE GUTTER ANGLE (FOLD-AROUND METHOD)

These images are of a completed internal quad-profile gutter after following these steps (sealant omitted for clarity).

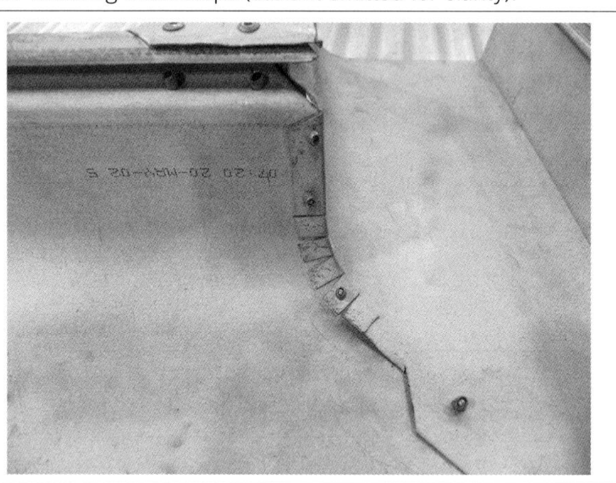

Step 1

a Measure the position of the internal corner of the fascia and mark it on the rear-bottom edge of the gutter. Project this centreline across the gutter *(1)*. This point is also the vertex of the angle you are about to make.

b Add the width of the gutter *(a)* on either side of the centreline *(2)*.

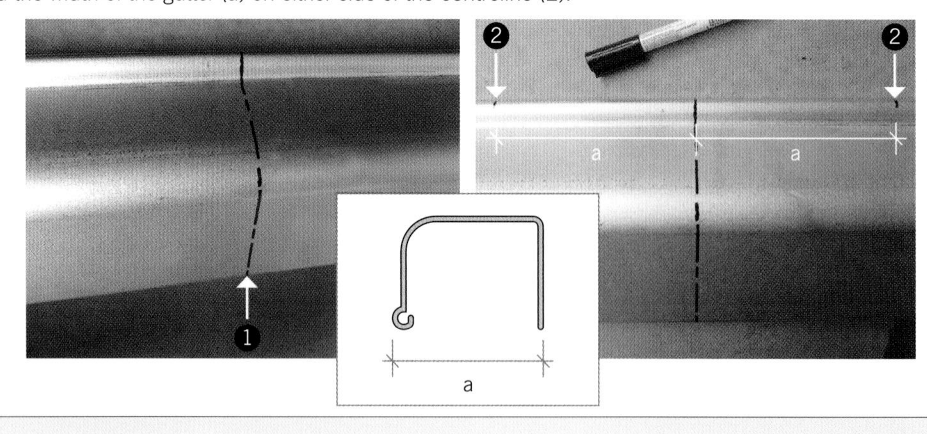

>>

>>

Step 2 Using a 45° square, project angle lines from the bottom-rear of the gutter to the front leading edge of the bead *(1)* on either side. The angle projections should look like this when done *(2).*

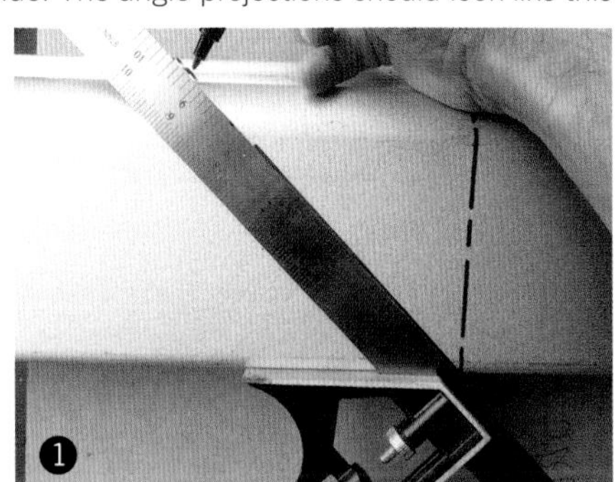

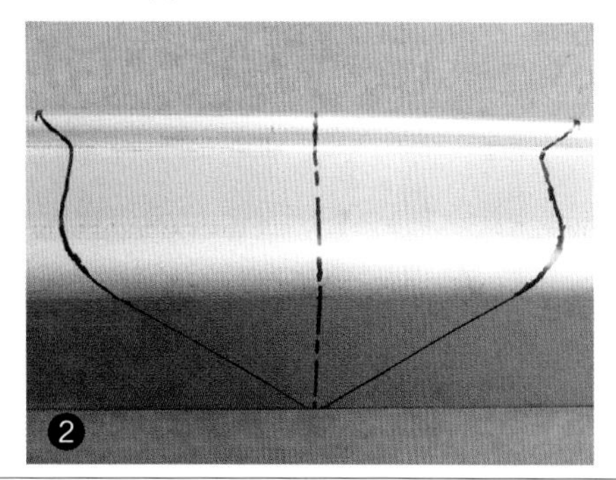

Step 3 Turn the gutter over and, using a 45° square, mark 2 × 45° lines for either side of the angle from the front face of bead, across and to the back of the bead *(1).* The angle projections should look like this when done *(2).*

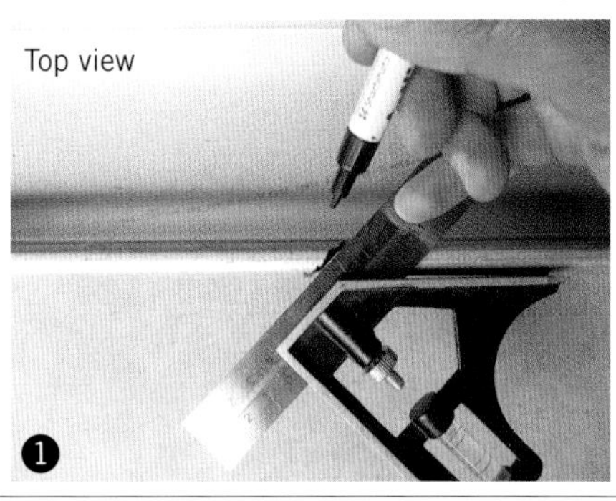

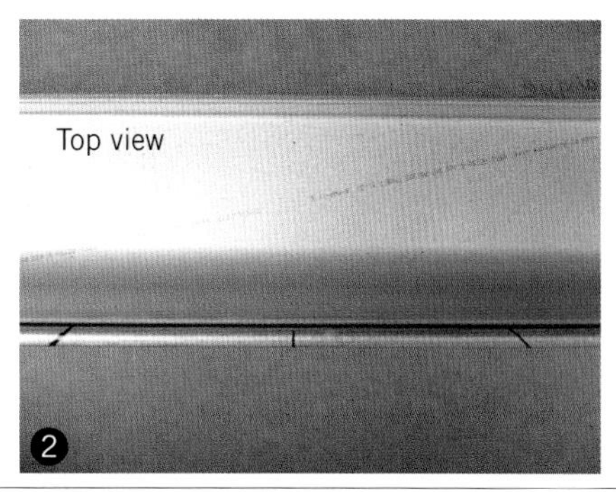

Step 4

a. Confirm the direction of water flow, then mark 25-mm laps on the upstream section. The slotted lugs around curve should be around 8 mm *(1).*

b. With a marker pen, cross-hatch the material to be removed *(2).* (In this diagram, the shaded areas show which section is to be removed.)

Note: The back of the gutter at the vertex point is folded around, not cut.

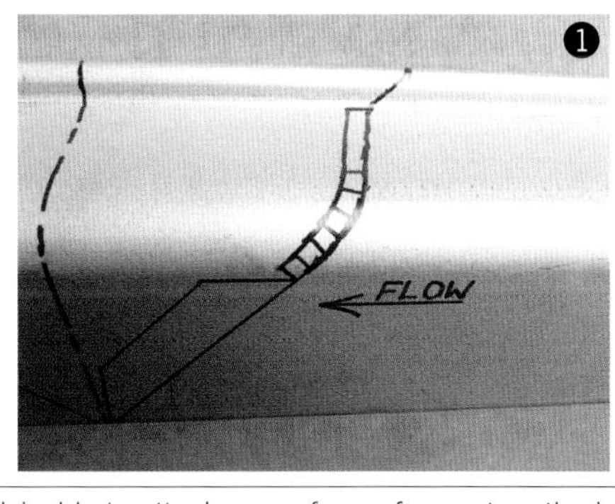

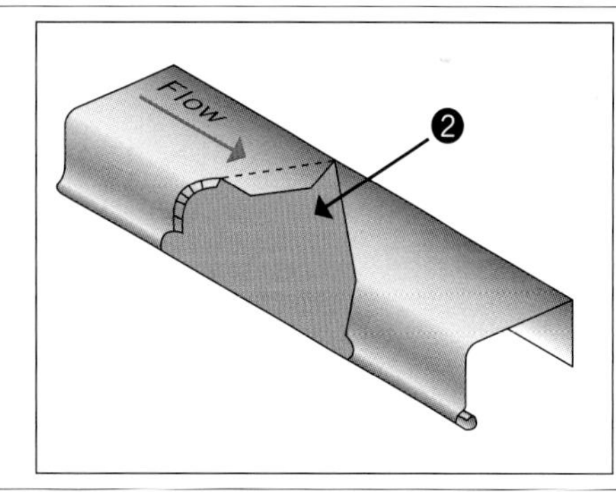

Step 5

a. It is advisable to attach some form of gusset on the internal angles *(1).* The gusset provides strength and also hides any imperfections in the bead cuts.

b. The gusset is simply made from a strip of scrap material *(2),* riveted underneath the bead then folded over on to the top of the bead, where it is trimmed to length and riveted once again.

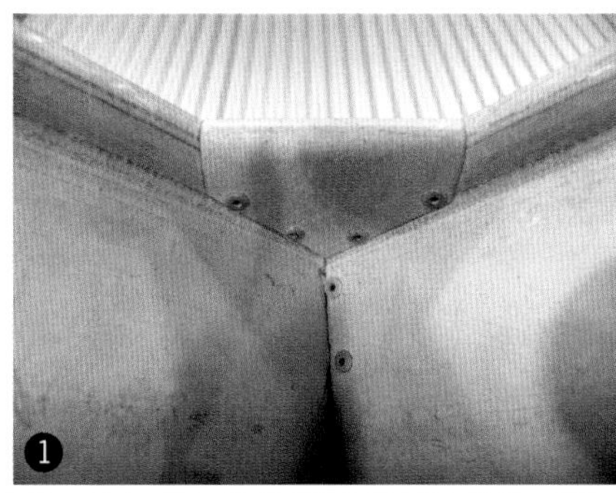

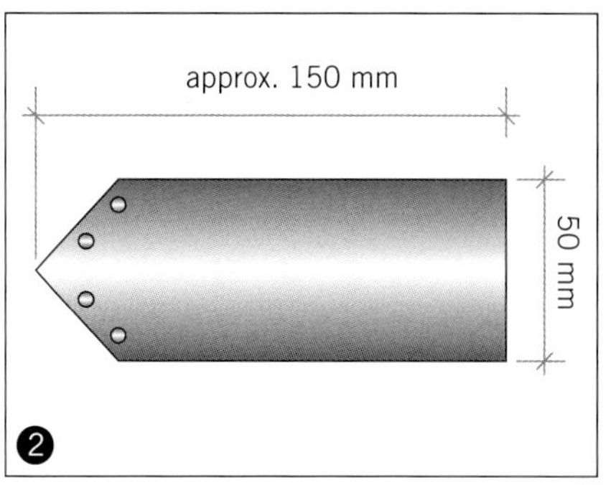

HOW TO

MARK OUT A RETURN STOP-END FOR A QUAD-PROFILE GUTTER

These images are of a completed quad-profile gutter after following these steps.

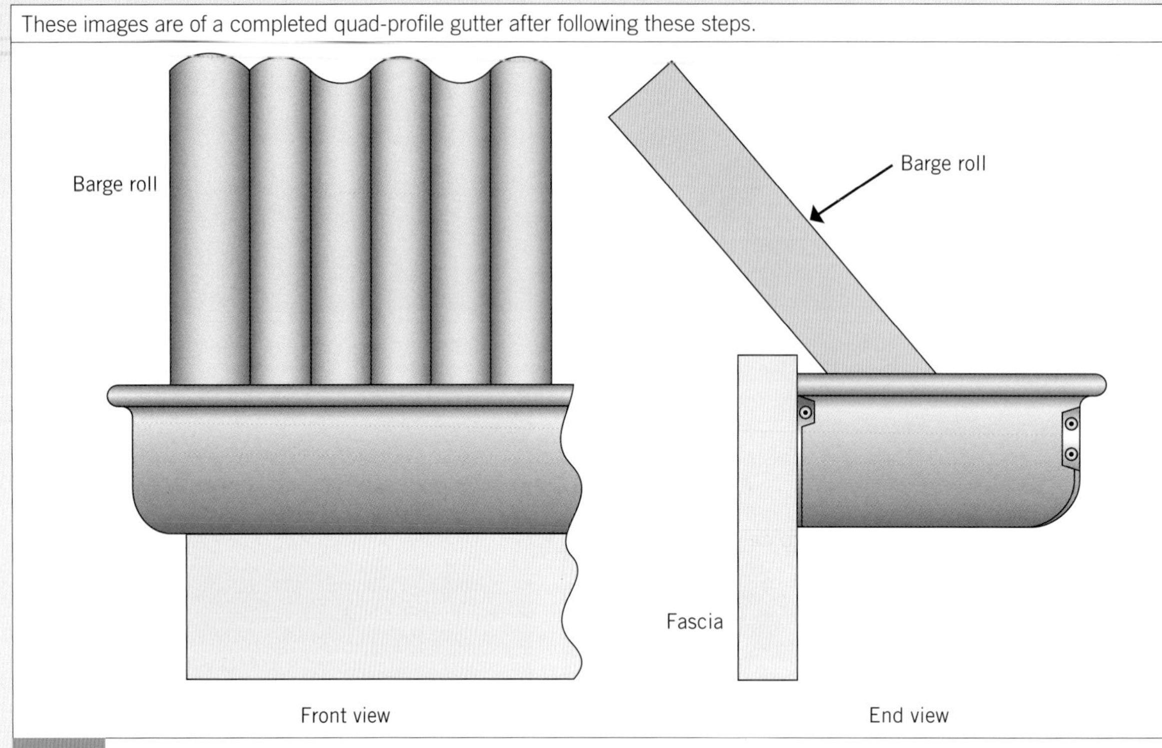

Step 1

a From the end of the length, measure back the same distance as the width of the gutter *(a)*. This is point *(1)*.
b Project the centreline from point *(1)* across the bottom, back and up the front to the top of the bead *(2)*.
c From the centreline *(1)*, measure downstream the same distance as the width of the gutter *(a)* – see point *(3)*.

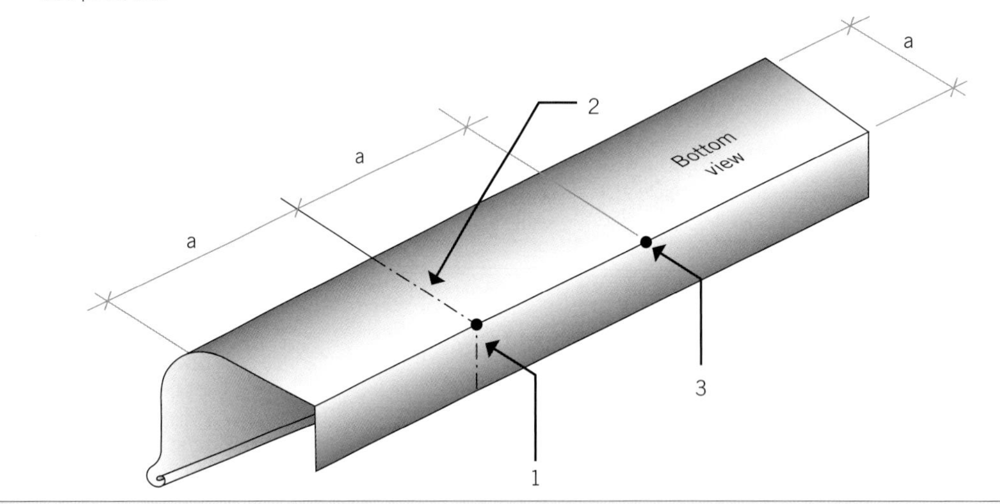

>>

>>

Step 2

a Using a 45° square, project angle lines from the back of the gutter to the front leading edge of the bead.
b With the square in place, position your view directly over the centreline on the front curve *(1)*.
c Without moving your head, position the very tip of the marker pen to follow the line of the square from the front edge of the bead and up the curve.
d Do the same on the other side *(2)* and both sides should look the same.
e This method takes some time to master but it is ultimately quicker and more accurate than others. Your teacher may show you another method using a pre-bent strip of hoop iron.

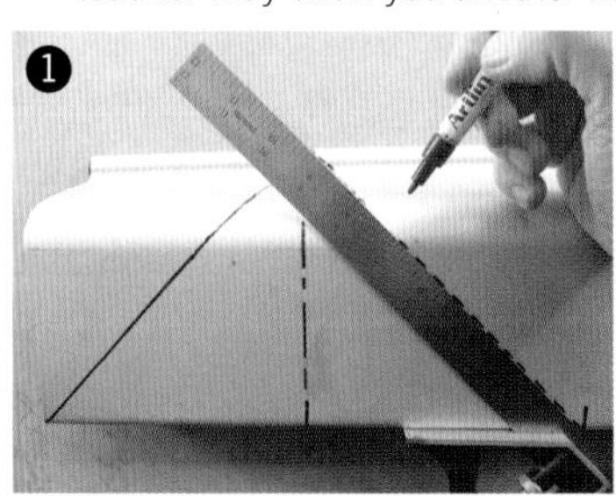

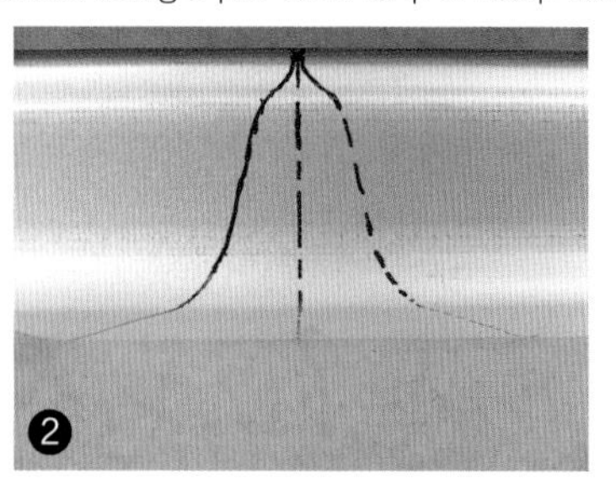

Step 3

a Turn the gutter over and, using a 45° square positioned on the centreline, mark 2 × 45° lines from the front face of the bead, across and to the back of the bead *(1)*.
b Approximately halfway between the centre and distance *(a)*, mark a cut-line from the 45° angle to the back of the gutter *(2)*.
c Mark another line back to centre as shown in *(3)*.

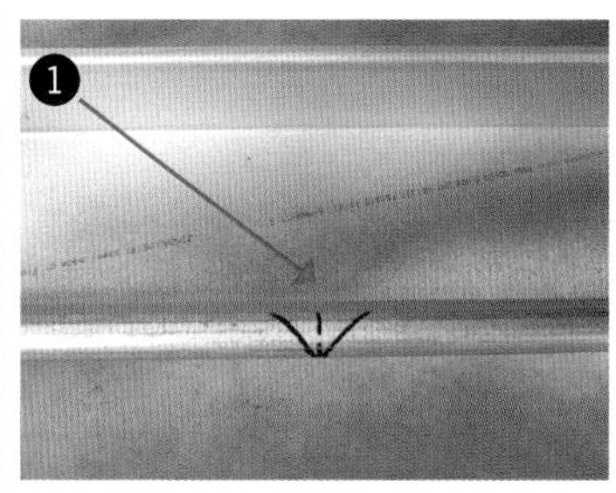

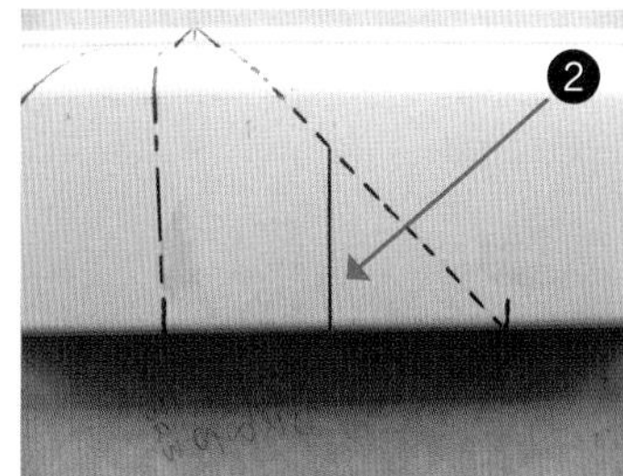

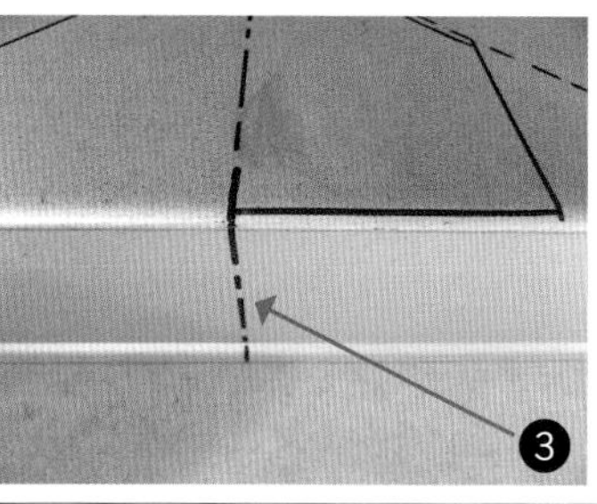

Step 4

a On the downstream piece, mark an 8-mm lug and a 3-mm lap from the front of the gutter down the curve *(1)*.
b On the upstream piece, mark a cut line showing where the excess pointed waste on the end-piece is removed *(2)*.

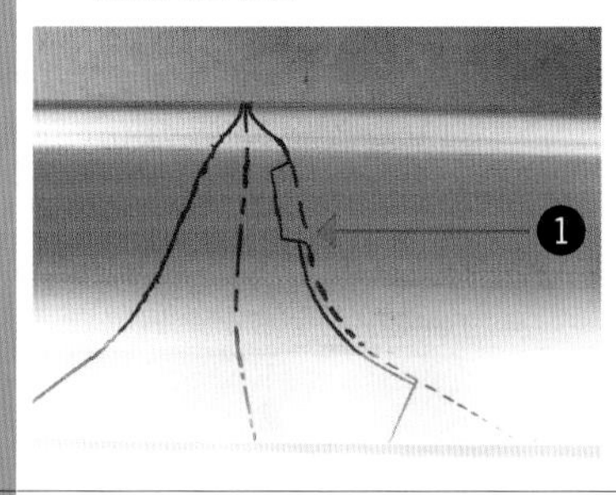

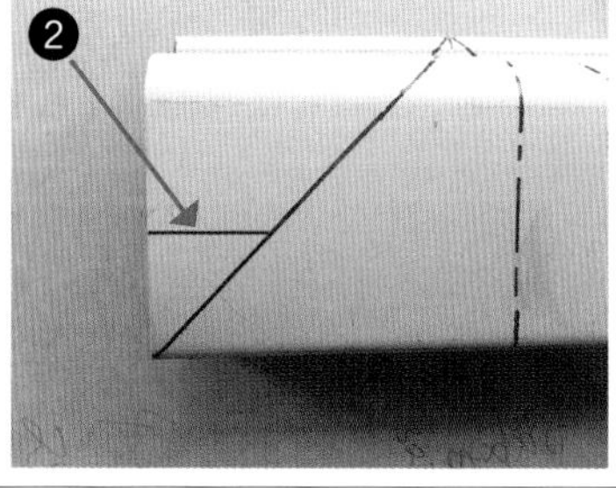

Step 5

a With a marker pen, cross-hatch the material to be removed *(1)*. (In these photos, the green shaded areas show which section is to be removed.)
b The back curve mark-out can be completed only once the end-piece has been cut from the gutter and fitted into place. A simple 8-mm lug and 3-mm lap are dressed around once riveted *(2)*.

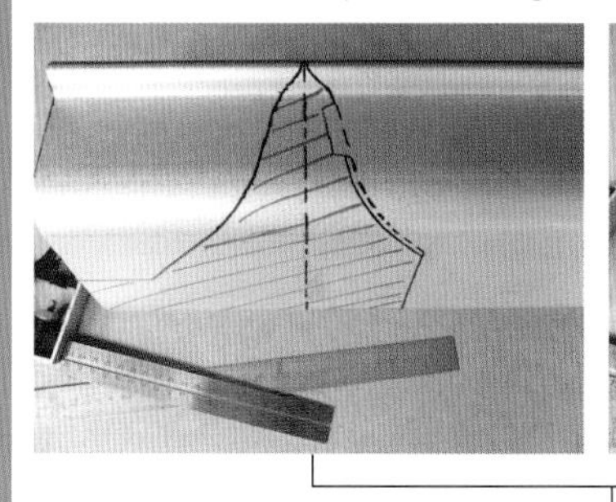
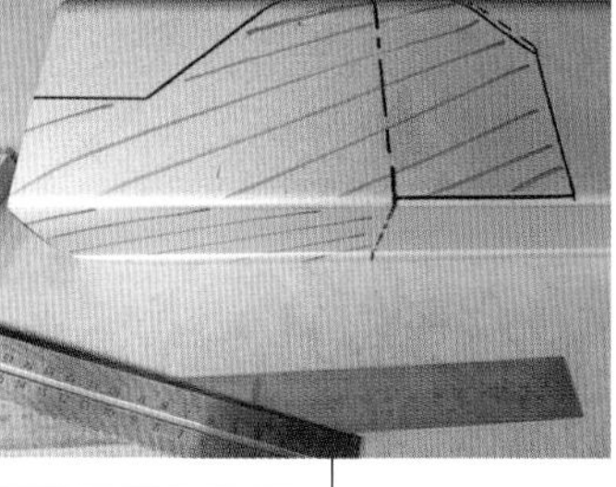
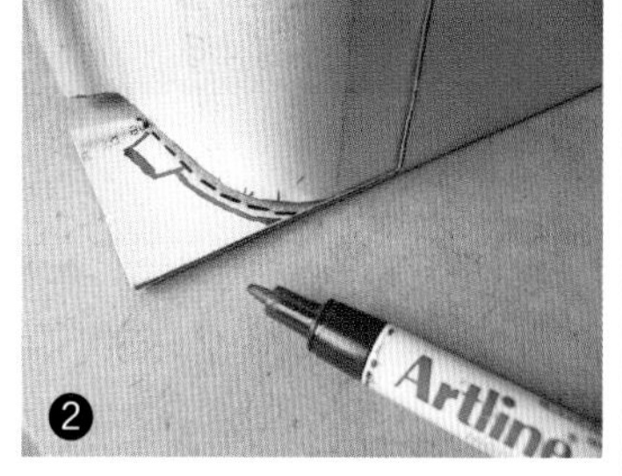

1

Moulded corner brackets

Internal and external moulded corner brackets (see Figure 10.19) are available for profiles such as quad and LYSAGHT OGEE® gutters and, depending upon the job, may provide an adequate and fast solution, particularly where roofing is not your core business.

The gutter is simply cut to the required angle and installed in position, ensuring that sufficient silicone is used to seal the joint.

Gutter profiles such as quad come in a range of sizes and variations to curve and bead. Therefore, when selecting these products you need to ensure that they will match the particular manufacturer profile.

Downpipe angles

While solvent cement jointed plastic downpipes are in common use, there is still considerable demand for skills in the fabrication of sheet-metal square, rectangular and round downpipe.

To create a specific downpipe angle, it is necessary to calculate the mark-out allowance for the angle. If necessary, revise the procedure reviewed earlier in this chapter before commencing this section.

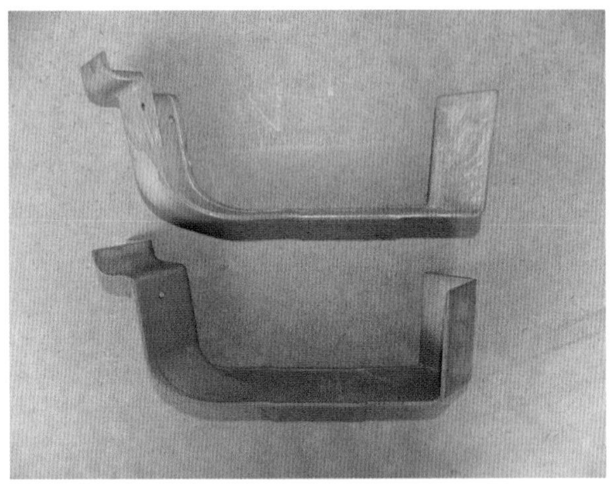

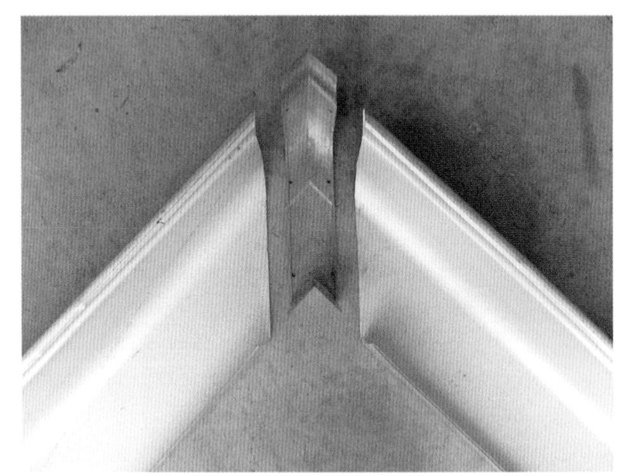

FIGURE 10.19 Internal and external moulded angles

HOW TO

MARK OUT A RECTANGULAR DOWNPIPE ANGLE

These images are of a completed rectangular downpipe angle after following these steps.

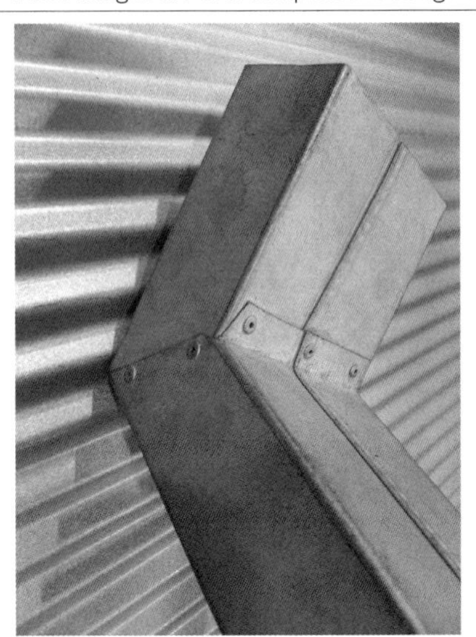

>>

>>

Step 1	a Before beginning, ensure that: i the small end of the length is downstream ii your mark-outs and fabrication will place the lock-seam out of sight and, where it is graded, uppermost. b Measure and mark the fold point where required and, using a square, project a centreline around the downpipe *(1)*. c Having measured the required angle and confirmed the mark-out allowance *(2)*, mark these lines each side of the centreline *(3)*. 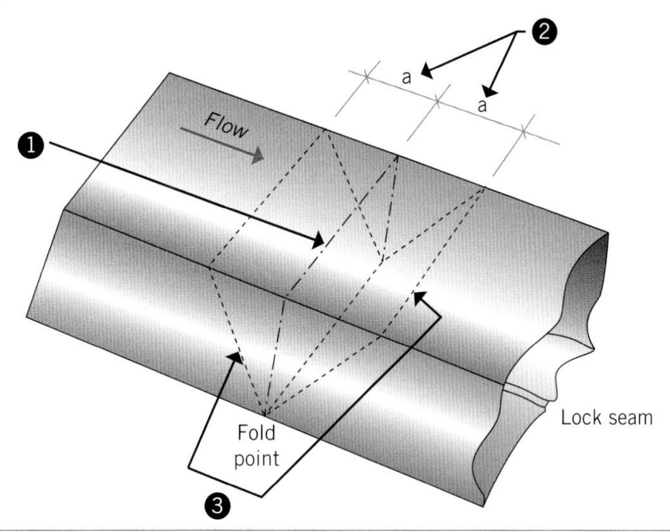
Step 2	a Double check the direction of flow and mark 10-mm laps on both the sides and the front of the upstream piece *(1)*. b (Optional) On the downstream piece, a 10-mm external lap can be included on the front *(2)*. c With a marker pen, cross-hatch the material to be removed *(3)*. (In this diagram, the shaded areas show which section is to be removed.) 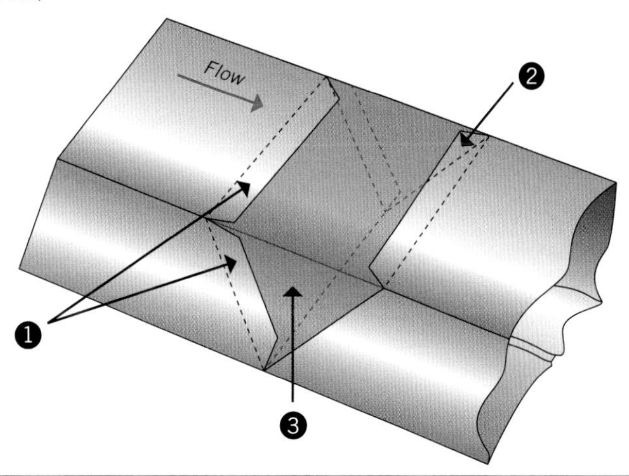
Step 3	For clarity, this diagram is shown with material removed and the optional downstream front lap turned outwards *(1)*; see inset *(2)*. Upstream laps go into the lower section. 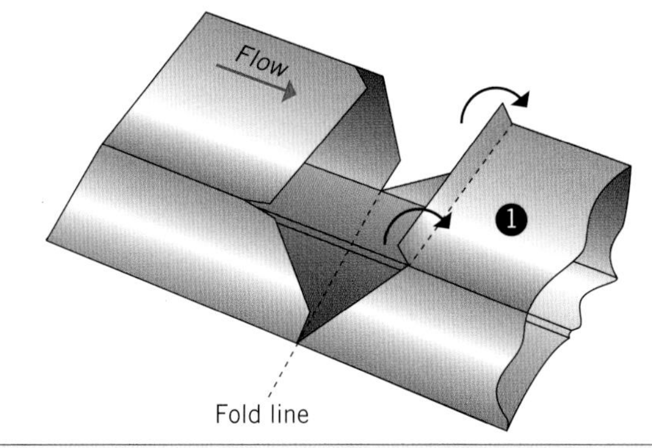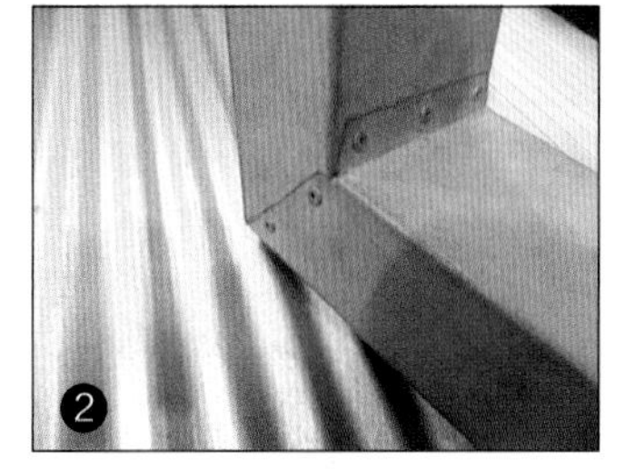

HOW TO

MARK OUT A ROUND DOWNPIPE ANGLE

These images are of a completed round downpipe angle after following these steps.

Step 1

a Before beginning, ensure that:
   i the small end of the length is downstream
   ii your mark-outs and fabrication will place the lock-seam out of sight and, where it is graded, uppermost.

b Sketch the angle and determine the mark-out allowance *(1)*.

c Measure and mark the fold point where required and, using a square, sight a centreline around the downpipe *(2)*.

d Having measured the required angle and confirmed the mark-out allowance *(1)*, project half of the mark-out allowance downstream of the centreline *(3)*.

e To allow the upstream piece to effectively sit inside the lower section, slightly more of the mark-out allowance needs to be removed. Therefore, add approximately 5–10 mm extra and plot the cut line *(4)*.

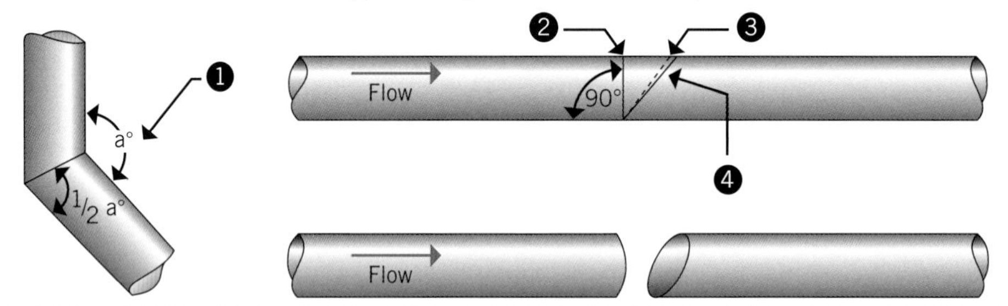

Step 2

a Fit the pieces together, ensuring that the lock-seams align, and then use a bevel to set the correct angle *(1)*.

b Hold the pieces together carefully so they do not move, then scribe a mark around the upstream section at the intersection of both pieces *(2)*.

c Separate the pieces, then add approximately 5–7 mm to this line to allow for lap *(3)*. Remove the waste material below this line (shaded).

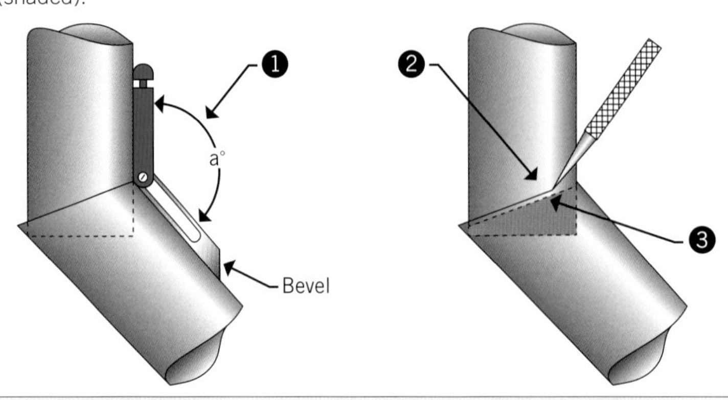

>>

>>

Step 3
a To create the upstream lap, use multigrips or pliers to gradually stretch and turn the throat lap to the correct angle *(1)*. This lap provides strength to the joint. Do not tear the material or cut slots!
b Thicker and more malleable galvanised steel can be dressed around with the handle of your Gilbows *(2)*.
c When fitted together, the lap should tightly grip to the throat of the lower piece *(3)*. Trim the lower piece as required.

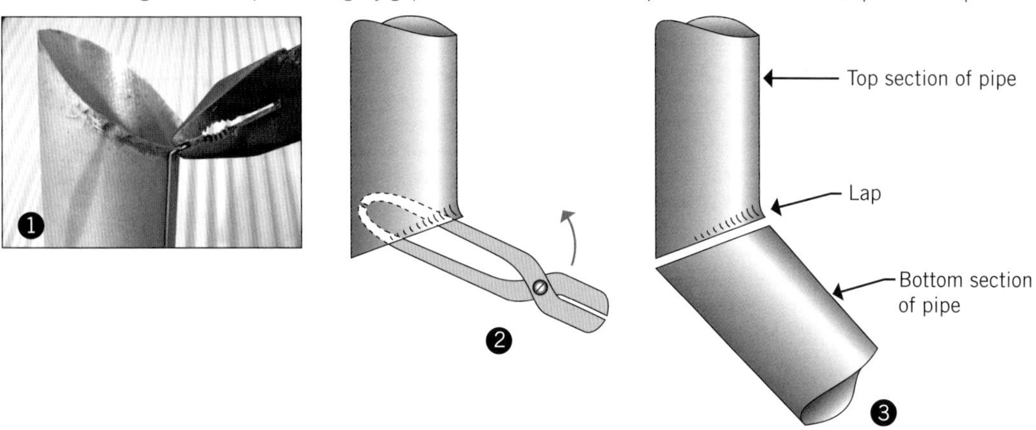

Step 4
a To create the downstream lap, fit the pieces together to the correct angle *(1)*.
b Where the lower piece protrudes past the back of the upper piece, scribe a mark to create a lap approximately 5 mm wide *(2)*. This lap locks the joint together. Any more than 5 mm will tend to crimp and create a weak joint. The goal is to stretch the metal into position.
c In a swiping motion, dress and stretch the back lap up to the required angle *(3)*. The joint should be tight and not move.

Note: Since the general demise of soldering, it is now accepted practice to also add a pop rivet to either side of the joint. Allow enough space and side lap as required.

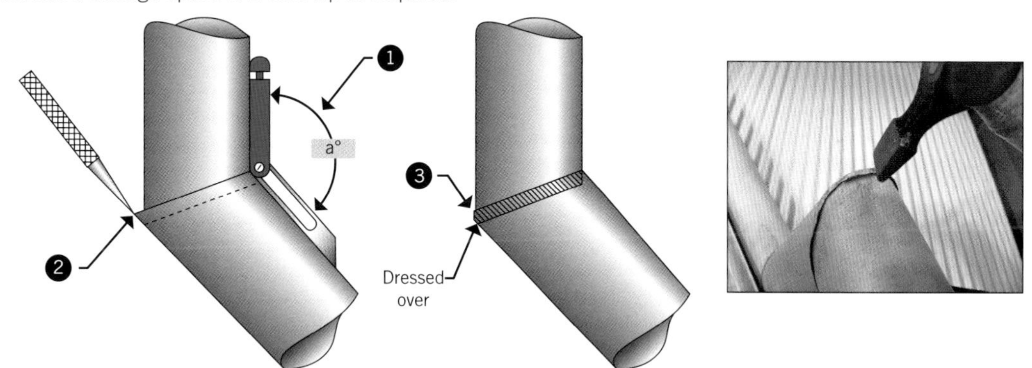

LEARNING TASK 10.4

1 What is the recommended width of the lap on the upstream section of square profile gutter?
2 Why might you attach a gusset on the internal angles of a quad profile gutter?

Fabricating rainwater goods

Preparation

Having successfully marked out your item, you need to start making the product. The fabrication of most rainwater goods will generally require the following tools:

- hacksaw
- straight snips
- aviation-style left- and right-hand snips
- square
- rivet gun
- drill and correctly sized bits
- self-locking C-clamps
- multigrips
- bevel
- cross-pein hammer
- sealant gun
- hand folder
- rags and spatula
- fine marker pen
- relevant personal safety equipment.

Your first attempt at a quad gutter angle or round downpipe may be difficult. Don't worry; it takes time and practice to develop the skills necessary to become proficient. The following points should be considered:

- Measure twice and cut once.
- Don't rush.
- Be accurate with each cut.
- Quality comes first; speed will follow in time.

Fabrication hints

The following sections are a brief series of hints and suggestions that may make your fabrication work easier.

Cutting a length of gutter

Cutting a length of gutter (particularly quad) with a hacksaw can be difficult at first. The blade seems to jam and at times will break. Many apprentices will turn to making two side by side cuts using left- and right-hand snips, thinking that this will be easier. However, cutting compound curved profiles with snips can be just as frustrating and no matter how fast you get with snips, it is always slower than using a hacksaw (once you have mastered the technique). Practise the steps in the following 'How to' box and see how you go. With a bit of practice you will be able to turn out quick, accurate cuts through any gutter profile with little trouble. The edge of any exposed slip joint can be easily and quickly trimmed up with a pair of snips if required.

Using self-locking clamps

The neat and efficient fabrication of rainwater goods requires that pieces be test-fitted and assembled exactly in position. However, sometimes it is easy to run out of hands!

Your work quality will improve dramatically once you have a couple of self-locking clamps on hand, as these make fabrication much easier. These come in various sizes and shapes and can be used to hold pieces in place while you trim edges or rivet (see Figure 10.20).

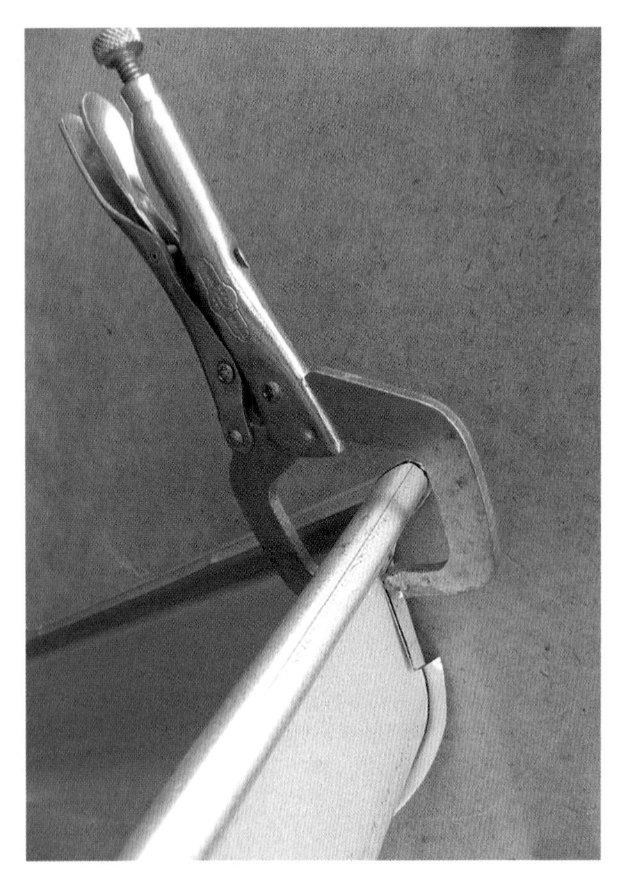

FIGURE 10.20 Self-locking C-clamps

Using a cold chisel to make sharper folds

Folding the seam on the back of a length of downpipe or the profile ridges found on square-profile gutter can result in an unsightly bend and deformation of the product. To get a sharper and more clearly defined fold, mark the position and then using a cold chisel give it a sharp hit with a hammer (see Figure 10.21).

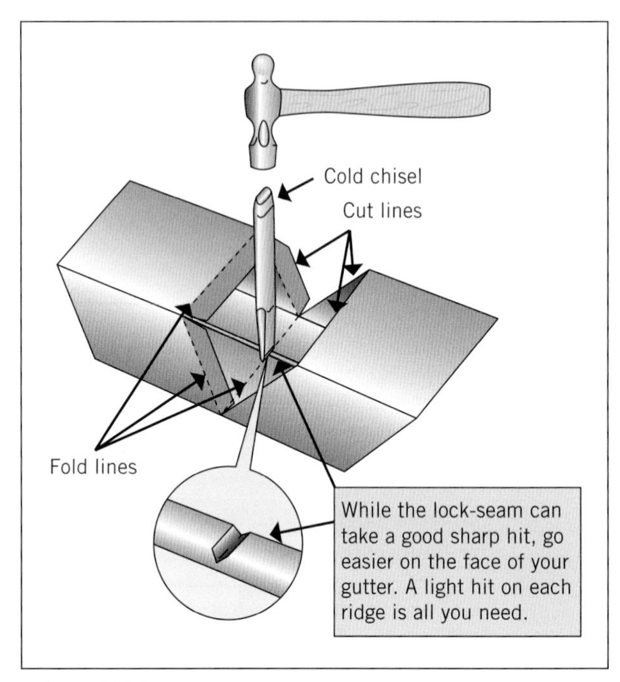

FIGURE 10.21 Use of a cold chisel to help create a sharper fold

Do I really need to use gussets?

Well, the quick answer is 'Yes'. The HB39:2015 clearly states that bead gussets should be used on all eaves gutter angles in order that the size, shape and angle be maintained.

HB39:2015 SECTION 5, CLAUSE 5.6(B) 'INTRODUCTION AND SCOPE'

Despite this, the use of gussets has become less common in recent years, with many installers simply using a single rivet on top of the bead.

While this certainly saves time in fabrication, a single fastener does nothing to maintain the true shape of the profile or angle of the product. This is commonly seen with square-profile gutter return-ends where, with no bracing, the return simply flexes out of square. As detailed below, in addition to the direction from the HB 39, there are good reasons to still include gussets as part of your fabrication process.

Gussets strengthen both internal and external angles, and this can be important for fitting repair angles and replacement work on existing dwellings (see Figure 10.22).

HOW TO

CUTTING QUAD GUTTER EFFICIENTLY

Step 1	a Make sure you are using a quality 32 TPI hack-saw blade, and have it correctly tensioned in the hacksaw frame with the teeth facing forward. b Brace the gutter in some way (normally with your foot or on some form of bench) so that it does not move around during the cut. c Before you start, point your index finger along the length of the hacksaw. It helps to guide the tool in a constant direction. Let the tool do the work, with an easy sawing motion. *Don't wrestle with it!*	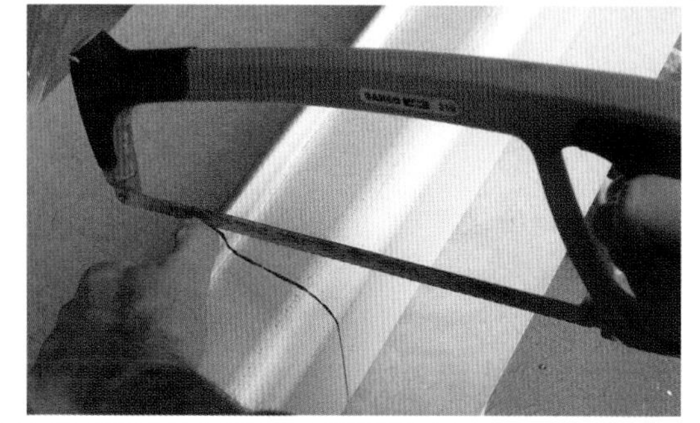
Step 2	a Cut through the bead and work your way down the front of the gutter. b The blade will begin to jam because the two pieces of sheet metal are close together, moving and catching on the blade teeth. c To stop this, place your thumb through the hacksaw frame and press each section apart. You will find that the blade instantly moves more easily.	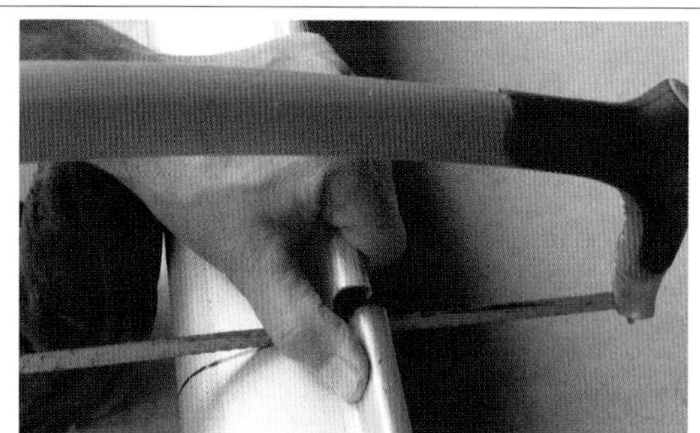
Step 3	Once you are past the curve, the blade will tend to catch on the front of the gutter as well as the bottom. Slightly rotate the work, keep separating each piece and also slightly turn the hacksaw on an angle (note the position of the blade in the diagram). As long as you keep following the line, the hacksaw will still cut straight, despite being on this angle.	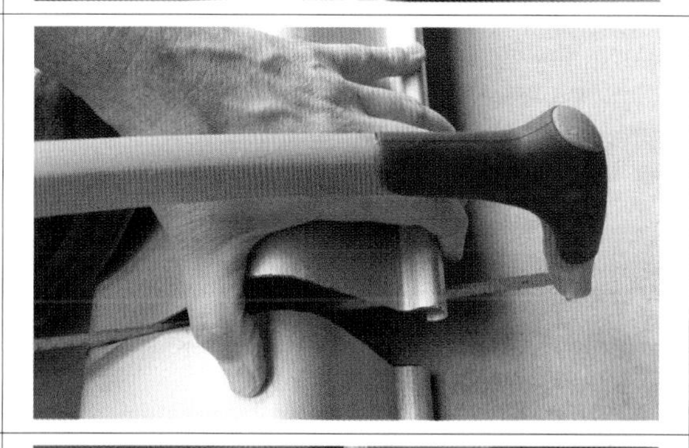
Step 4	Having cut through to the bottom edge of the gutter, finish the job by simply cutting up the back with your snips. Remove any hacksaw swarf by running the edge of your 'Gilbows' at right angles along the cut. If necessary, trim up any rough exposed edge with snips.	

FIGURE 10.22 Bead gussets used to strengthen internal and external angles

Unlike the soldering days of decades past, small mistakes in bead alignment and minor gaps can't be filled with solder and hidden. Standard translucent silicone sealant allows the light through and makes a minor overcut stand out. A gusset hides these minor faults and adds a final tradesperson's finish to the job at the same time.

Work on heritage properties is an opportunity to showcase your skills and at the same time improve the level of authenticity for the project. Well-made gussets are almost mandatory to effectively finish off your galvanised LYSAGHT OGEE® Gutter and quad gutter angles.

COMPLETE WORKSHEET 2

LEARNING TASK 10.5

1 Which hand tool can we use to minimise unsightly bends and deformation of the product due to the seam when folding?
2 What is a technique you can adopt to minimise the blade of the hacksaw catching when cutting quad gutter?

Installation

In the last part of this chapter, you will be introduced to specific knowledge and skills related to the actual installation process. It is not intended that this text list every installation step required, as the process of installation is what you do on the job every day, and with a good employer you will learn it more effectively in practice than you would reading about it in a book. Instead, this section will focus on some of the reasons behind installation practices and at the same time expose you to more of the requirements of HB39:2015 and AS/NZS 3500.3.

HB39:2015 AND AS/NZS 3500.3 'SCOPE AND GENERAL'

Expansion and contraction

Though it may not be obvious to the eye, when metal is heated it expands. As it cools down, it contracts. While you may not have seen this happen, you have probably heard a steel-clad roof creak and crack when the sun suddenly comes out on a cloudy day. These are the sounds caused by movement against screws, battens and joints.

When heated, metal will expand in all directions, but more so along its length than its width. Certain materials will expand more than others, and the greater the material length, the more obvious is the change.

Why does a roof plumber need to know this? Expansion and contraction of materials is a natural phenomenon: it *will* happen and you can't stop it. If you do not account for this in your roof installations, the performance of your roofing materials may be affected.

EXAMPLE 10.7

Imagine that you installed a new box gutter tightly between two parapet walls. With a significant rise in temperature, the gutter will try to expand at either end. If it cannot do so, it may start to buckle in the middle or apply stress to joints and fasteners. Joint leakage may be the result.

At any point where you are working between fixed points, or are installing long runs of gutter or flashing, you need to make some allowance for expansion and contraction.

The rate of expansion varies for different types of metal and can be expressed mathematically as the coefficient of expansion. This relates to the change in size when 1 metre of material is raised in temperature by 1°C. The coefficient of expansion for some materials relevant to roof plumbing can be seen in Table 10.1.

TABLE 10.1 Coefficient of expansion for common roofing materials

Material	Coefficient of expansion
Copper	0.0000167
Aluminium	0.000024
ZINCALUME® steel	0.000001

The actual expansion of any given material can be determined by inserting these figures into a simple formula. To determine the total expansion rate, the following applies:

$E = L \times T \times C \times 1000$

where

E = total expansion in mm

L = length of material in m

T = temperature difference in degrees Celsius

C = coefficient of lineal expansion

The important point to make here is that you need to account for expansion and contraction in your installations. Fortunately, code requirements take some of the complexity out of this by detailing at what point provision for expansion and contraction must be accommodated.

Eaves gutters

Eaves gutter installations must not exceed a length of 20 m without provision being made for expansion. An example of how this may be achieved can be seen in Figure 10.23.

Two stop-ends can be installed and sealed inside the gutter on either side of a slip joint. The joint itself is not fastened and remains free to expand and contract as necessary. Where this joint occurs at a low point in the gutter system, a downpipe outlet must be provided.

EXAMPLE 10.8

An aluminium eaves gutter is to be installed along the back of a commercial building. The run is 30 m long. The area in which the building is located can experience sheet-metal surface temperature changes of up to 60°C between night and day.

How much would such a length of gutter expand over the course of a day and night?

$E = L \times T \times C \times 1000$

$= 30 \times 60 \times 0.000024 \times 1000$

$= 43.2$ mm

Movement of this amount may cause buckling and damage to the box gutter joints if no allowance for expansion is made.

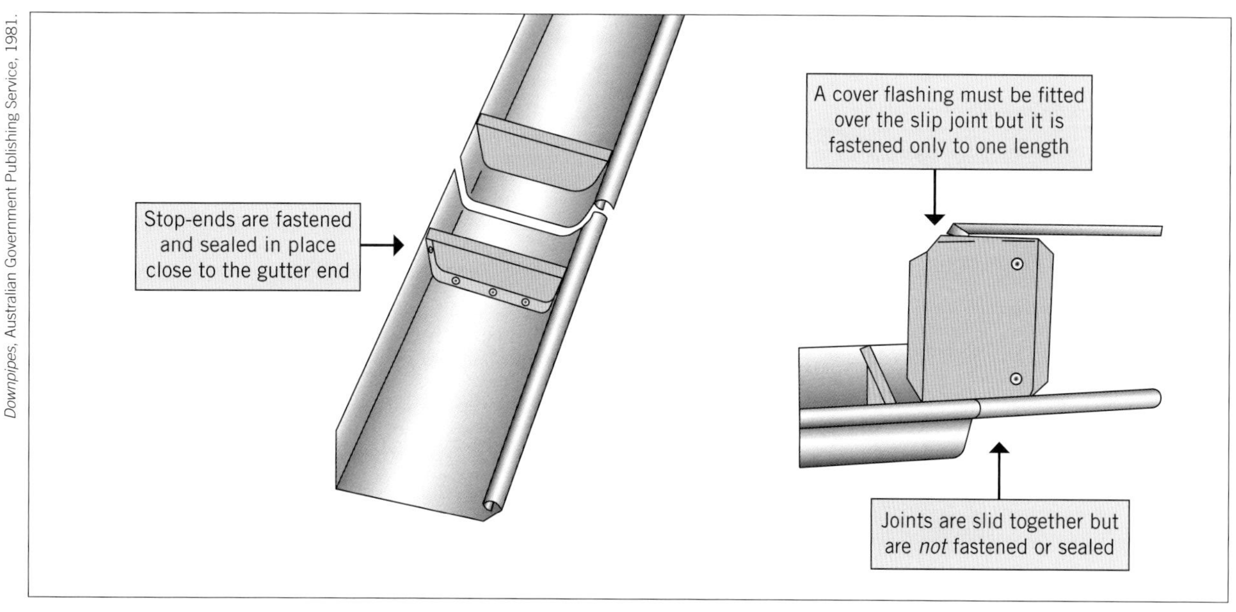

Source: Based on Basic Training Manual 12-1 *Roof Plumbing, Introduction and Downpipes*, Australian Government Publishing Service, 1981.

FIGURE 10.23 A simple eaves gutter expansion joint and flashing cap

Box gutters

Box gutters must have expansion joints included at various lengths, subject to the type and thickness of the material being used. This is detailed in your code book.

See Figure 10.24 to see how this is calculated in accordance with HB39:2015.

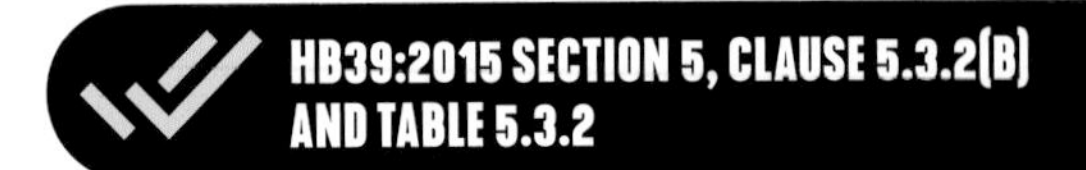

Determination of fall

Calculation of fall in a length of gutter is an important skill for any roof plumber. The roofing code details the minimum fall ratios for both eaves and box gutters, and you must know that what you are installing is meeting these requirements and performing well. Guessing is not good enough.

The house in Figure 10.25 was only a couple of years old when the photo was taken. The lack of sufficient fall in this section of gutter has allowed airborne dust and silt to build up to a point where it is now deep enough to grow grass.

This problem can occur when the installer checks fall only intermittently with a short spirit level instead of calculating overall fall from the highest to lowest point.

The roofing code quotes fall requirements as a ratio, such as 1:500, 1:200 or 1:25. A ratio of 1:200 means that there must be a fall of at least 1 unit for every 200 units of run. This is all very well, but in a practical sense, you need to know what the fall is in millimetres over a given length. The following formula allows you to convert fall expressed as a ratio into fall in millimetres:

$$\text{Fall(mm)} = \frac{\text{L(m)}}{\text{Grade}} \times 1000$$

Follow through with the following example to see how this may be applied in a practical way.

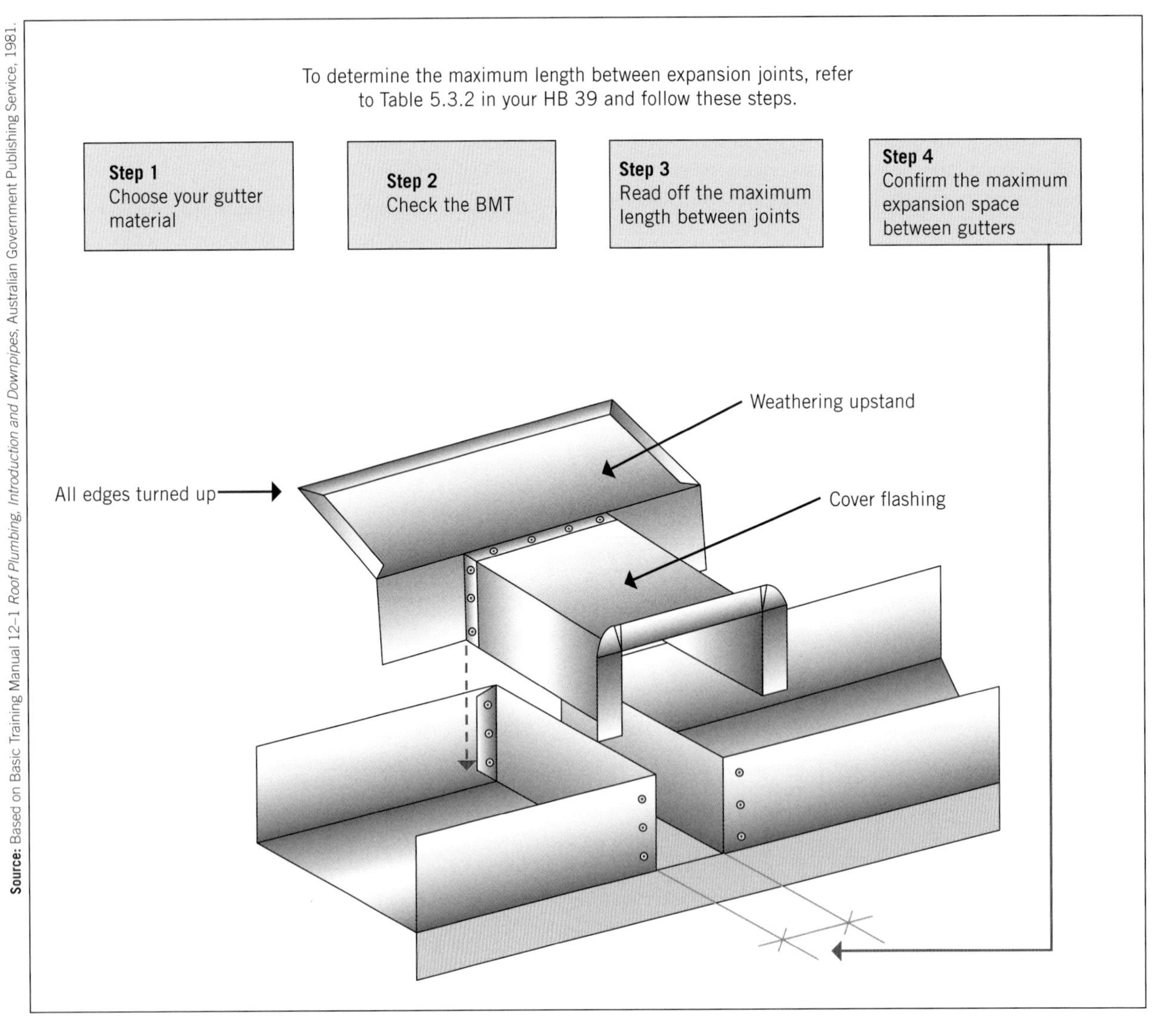

FIGURE 10.24 Box gutter expansion joint, cover and weathering upstand flashing

FIGURE 10.25 Grass growing in an incorrectly installed gutter

EXAMPLE 10.9

Calculation of fall

You are installing a straight run of eaves gutter with a constant fall of 1:500 over a distance of 18 m.

In setting out your brackets, you need to calculate the fall in mm from the highest point to the lowest.

$$F(mm) = \frac{L(m)}{Grade} \times 1000$$
$$= \frac{18}{500} \times 1000$$
$$= 0.036 \times 1000$$
$$= 36 \text{ mm}$$

Determination of gradient

Sometimes you may already know the fall on an existing gutter length, but need to check that it is compliant with the roofing code. To find out the gradient of an existing run of gutter expressed as a ratio, use the following formula:

$$\text{Gradient (ratio)} = \frac{L\ (m)}{F(m)}$$

Follow through with the next example to see how this may be applied in a practical way.

Hint: Formulae are not much help if you can't remember them on the job. A good way around this is to engrave or otherwise permanently mark each formula on the side of your spirit level for quick reference. Alternatively, ensure you know where to search for each formula with your mobile phone as and when you require it.

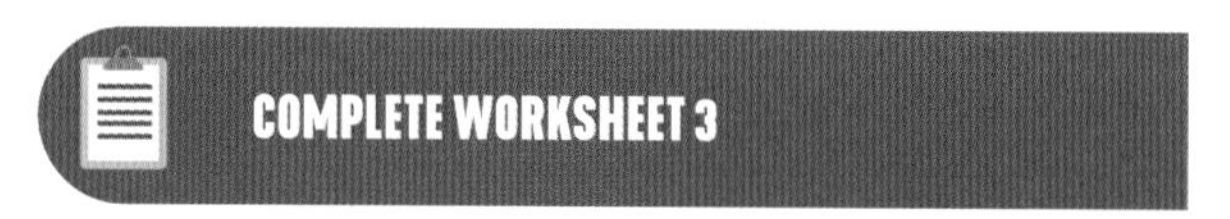

Metal fascia

It may not be obvious at first, but modern metal fascia is a form of flashing, as it protects the end of the trusses, preventing water and wind from entering the space around the eaves and greater roof area. Metal fascia and matching gutter systems are used for the majority of domestic installations (see Figure 10.26).

FIGURE 10.26 Metal fascia being installed

EXAMPLE 10.10

Calculation of gradient (ratio)

You have been asked to install a new box gutter on an old factory roof. You know that the roofing code requires that box gutters must have a minimum fall of 1:200.

Using an automatic builder's level, you determine that the fall in the existing gutter is 100 mm over 30 m, from the highest point to the lowest.

What is the existing grade of the gutter support? Are you able to lay your new box gutter on these existing sole boards?

$$G\ (ratio) = \frac{L(m)}{F(m)}$$
$$= \frac{30}{0.1}$$
$$= 300$$
$$= 1{:}300$$

A fall of 1:300 is flatter than the minimum permissible in the roofing code and therefore you will have to modify or replace the existing gutter support system in order to comply.

Installation process

The metal fascia system and gutters are installed before the roof sheets go on; this is one of the first jobs to be done once a frame has been braced and 'signed-off' by the building surveyor or local authority. The metal fascia itself incorporates a rebate that also holds the soffit lining in place. Before starting the job, you will need to confirm with the builder at what level the lining is to be positioned.

Compare the soffit level (normally at the top of the windows) with the fascia profile depth and check that the roof sheets will still be able to project into the gutter without the top of the fascia being too high. Once you are confident that the fascia dimensions meet the building specifications, it is time to get stuck into it. The steps in the 'How to' box show a basic sequence of installation that would be similar to most systems on the market.

HOW TO

PROCEDURE FOR INSTALLING METAL FASCIA

This image is of a completed metal fascia after following these steps.

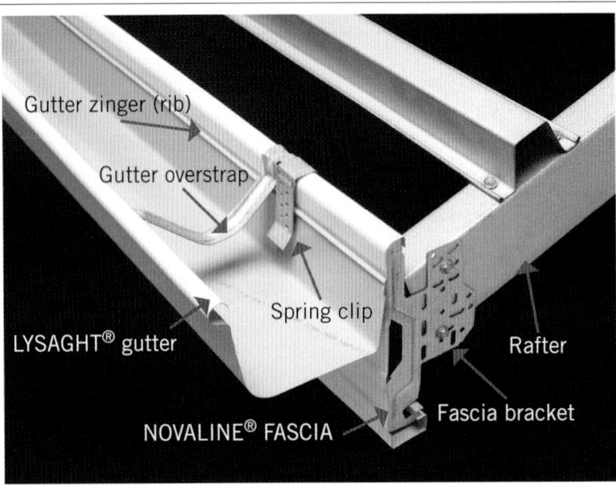

Step 1	Attach a rafter clip at each end of the run, and then run a stringline from end to end of the run. Attach intermediate rafter brackets at the mid-region of the run, on level and in line with the stringline.	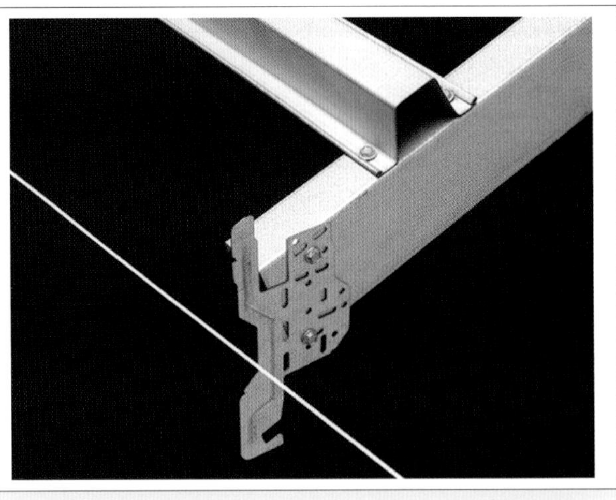

>>

>>

Step 2	Hook the fascia horizontally under each end clip, then flip the fascia forwards and lever it over the top of the bracket into the vertical position.	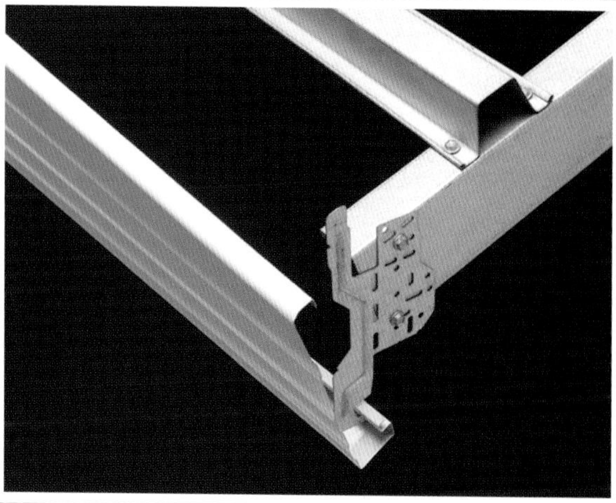
Step 3	Adjust the fascia to the correct position by sliding it along the brackets, then secure the remainder of the intermediate brackets by twisting into the fascia first and then connecting to the rafters.	 Notching, holing or distortion of the NOVALINE® FASCIA is not recommended.
Step 4	Complete the job by fitting the relevant accessories.	**NOVALINE® FASCIA accessories** 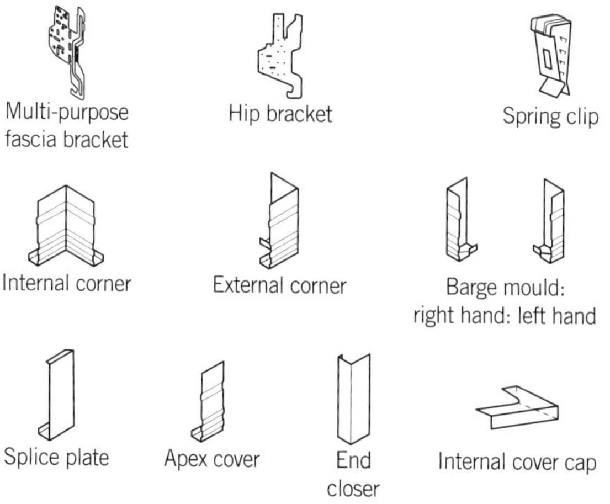 Multi-purpose fascia bracket; Hip bracket; Spring clip; Internal corner; External corner; Barge mould: right hand: left hand; Splice plate; Apex cover; End closer; Internal cover cap

Source: © BlueScope LYSAGHT®, www.lysaght.com

Eaves gutter installations

Eaves gutters must be sized and located to accommodate the volume of water flowing from the roof and effectively direct it to the downpipes. The approved plans for the project will detail gutter requirements, and you must ensure these are adhered to so as to guarantee design performance.

Specific code requirements relate to eaves gutter installations, and these include the following:

- Gutter brackets (see Figure 10.27) should be installed in a position that will allow them to provide a uniform and *minimum fall of at least 1:500* from each high point to the outlets. The AS/NZS 3500.3 actually allows for flatter installation in particular circumstances but only if the gutter system is sized to match.
- Lengths should not exceed 20 m without provision for expansion and contraction.
- The back of the gutter must be cleated by the bracket cleat or secured with a turned-down galvanised nail.
- Lap joints must be not less than 25 mm and be lapped in the direction of flow.
- Gutter brackets must be installed at stop-ends and at intervals complying with manufacturer's recommendations. The interval should not exceed 1200 mm.

During the preparation for your installation, some or all of the following additional points should also be considered:

- *Metal fascia systems* – where an eaves gutter is being installed as part of a matching metal fascia system, confirm with the builder the level at which the fascia is to be set in relation to window lintels. Usual practice is that eaves boards, held in place by the fascia, must match or be slightly higher than the top of the window line.
- *Downpipe locations* – confirm the location of the approved discharge points and matching downpipe locations, ensuring that these match the points detailed in the approved plans.
- *High points* – compare the downpipe locations to the proposed gutter layout and identify your high points.
- *Expansion joints* – identify any sections of gutter run that will require the installation of an expansion joint, and where that expansion joint is located in relation to high points and the nearest downpipe.
- *Sagged fascia* – when replacing gutters on an older building, be suspicious of any section that has rusted away through lack of fall. While the gutter may have been installed badly in the first place, it may also be due to sections of the building itself that have sagged over time. Some cases are so bad that it is difficult to get enough fall in the new gutter, and an additional downpipe may have to be considered. Where this becomes apparent you must contact your employer right away, as you cannot simply run another downpipe to ground without there being an approved point of discharge nearby.
- *Rotten fascia* – removing old gutters may reveal sections of timber fascia that have rotted away. These sections will need to be replaced before new gutters can be installed.

Considerations relating to gutter fall

The goal is to aim for a smooth gradient in your gutter that will not cause permanent ponding while at the same time ensuring that the fall is not too excessive or obvious.

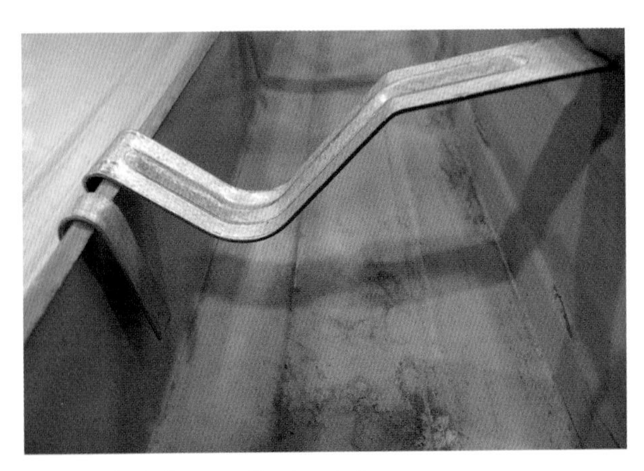

FIGURE 10.27 Eaves gutter brackets

The highest potential point of any gutter section is the point at which the bracket and/or back of the gutter is at its highest position on the fascia while complying with minimum continuous overflow requirements. This is a calculated gutter overflow point to prevent water from flowing over the fascia and into the building. Your supervisor will inform you of this measurement for each application. Note that it is not permissible to cut the back out of the gutter to achieve more fall.

From the high point for each section of gutter, try to achieve adequate fall but be careful that you do not apply too much. Excessive fall stands out against horizontal levels such as fascia and rooflines. It may also expose the sheet ends, which is unsightly and also allows possible water ingress from wind-driven rain.

Box gutter installations

Box gutters sit within the building line and are drained towards a sump or rainwater head.

It is therefore critical that they are sized and designed in accordance with the AS/NZS 3500.3 'Stormwater drainage'. Your box gutter installations must be installed so that the risk of expensive water damage inside the building is eliminated.

For all box gutters up to 600 mm in width, the following specific requirements apply:

- Box gutters must be installed with a maximum fall of 1:40 and a uniform minimum fall of 1:200 towards the outlet.
- Commercial box gutters must be a minimum of 300 mm wide and 75 mm deep at the upstream end.
- Lap joints must be not less than 25 mm and must be lapped in the direction of flow.
- Box gutters must not exceed specified lengths without provision for thermal movement.
- All box gutters must incorporate an overflow system that can discharge all water to atmosphere clear of the building in the event of a total blockage.

Box gutter support

Box gutters must be supported correctly to ensure structural soundness and a uniform fall to outlets. Up to a box gutter size of 450 mm, supports may be:

- continuous – this means that the full length and no less than 25% of the width of the sole is in continuous contact with the gutter support. This is normally achieved by laying a multi-ribbed roof sheet sole support (see Figure 10.28) inside

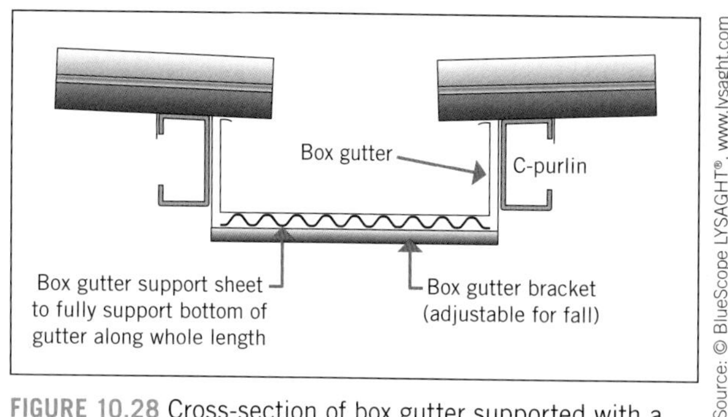

FIGURE 10.28 Cross-section of box gutter supported with a continuous support system

  adjustable box gutter hanging brackets (LYSAGHT SPANDEK® is most commonly specified)
- discontinuous – in this system the gutter is supported by adjustable brackets spaced at minimum intervals of only 750 mm.

For box gutter sizes with widths greater than 450 mm, a continuous support system is required.

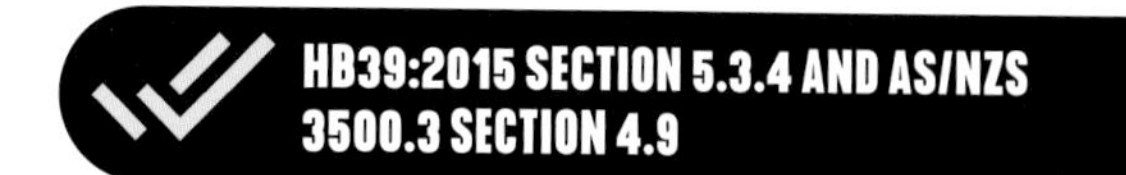

Sumps and rainwater heads

Eaves gutters may discharge into rainwater heads and box gutters into both sumps and rainwater heads. The design of sumps, rainwater heads and associated overflows is vital to the performance of the roof drainage system. During the installation phase you must be aware of and accommodate specific code and standards requirements in your work. These are covered here in brief.

Sumps

When installing sumps, you must be aware that there are requirements related to their minimum size, downpipe connection location and overflow provision:

- Sumps must be sized not less than 400 mm in length.
- The width of a sump should match the width of the box gutter.
- You cannot connect a downpipe just anywhere in the sump! The downpipe must be installed in the sump so that its centre point is no further from the side of the sump than:
  - the diameter of the circular downpipe
  - the average of the two side dimensions of a rectangular downpipe.

EXAMPLE 10.11

Box gutter sumps and overflows

A roof plumber is regularly required to make repairs to existing installations. Look at the image in Figure 10.29. If your employer sent you out to look at a reported rusted and leaking sump such as the one shown in the photograph, what would you identify as being incorrect? What would need to change to make the installation compliant? Who would design a new solution?

Discuss these questions with your teacher or supervisor.

FIGURE 10.29 A non-compliant box gutter sump

A sump *must* be designed and fitted with one of two overflow options that must not have a design flow less than the gutter outlet itself. These are either:

- a side outlet overflow, or
- a high-capacity overflow device.

Overflows must discharge to atmosphere (see Figure 10.30). Note that these overflows are designed to ensure that all water is effectively discharged in the event of a total blockage. You must therefore fit these as specified in the approved plans.

FIGURE 10.30 Bottom view of a sump overflow outlet

HB39:2015 SECTION 5.7.2 AND AS/NZS 3500.3 SECTION 3.7 AND FIGURE 3.7.3

Rainwater heads

A rainwater head allows a box gutter or eaves gutter to overflow outside the building. Rainwater heads are supplied off the shelf or to custom order in a wide range of styles and sizes (see Figure 10.31). However, during the installation you will be responsible for the correct fitting and modification of rainwater heads so that they meet the overflow and connection requirements detailed in the code and standards.

The basic requirements for rainwater head installations include the following:

- The weir of the overflow must be a minimum of 25 mm below the sole of the box gutter.
- The width of a rainwater head shall be at least the same width as the box gutter.
- The downpipe must be installed in the rainwater head so that its centre point is no further from the side of the sump than:
  - the diameter of the circular downpipe
  - the average of the two side dimensions of a rectangular downpipe.
- Overflows must discharge to atmosphere.
- Refer to the Standard for the only approved DTS rainwater head design

HB39:2015 SECTION 5.7.3 AND AS/NZS 3500.3 SECTION 3.7 AND FIGURE 3.7.3

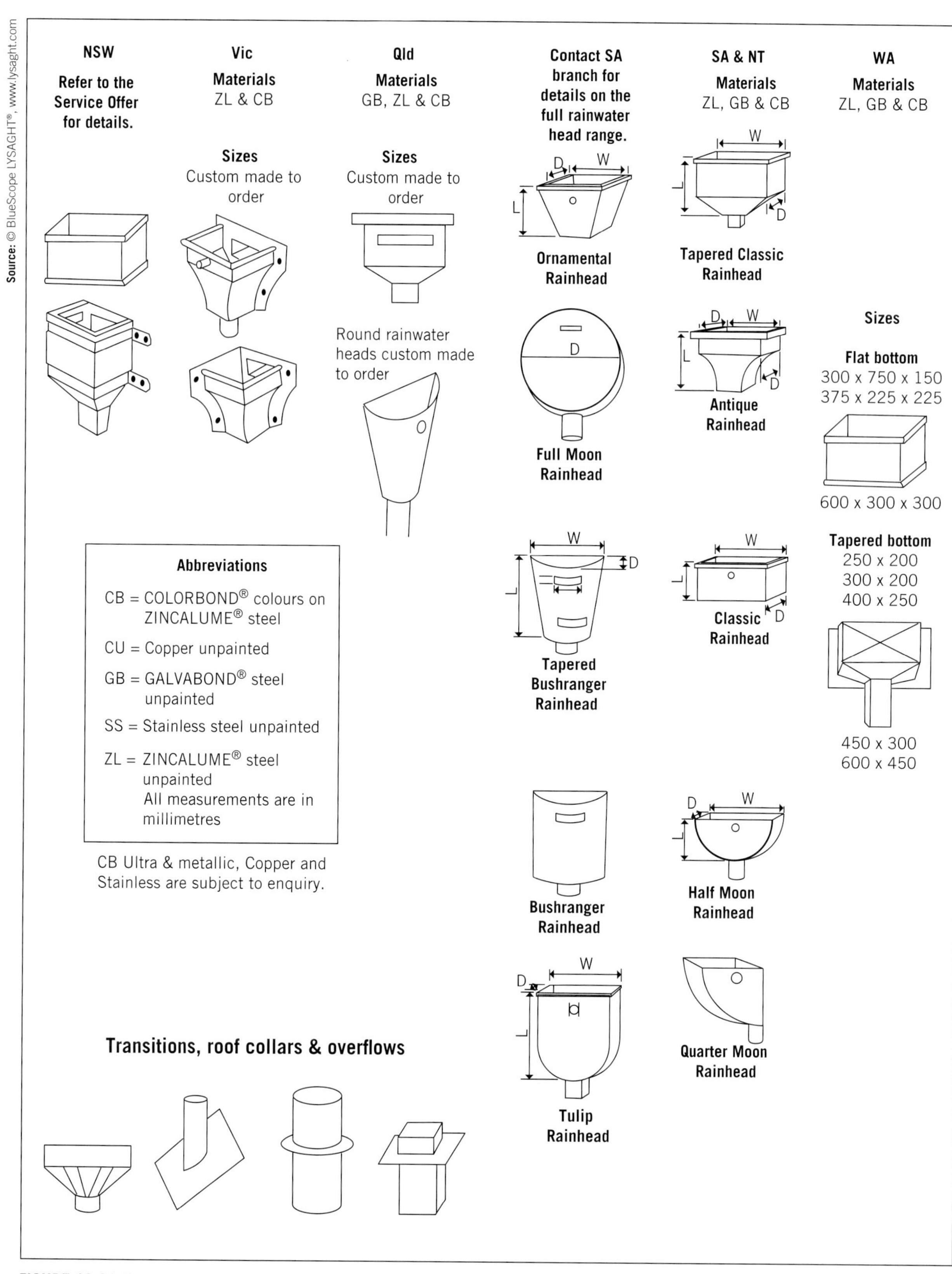

FIGURE 10.31 Example of the range of rainwater heads available to meet individual job requirements

Valley gutter installations

The valley gutter is installed between two inclined roof slopes and runs all the way from the ridge down to the eaves gutter. Valley gutters can be fabricated in your workshop to suit particular needs, though today it is far more common simply to order the required profile as a pre-fabricated continuous length. However, it is important to note that valley gutters must also be sized in accordance with the AS/NZS 3500.3. Do not assume that the supplier will always provide you with a compliant product.

Some points relating to valley gutters are as follows:

- The maximum catchment area above any valley gutter is not to exceed 20 m^2
- Valley gutters are not to be installed on roof slopes less than 12.5° (at slopes less than 12.5° the valley must be sized as a box gutter)
- The minimum valley gutter discharge width between overlapping sheets is 150 mm

See Figure 10.32 for further installation details.

Downpipe installations

Downpipes direct the water from the roof and gutters to an approved point of discharge. To do this effectively, downpipes are sized in accordance with the carrying capacity of the gutter profile as measured by the effective cross-sectional area in square millimetres. The number of downpipes is then calculated to match the rainfall catchment area of the roof. Building designers and/or your employer must ensure that the sizing is correct, and during your installation it is very important that you comply with the requirements of the approved plans in regards to location, size and termination point.

The basic installation requirements include the following:

- Downpipes must be securely fastened to structures with compatible brackets/clips spaced at a maximum of 1-m intervals on grade and 2 m vertically (see Figure 10.33).
- Downpipe brackets must be sized to allow for thermal expansion and contraction.

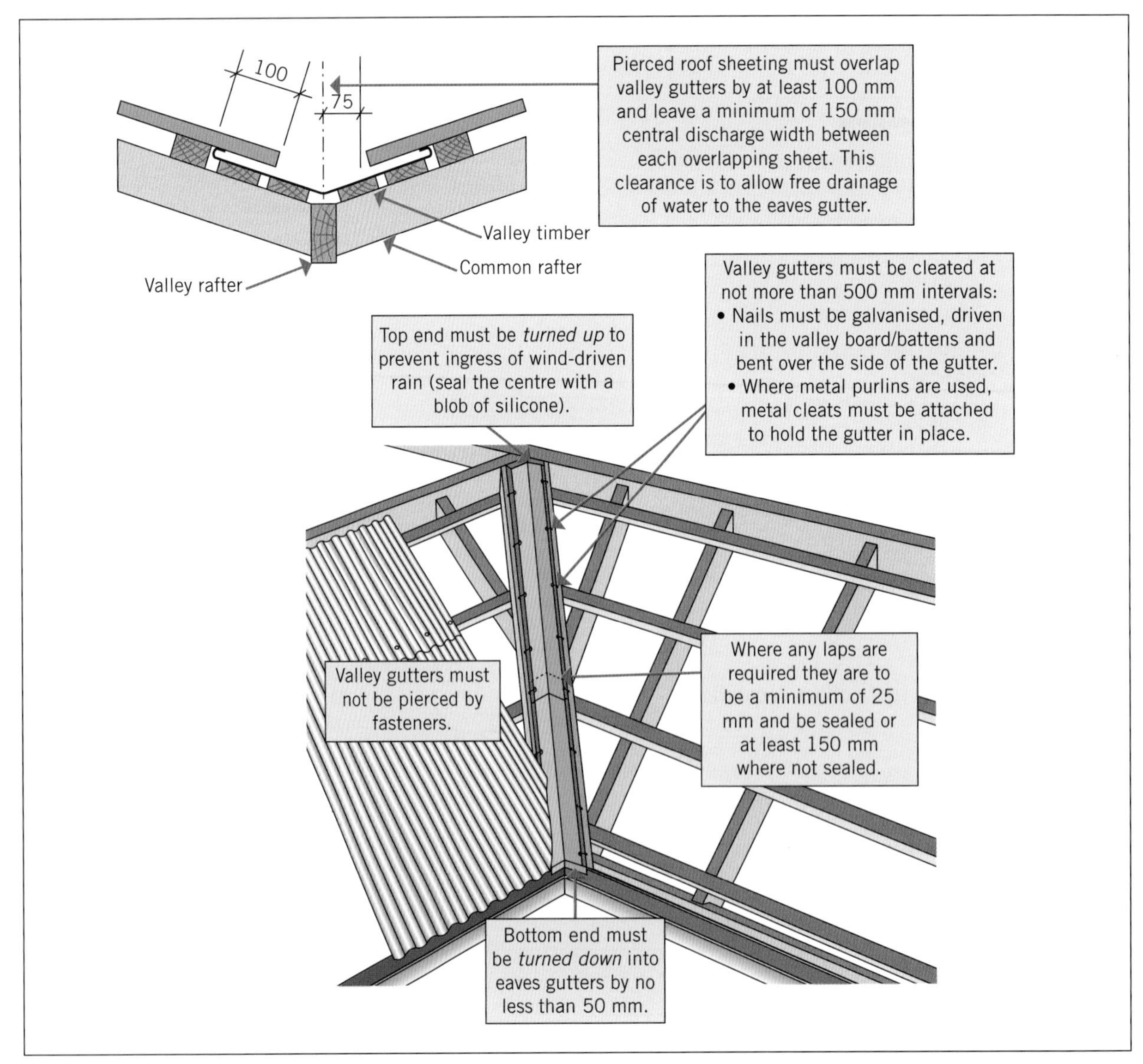

FIGURE 10.32 Basic valley gutter installation requirements

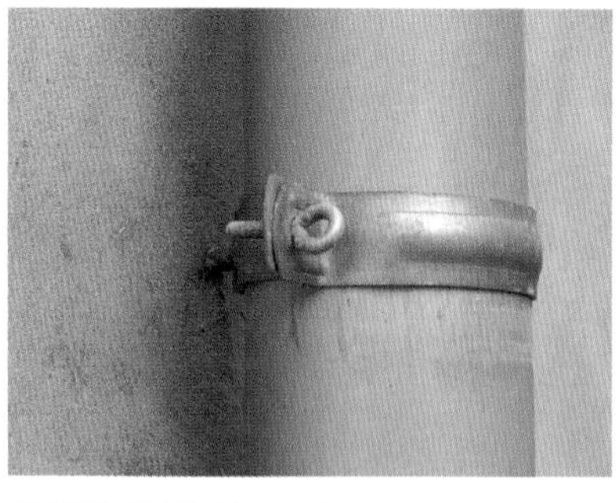

FIGURE 10.33 Downpipe supports

- PVC-U downpipes must include an *expansion joint* at no greater than 6-m intervals.
- Seams should be positioned uppermost on graded systems and be concealed on vertical sections.
- Graded sections should have a minimum fall of 1:25 (approximately 2.3° from the horizontal plane).
- All metal downpipe joints should be lapped not less than 50 mm in the direction of flow, and fully sealed if on grade.
- Downpipes directly connected to stormwater drains must have overflow protection – this means that some form of surcharge pit must be installed in the stormwater line. Alternatively, the downpipe can be 'disconnected' by terminating over an open stormwater grate.
- Downpipes should be at least 100 mm clear of any gas or electrical cable and at least 50 mm from any other service.

Spreaders

Some buildings have one roof situated higher than another. Sometimes it may not be possible to drain the higher roof directly to an approved point of discharge. In this case, part of this higher roof may have to discharge onto the lower roof first. To do this you must use a spreader to avoid a high volume of water from discharging at one point (see Figure 10.34). This will ensure that sheet laps, downstream gutters and flashings are not swamped, resulting in leakage and possible damage.

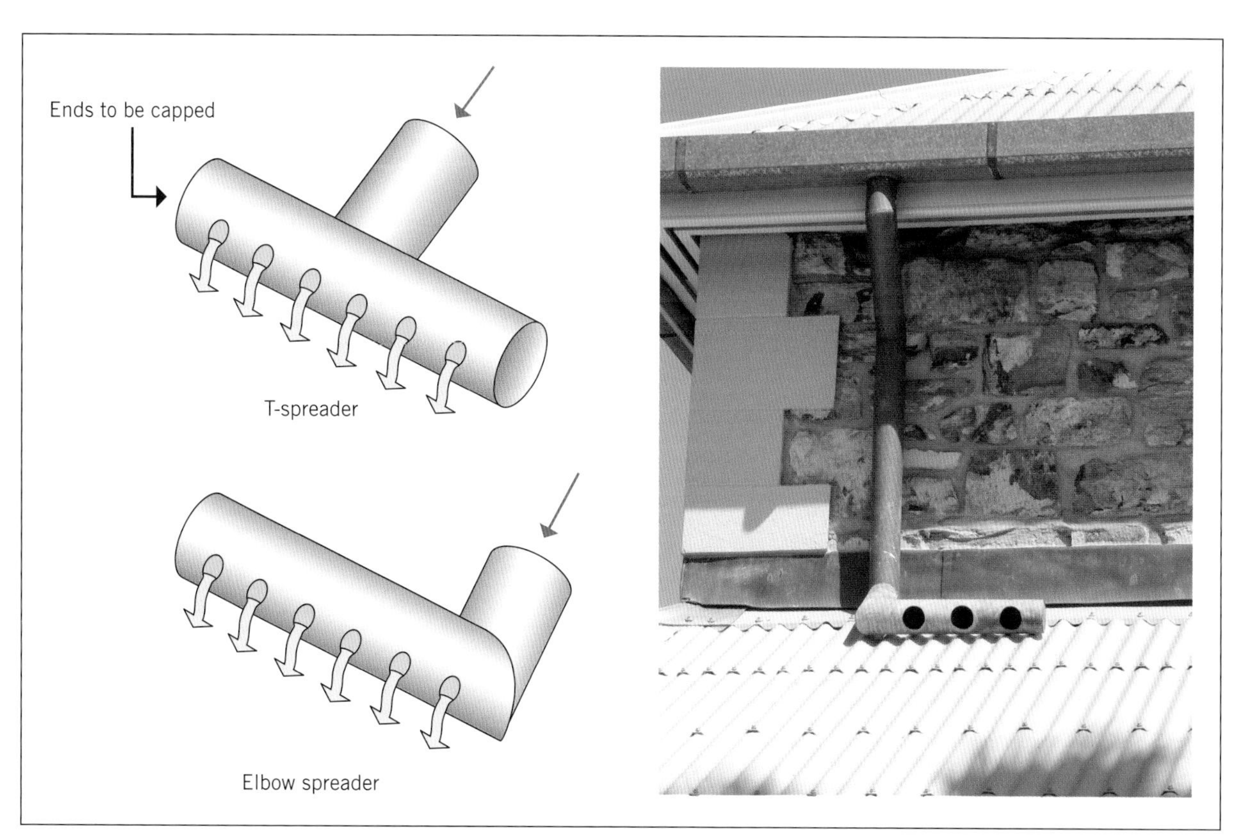

FIGURE 10.34 Spreaders

Some basic requirements apply:

- The roof catchment area being drained with each spreader must not exceed 15 m^2.
- The water must not be discharged directly over roof sheet side laps, ridge tiles, mortar jointed tiles, flashings or timber fascia.
- The downstream drainage system should be sized to handle the additional flow.
- The ends of the spreader should be capped to promote even build-up and flow from all discharge holes.

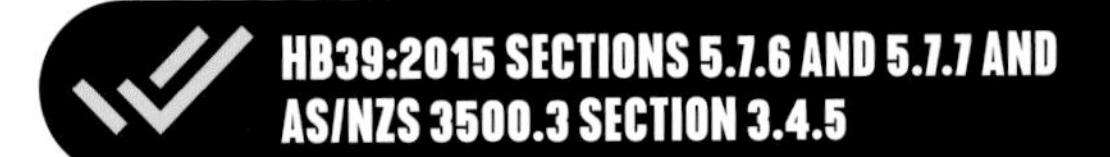

HB39:2015 SECTIONS 5.7.6 AND 5.7.7 AND AS/NZS 3500.3 SECTION 3.4.5

Testing the roof drainage system

Having completed your installation, it is necessary to check all component parts and carry out a performance test. For all sections outside the building, this may entail basic leak and flow testing, checking joints for soundness and looking for any sections that may be causing ponding or restrictions.

The testing of downpipes inside buildings is more involved, and this is carried out with a hydraulic or air test in accordance with AS/NZS 3500.3. During times of water restrictions, a hydraulic test may be undesirable. Discuss these local requirements with your employer.

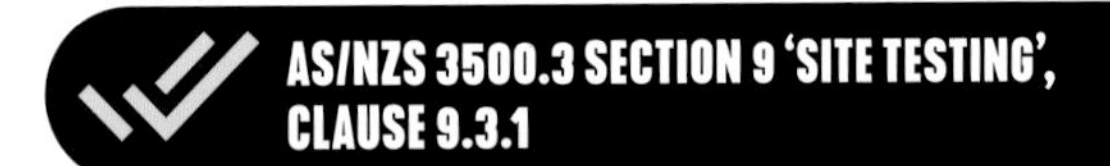

AS/NZS 3500.3 SECTION 9 'SITE TESTING', CLAUSE 9.3.1

LEARNING TASK 10.6

1. What is the mathematical term used to express the expansion of a material by a temperature rise of 1°C?
2. What is the minimum length for sumps?
3. A sump must incorporate one of two overflow options. What are they?
4. Why should the ends of the speader be capped?

Clean up

The clean-up process is unfortunately one aspect of an installation that is often not done satisfactorily. However, unless the clean-up is thoroughly carried out, damage to both the installation and your reputation may follow. The following requirements are therefore important:

1. *Remove all installation swarf and debris.* During the installation process, swarf, offcuts and rivet shafts invariably accumulate on the job. It is vital to the quality of your work that such contaminants are hosed or swept from the full roof drainage system every day where possible, and at least weekly.

   Also ensure that strippable plastic films are removed and disposed of according to site protocol.
2. *Comply with correct and sustainable waste removal.* Larger offcuts and waste building materials must be disposed of in the correct skip in accordance with site requirements. Such requirements are normally detailed during the induction process.

   In most municipalities today, old gutter and roof sheets can be disposed of at a waste transfer station where all metal products are isolated from general waste for recycling.
3. *Refurbish tools and equipment.* All tools and equipment should be maintained and checked for serviceability before returning to the site store or workshop. Problems must be reported to the relevant supervisor.

LEARNING TASK 10.7

1. Which quality assurance processes might apply to the clean-up?

SUMMARY

In this chapter you have been introduced to basic requirements related to the fabrication and installation of roof drainage components. The correct identification of components and how they are applied as part of the broader roof drainage system is vital to your work as a roof plumber. You have also started to gain an understanding of how important adherence to relevant codes and standards is for roof plumbing.

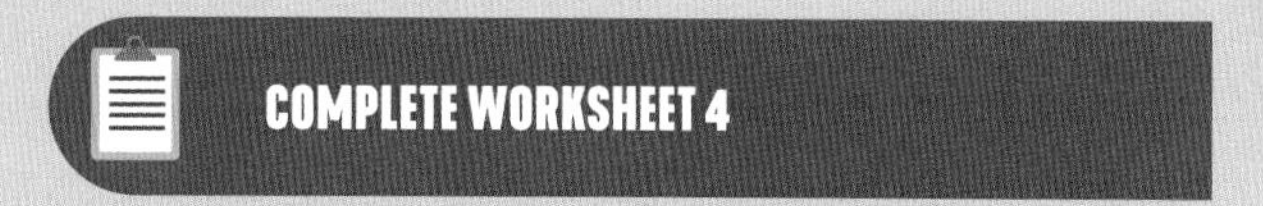

WORKSHEET 1

To be completed by teachers

Satisfactory ☐

Not satisfactory ☐

Student name: ______________________

Enrolment year: ______________________

Class code: ______________________

Competency name/Number: ______________________

Task: Working with your teacher/supervisor, refer to this text, your code/standard and any other relevant resource to answer the following questions.

1. What are two reasons for the use of downpipes and connection to an approved point of discharge?

2. What are two purposes of a rainwater head?

3. Where is a valley gutter situated?

4. Where would you find a sump installed?

5. Draw and name at least three gutter profiles commonly used in your local area.

6 Provide three examples of what is known as an approved point of discharge.

7 In the picture below you can see a galvanised gutter draining rainwater off a painted steel roof. What is the name given to the condition that causes such corrosion?

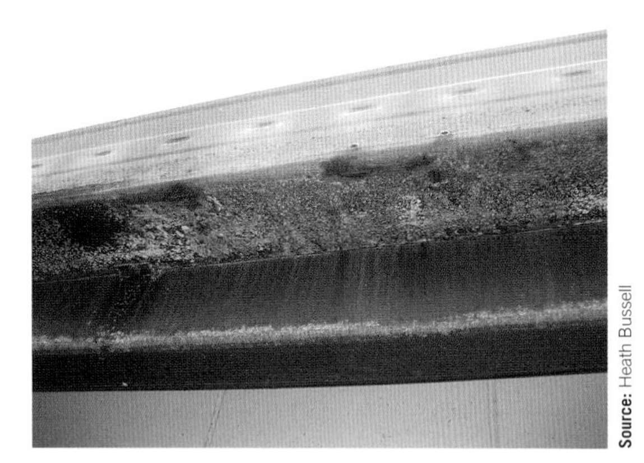

Source: Heath Bussell

Source: Heath Bussell

8 Are COLORBOND® steel products and plain ZINCALUME® steel considered compatible with each other?

9 What type of silicone sealant is recommended for roof and gutter applications?

10 What is the purpose of a box gutter overflow?

11 Which Australian Standard is used to determine the size of rainwater drainage systems and products?

WORKSHEET 2

To be completed by teachers	
Satisfactory	☐
Not satisfactory	☐

Student name: ____________________

Enrolment year: ____________________

Class code: ____________________

Competency name/Number: ____________________

Task: Working with your teacher/supervisor, refer to this text, your code/standard and any other relevant resource to answer the following questions.

1 To which code and standard should you routinely refer for all roof drainage installations?

____

____

____

2 What are two reasons for reducing waste through the efficient selection and ordering of materials?

____

____

____

3 For what additional information would you access manufacturers' websites in preparation for a gutter installation? List at least four items.

____

____

____

4 What would be two limiting factors when handling long lengths of gutter?

____

____

____

5 Use Pythagoras' theorem to calculate the incline for a valley gutter where the rise is 2.4 m and the run is 4 m.

____

____

____

6 Why is it not recommended to surround drill holes with a circle of silicone sealant?

____

____

____

7 If you wanted to make a 135° angle to go around a bay window, how many degrees would be on each side of the vertex or fold point?

____

____

____

8 What are five goals that you should aim for in order to fabricate quality rainwater goods?

____

____

____

____

____

9 List at least 10 tools that you would use to fabricate general rainwater goods.

____

____

____

____

____

10 For general cutting of downpipe and gutter using a hacksaw, what sort of blade should you use? Circle the correct answer:

18 TPI 32 TPI 24 TPI

WORKSHEET 3

To be completed by teachers	
Satisfactory	☐
Not satisfactory	☐

Student name: ______________________

Enrolment year: ______________________

Class code: ______________________

Competency name/Number: ______________________

Task: Working with your teacher/supervisor, refer to this text, your code/standard and any other relevant resource to answer the following questions.

1 What would be the expansion in a 20-m copper eaves gutter where an 80°C rise in temperature is measured?

____

____

____

2 At what maximum interval should an expansion joint be allowed for in a steel box gutter with a base metal thickness (BMT) of 0.42? Quote a code reference.

____

____

____

3 What is likely to happen if a gutter is installed between two fixed points with no provision for expansion?

____

____

____

4 What is the maximum length that an eaves gutter cannot exceed without provision for expansion? Quote a code reference.

____

____

____

5 What is the minimum expansion space required at the end of an aluminium box gutter? Quote a code reference.

____

____

____

6 Calculate the fall in millimetres of an eaves gutter run of 13 m at 1:500.

7 Calculate the fall in millimetres of a box gutter run of 35 m at 1:150.

8 What is the grade of a box gutter that has a 125 mm fall over 20 m? Is this fall permissible?

9 What is the grade of a box gutter that has a 110 mm fall over 35 m? Is this fall permissible?

WORKSHEET 4

To be completed by teachers	
Satisfactory	☐
Not satisfactory	☐

Student name: ______________________________

Enrolment year: ______________________________

Class code: ______________________________

Competency name/Number: ______________________________

Task: Working with your teacher/supervisor, refer to this text, your code/standard and any other relevant resource to answer the following questions.

1 What is the minimum fall required for box gutters of up to 600 mm in width? Quote a code reference.

____

____

____

2 From your HB39:2015, complete the following sentence in relation to box gutters, and quote a code reference:

'Lap joints of box gutters to have ____________ mm laps sealed and fastened in the ____________ fall'.

____

3 Complete the following, and quote a Standards reference:

Valley gutters shall not be used on a roof pitch of less than ____________ degrees.

____

4 What is the minimum fall recommended when installing an eaves gutter? Quote a code reference.

____

____

____

5 At what maximum interval must you secure a valley gutter? Quote a code reference.

____

____

____

6 What two components are missing from the box gutter installation in this image?

____

____

____

Source: Heath Bussell

7 Your boss asks you to install a series of brackets for a box gutter discontinuous support system. What is the maximum distance permitted between each bracket? Quote a Standards reference.

8 You notice that your mate cuts a valley gutter so that it is turned down into the eaves gutter around 30 mm. Is this OK? Quote a code reference.

9 What is the minimum effective discharge width required between roof sheets in the middle of a valley gutter? Quote a code reference.

10 You need to grade a downpipe system across the side of a commercial building, picking up four vertical drops. What is the minimum grade permitted? Quote a code reference.

11 You are about to cut a circular outlet hole in the bottom of a box gutter sump. Where should the hole be positioned? Quote a code reference.

12 What is the maximum interval between downpipe clips on both vertical and graded sections? Quote a code reference.

____

____

____

13 Downpipes must maintain a minimum distance from adjacent services. Complete the following minimum distances, and quote a standards reference:

From gas and electrical cables ____________ mm.

From all other services ____________ mm.

14 Complete the following, and quote a standards reference:

A spreader is permitted to drain a roof area of no greater than ____________ m^2.

15 How frequently should the roof and drainage system be cleaned down during work?

____

____

____

16 What special testing requirement relates to downpipes installed inside buildings?

____

____

____

17 What is the maximum catchment area permitted above a valley gutter? Quote a code reference.

____

____

____

RECEIVE ROOFING MATERIALS

The profitability and safety of an installation can be at risk even before the job is started. Whether or not a roof installation is efficient and profitable can depend on how well you plan for the delivery of materials.

Overview

Time spent on planning is never wasted time. All tasks require a focused and methodical evaluation before commencement so that expensive mistakes and unnecessary repetition can be avoided. More broadly, poor delivery planning can lead to:

- lost time
- damaged materials
- lost or stolen materials
- dangerous storage or handling practices
- structural damage
- inefficiency, causing delays for other trades
- loss of profitability.

Planning for and receiving materials onsite is not just the responsibility of your employer or supervisor. All roof plumbers have a role to play in the efficient and safe receiving of roof materials. Getting it wrong can be dangerous and expensive.

In this chapter you will be introduced to many of the considerations relating to the delivery and receipt of roofing products on the job site. The differences between small domestic re-roofing jobs and larger building projects are identified as and when required.

The four stages of materials delivery and handling

You have won a roofing quotation or tender. What now? What needs to happen between this point and when you begin to install products on the job?

Delivery of materials can be as simple as a single drop-off of roofing products for a small domestic carport, or as complex as a multi-stage, large industrial installation requiring numerous deliveries, onsite storage and mechanical handling of sheets at height. However, regardless of the job size or complexity, you can break down the receiving of materials into four main stages, as shown in Figure 11.1.

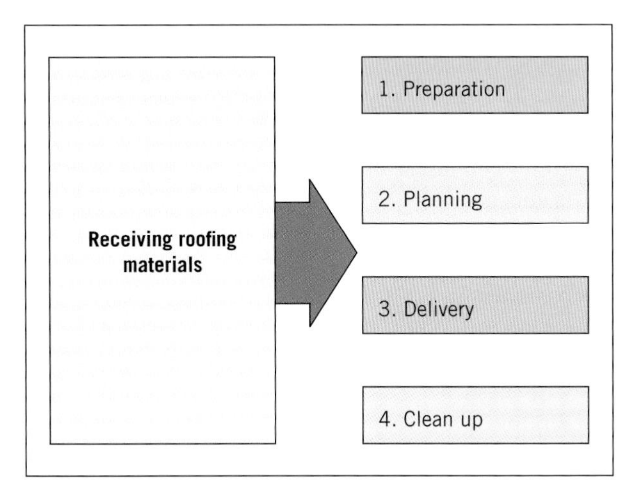

FIGURE 11.1 The four stages of materials delivery

Preparation: initial inspection and consultation

At the quotation stage you may have only seen plans or visited the job site before work began. Many things may have changed in the meantime, so the preparation stage is all about visiting the job to check and confirm the basic characteristics of the site, work health and safety (WHS) issues, quality assurance requirements, special tools/equipment needs and the suitability of the materials delivery area.

Obtain the approved plans and specifications

Approved plans and specifications are stamped and dated as part of the local government building approval process (see Figure 11.2). You should work off only these documents. These are kept onsite at all times and copies are available for subcontractors as required.

For normal domestic re-roof and re-gutter tasks, there is unlikely to be an overall set of plans as normally found in new construction. However, planning and sketching is still required to organise a work sequence, calculate quantities or design customised roof drainage components such as rainheads and sumps by your sheet metal shop.

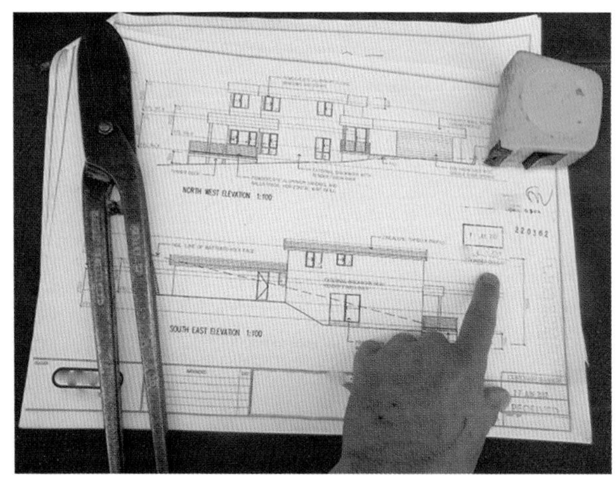

FIGURE 11.2 Checking that the plans and specifications are stamped as approved

Review the site WHS plan

All construction sites must have some form of WHS plan in place. Hazard identification and risk control are a mandatory and standard part of every job. (More detail on this subject is provided in Chapter 3 'Basic roof safety' of this text.) For issues relating specifically to materials delivery, look for some of the following:

- site induction requirements
- overhead power lines near the delivery, handling and working areas
- steep slopes
- any forecast adverse climatic conditions during the planned delivery, such as heat, wind, glare, dew and rain
- height access and fall protection requirements.

Compliance with company quality assurance protocols

On many jobs, work processes, controls and protocols must meet specified quality assurance requirements detailed within the tender process. As a subcontractor, you need to be aware of and possibly document your compliance with such requirements throughout the duration of the job. In addition to WHS issues, considerations may include:

- materials handling controls
- compliance with materials warranty requirements
- specific training requirements
- recycling and waste management
- job-site 'housekeeping' requirements.

Tools and equipment

The basic tools and equipment related to materials delivery may include those shown in Table 11.1. This list must be refined and added to as the specific planning details of the site dictate.

TABLE 11.1 A basic tool and equipment list for materials delivery

Tools	Personal protective equipment	Equipment
Knife for opening packs and cutting straps Snips for cutting steel strapping Crow bar or pinch bar for levering packs into position	Broad-brimmed hat Gloves for handling sharp edges Hardhat Sunscreen and clothing appropriate to the climate High-visibility ('hi-vis') clothing Approved boots, glasses and hearing protection Personal fall arrest equipment as required	Barricades Signage Traffic 'Stop/Slow' bats Trolleys, work platforms, forklifts and cranes Tarpaulins Securing straps Stacking timbers Two-way radio

Location and suitability of delivery area

Check that the delivery area is in a location that is suitable for the delivery of roofing materials. Points to look for include the following:

- the proximity to the roof structure
- whether the area is clear of debris and other materials
- whether the slope of the land is suitable for safe and efficient unloading
- whether the area will permit the delivery of long material lengths
- whether the area is isolated from overhead power lines or, if they are close, that they are adequately identified
- if required, whether an adequate storage area is identified.

Basic preparation allows you to then move onto the detailed planning required for the efficient and safe delivery of the materials onsite.

LEARNING TASK 11.1

1 You need to have a scissor lift delivered on site to work on some rusted rainheads. What are at least three site WHS considerations you need to identify?
2 What are four consequences of failing to plan effectively for a roofing job?
3 Why is it important to have a site inspection?
4 Why are commercial plans and specifications stamped and dated?

Planning and delivery – getting the details right

Just as each job is different, so too are the requirements for material delivery. In this section, you will see how attention to detail can mean the difference between safe and unsafe practices and between job profitability or loss.

Confirm quantities

From the plans or existing roof, calculate and confirm the quantities of all required materials. Depending upon the job, items may include:

- gutter and fascia
- fasteners
- downpipe
- roof sheets
- flashings
- fixings
- insulation
- wire mesh
- sarking
- sealants
- roof components (e.g. skylights, ventilators)
- tool consumables (e.g. blades, discs, drill bits).

You will be introduced to more details relating to quantity calculations in subsequent chapters.

Delivery of materials

The delivery stage includes important decisions and considerations relating to job progression, delivery costs and the amount of work in each section of the project.

Sequencing

Although every job involves different considerations, it would be unusual for all materials to be delivered at once, apart from for small installations, re-roofs and repairs. Unless adequate and secure storage facilities are available, you would normally plan your deliveries around some of the following factors:

- *Size and stages of work*. For example, the installation of safety mesh over the roof frame may take two full days and form a complete stage in your planning. You may then plan for the delivery of insulation blanket and roof sheeting on the third day. The installation of fascia and gutter is another common part of the job for which only those materials required for this stage of work are ordered and installed before the next stage of the job is begun.

- *How much work you and/or your team can complete in a full working day*. If you had to sheet a 10 000 m^2 factory roof, you would obviously need more than one day! Therefore, you would generally plan your deliveries of materials to match the time expected to complete each section of the job.
- *Coordination with other trades*. There is little point in ordering certain materials where you need to wait for other trades to complete a section of work. For instance, you would not order your downpipes to arrive with the fascia and gutter delivery if the wall cladding or brickwork is not to be completed for another six weeks. Where a builder or project manager is in charge of the job, you will need to consult closely to determine when you will be required or able to carry out work during the various stages of the project.
- *Cost of delivery*. In some areas or for particular materials, the cost of delivery could be expensive and you may choose to reduce delivery costs by increasing the amount of material delivered. If you have access to secure storage onsite, using this may be a viable option.
- *Cost of mechanical handling*. Where you need to hire a crane to lift materials onto the roof, you will need to balance hire costs against convenience. While you might normally sequence a series of deliveries for certain jobs, there may be instances where the cost of crane hire is prohibitive. In other situations, however, safety considerations may require that a crane is used multiple times, and in this case the cost will need to be built into the price of the job.

The smooth progress and completion of any job depends upon efficient delivery sequencing. Worksheet 1 includes some exercises that provide an opportunity to think more about delivery considerations.

COMPLETE WORKSHEET 1

Load handling

Once you have determined *what* you are going to deliver and *when*, you now need to think about *how* you are going to handle the materials. Materials will arrive by truck and will be packed in various ways. Some long items, such as roof sheets or lengths of flashing, will be bundled together with steel or nylon straps. Other items, such as sarking rolls, screws and fixings, will come in cartons.

Some jobs may allow you to break up packs and cartons into individual items so that you can safely position them where required for installation or storage. Other circumstances may dictate that larger and heavier items must be manoeuvred by mechanical means for both safety and efficiency.

Regardless of handling method, always plan the process with the following in mind:

- The method of handling must be safe.
- The method should avoid double-handling wherever possible. Considerations include:
  - proximity to the point of use
  - the sequence of delivery and use
  - the alignment of sheets.

Manual handling

The following guidelines should be considered when moving materials by hand:

- *How heavy is too heavy?* Unfortunately, there is no simple answer to this question, as it often depends upon the material's overall dimensions and an individual person's strength, size and experience. In general, weights above 20 kg will often require some form of assistance, particularly when they are bulky and/or repetitive. Perhaps more relevant is the fact that injuries can occur when lifting almost any weight if the lift is completed incorrectly.
- *Bend at the knees*. You have probably heard this many times before, but it is good advice! Your back is not designed for bearing all the weight of lifting heavy objects, so it is important to use your legs to take the load instead. Avoid immediate and future damage by following the simple recommendations outlined in Figure 11.3.
- *Wear gloves*. The edges and corners of sheet metal can be extremely sharp. Sudden gusts of wind, a loss of grip or unexpected movements made by another person may result in severe lacerations to the skin, muscles, tendons and ligaments that can leave you temporarily or permanently incapacitated. Gloves must be suitable for the purpose and fit your hands well. Poorly selected gloves can be more dangerous than no gloves at all!
- *What is the team lift protocol/sequence?* When lifting larger objects with other workers in a team, always discuss some form of lift protocol before lifting. Unless the lift is coordinated with clear directions (as seen in Example 11.1), injuries may occur when one person lifts or moves before another person is ready. Before lifting begins, someone should be nominated to control the sequence by giving clear directions (see Figure 11.4).

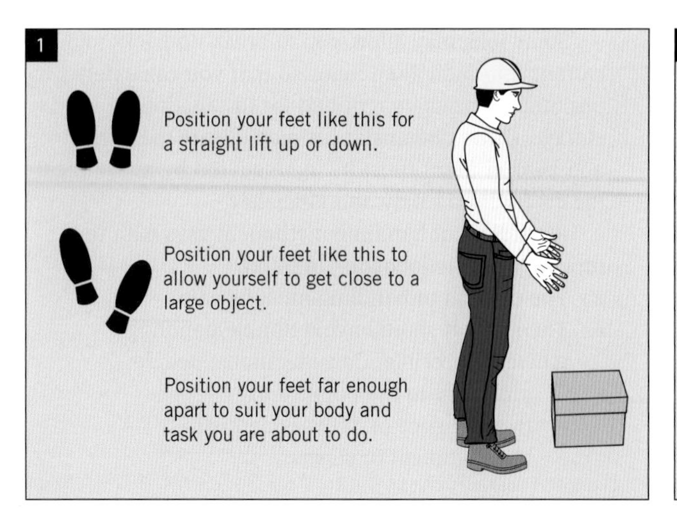

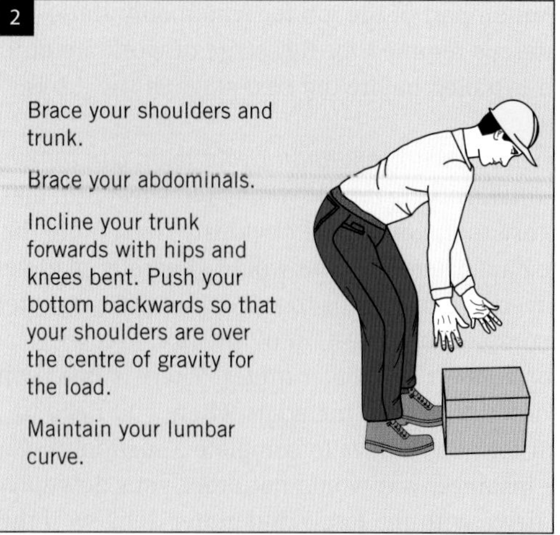

3

Test the load weight. If it is too heavy get help.

Keep your head and chest up.

Keep the load close to your body, with your back, abdominals, shoulders and elbows braced.

Extend your hips and knees to straighten up. Change direction with your feet instead of twisting your body while bent or loaded.

4

To lower, reverse this procedure.

FIGURE 11.3 The safe way to lift a heavy object

Source: Based on Workplace Standards Tasmania, WorkSafe Tasmania.

FIGURE 11.4 Lifting as a team

EXAMPLE 11.1

The following example is one method you can use as a lift sequence:

1. Team members position around the load.
2. When ready, the nominated person clearly says: 'Prepare to lift' (all team members take up the lift position).
3. The nominated person says: 'LIFT!' (all team members lift at once).

- *Keep sheets low as possible and stay on the windward side.* When lifting roof sheets or lengths of gutter, try to keep the item relatively low to lessen the chance of it being caught by sudden wind gusts. In addition, position yourself on the side of the sheet from which the wind is blowing, so that if a sudden gust of wind catches the sheet you can drop it instead of being knocked over by the sheet.
- *Take care when passing materials onto a roof by hand.* Great care must be taken when lifting long material lengths onto a roof. You must ensure that a well-understood procedure is in place to avoid any possibility of materials being dropped onto the person below. Avoid damaging gutters and flashings by dragging lengths over the edges of the roof. Full lengths of sheets and gutter must be supported so that the material does not buckle.

Moving sheets and other materials around the job site and particularly up on the roof itself must be done in a planned and methodical manner to avoid injury and death.

- *Check the alignment of roof sheets before handling them into position.* When roof sheets are secured to a roof frame, each sheet is lapped over the one preceding it. Each sheet has a specific under-lap and over-lap side. Placing your sheets in position around the wrong way will require each sheet to be rotated on the roof before securing. This is unsafe and inefficient.
- *Secure all materials in place.* Once materials have been placed into the desired position, you need to ensure that they are well secured to prevent them from being dislodged by the wind or site operations. Ratchet-type straps, ropes and chains may be required to do this effectively.

Mechanical handling

Where the size, quantity or WHS requirements of a site dictate that mechanical handling must be used, you must consider some of the following points:

- Can the size of crane that you require manoeuvre onsite?
- Is the site relatively level and the ground surface stable?
- Will the crane be able to lift materials directly from the delivery trucks or storage site?
- Will the road or the site access be impeded by the crane during the lift?
- Will traffic control be required?
- Are there any electrical hazards within the proposed handling area?
- Will the operations of other trades be affected by the lift?
- Will the lift process comply with site WHS requirements?

It is important for you to realise that crane operation and lifting is a specialised job that requires the operator to be suitably qualified and to hold a relevant high-risk work licence. It is highly recommended that you meet the crane operator, rigger or company representative onsite to discuss all details of your lift ahead of time. A reputable company will supply all the rigging and associated lift equipment suitable for the task.

As materials are lifted onto the roof or storage location, you will often be required to provide directions to the crane operator as the load is lowered into place. The final position of the load is not generally visible to the operator, so you need to be clear and concise with your directions. Before the lift, ask the operator what verbal and hand signals you are to use so that there is no confusion during the task.

Site storage

The requirements and size of most installations will mean that some quantity of material will need to be stored onsite. Smaller items such as brackets, sarking rolls and boxes of screws can often be stored in site sheds/containers or vehicles until required. Where possible and practical, position packs of sheeting at appropriate intervals around the job.

Wet storage corrosion

If long roofing sheets must be stored outdoors for a period of time, you must take particular care of how they are stacked, in order to prevent corrosion damage. Roofing sheets are delivered in bound packs in which each sheet is placed closely on top of another. Although this is necessary for efficient transportation, it makes the sheets susceptible to capillary action if they are allowed to become wet (see Chapter 7 'Capillary action').

FROM EXPERIENCE

From the moment that materials arrive onsite, you are responsible for their care and condition. Packs of roofing sheets can suffer accelerated corrosion if they are not stored correctly.

Roofing products must be kept dry during storage. Failure to do so may result in costly damage to sheets that become unusable as a result. If packs are delivered wet and cannot be used immediately, they must be separated and stacked so that they dry out. Your HB39:2015 provides some simple guidelines on how to store sheets onsite.

Failure to store sheets correctly could result in the loss of product warranty, with you being held responsible for replacement.

You should also be aware that corrosion and stain damage from wet storage can occur from simple condensation alone. Therefore, even in dry weather, ensure that sheets are stored according to the manufacturer's recommendations. As a minimum, these should include the following:

- Keep sheets off the ground.
- Separate the packs with cross-timbers in order to allow airflow.
- Cover them with a loose-fitting but well-secured tarpaulin that allows adequate air circulation.
- Ensure that sheets are stored in a location in which they will not be walked on or suffer impact damage of any kind.

More information on the problems caused by wet storage can be found in the most current BlueScope Steel technical bulletin on the subject: https://cdn.dcs.bluescope.com.au/download/tb-07

(Note: over time, the link above may become out of date. The most current version of this document can be found on the BlueScope website by conducting an internet search for: 'Care of BlueScope coated steel products during transport and storage'.)

Domestic site storage: occupied dwellings

The storage of materials on job sites where the building is occupied during the work needs particular mention at this point. In most cases where roof work is to be carried out on an existing dwelling, the roof plumber is the only contractor and is entirely responsible for the safety of occupants during the work. The possibility of children being present on the site makes secure storage absolutely vital (see Figure 11.5). Points to consider include the following:

- New material should be stored onsite only when immediate installation after delivery is not possible.
- Ensure that all materials are securely barricaded.
- Ensure that no long sheet or gutter lengths protrude from the storage area.
- Ensure that all waste is disposed of directly into a skip or other container.
- Where children are present, liaise closely with their parents concerning supervision and safety around storage areas and work zones.

FIGURE 11.5 Incorrect storage on a domestic site could lead to a serious injury.

Spot loading (point loading)

'Spot loading' is a term that describes the loading of roofing sheet packs and other materials directly on top of the roof structure at predetermined intervals. Each pack contains the correct number of sheets required to cover the space between each load pack. These packs are usually placed with the aid of a crane as part of the delivery process. With correct planning, spot loading allows the job to progress faster due to a reduction in manual handling.

While spot loading is particularly relevant to larger commercial and industrial installations, it is also used on domestic jobs where safety considerations, the roof size or the height of the roof makes it a viable option.

However, before you decide to spot load your sheets around the roof structure, you must make some very important checks, which are detailed below.

Is the structure ready and suitable for spot loading?

Do not assume that a particular structure is ready for spot load weights to be applied. Many a job has suffered damage when materials loads have been placed on the structure prematurely. For example, even a small pack of 20 × 8 m lengths of 0.48 mm corrugated sheets weighs approximately 600 kg! One or more packs of such weight may cause the frame to collapse, twist or 'corkscrew' if not correctly secured. Therefore, before loading, you *must* ask the builder or engineer to confirm the following:

- Are all of the required frame bracing and tie-downs in place and secured?
- Are all bolts and screws in place and tight?
- Are all of the required structural plate welds completed?
- Has the building surveyor/engineer 'signed-off' on the frame?

In addition, *be wary of roof pitch*. Steep-pitched roofs are obviously unsuitable for placing a roof sheet pack without securing, but lower pitched roofs are harder to gauge. There is a limit to the angle of an incline plane on which any object can rest without sliding down. For granular products, such as sand or gravel, this is commonly known as the 'angle of repose'. The same principles apply to the placement of roofing sheets and materials on a sloping roof.

The limit itself depends upon the object's mass, the friction between surfaces and the degree of incline. However, even below this point, objects may begin to slide if they are moved in any way. Where the roof pitch is greater than approximately 5°, there is a risk that an otherwise stable sheet pack may begin to creep down the roof due to vibration or strong wind gusts.

If you have any doubts about the pitch of a roof, it is best to be cautious and prevent roof sheet packs from sliding by fitting a secure 'stop' at the lower end of the pack in addition to any straps across the pack.

If, after these considerations, you are satisfied that the structure is suitable and has been completed to the appropriate stage, you are ready to plan your load points.

What are the spot load limits?

Roof structures are generally designed to bear the weight of only the roof itself, the roof components and the maintenance workers, plus a certain margin for safety. The structure is not necessarily capable of supporting the concentrated weight of multiple sheets at a single point. Having determined that the frame is correctly braced and ready for loading, you must ask the following questions of the builder or engineer:

- From which direction does the prevailing wind blow? It is standard trade practice to always start laying your sheets from the side of the roof that is opposite to the direction of the prevailing wind, so that the sheet lap edges face away from the wind (see Figure 11.6). The position of your packs will be first determined from this information.
- Where can loads be positioned on the roof structure? For example:
  - What is the maximum weight permitted over the load-bearing walls/support?
  - What is the maximum weight permitted on purlins between the load-bearing walls/support?

The answers to these questions will assist you in calculating where each pack should be placed.

Where should sheet packs be placed?

Your goal in spot loading is to place material packs onto the roof structure at the most advantageous interval so that installation speed, efficiency and safety are maintained or enhanced. Determination of where the packs shall be placed has two stages:

1. Decide where you would *like* to put the material packs based on work sequence and sheet *effective cover*. To do this you need to know how to do the following:
   a. Determine the number of sheets required.
   b. Determine the possible pack locations based on the number of sheets per pack.
2. Calculate the weights and confirm the pack locations on the roof structure.

The effective cover of any roof sheet is the width of roof covered by the sheet. Sheet laps are *not* included in this measurement. For example, the full width of a sheet of LYSAGHT KLIP-LOK® 406 is approximately 436 mm. However, its *effective cover* is actually 406 mm.

Determine number of sheets required

Dividing the effective cover of a roof sheet profile into the width of roof will tell you how many sheets you require to cover that area. Using LYSAGHT KLIP-LOK® 406 again, look at Example 11.2.

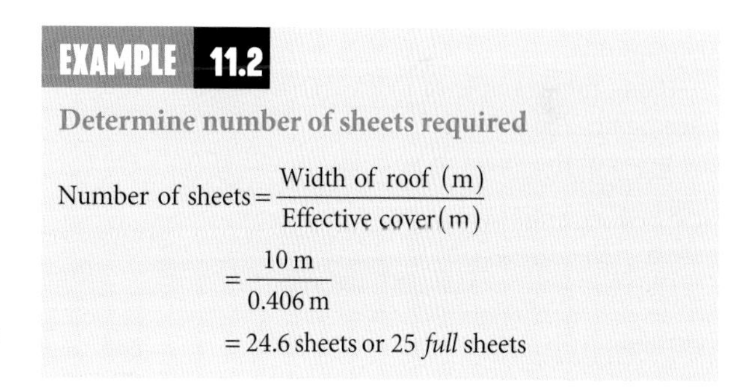

EXAMPLE 11.2

Determine number of sheets required

$$\text{Number of sheets} = \frac{\text{Width of roof (m)}}{\text{Effective cover (m)}}$$

$$= \frac{10\text{ m}}{0.406\text{ m}}$$

$= 24.6$ sheets or 25 *full* sheets

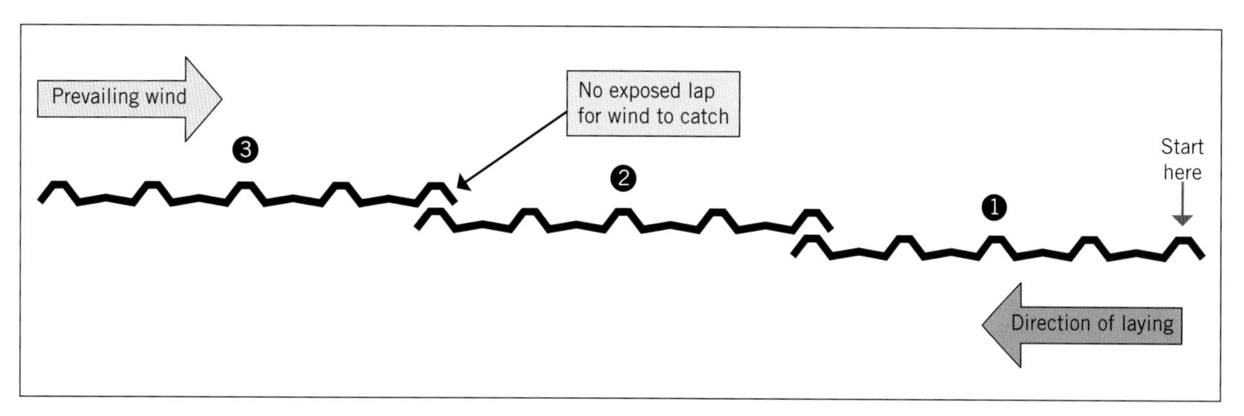

FIGURE 11.6 Securing roof sheets so that the laps face away from the prevailing wind

Determine how much roof each pack will cover

Multiplying the number of sheets in a pack by the effective cover of the sheet profile will tell you how many metres of roof each pack will cover. For example, if you knew that each sheet pack of LYSAGHT KLIP-LOK® 406 contained 20 sheets, you might decide to place each pack at approximately 9-m intervals (see Example 11.3).

Fortunately, manufacturers provide product information brochures that include basic sheet coverage tables so that, regardless of what profile you are using, an answer is readily available. An example of such a table can be seen in Figure 11.7.

Sheet coverage																				
Width of roof (m)	3	4	5	6	7	8	9	10	11	12	13	14	15	16	17	18	19	20	30	40	50
KLIP-LOK® 406 Number of sheets	8	10	13	15	18	20	23	25	28	30	33	35	37	40	42	45	47	50	74	99	124

Source: © BlueScope LYSAGHT®, www.lysaght.com

FIGURE 11.7 Example of a manufacturer's sheet coverage table

EXAMPLE 11.3

Determine how many metres each pack will cover

Pack interval = Number of sheets/pack × Effective cover of profile
= 20 × 0.406 m
= 8.12 m (rounded up to 9 m).

Calculate pack weights

Packs of roofing sheets can place significant weight on a roof structure, and you need to know for sure that you are not overloading any section of the frame. To check this, you need to refer to the manufacturer's product information, which is available for every roof sheet profile.

Using the example of LYSAGHT KLIP-LOK® 406 once again, the product brochure will generally provide the mass of the sheet in both kg/m and kg/m^2 (see Figure 11.8).

KLIP-LOK® 406 mass			
BMT (mm)	kg/m	kg/m^2	m^2/t
0.48 **ZINCALUME®**	2.28	5.62	178
0.48 **COLORBOND®**	2.32	5.71	175

Source: © BlueScope LYSAGHT®, www.lysaght.com

FIGURE 11.8 The mass of a sheet profile as shown in a product brochure

To determine the weight of a pack of roof sheets, you need to first find out how much one sheet weighs. The mass of one sheet is found by multiplying the length of each sheet by its mass in kg/m, as detailed in the product brochure. Then you multiply this answer by the number of sheets. This process is shown in Example 11.4.

EXAMPLE 11.4

Determine the weight of a pack of roof sheets

- Number of sheets per pack: 20 sheets
- Material: 8 m LYSAGHT KLIP-LOK® 406 ZINCALUME® steel with a base metal thickness (BMT) of 0.48 mm

Pack weight = Length of sheet × mass in kg/m × number of sheets per pack
= 8 m × 2.28 kg/m × 20 sheets
= 364.8 kg.

Once you know the weight of each roof pack, you can then compare this figure to the spot load limits that you previously sourced from the builder or engineer during the planning stage. If the weight exceeds the spot load limit, you will need either to put the pack in a more suitable area or to break the pack up into a smaller number of sheets and reduce the interval between each pack on the roof.

Figure 11.9 shows how this process might be applied.

LEARNING TASK 11.2

1. How many sheets of LYSAGHT KLIP-LOK® 406 would you require to cover a roof 12 m wide?
2. What is the type of corrosion that roof sheets suffer if bound sheet packs are allowed to become wet while stored on site?

LEARNING TASK 11.3

1. When positioning your sheet packs, why would you need to know the direction of the prevailing wind?
2. What are three important considerations when passing materials onto a roof by hand?
3. What other form of communication might be required when directing the crane operator?

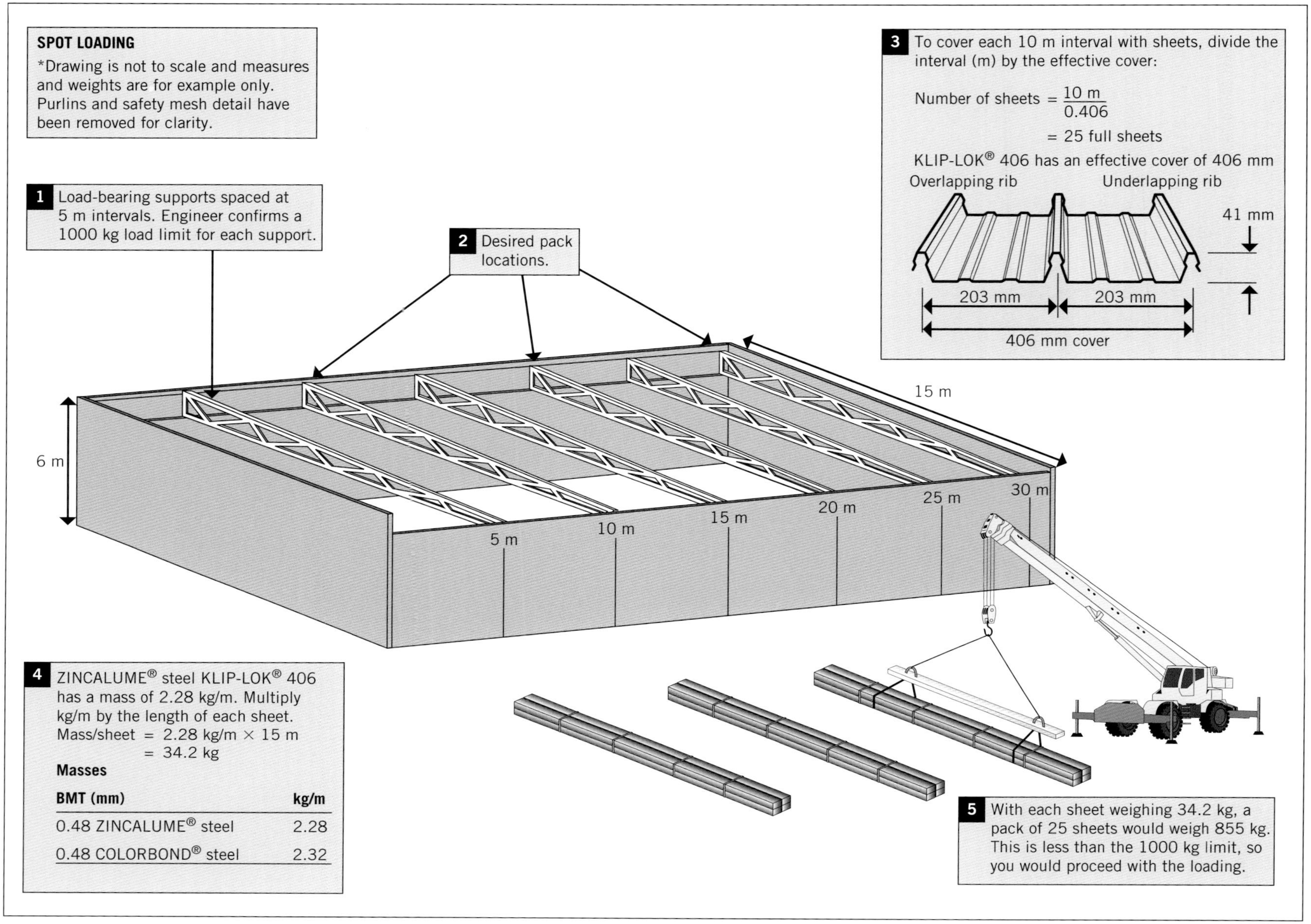

SPOT LOADING

*Drawing is not to scale and measures and weights are for example only. Purlins and safety mesh detail have been removed for clarity.

1 Load-bearing supports spaced at 5 m intervals. Engineer confirms a 1000 kg load limit for each support.

2 Desired pack locations.

3 To cover each 10 m interval with sheets, divide the interval (m) by the effective cover:

$$\text{Number of sheets} = \frac{10\text{ m}}{0.406} = 25 \text{ full sheets}$$

KLIP-LOK® 406 has an effective cover of 406 mm

Overlapping rib

Underlapping rib

41 mm

203 mm

203 mm

406 mm cover

15 m

6 m

5 m

10 m

15 m

20 m

25 m

30 m

4 ZINCALUME® steel KLIP-LOK® 406 has a mass of 2.28 kg/m. Multiply kg/m by the length of each sheet.

Mass/sheet = 2.28 kg/m × 15 m
= 34.2 kg

Masses

BMT (mm)	kg/m
0.48 ZINCALUME® steel	2.28
0.48 COLORBOND® steel	2.32

5 With each sheet weighing 34.2 kg, a pack of 25 sheets would weigh 855 kg. This is less than the 1000 kg limit, so you would proceed with the loading.

FIGURE 11.9 Determining spot load points

Clean up

Most tasks on a construction site require that you clean up all waste and pack away tools and equipment in a serviceable condition.

GREEN TIP

Receiving material deliveries invariably ends up with considerable waste packaging left over. All recyclable material and other waste must be disposed of in the appropriate manner in accordance with job-site protocols

Considerations include the following points:

- Dispose of all waste in an approved manner.
- Once they are no longer required, remove and store all barricades and signs.
- Clean and maintain all tools and equipment and return them to the correct storage location.
- Report any equipment faults.
- Refurbish all hire equipment in accordance with the hire company's instructions.
- Check and maintain your own tool kit and personal protective equipment (PPE).
- Complete any site and/or employer documentation required at the end of the task.

LEARNING TASK 11.4

1. Why is recycling excess material important?
2. Which quality assurance processes might apply to the clean up?

SUMMARY

In this chapter you have been introduced to the processes involved in the application of a formal planning process. Only through detailed and methodical planning are you able to run safe and profitable jobs.

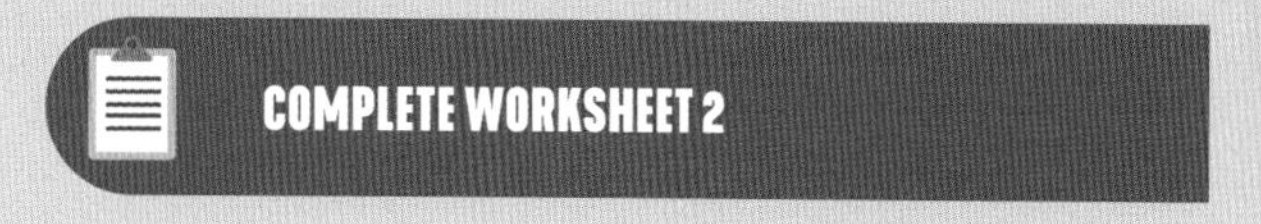

To be completed by teachers	
Satisfactory	☐
Not satisfactory	☐

Student name: ____________________

Enrolment year: ____________________

Class code: ____________________

Competency name/Number: ____________________

Task: Working with your teacher/supervisor, refer to this text, your HB39:2015 and any other relevant resource to answer the following questions.

1 List at least five problems that may result from poor delivery planning.

____

____

____

____

____

2 During a site visit to confirm your delivery arrangements, what WHS issues would you be looking for?

____

____

____

____

____

3 Identify at least three considerations to confirm the suitability of the proposed delivery area.

____

____

____

____

____

4 In the table you can see a list of common roofing materials. In this scenario, you are planning the delivery of materials for installation on a standard hip-roofed, low-set, brick-veneer, four-bedroom home.

From the framing stage right through to completion, think about how many deliveries you might require over the course of the job. Number the delivery of each group of materials in order: D1, D2, D3 and so on.

Because there are so many variables on construction sites, there can be no single right answer. However, there are some natural groupings, and the solution is worth discussing with your supervisor or teacher, who may have other scenarios for you to consider.

Domestic job	
Materials	**Delivery sequence: D1, D2, etc.**
Valley gutter	
Gutter brackets	
Metal fascia	
Roof screws	
Ridge capping	
Sarking	
First-flush diversion valve	
Coloured rivets	
Polyethylene rainwater tank	
Skylights	
Gutter lengths	
Roof sheets	
Downpipes	
Downpipe outlets	
Roof ventilators	
Downpipe brackets	
Box silicone sealant	

5 In the table below you can see a list of common roofing materials. In this scenario, you are planning the delivery of materials for installation on a 40 m × 15 m tilt-slab commercial warehouse with a single, low-pitch roof draining to a box gutter on one side.

From when the metal roof frame is installed right through to completion, think about how many deliveries you might require over the course of the job. Number the delivery of each group of materials in order: D1, D2, D3, and so on.

Because there are so many variables on construction sites, there can be no single right answer. However, there are some natural groupings, and the solution is worth discussing with your supervisor or teacher, who may have other scenarios for you to consider.

Commercial job	
Materials	**Delivery sequence: D1, D2, etc.**
Concealed fixing fasteners	
Gutter lengths	
Safety mesh rolls	
Box gutter support brackets	
Parapet capping	
Insulation blanket	
Box gutter sole sheets	
Box silicone sealant	
Roof sheets	
Box gutter lengths	
Downpipe brackets	
Roof sheet concealed fixings	
Soaker flashing lengths	
Downpipe lengths and bends	
Box gutter sumps	
Translucent sheets	

WORKSHEET 2

To be completed by teachers	
Satisfactory	☐
Not satisfactory	☐

Student name: ______________________

Enrolment year: ______________________

Class code: ______________________

Competency name/Number: ______________________

Task: Working with your teacher/supervisor, refer to this text, your HB39:2015 and any other relevant resource to answer the following questions.

1 What are three considerations that you need to take into account to avoid double-handling materials?

____

____

____

2 Is the method for the manual handling of a 20-kg box on the ground the same as handling the same box up a ladder? Explain your answer.

____

____

____

3 When handling long sheets around the site and on the roof frame, should you place yourself upwind or downwind of the sheet? Why?

____

____

____

4 Explain why roof sheet alignment is so important when storing and locating packs.

____

____

____

5 You need to organise a crane to lift multiple packs of roof sheets onto the frame. What are at least five considerations that you need to take into account before ordering the machine?

____

____

____

6 An unavoidable delay in a building project has meant that you need to store roof sheets onsite. How should the sheets be stored until you are able to recommence work?

---

7 Before spot loading on a roof frame, what key questions should you have answered by the builder or engineer?

---

8 In the following photo, the direction the prevailing wind blows from is marked with an arrow. Place a cross on the end of the building from where you would commence laying the roof sheets.

Note: Questions 9–12 refer to information found on the following links or in the product literature supplied by your teacher or supervisor:

- LYSAGHT CUSTOM ORB®: **http://professionals.lysaght.com/products/custom-orb**
- LYSAGHT TRIMDEK®: **http://professionals.lysaght.com/products/trimdek**
- LYSAGHT KLIP-LOK® 406: **http://professionals.lysaght.com/products/klip-lok-406**

9 What is the effective cover of LYSAGHT CUSTOM ORB® corrugated roof sheets? How many sheets would you need to cover a roof that is 16 m wide?

10 You are installing a ZINCALUME® steel LYSAGHT TRIMDEK® clad roof on a commercial building. The product has a base metal thickness of 0.48 mm and the sheets are 9 m long. The engineer has confirmed that your roofing packs can be placed only on the frame above rafters located at 10 m intervals, and must not weigh more than 600 kg.

Using your product information brochure, determine the following:

i How many sheets of LYSAGHT TRIMDEK® would cover the interval between each 10-m spot load point?

ii What would be the mass of each pack?

11 The builder tells you that your proposed load point for a roof sheet pack of COLORBOND® steel LYSAGHT KLIP-LOK® 406 can take a weight of no more than 500 kg. You had wanted to place a pack of 30×10 m sheets on that spot. Would this be permissible?

12 List at least four tasks that may need to be completed when packing up after a delivery of materials.

SELECT AND INSTALL ROOF SHEETING AND WALL CLADDING

12

The chapter deals with the use of metallic and non-metallic profile sheeting to cover building roofs and walls. A building depends on well-installed roofs and walls for watertightness, wind protection, thermal protection and security. As the roof and external/internal walls comprise such a visible portion of the overall building, it is also important that the product selection and installation quality are aesthetically pleasing.

Before proceeding with this chapter it would be advisable to revise Part A of this text with particular attention to Chapter 7, while at the same time pre-read and cross reference Chapter 13 and Chapter 14.

Overview

To avoid unnecessary repetition, this chapter will have as its primary focus the installation of concealed-fixed roof and wall systems, though many aspects will equally apply to pierce-fixed roofing profiles. This chapter should therefore be reviewed in conjunction with Chapter 14, where pierce-fixed roofing is covered in more detail.

Types of roof and cladding systems

The wide range of profiled roofing/cladding sheet products available for roof and walling installation can be broadly separated into two specific categories based on the way they are fastened to the roof or wall. These are *pierce-fixed systems* and *concealed-fixed systems*.

Each of these systems provides designers, builders and roofing contractors with a range of characteristics and features that can be matched to specific project needs, such as the metallic profile roof and wall cladding shown in Figure 12.1.

FIGURE 12.1 Example of metallic profile roof and wall cladding

Pierce-fixed systems

Pierce-fixed sheets (see Figure 12.2) are secured to roofs and walls with fasteners that pass through the sheet and into the supporting purlin, batten or other structural member. When secured to roofs, the fasteners are pierced through the crest or rib of each profile. The sheets used for wall cladding are fastened through the pan of the profile.

Fasteners for pierce-fixed systems must be of the size, type and coating recommended by the manufacturer.

Corrugated sheet steel is a pierce-fixed roofing and cladding system and one of the most widely used profiles in Australia.

Concealed-fixed systems

At first glance, some roof sheet profiles do not seem to have any fasteners visible at all. These sheet profiles are grouped under the term concealed-fixed systems. Such sheeting and cladding profiles use a system of

FIGURE 12.2 Example of pierce-fixed roof sheets and wall cladding

concealed clips that are located beneath the sheets and fastened to the support purlin (see Figure 12.3). The clip is shaped in such a way as to lock into the rib profile of the sheet, effectively holding it in place without actually piercing it.

The installation of these systems requires the use of proprietary clips from the manufacturer and strict adherence to the recommended installation process. Concealed-fixed roofing and cladding products are especially popular for use on commercial and industrial buildings.

When installed correctly, concealed-fixed products offer a strong, watertight roofing and cladding solution.

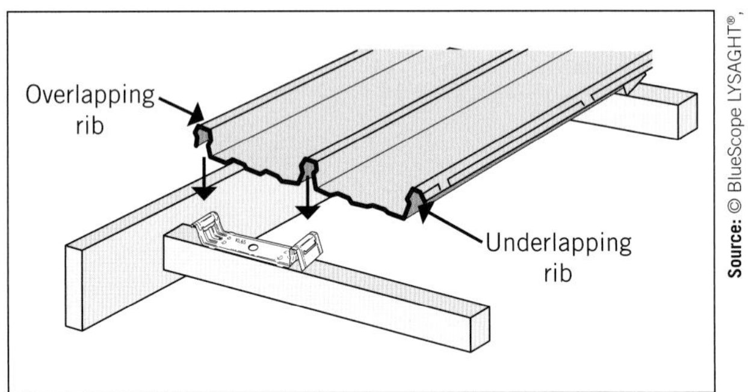

FIGURE 12.3 Concealed-fixed roof sheet profiles are secured with clips fastened beneath the sheet

Source: © BlueScope LYSAGHT®, www.lysaght.com

Profiles, effective cover and minimum pitch

Each roofing and walling product has a particular cross-sectional shape that forms a series of ridges and pans across the width of the sheet. This shape is known as the profile. While providing rigidity to the length, the profile may also be used to satisfy other requirements such as:

- aesthetic difference
- better holding capability
- different pitch applications
- suitability for curving
- ease and speed of installation.

Profiles are known by their respective trade names and are also grouped under particular profile categories. It is important that you become familiar with these category names as they are regularly referred to in codes, plans and specifications:

- corrugated
- trapezoidal
- close-pitched trapezoidal
- wide rib
- narrow rib.

FROM EXPERIENCE

On new projects or larger commercial installations, the plans and specifications will provide all the basic product selection and design information for the job. However, this is not the case for most re-roof and maintenance jobs. It is often the roof plumber alone who must advise the customer and select fit-for-purpose products for each situation. You must get in the habit of regularly researching all roofing products and materials.

Table 12.1 shows some examples of corresponding LYSAGHT® products.

Effective cover

The width of every roof and walling sheet has two measurements with which you need to be fully familiar. These are *actual width*, which is the measurement from one side of the sheet to the other, and *effective cover*, which is the measurement of the amount of cover provided on the roof less the lap on one side. This measurement enables you to work out how many sheets you need to cover a particular width of roof area.

A selection of roof sheet and wall cladding profiles comparing the difference between their respective actual widths and effective cover is shown in Table 12.1.

Minimum pitch

Not all roof sheet profiles are suitable for use at any pitch, and their suitability will vary according to the rib and lap height. Use of a sheet designed with a minimum pitch of 5° on a roof sloping at only 2° may result in water not flowing fast enough to the gutter and overflowing past the lap between each sheet.

You need to check the plans and, if necessary, use a pitch gauge to confirm the actual roof pitch in degrees. Compare this pitch with the products offered by the manufacturer and select a suitable profile. Table 12.2 shows some of the variations in minimum pitch between different profiles.

TABLE 12.1 Comparison between actual sheet width and effective cover of various roof sheet and wall cladding profiles

Roof sheet profiles	Cross-section	Actual width (mm)	Effective cover (mm)
KLIP-LOK® 406 (wide rib)		432	406
KLIP-LOK® 700 (wide rib)		710	700
LONGLINE® 305 (narrow rib)		310	305
Wall cladding profiles	**Cross-section**	**Actual width (mm)**	**Effective cover (mm)**
EASYCLAD® (wide rib)		330	300
MINI ORB® (corrugated)		841	820
TRIMWALL® (trapezoidal)		816	762

Source: © BlueScope LYSAGHT®, www.lysaght.com

TABLE 12.2 Minimum roof pitch of different profiles

Profile	Cross-section	Minimum roof pitch
KLIP-LOK® 406 (wide rib)		1°
KLIP-LOK® 700 (wide rib)		2° (BMT* 0.42 mm) 1° (BMT* 0.48 mm)
LONGLINE® 305 (narrow rib)		1°
SPANDEK® (close-pitched trapezoidal)		3°
CUSTOM ORB® (corrugated)		5°

*BMT = base metal thickness

Source: © BlueScope LYSAGHT®, www.lysaght.com

Source: Heath Bussell

Source: Heath Bussell

FIGURE 12.4 This new roof has been installed below the minimum allowed pitch of 1° degree

Translucent sheeting

Translucent sheets are designed to provide natural light into buildings, such as the shed shown in **Figure 12.5**. They are available in a range of products, materials and colour tints to suit various applications and roofing profiles. Always ensure that translucent sheets are selected in accordance with an Australian Standard suitable for the application. In most cases you will look for the following: AS 1562.3 'Design and installation of sheet roof and wall cladding – Plastic'.

FIGURE 12.5 Typical use of translucent sheeting on a commercial shed

Some specific installation requirements relate to the use of translucent sheeting and these are covered in more detail later in this chapter.

AS 1562.3 'DESIGN AND INSTALLATION OF SHEET ROOF AND WALL CLADDING – PLASTIC'

LEARNING TASK 12.1

1 The installation of concealed-fixed systems requires the use of proprietary clips. Explain the term *proprietary*.
2 The width of every roof and walling sheet has two measurements. How are these measurements described?

General planning and preparation

Efficient and accurate planning and preparation is central to all successful jobs. You need to source the following basic requirements and documentation:

- scope of work
- approved plans and specifications (see Figure 12.6)
- manufacturer's product information.

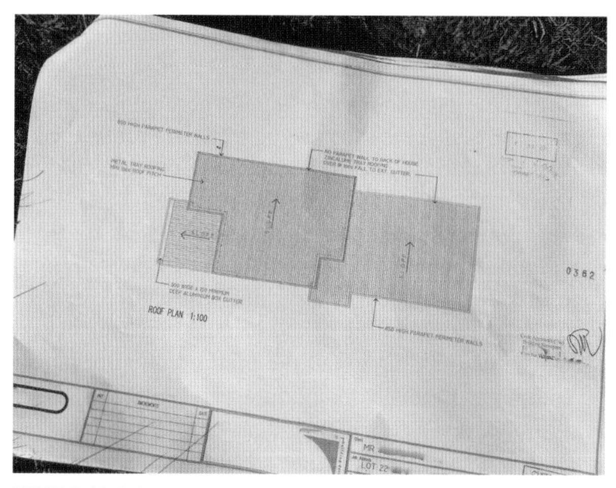

FIGURE 12.6 Approved plans showing plan view of roof layout

More detail on these steps is covered in Chapter 14. In this chapter regular reference will be made to the following two sources of manufacturer information:

- LYSAGHT® product range, available at http://www.lysaght.com > products > roofing and walling
- LYSAGHT® *Roofing & Walling Installation Manual*, available at http://www.lysaght.com > installation and maintenance > roofing and walling.

These and similar publications are available for free download from the relevant company websites.

Roofing requirements in areas subject to bushfires

In recent years it has been legislated that all new construction must be subject to a Bushfire Attack Level (BAL) assessment as part of the planning process. The objective of the evaluation is to determine the level of risk that each building is exposed to.

Based upon a combination of ground slope, vegetation type and proximity to the building, a property will be assigned one of the following BAL ratings:

- BAL – Low
- BAL – 12.5
- BAL – 19
- BAL – 29
- BAL – 40
- BAL – FZ (Flame Zone)

AS 3959:2018 FIGURE 1.1

Very specific design and construction requirements relate to the installation of ventilators from BAL 12.5 upwards. On new construction, the BAL rating of the property will be noted clearly in the plans and specifications, and it is your responsibility to ensure that all plumbing and roofing components are rated and installed in accordance with the relevant assigned category.

BAL rated components

Once you have determined the BAL rating of the property from the plans and specifications, you must ensure that you purchase components that are certified to meet or exceed that rating. For example, if the plan states that BAL 29 has been assigned to the property, a component meeting BAL 12.5 would not be suitable, while one meeting BAL 40 would be fine.

BAL rated installation

Not only must you ensure your components match the minimum BAL rating, you must also follow strict installation instructions. Particularly in relation to flashings, these include the use of mesh screens and non-combustible seals, including mineral wool blanket and fire rated silicones and polyurethane gap fillers.

A resource for roof-plumbing specific solutions matching each BAL rating can be found on the Bradford Insulation website: https://www.bradfordinsulation.com.au/

Insulation requirements

Increasingly, the installation of different forms of insulation is a significant part of a roof plumber's responsibility. For construction approval to be granted, building codes and standards require minimum thermal efficiency standards to be achieved. To meet these standards, the roof plumber will regularly be called on to install various insulation products as a normal part of most installations.

The role of insulation

Insulation products and their appropriate use and installation are a major feature of all modern construction.

GREEN TIP

The correct specification and installation of building insulation is a prime example of how the plumbing industry is engaged in achieving sustainable construction solutions for the community.

Insulation products can fulfil a number of roles:

- reducing heat flow into a building
- reducing heat loss from a building
- preventing the formation of condensation in roof and wall spaces
- suppressing external and internal noise (e.g. rain, traffic, work activities)
- reducing overall load and running costs from heating and cooling appliances
- acting as a fire retardant
- reducing the occurrence of thermally induced metallic creeping and popping sounds caused by roof sheets moving against supports.

HB39:2015 SECTION 6.2 'INSULATION OF STEEL ROOFING'

The performance and efficiency of these products will depend to a large extent on adherence to all thermal and BAL installation guidelines.

R-value

The performance capability of any insulation product is quoted as its 'R-value'. The R-value is a measure of thermal resistance and is related to the thickness of the insulation material and its inherent thermal characteristics. The higher the figure, the better is its potential performance.

For example, a reflective foil blanket of 60-mm nominal thickness might have a base R-value of R1.3, while a thicker 100-mm blanket may be rated at R2.5. Typical wall insulation might be R1.5, while ceiling batts could be R3.5 or higher.

HB39:2015 SECTION 6.4 'RESISTANCE TO HEAT TRANSFER'

Types of insulation products

Roof and wall cladding installation will require that you become familiar with a range of insulation products. These products may be manufactured from fibreglass, rockwool, natural wool and loose fill. Most commonly you will encounter the following:

- reflective foil laminate (RFL) – a basic reflective and vapour membrane installed beneath roof sheets
- foil-faced blanket – a bulk insulation blanket installed directly between the roof sheets and the purlins, which incorporates a layer of attached RFL
- ceiling batts – plain bulk insulation installed directly above the ceiling lining within the roof space
- wall batts – bulk insulation installed between wall studs
- mineral wool rolls – BAL rated 300-mm wide rolls of non-combustible glasswool fibre installed at gutter lines, valley gutters and under ridge capping.

HB39:2015 TABLE 6.4(B) 'INSULATION TYPES AND APPLICABILITY

Specification of insulation

The approved plans and specifications will detail the minimum R-value required for the insulation in the walls and ceilings. It will be your responsibility to source a product that will meet these minimum requirements and also be compatible with the roofing system that you are using.

Increased thermal performance requirements for buildings have led to higher R-values and, therefore, thicker insulation material. This will have an impact on the selection and use of roof system fasteners, as follows:

- For *pierce-fixed* roofing systems where an insulation blanket is specified for use beneath the roof sheets, you will need to accommodate the thickness of this extra material by selecting longer screws.
- For *concealed-fixed* roofing systems, the compression of the insulation blanket between the roof sheet, fasteners and purlins will reduce the overall thermal rating of the installation. This may cause the building to fail the minimum performance measures. In this case, manufacturers can provide proprietary spacers that sit between the purlins and roof sheet and provide a gap that prevents insulation compression.

More detail on use of insulation is included in the installation section of this chapter.

Site inspection

Never assume that a site or building frame will be exactly as detailed in the plans and specifications.

Changes, inaccuracies and product variations are common, and you always need to check and confirm all details by regularly inspecting the site. Examples of what you will be doing during a site visit include:

- consulting with the site foreman, WHS representative, builder and/or owner
- checking the site access
- checking the storage areas
- checking for overhead electrical hazards
- checking roof access requirements
- confirming all measurements
- confirming the intended direction of sheet laying.

Safety

All aspects of the job must be planned for and carried out with constant observance of all safety considerations. Review Chapter 3 for more details.

Task sequencing

Task sequencing involves the efficient staging of each aspect of the installation while also fitting in with the activities of other trades and the overall project timeframes. More details on this are included in Chapter 14.

Tools and equipment

The tools and equipment required for roof sheeting and wall cladding include all of those items listed in Chapter 9. In addition to these tools, you will need to use various types of pan folders when installing concealed-fixed roof sheet profiles. Examples of these folders are shown in Figure 12.7.

Folders are used to turn up the high-end of the roof sheet pan to prevent water from being blown back past the end of the sheet. They are also used to turn down the lower end of the sheet to prevent water from running back under the sheet due to capillary action. You may choose to fabricate your own folders or purchase them directly from the manufacturer. Application of these tools will be covered in more detail later in this chapter.

LEARNING TASK 12.2

1 Why is it important to keep up with legislation?
2 The performance capability of any insulation product is quoted as what type of value?

Product selection

Product selection has never been simpler, and a range of sources of information are available that include:

- knowledgeable suppliers' representatives
- manufacturers' sales and product representatives
- hardcopy product information
- internet searches for installation and product guides.

The number of products that you need to select will vary, depending on the type and size of the work, but there is always some level of decision making required. This section gives some examples of how to select certain installation products.

Selection of roof sheets

Roof sheets must be selected to match the specifications and meet all requirements related to pitch, profile, fastener and support system. For example, information to support your concealed-fixed roof and cladding installations is available via simple internet searches, as shown in Figure 12.8.

Selection of fasteners

The choice of fasteners is determined by accessing two primary sources of information:

- the roof sheet manufacturer's recommendations for size, gauge and coating
- the fastener manufacturer's recommendations for products meeting these size, gauge and coating requirements.

In relation to concealed-fixed roof sheet and wall cladding requirements, you will need to select fasteners that are suitable for securing the clips to metal purlins

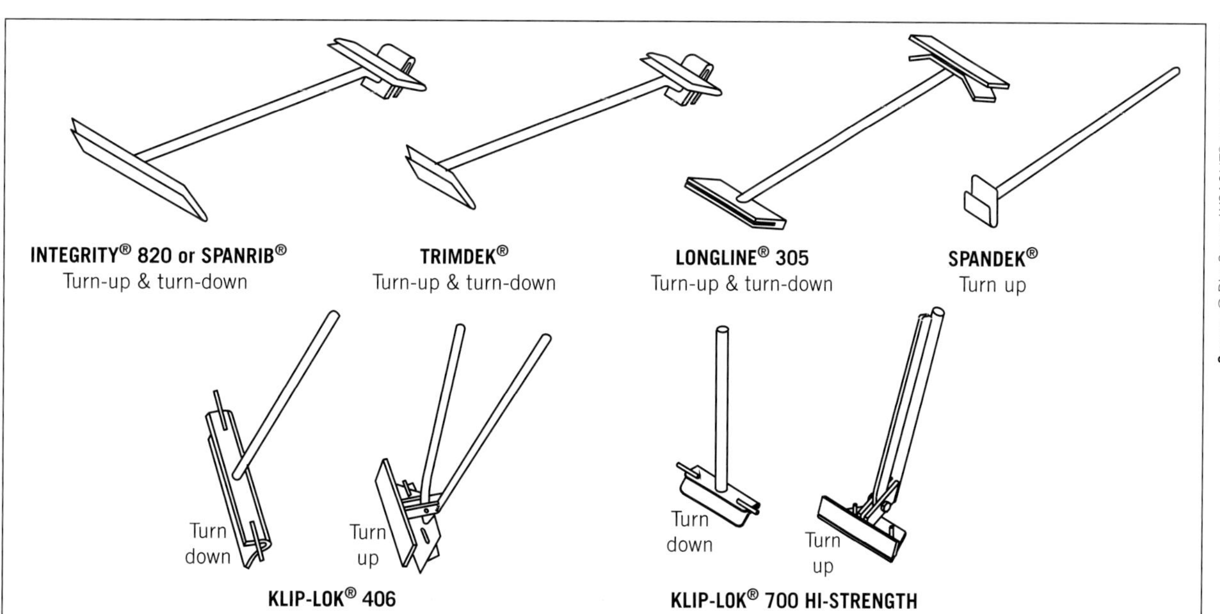

FIGURE 12.7 A selection of folders is available to match specific sheet profiles.

Ask the question:
Which roof sheet profiles match the specification requirements?

Internet search
Type keywords or phrase: installing concealed-fixed cladding
Select documents from search result

[PDF] roofing & walling installation manual - Lysaght
www.lysaght.com/sites/default/.../LysaghtRoofingWallingInstallationManualJul2015.p...
8.4 Pierce-fixing on valleys (for walling only). 36. 8.5 Pierce-fixing on side-laps. 36. 9.0 **Installing concealed-fixed cladding**. 37. 9.1 Installing KLIP-LOK® roofs.

Download document/s
In this instance a pdf file from the manufacturer
Lysaght is chosen:

9. Installing concealed-fixed cladding

Concealed-fixing is the method of fixing sheets using fasteners which do not pass through the sheet. Instead, the cladding is held in place with clips. This is different from the alternative method called pierce-fixing (Chapter 8). The method of fixing you use is determined by the cladding profile you are using.

Concealed-fixing is used for:

- KLIP-LOK 700 HI-STRENGTH®
- KLIP-LOK CLASSIC® 700
- KLIP-LOK® 406
- EASYCLAD®
- LONGLINE 305®

VERY STEEP PITCHES

To prevent concealed-fixed cladding from sliding downward in the fixing clips, on very steep pitches, you should pierce-fix through each sheet under the flashing or capping, along the top of the sheets, but not less than 25mm from the ends of a sheet.

9.1 INSTALLING KLIP-LOK ROOFS

Use the same general procedure described in Section 8.1 (General installation procedure). However, at the start of installing KLIP-LOK® 406, KLIP-LOK CLASSIC® 700 (KL-CLASSIC) or KLIP-LOK 700 HI-STRENGTH® (KL-700HS) a row of clips is fixed to the supports before the first sheet is located over them and locked in position.

CLIPS

Figure 9.1
KLIP-LOK® clips.
Direction of laying
KLIP-LOK CLASSIC® clip (For KLIP-LOK CLASSIC® 700)
KL-700 HS fixing clip for KLIP-LOK 700 HI-STRENGTH®
Shortened first clip

Answer
Review of the document reveals the following profiles suitable for concealed-fixed installation

- KLIP-LOK® 700
- KLIP-LOK® 406
- LONGLINE® 305
- EASYCLAD®

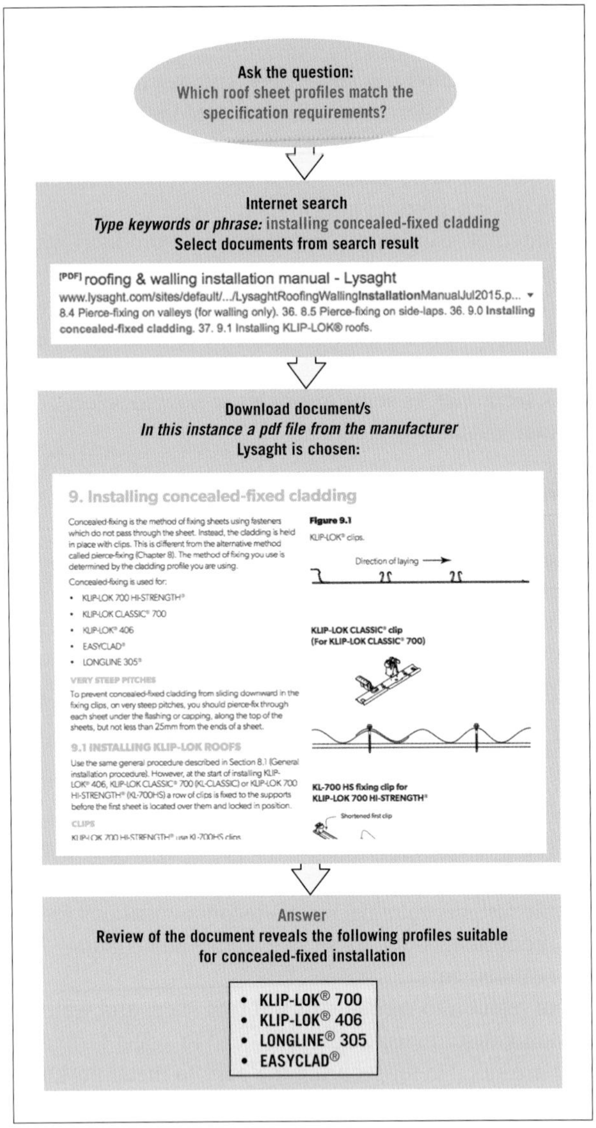

FIGURE 12.8 Using an internet search to find information for concealed-fixed roof and cladding installations

or timber battens, subject to the requirements of your job. To find a solution, you need to determine the type of fastener recommended by the roof/wall product manufacturer and then find a matching product from the fastener manufacturer.

The example in Figure 12.9 demonstrates some steps in the process of fastener selection.

LEARNING TASK 12.3

1 What are two sources of information that can help with the selection of product?
2 What are the three steps required when selecting the correct fasteners from manufacturers' documentation?

Quantity calculations

Quantity calculations are based on the comparison between product characteristics or application and the dimensions of the building in which the installation is to take place. The examples in this section are based on a single commercial building that is 30 m long with a nominal roof pitch width of 10 m.

Safety mesh quantity

Many jobs will require the installation of safety mesh (see Figure 12.10) over the full roof area to provide fall protection to workers on the roof. The approved plans and specifications will detail the type of mesh, wire gauge and coating required, and the relevant Australian Standard. Your job will be to order the correct amount.

EXAMPLE 12.1

Calculate number of rolls of safety mesh

You need to work out how much safety mesh to order for a skillion roof area that measures 30 m with a pitched width of 10 m from bottom purlin to top edge purlin. The job specification document says the mesh must meet the following requirements:

- conform to AS/NZS 4389
- 300 mm × 150 mm mesh dimension
- 2 mm wire gauge

You research these requirements and find a suitable product that has a full width of 1.8 m and is run down the roof pitch with an overlap of each other run by 150 mm. The product literature says that a roll of 1.8-m wire will cover 82.5 m² when used with a single 150-mm overlap.

$$\text{Number of 50 m rolls} = \frac{\text{Roof in square metres}}{\text{Square metres per roll}}$$

$$= \frac{30 \times 10}{82.5}$$

$$= \frac{300}{82.5}$$

$$= 3.64$$

$$= 4 \text{ rolls (answer is rounded up to the next highest)}$$

Some suppliers may be able to work out the quantity for you and cut mesh to length. You would need to have the overall dimensions of the job ready when ordering.

Safety mesh may also be determined in simple lineal metres. This is based on the effective cover of each run divided into the overall width of the roof. You must remember to allow at least a metre extra for each

LYSAGHT

Fasteners: Screw to steel–AS3566 Class 4 Roof & Wall cladding - non-cyclonic

2 Select the type of support material.

Roofing - non-cyclonic

LYSAGHT PROFILE	CREST			PAN		
	Single & Total Lapped ≥0.55 up to 1.0mm BMT	Single ≥1.0mm up to 3.0mm BMT	Lapped ≥1.0mm up to 1.9mm BMT	Single & Total Lapped ≥0.55 up to 1.0mm BMT	Single ≥1.0mm up to 3.0mm BMT	Lapped ≥1.0mm up to 1.9mm BMT
CUSTOM ORB	M6-11X50 ROOFZIPS	#12-14X35 HH HG/TG M5.5-14X39 AUTOTEKS	#12-14X35 HH HG/TG M5.5-14X39 AUTOTEKS	#10-16X16 HH M5-16X25 DH M6-11x25 ROOFZIPS	#10-16X16 HH M5-16X25 DH	#10-16X16 HH
CUSTOM BLUE ORB	M6-11X50 ROOFZIPS	#12-14X35 HH HG/TG M5.5-14X39 AUTOTEKS	#12-14X35 HH HG/TG M5.5-14X39 AUTOTEKS	#10-16X16 HH M5-16X25 DH M6-11x25 ROOFZIPS	#10-16X16 HH M5-16X25 DH	#10-16X16 HH
TRIMDEK	M6-11X50 ROOFZIPS	#12-14X45 HH HG/TG #12-14X55 HH HG/TG M5.5-14X50 AUTOTEKS	#12-14X45 HH HG/TG #12-14X55 HH HG/TG M5.5-14X50 AUTOTEKS	#10-16X16 HH M5-16X25 DH M6-11x25 ROOFZIPS	#10-16X16 HH M5-16X25 DH	#10-16X16 HH
SPANDEK	M6-11X50 ROOFZIPS	#12-14X45 HH HG/TG #12-14X55 HH HG/TG M5.5-14X50 AUTOTEKS	#12-14X45 HH HG/TG #12-14X55 HH HG/TG M5.5-14X50 AUTOTEKS	#10-16X16 HH M5-16X25 DH M6-11x25 ROOFZIPS	#10-16X16 HH M5-16X25 DH	#10-16X16 HH
INTEGRITY 820	M6-11X65 ROOFZIPS	#12-14X68 HH HG/TG	#12-14X68 HH HG/TG	N/A	N/A	N/A
FLATDEK	N/A	N/A	N/A	#10-16X16 HH #12-14x20 HH M5-16X25 DH M6-11x25 ROOFZIPS	#10-16X16 HH #12-14x20 HH M5-16X25 DH	#10-16X16 HH #12-14x20 HH
KL406 CLIP FIXED	N/A	N/A	N/A	#10-16X16 WH #10-16X22 WH	#10-16X16 WH #10-16X22 WH	#10-16X16 WH #10-16X22 WH
KL700HS CLIP FIXED	N/A	N/A	N/A	M6-11x25 ROOFZIPS #15-15x25 HH	#12-14x20 HH #12-14x30 HH	#12-14x20 HH #12-14x30 HH
KL700 CLASSIC CLIP FIXED	N/A	N/A				
LONGLINE 305	N/A	N/A				

1 Select the profile

3 Read off the fastener type, then match to the manufacturer

Wafer Head

Application
For jobs needing a low profile head
- metal deck clips - electrical fittings - signs - sheetmetal.

Gauge	T.P.I.	Length	Pack	Carton	Part Number	Pack Type	Finish
8	18	12	1000	8	6-311-0009-2	Bulk	◇
10	16	16	100	50	1-311-9021-8	Hang	◆
			100	50	1-311-9150-5TPMP	Trade	◆
			1000	8	6-311-0150-6	Bulk	◇
			1000	8	6-311-0150-6MP	Bulk	◆
10	24	16	1000	8	6-311-0021-3MP	Bulk	◆
10	16	22	50	50	1-311-7025-3	Hang	◇
			1000	6	6-311-0179-9	Bulk	◇
			1000	6	6-311-0179-9MP	Bulk	◆
10	24	22	1000	6	6-311-0025-6MP	Bulk	◆
10	16	30	1000	3	6-311-0535-7MP	Bulk	◆
10	24	40	1000	3	6-311-0312-9MP	Bulk	◆

Source: © BlueScope LYSAGHT®, www.lysaght.com and courtesy of Buildex®.

FIGURE 12.9 The process of choosing fasteners that match the manufacturer's requirements

FIGURE 12.10 Rolls of safety mesh for covering a roof area

run in order to tie off on the top and bottom purlins. Your teacher and/or supervisor will be able to advise you on what is the common practice in your area.

Insulation quantity

The specifications for this job also require the installation of R1.3 foil-backed insulation blanket. This product is available with a nominal coverage of 18 m^2 per roll. How many rolls do you need to cover this skillion roof?

EXAMPLE 12.2

Calculate number of rolls to cover a skillion roof

$$\text{Number of insulation rolls} = \frac{\text{Roof in square metres}}{\text{Square metres per roll}}$$
$$= \frac{30 \times 10}{18}$$
$$= \frac{300}{18}$$
$$= 16.7$$
$$= 17 \text{ rolls (answer is rounded up to the next highest)}$$

Roof sheets and wall cladding quantity

Once you know the effective cover measurement, you can calculate the number of sheets required for either roof sheeting or a wall cladding job. This is determined by dividing the total width of the roof or wall by the effective cover of the particular sheet profile:

$$\text{Number of sheets required} = \frac{\text{Roof total width}}{\text{Effective cover}}$$

EXAMPLE 12.3

Calculate number of sheets

You need to sheet a commercial skillion roof that has an overall width of 30 m. The specifications require the use of LYSAGHT KLIP-LOK® 406 sheets for the job.

$$\text{Number of sheets required} = \frac{\text{Roof total width}}{\text{Effective cover}}$$
$$= \frac{30}{0.406}$$
$$= 73.89$$
$$= 74 \text{ sheets (rounded up to whole sheet)}$$

You will notice that in this calculation the final answer has been rounded up to the next whole sheet. Always round up your answer to ensure that you have full coverage of the roof.

Manufacturers also provide tables for each profile that have been pre-calculated to match simple roof widths to numbers of sheets (see Figure 12.11). You simply round up to the next full metre and then read off the number of sheets. However, if the roof is wider than indicated or precise measurements are required, you will need to work it out manually.

Clips and fasteners quantity

Clips for concealed-fixed roof sheeting and wall cladding will generally come in boxes containing a set number of clips. Some suppliers will be able to give you the exact quantity and others will supply you with the required number of clips rounded up to the next box size. Each profile will have a specified number of clips and fasteners per purlin.

EXAMPLE 12.4

Determine the number of clips and fasteners

In the example of our 30 m × 10 m skillion shed, there are 8 purlin runs across the width of the building. A review of the manufacturer's website reveals that a sheet of Klip-Lok® 406 requires 1 clip per support. You have previously worked out that there are 74 sheets required for the job. Therefore, the number of clips is determined as follows:

Number of clips = number of purlins × number of sheets
= 8 × 74
= 592

Once you know the number of clips, it is then easy to determine the number of fasteners needed. The product literature for LYSAGHT KLIP-LOK® 406 also states that the required fasteners for screwing clips onto metal purlins would be 2 × #10–16 × 16 wafer head screws per clip.

Number of screws = number of clips × 2
= 592 × 2
= 1184 screws (round up to the next size box/packet)

Screws may come in cartons of 1000, with smaller quantities made up with small packets. In this case you might get 1 carton of 1000 plus 2 packets of 100 screws each.

COMPLETE WORKSHEET 1

LYSAGHT KLIP-LOK® 406 sheet coverage																				
Width of roof (m)	3	4	5	6	7	8	9	10	11	12	13	14	15	16	17	18	19	20	30	40	50
Number of sheets	8	10	13	15	18	20	23	25	28	30	33	35	37	40	42	45	47	50	74	99	124

FIGURE 12.11 Example of a manufacturer's table for calculating sheet coverage

Source: © BlueScope LYSAGHT®, www.lysaght.com

Installation of a concealed-fixed roof

In this section you will be introduced to the basic procedures required to lay a standard concealed-fixed roof. The installation can be broken up into a number of stages. These are:

1. Install rainwater goods.
2. Lay safety mesh.
3. Lay insulation.
4. Fix sheets.
5. Install flashings.

Install rainwater goods

Rainwater goods include valley gutters, eaves and box gutters, and they will generally need to be installed before the roof sheets. More details dealing with the installation of these parts of the roof system are covered in Chapter 10.

Lay safety mesh

Where required, the installation of safety mesh is vital for the safety of all roof workers during the installation and also, at a later date, for maintenance workers where translucent sheeting is used. Safety mesh must be manufactured in compliance with AS/NZS 4389 'Roof Safety Mesh', and it is generally found as a 2 mm gauge wire with a rectangular 300 mm × 150 mm mesh. The longitudinal wires (down the roof pitch) are 150 mm apart and the transverse wires (across width of roof) are 300 mm apart.

AS/NZS 4389 'ROOF SAFETY MESH

Depending upon site height access options, you will work with an assistant with each person standing on scaffolding on either side of the building. The roll of mesh will be rolled out from one side to the other with the use of rope/s tied to each roll. The common installation procedure is as follows:

1. *Roll out mesh.* On one side of the building, the wire is prepared for rolling out from the scaffolding platform. If possible, insert a pipe through the middle of the wire, enabling it to be rolled out more easily.

   Ensure that the transverse wires are uppermost and the longitudinal wires are facing downwards. A rope is secured to the wire and then carefully pulled over the pitch of the roof by a worker on the other side of the building. Where this is not possible, the first roll of wire will need to be placed by a roof worker who is equipped with full fall arrest equipment.

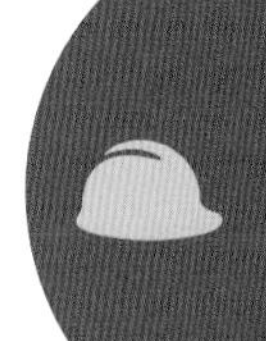
Caution! When rolling out and cutting mesh, be especially careful of wire 'spring-back'. You could easily lose an eye or suffer severe puncture wounds if wire is released suddenly. Wear all appropriate personal protective equipment and communicate with workmates so that everyone knows what their role is.

2. *Secure mesh.* The end of the wire should be secured to the lowest purlin in accordance with the relevant code and the manufacturer's requirements (see Figure 12.12). Once the end is tied off, the other end can be pulled to length and cut. Pull it tight but allow a degree of natural sag to avoid difficulty when fixing sheets later. Allow at least 300 mm extra to enable the wire to be tied off correctly.
3. *Lap next length.* Position the next length of mesh so that it overlaps the previous length by the amount detailed in the specifications or the manufacturer's instructions. This will be at least a single lap of 150 mm (one mesh section). A double lap would be 300 mm (two mesh sections). Secure the side laps according to the manufacturer's requirements. This may be as follows:
   a. For purlin spacing less than 1700 mm, use a 2-mm ring fastened or twitched at maximum 3 m centres.
   b. For purlin spacing greater than 1700 mm, use a 2-mm ring fastened at maximum 900 mm centres.

HB39:2015 SECTION 4.4 'SAFETY MESH'

Wherever possible, try to ensure that the entire roof work area or stage is fully wired before beginning the rest of the installation.

Lay insulation

The following details relate to the use of insulation blanket and reflective foil laminate (RFL).

Insulation blanket

The basic considerations for installing insulation blanket are as follows:

- Using a stringline, check that all purlins are in plane so that all sheets can be fixed with an even grade. Once the insulation is on you will not be able to do this.
- Lay the blanket directly on top of the safety mesh with RFL facing downwards as seen in Figure 12.14. (Note, however, that blanket used in tropical areas

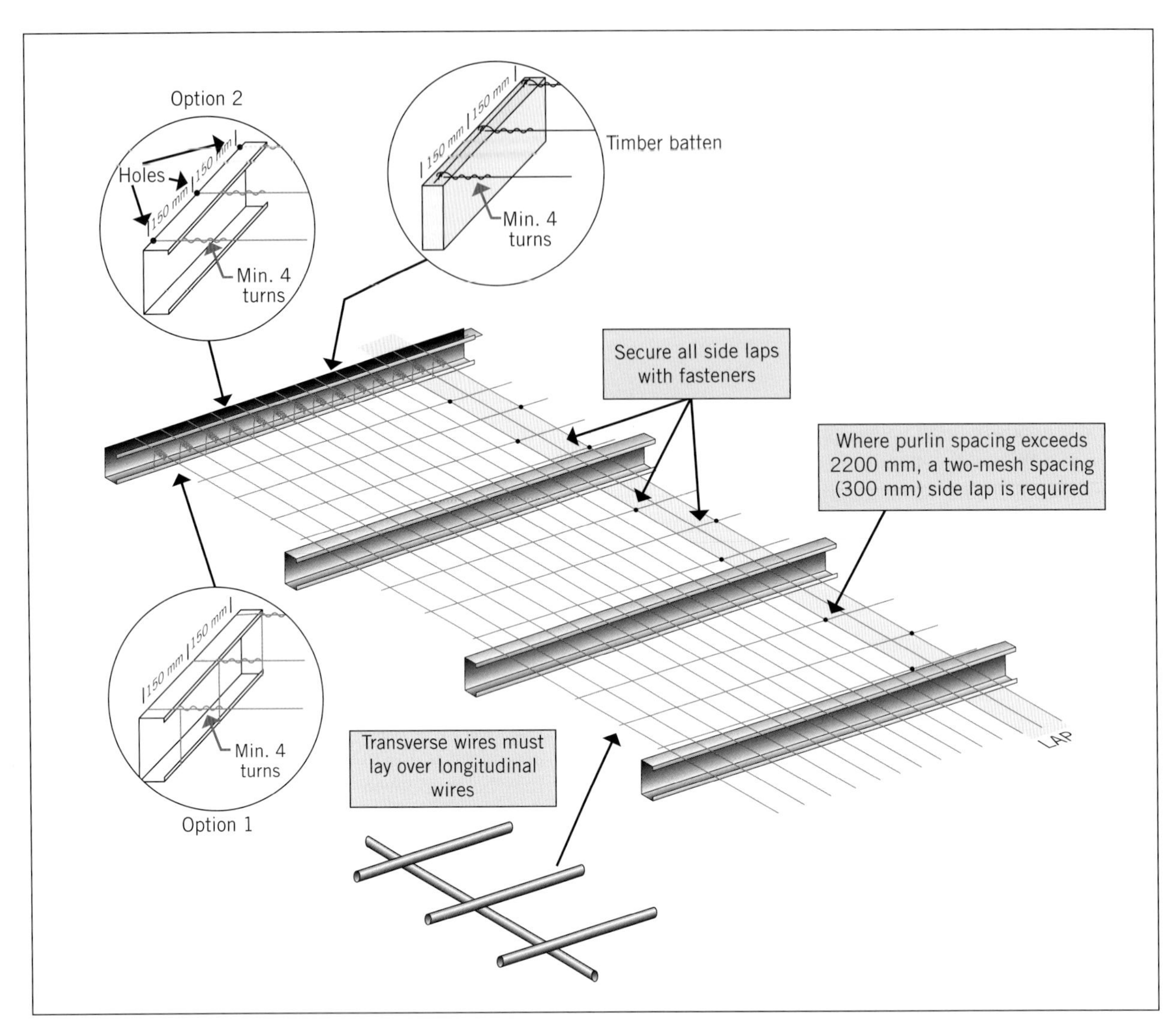

FIGURE 12.12 Lay safety mesh in strict accordance with manufacturer's instructions and the AS/NZS 4389

may need to be installed with the RFL facing up or, alternatively, installed in a conventional manner but with a second layer of RFL laid on top. Check with the manufacturer.)

- Lay the blanket down the pitch of the roof and overlap the preceding roll with the 150-mm side lap (subject to purlin spacing).
- Seal each lap with the adhesive tape specified by the manufacturer.
- Fit spacers where necessary.
- Lay only as much insulation as you can soon cover with sheets.
- Prevent exposed insulation from becoming wet.
- Prevent exposed insulation from flogging in a high wind.
- Review any BAL rating requirements that relate to the laying of the insulation, particularly at the end of sheets

Spacers

For some installations, the fixings are screwed into the purlins with the insulation blanket sandwiched between. On some buildings this may be fine, but where the insulation is squashed, it will not perform to standard and the thermal performance rating of the building may be compromised.

Provisions within the National Construction Code now require that for some commercial building classes and climate regions any insulation blanket used between purlins and the roof sheets must maintain its thickness and position.

To achieve this requirement, some insulation manufacturers supply proprietary spacers that provide a fixed space between the sheets and the purlins for the insulation to maintain its full thickness and thermal properties. The standard roof sheet clips can be fixed to the top of the spacer before or after installation.

Not all roof profile clips are suitable for each brand or type of spacer. The insulation manufacturer will provide advice on the availability and use of each product to suit your requirements. More information on insulation spacers can be found at the following link: Fletcher Insulation – http://www.insulation.com.au > products > roofing > tapes and accessories.

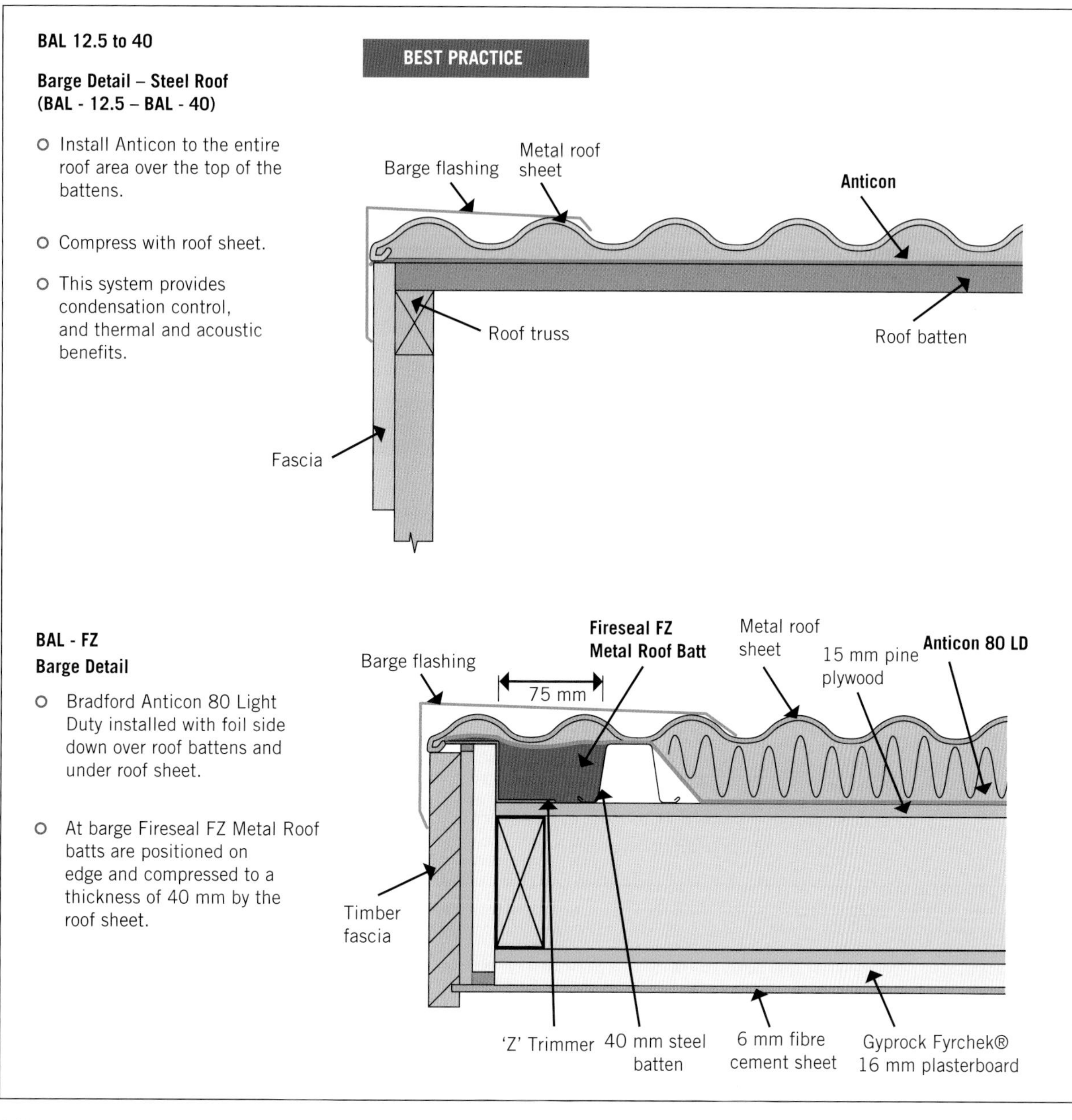

FIGURE 12.13 Comparison of installation requirements of barge detail at BAL 40 and BAL FZ

FIGURE 12.14 Insulation blanket and safety mesh

Reflective foil laminate

Reflective foil laminate should be laid in a similar manner to insulation blanket, and the following points should be considered:

- Ensure that RFL is sagged between each batten. The air gap between the membrane and the roof sheet is necessary for effective condensation control (see Figure 12.15). Do not pull the RFL tight.
- Lap each length by at least 150 mm.
- Seal each lap and end-join with the adhesive tape recommended by the manufacturer.
- Do not allow excessive RFL to hang into the gutter.
- Prevent RFL from flogging in the wind.

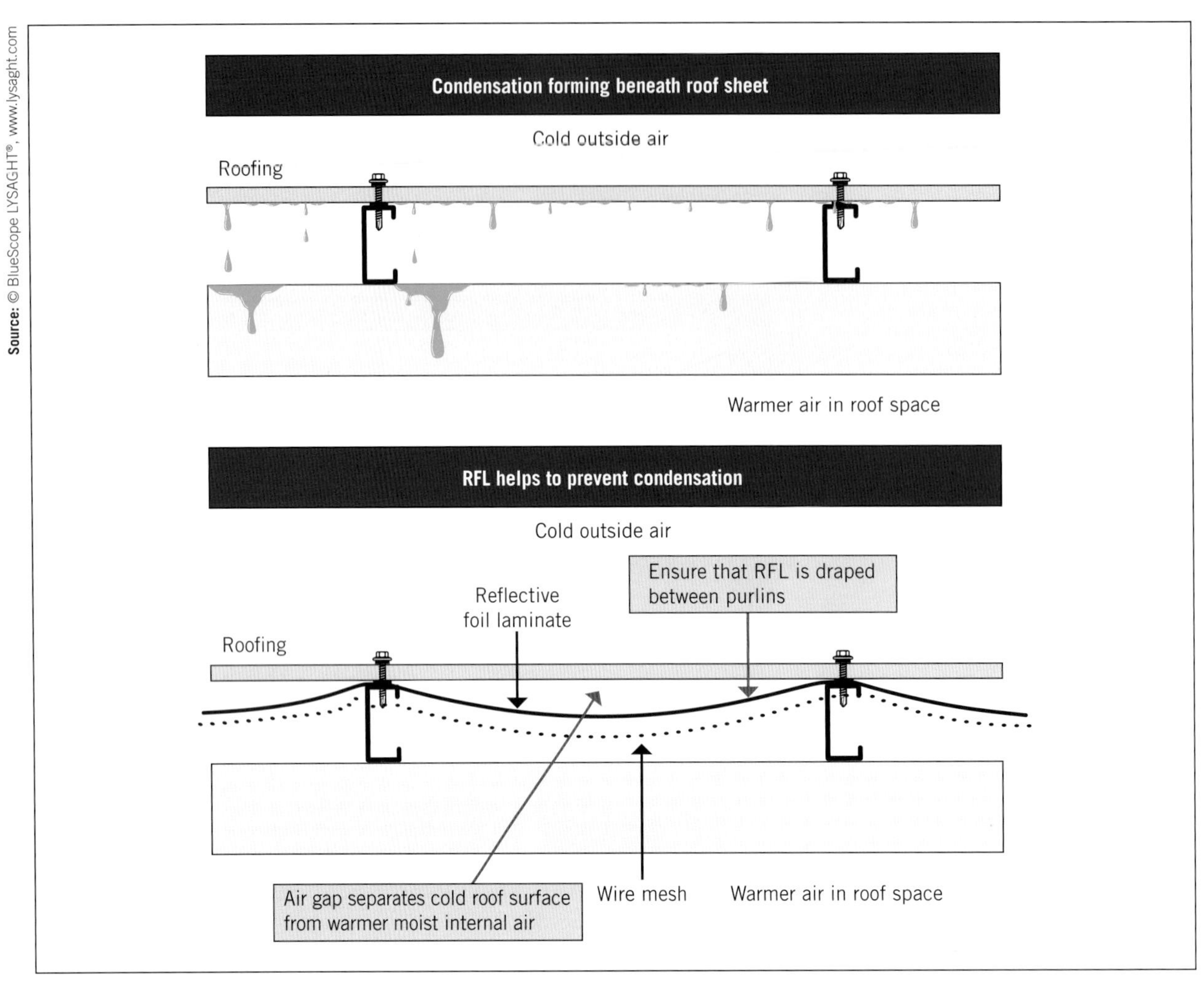

FIGURE 12.15 Reduce condensation by ensuring there is a sufficient gap between RFL and the roof sheet.

Fix sheets

The following steps relate to the fixing of the roof sheets to the structure.

Mark out roof for sheet alignment

The first sheet that you secure on any roof must be correctly aligned and square with the frame. Failure to check and mark out your sheet positions can result in all of the remaining sheets being out of square to the fascia or gable ends. At times, this is so bad that it is visible from the ground, and it is evidence of extremely poor installation.

Although it is possible to correct misaligned pierce-fixed sheets gradually by 'creeping' them till the measurement is correct, this is more difficult and sometimes impossible with concealed-fixed sheets. It is better to spend more time in getting the set-out correct in the first place than trying to do it halfway into the installation.

On a *gable and skillion roof* you need to check that the end rafters are square with the fascia.

On a *hip roof* the first sheet is normally laid to intersect with the highest point of the hip. In this instance you need to use the fascia as a base line and establish a square set-out to the top of the hip.

The simplest way to confirm alignment is to triangulate the area. Review Chapter 14 and refer to Figure 12.16 to see how the 3:4:5 rule is applied.

Setting the first sheet (concealed-fixed)

There are some minor variations between different profiles and products, but the following procedure is a general guide:

1 Use the 3:4:5 rule to confirm that the fascia is square with both end-wall/barge boards at each end of the roof area. If not, determine the amount of error so that the first sheet can be positioned to accommodate the inaccuracy. This is not normally a problem on a new construction, but older buildings will often have considerable variation in the frame.
2 Position and secure the first clip on the lowest purlin just above the gutter, ensuring that it is in the correct position relative to the end-wall and sheet width (see Figure 12.17).
3 Use a stringline up the pitch of the roof to position and secure the clips on all other purlins.
4 Fit the first sheet into position.

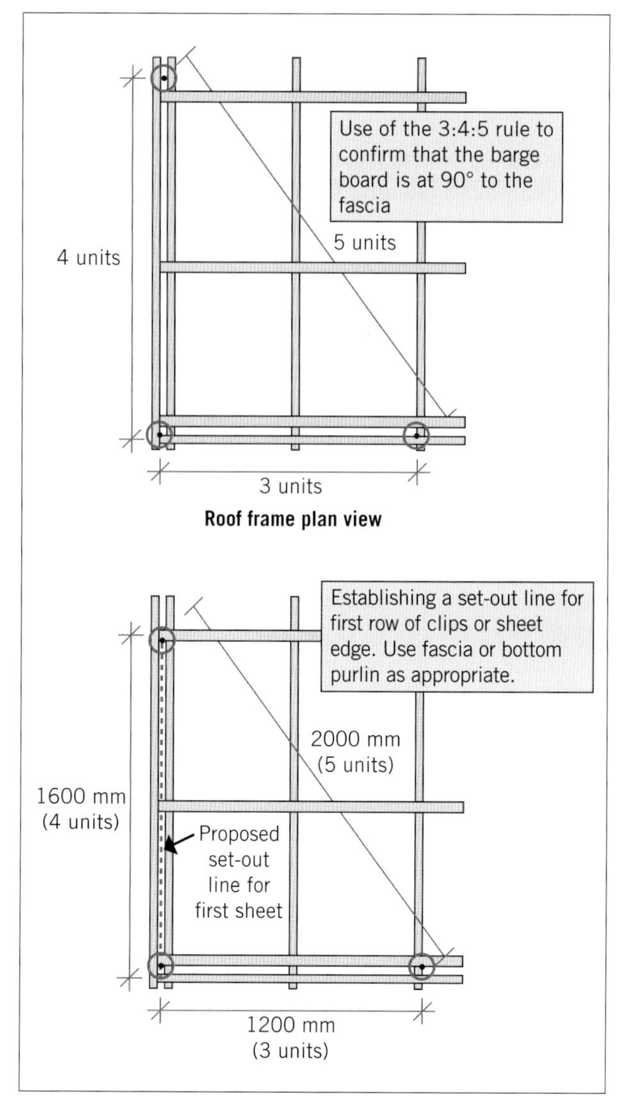

FIGURE 12.16 Applying the 3:4:5 rule

Turn up trays

Turn up all trays in the manner recommended by the sheet manufacturer in order to prevent wind-driven rain from passing the end of the sheet (see Figure 12.18).

Begin laying sheets

Once the first sheet is secured in place, you are ready to begin laying the remaining sheets in accordance with the manufacturer's instructions. Concealed-fixed sheets generally do not drift too far out of alignment, but it is still good practice to check them regularly, just in case.

Regular alignment checks

You will have identified any inaccuracies between the fascia line and the end-wall/barge board when you were setting out the alignment of the first sheet. Your next goal is to ensure that when you have finished sheeting that particular section, the sheets are still parallel with the end-wall/barge board and perpendicular to the fascia line. The following method is one way of ensuring that this is achieved (see Figure 12.19):

1 Once the first sheet is in position and parallel to the far end-wall/barge board, measure back from the completion point and mark off a series of regular checkpoints on the top and bottom purlins.
2 The distance between each mark should be short enough so that one person can make a quick check while the sheets are being laid without having to waste time measuring all the way to the end of the run. A 1–2-m interval is usually fine.
3 On a regular basis, simply measure between the edge of the last sheet and the next purlin mark to check for any misalignment.

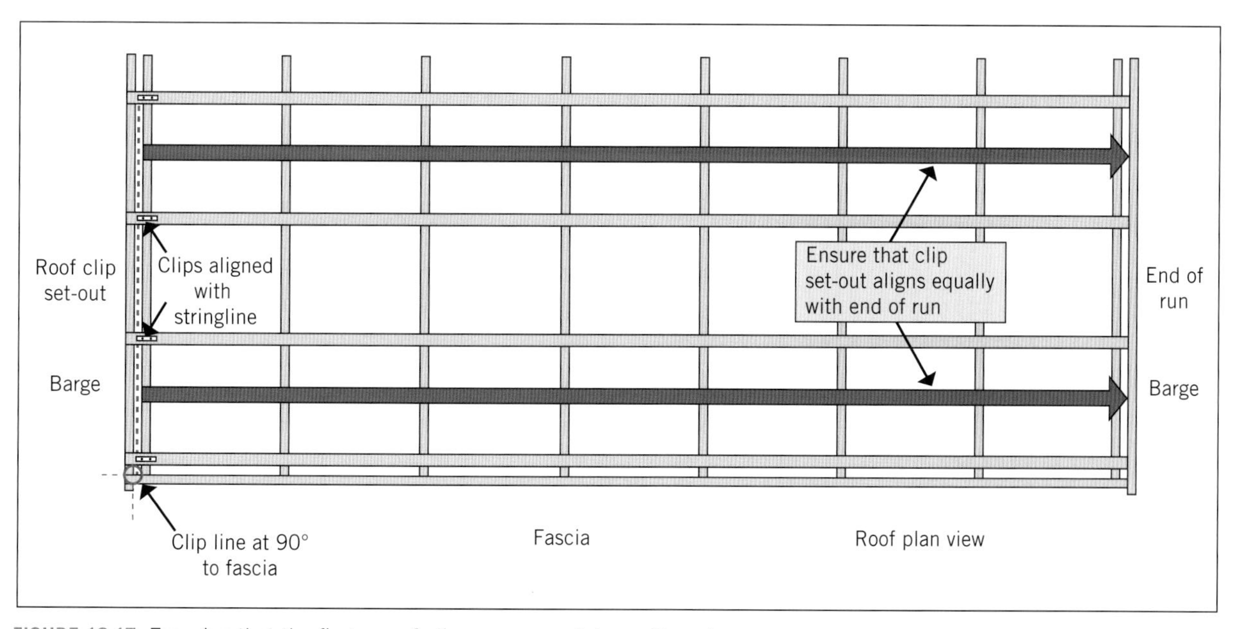

FIGURE 12.17 Ensuring that the first run of clips are accurately positioned

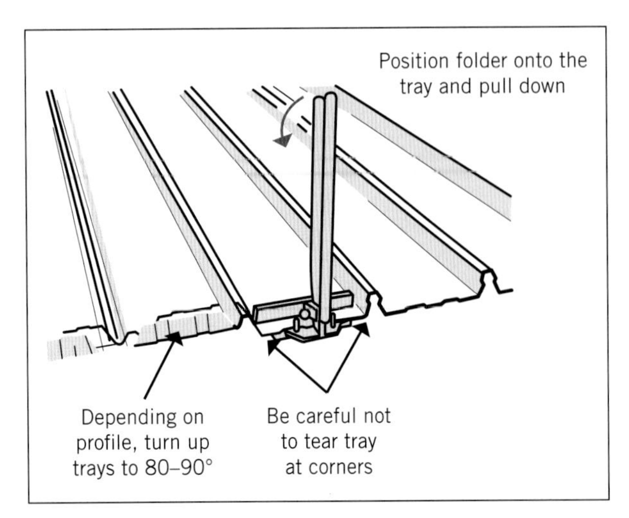

FIGURE 12.18 Procedure for weathering the trays of each sheet

Can't I just measure the sheet overhang past the fascia?

Some roof workers will measure each side of the overhang of the sheet past the fascia in an attempt to maintain sheet alignment. This method can't be recommended, particularly when long sheets are being installed. A 1-mm discrepancy at the bottom may translate into a 10-mm misalignment at the top of the sheet. You can only maintain true alignment by *regularly checking both the bottom and top of the sheet against a perpendicular base line from the fascia.*

Correction of misaligned sheets

Where concealed-fixed sheets have drifted out of alignment, the method you use to correct this problem will depend on the type of clipping system employed. Consult the manufacturer's installation instruction for the recommended procedure.

Regardless of the method used, it is most important to ensure that nothing is done that will compromise the holding capability of the clips and that the correction is gradual and made over a number of sheets.

What to do where the frame is badly out of square

Where an older building frame is significantly out of square and it is not practicable to fix the problem, you will need to determine how you will accommodate the misalignment in the lay of the sheets. In such a case, the following should be considered:

- Regardless of the solution you choose, ensure that your work still remains watertight and compliant.
- Try to 'take up' the misalignment by allowing the sheets to 'saw-tooth' into the gutter where they can be later trimmed, or at the very least kept from view.
- Ensure that the intersection between flashings and roof sheets is visually even wherever possible, so that the job still looks aesthetically sound.

Install clips and fasten to purlins

As a wide range of clipping systems are used for concealed-fixed profiles, there is no single procedure for carrying out the installation. However, you need to be aware of the following points:

- Read the manufacturer's instructions on how to install the clips.
- Identify the directional arrow so that the clip is facing the right way.
- Use only the type and number of fasteners recommended by the manufacturer.
- Ensure positive engagement of the clip to the previous sheet/clip and check that it does not skew out of line when fastening.

Figure 12.20 shows an example of clips used for concealed-fixed sheeting.

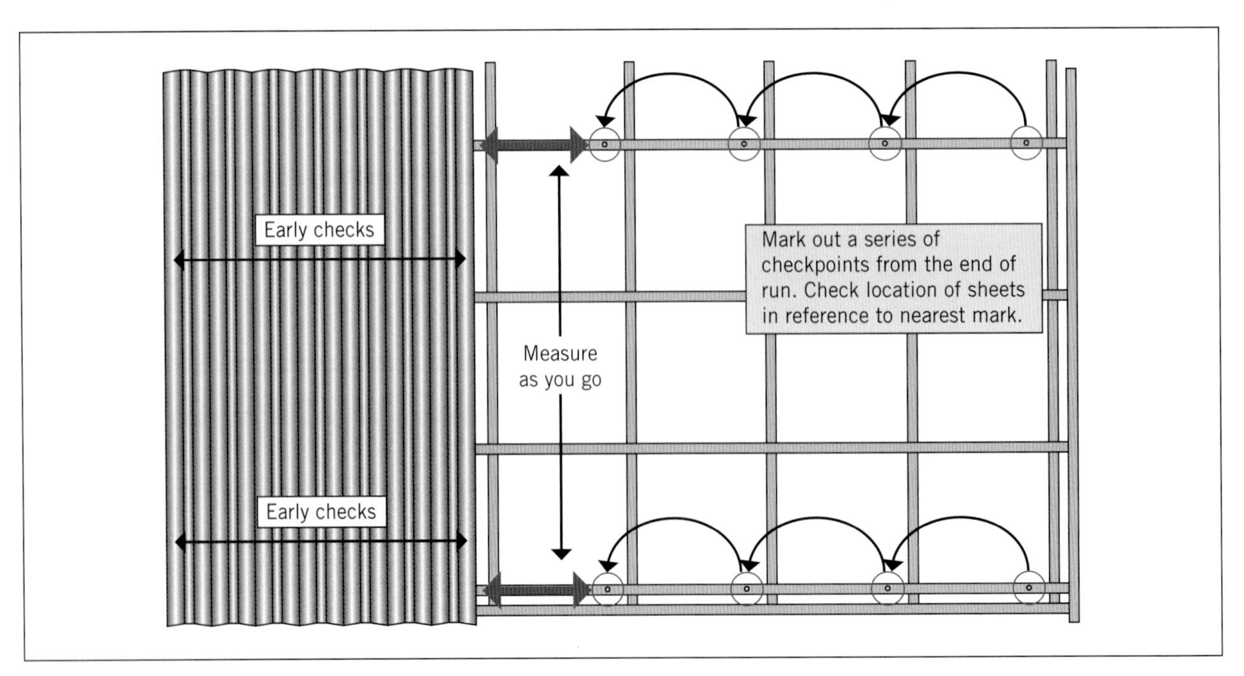

FIGURE 12.19 Simple checks to keep your sheets running true

FIGURE 12.20 Clips used for concealed-fixed sheeting

Fit sheets to clips

The method of fitting the sheet to the clips will be determined by the particular product requirements. In general, the following procedure applies (see Figure 12.21):

1. Lay the sheet into position.
2. Check that the sheet overhangs into the gutter by a minimum of 50 mm.
3. Working from the edge closest to the previously laid sheet, walk the new sheet onto each clip by applying vertical foot pressure to the top of the rib at each purlin location. Do not apply pressure in the tray, as this will not allow the rib to engage correctly.
4. Listen and feel for positive engagement of the sheet onto the clips. Use a rubber mallet if necessary to help the sheet lock into position.
5. If the profile requires some form of mechanical locking, use the tool as instructed to dimple the ribs onto the clips.

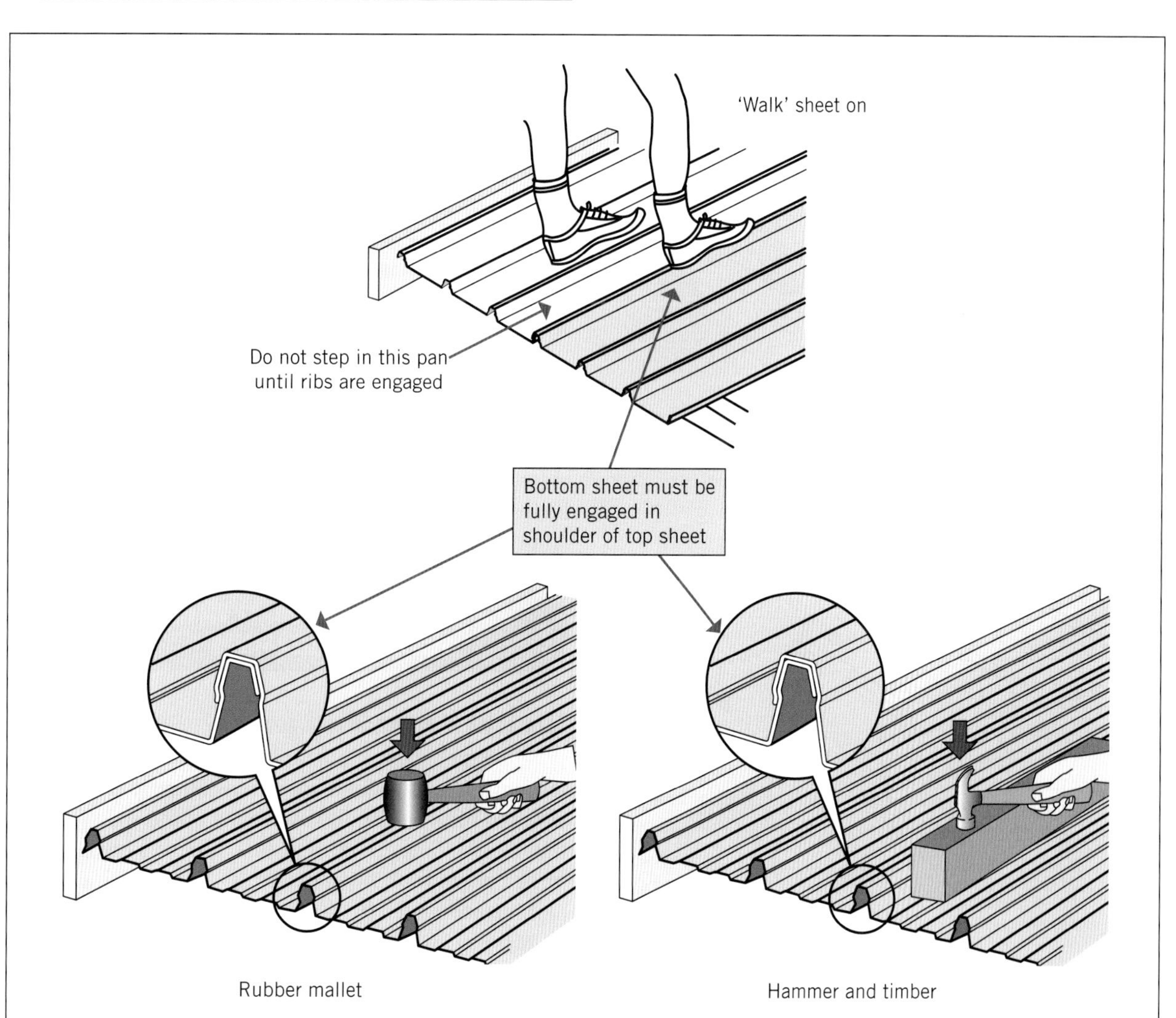

Source: © BlueScope LYSAGHT®, www.lysaght.com and Basic Training Manual 12-3 *Roof Plumbing, Deck Fixing and Materials*, Australian Government Publishing Service, 1981.

FIGURE 12.21 Ensuring positive engagement of sheet over each clip

6 Where concealed-fixed roof sheets have been installed on a steep pitched roof, it is possible for normal thermal expansion and contraction to cause the sheets gradually to slip down the roof. To prevent this, secure the very top of each sheet with a fastener located under the flashing.

The last sheet

In many instances, the last sheet will not be a perfect fit with the end of the run. Always terminate the run with a full rib upstand and cover it with an appropriate flashing.

Where the final space to be covered is *more than one pan in width*, cut the sheet along its length on the far side of the rib and secure this half-length with a single row of clips.

Where the space to be covered is *less than one pan in width*, the sheet may need to be secured by a series of 'over clips'. This is then covered with a flashing over the remaining space and the final rib.

Figure 12.22 gives some examples of how the final sheet can be secured.

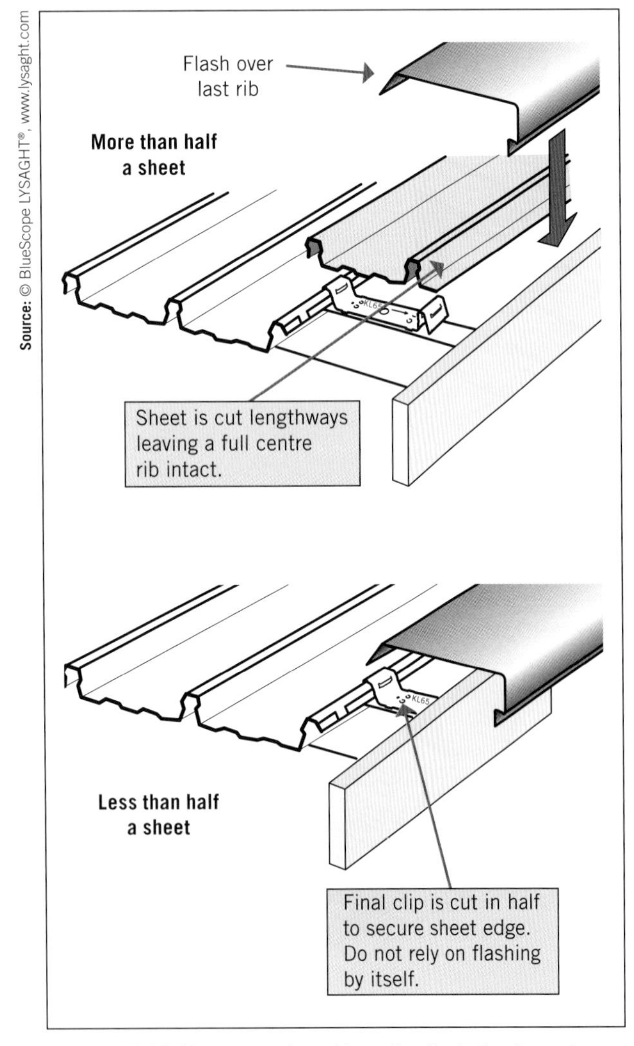

FIGURE 12.22 Two examples of how the final sheet can be secured

Take care when walking on roof sheets

When moving around the roof, prevent the kind of unnecessary damage shown in Figure 12.23 by observing the following simple guidelines:

- Tread carefully and distribute your weight evenly.
- Wear soft-soled boots and regularly check for trapped swarf, trimmings and stones caught in the tread.
- When moving across the roof, always try to walk on or close to the roof purlins.
- When moving up and down the roof on close-ribbed and corrugated roofing, walk on at least two ribs at once.
- When moving up and down on pan-type roofing, walk directly in the pans.

FIGURE 12.23 Unnecessary damage caused by careless movement on a roof

Rib closure

Rib closure is normally required to prevent access by birds, prevent bushfire embers from entering roof space and act as a further method of thermal control. Depending on the product and the application, metallic stops, mineral wool or closed-cell foam filler strips can be installed before or after the roof sheets have been laid (see Figure 12.24 for an example).

Turn down sheet ends into gutter

Where a roof sheet is laid at less than 10°, surface tension, capillary action and wind can act to cause water to run back on the underside of the sheet and into the building. To prevent this, the pans must be turned down by approximately 20° using a folder or tool recommended by the manufacturer (see Figure 12.25). Be careful not to tear the sheet when doing so.

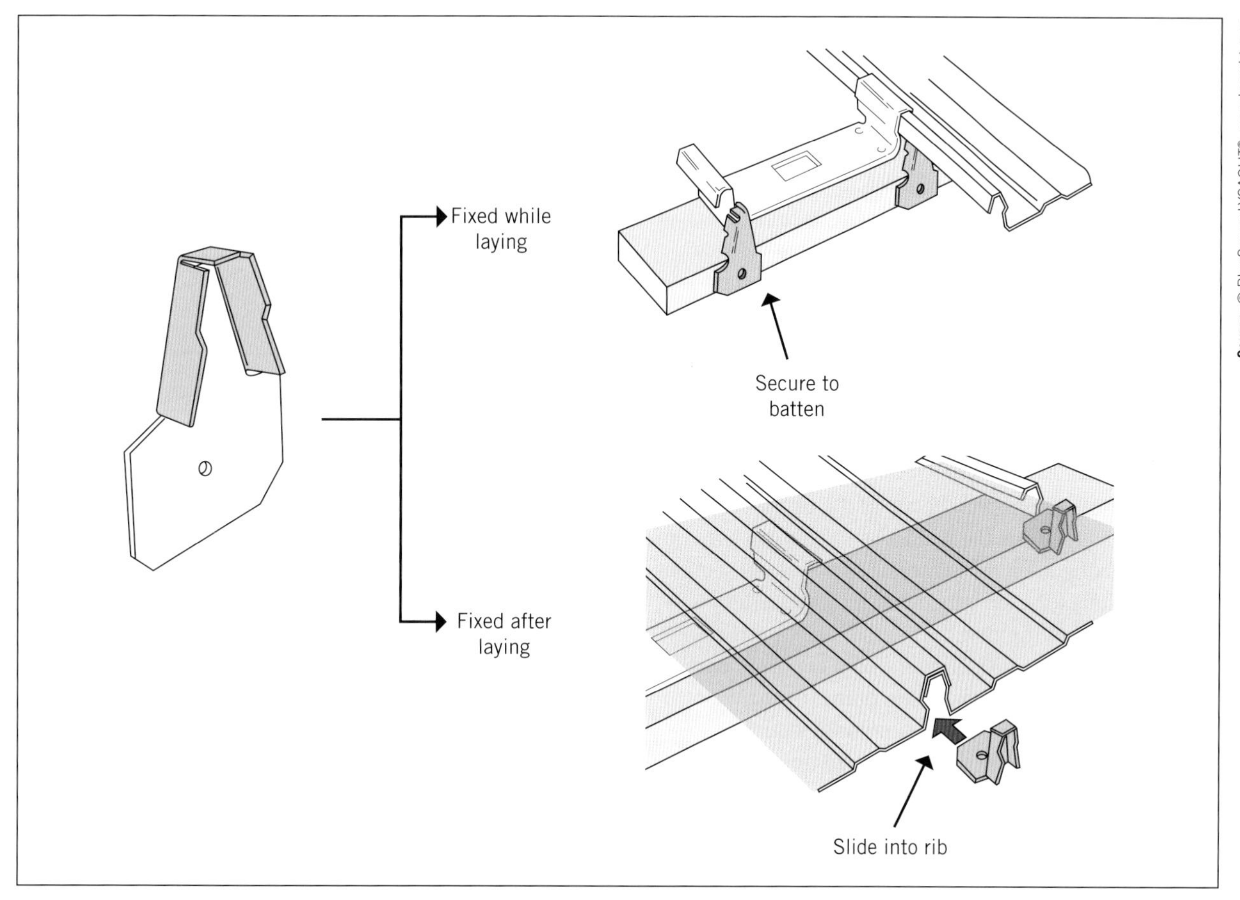

FIGURE 12.24 Example of how rib stop-ends can be installed before or after sheet installation

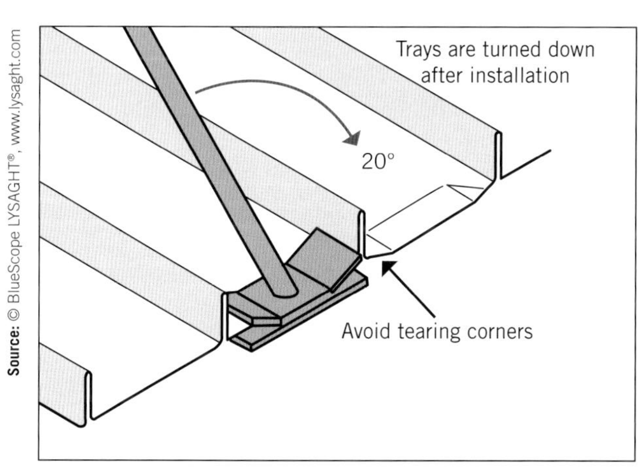

FIGURE 12.25 Turning down the sheets into the gutter

For roof sheet installation at a pitch of 5° or less you may need to also trim off the corner of the underlying lapped sheet to prevent capillary action from drawing water into the lap. Check manufacturer's instructions for relevant requirements for each profile.

Install flashings

With the roof sheets in place, you are ready to install all over-flashings to make the roof watertight. Flashing is covered in more detail in Chapter 13 and Chapter 15.

LEARNING TASK 12.4

1 What does the acronym RFL stand for?
2 Which code requires that, in some commercial building classes and climate regions, any insulation blanket used between purlins and the roof sheets must maintain its thickness and position?
3 Why would you regularly check both the bottom and top of the sheet against a perpendicular base line from the fascia?

Installation of wall cladding

In relation to the setting out and installation of metal profile sheet wall cladding, the general principles for installing roof sheets also apply to the use of wall cladding products. Details relating to specific wall cladding and flashing requirements are covered below.

Building wrap

In general practice, the builder of any new construction will wrap the frame in some form of RFL. During the installation of the cladding flashing and sheets, you will need to ensure that any RFL is lapped over any window and foot-mould flashings so that water condensation will run out and away from the building.

Any tears or cuts in the RFL should be sealed with the adhesive tape recommended by the manufacturer.

Setting position of first sheet

With any required under-flashings in place (see later in this section), the setting of the first sheet is as important for a wall as it is for roof sheets. The following points apply:

- For standard horizontal installations, the first sheet will normally be located at the lowest point of the wall so that each subsequent sheet is lapped over the one below it. However, where a series of windows are positioned at the same level along the wall, you may choose to position the top of the first sheet immediately below the window sill. This can save time by removing the need to measure and cut out sections in the sheet for each window (see Figure 12.26).
- When setting the first horizontal sheet position, ensure that the measurement between each end of the sheet and the completion point (near the soffit or the top of the wall) are nominally equal.
- Where sheets are to be fitted on an angle across the wall (45° is common), you must set a chalk line across the longest section of the wall (normally mid-wall) and use that as the base line from which your first sheet is set.

FIGURE 12.26 Fix the first sheet immediately below the window sill line.

Fastening of sheets

Sheets should be fastened to the wall using the method specified by the manufacturer. Some profiles are designed for a concealed-fixed clipping system, whereas other types are secured using pierce-fixed methods. In most instances, pierce-fixed wall cladding is secured through the pan/valley of the sheets (see Figure 12.27).

Rib closure

Ensure that the ends of sheet profiles are sealed off in the manner recommended by the manufacturer or any BAL installation detail. Closed-cell foam strips matching the shape of the particular profile are most commonly used. These strips help to prevent access by vermin and wind-driven rain and can assist in reducing thermal losses through the profile openings.

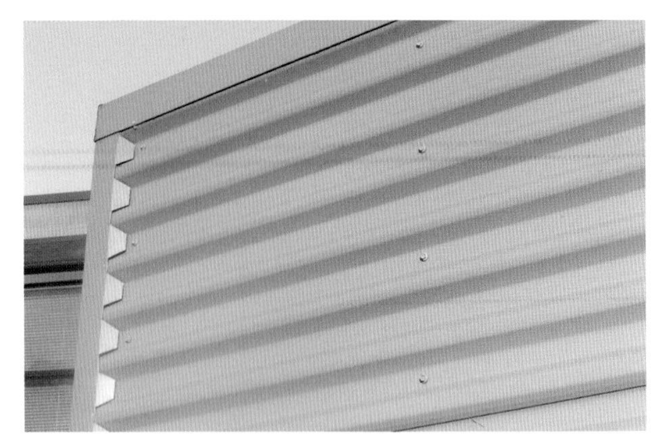

FIGURE 12.27 Wall cladding secured through the pans of the sheet and corner flashing notched into the profile

Flashings for wall and ceiling cladding

The use of sheet-metal cladding products on external walls and ceilings requires flashing solutions that will ensure that the building remains watertight when subject to wind-driven rain. Furthermore, in those areas where bushfires are a threat, flashings must also be installed in reference to any required BAL ratings.

Cladding can be applied horizontally, vertically and sometimes diagonally. This and the choice of cladding profile have an influence on the way your flashings are applied. The manufacturers of cladding sheets also provide a range of flashings to suit general applications. Where you need to have a special flashing made to suit different requirements, ensure that:

- the flashing provides sufficient width of cover/protection
- the internal edge is turned back on an angle where wind-driven rain is likely
- a safety fold is provided on the external edge in order to offer protection from cuts and to give rigidity to the flashing.

Window frames and other wall openings

One of the main problem areas for external cladding is around window frames. Although windows often come with flexible flashing around the frame, this is not always satisfactory and you may need to provide additional sheet-metal protection to do the job properly. You also need to ensure that any attached flashing and the building thermal/vapour wrap are correctly lapped in the direction of flow around your flashings. Where gaps are large between the flashing and the cladding profile, use specially made closed-cell foam fillers to close the space.

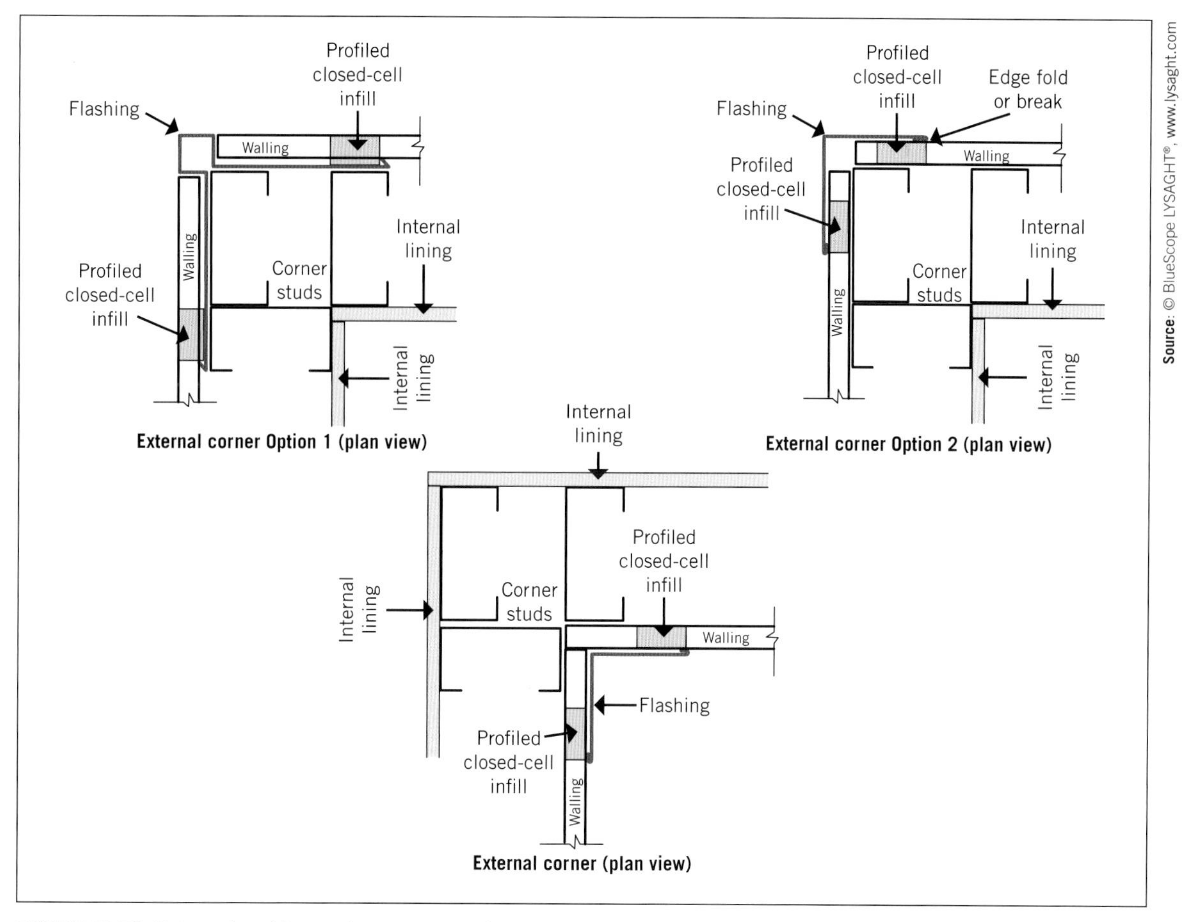

FIGURE 12.28 External and internal corners – horizontal cladding

The manufacturers of sheet-metal cladding products generally can provide good installation guidelines for you to follow.

A selection of possible flashing solutions is given below.

Corner flashing

Figures 12.28 and 12.29 show some options for external and internal corner flashing. Corner flashing should be notched into wide pan sheets that are fastened horizontally on the wall.

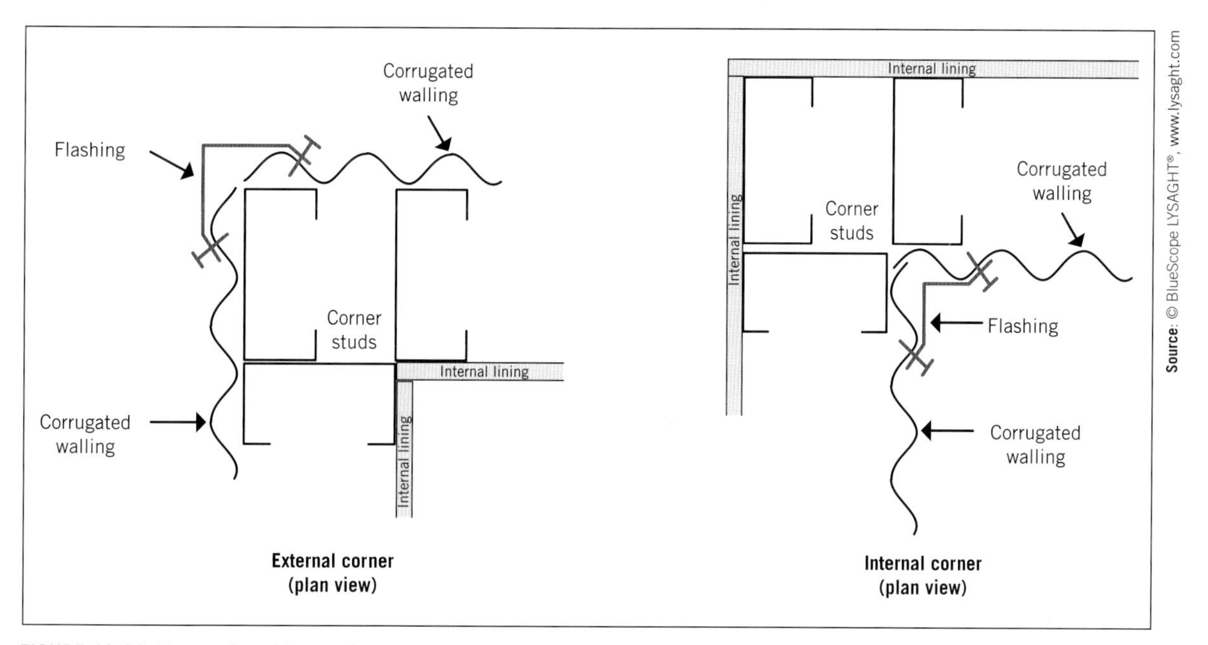

FIGURE 12.29 External and internal corners – vertical cladding

Window sill and header flashing

The term 'window sill' refers to the bottom horizontal section of the window frame. The upper horizontal section is called the 'header'. Correct flashing here is vital, and Figures 12.30 and 12.31 show some options.

Window jambs

The term 'window jamb' refers to the sides of the windows. The diagrams in Figure 12.32 are shown in both plan view and elevation to demonstrate the installation requirements for window jambs. One of the most important installation requirements is that any water that gets around the flashing must be drained *to the outside* of the cladding just below the sill.

Foot moulds

The foot moulding or bottom trim is the flashing used to finish off the bottom edge of the cladding in a neat, weatherproof and safe manner (see Figure 12.33).

Trim and cover flashing

Trims and covers ensure that the ends and joins in cladding are effectively flashed so that sharp edges are covered and any weather kept out (see Figure 12.34).

The author suggests that this chapter be cross-referenced as needed with the latest version of the 'Lysaght Architectural Detailing & Flashing Manual', which can be found at this link: http://www.professionals.lysaght.com > resources > user guides and manuals

LEARNING TASK 12.5

1. What is another term used to identify the bottom trim?
2. What are the two horizontal sections of a window that need flashing?
3. What are the vertical sections of a window that need flashing?

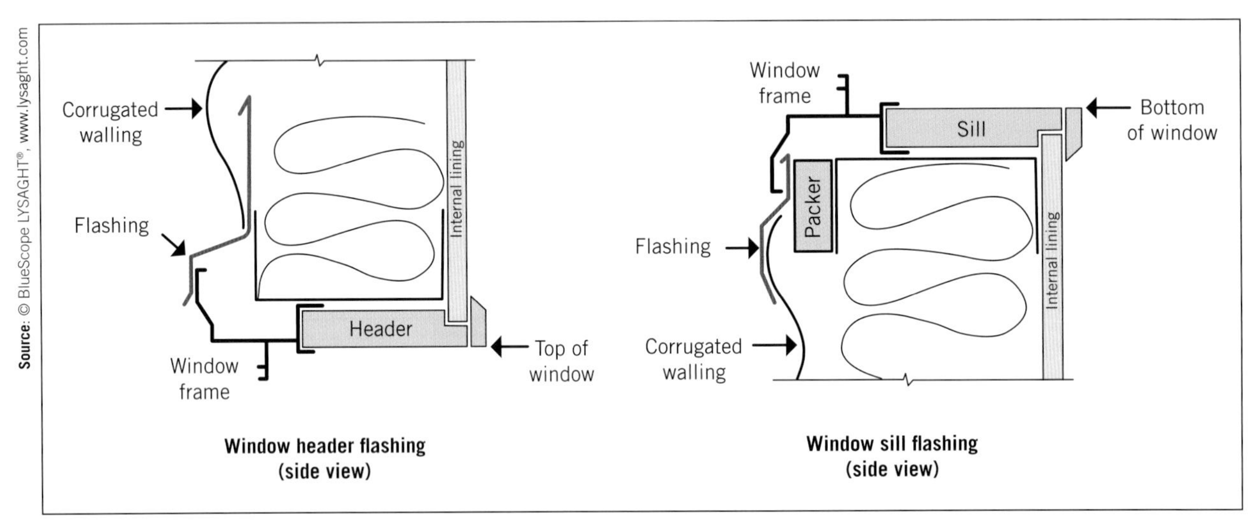

FIGURE 12.30 Window sill and header – horizontal cladding

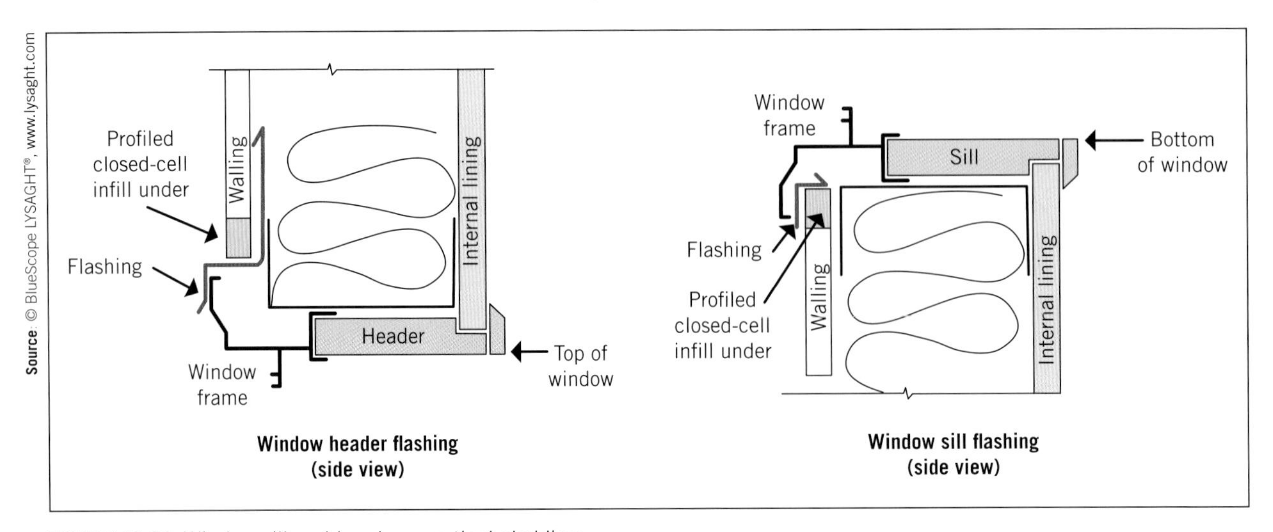

FIGURE 12.31 Window sill and header – vertical cladding

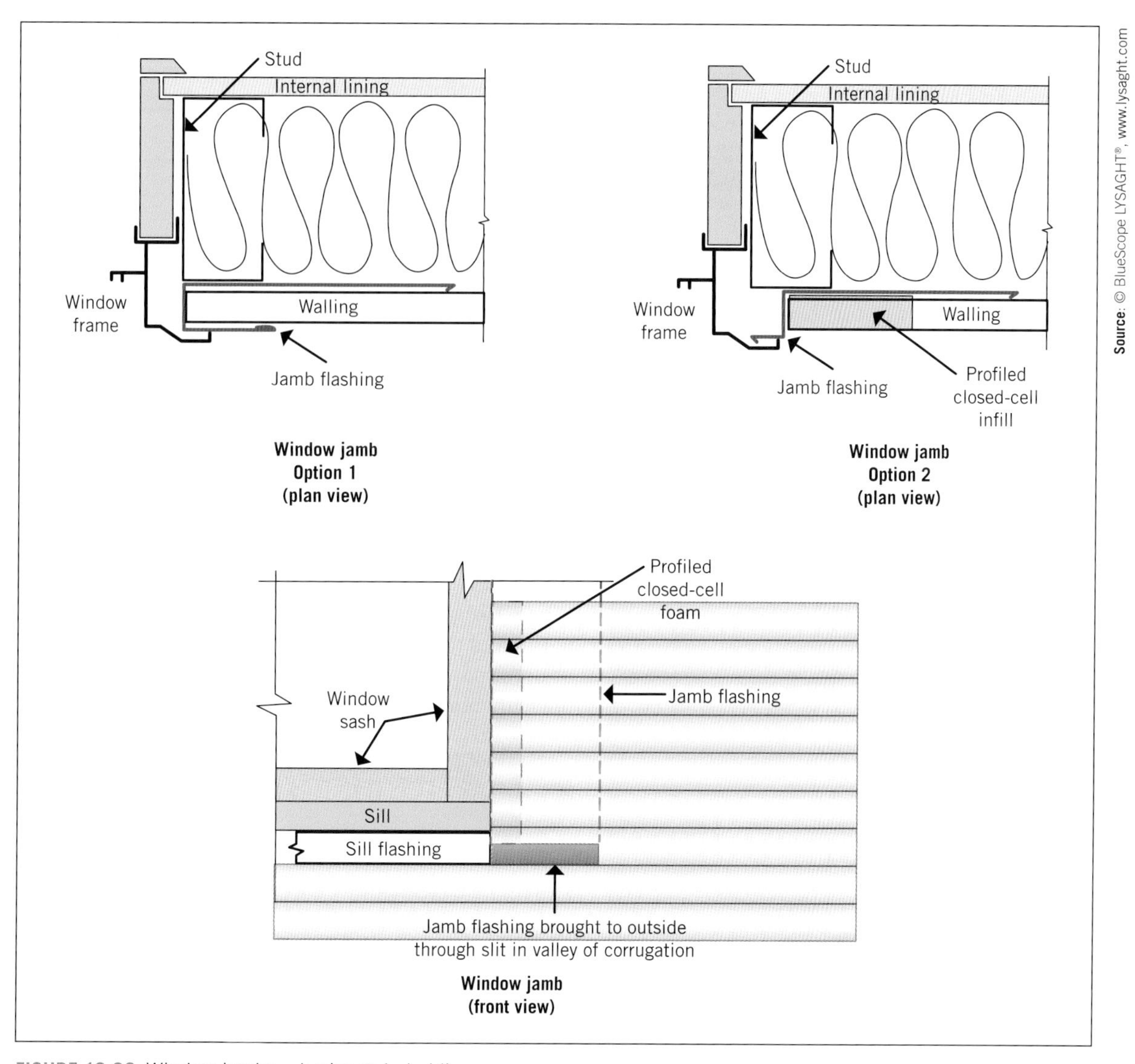

FIGURE 12.32 Window jambs – horizontal cladding

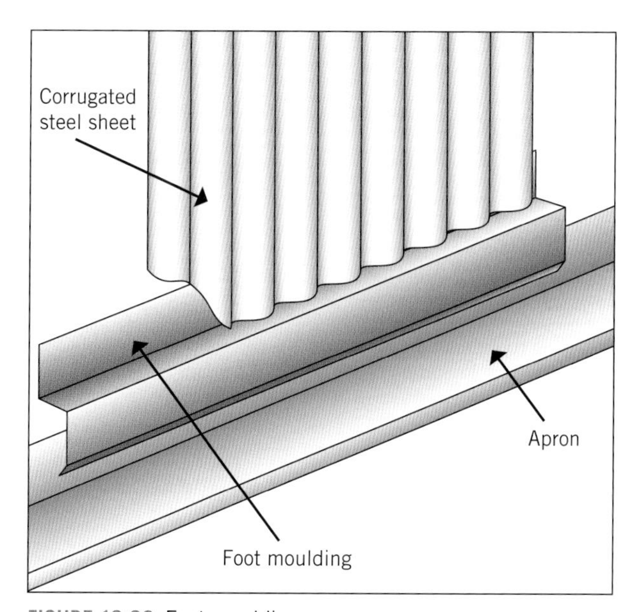

FIGURE 12.33 Foot moulding

Installation of translucent sheeting

Fibreglass and polycarbonate products are available in a range of profile shapes to match complementary metal roof sheets. Translucent sheets must be used in reference to specific code and installation requirements so that their design performance is maintained.

HB39:2015 SECTION 9.1 'TRANSLUCENT ROOF MATERIALS'

Safety mesh

In almost all cases, safety mesh must be installed beneath translucent sheets (see Figure 12.35). The safety mesh protects both the installation workers and future roof workers who would be at risk of falling through old and brittle translucent products. Never install translucent sheeting without approved safety mesh unless the specifications and applications permit this.

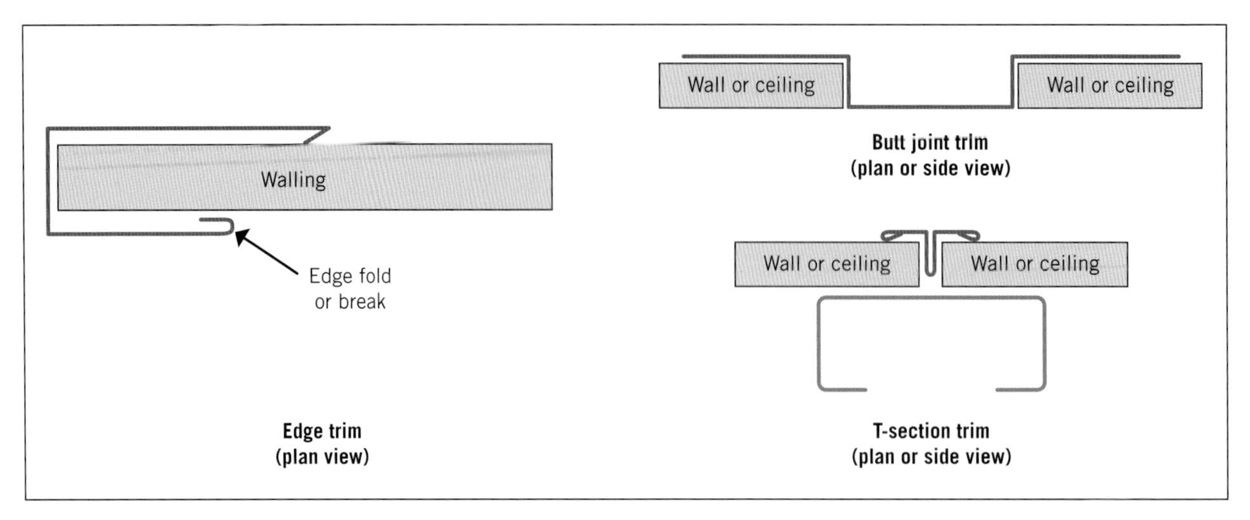

FIGURE 12.34 Trims and covers

FIGURE 12.35 Safety mesh used beneath translucent sheeting

Protection of sheets over purlins

Where the safety mesh is sandwiched between the purlin and the sheet, protection in the form of a foam or plywood strip must be secured over each purlin to prevent abrasion.

Fastener selection

All translucent sheeting must be pierce-fixed, including those profiles normally used for concealed-fixed applications. Fasteners suitable for use with translucent sheeting include special wide washers to improve holding power and flexibility (see Figure 12.36).

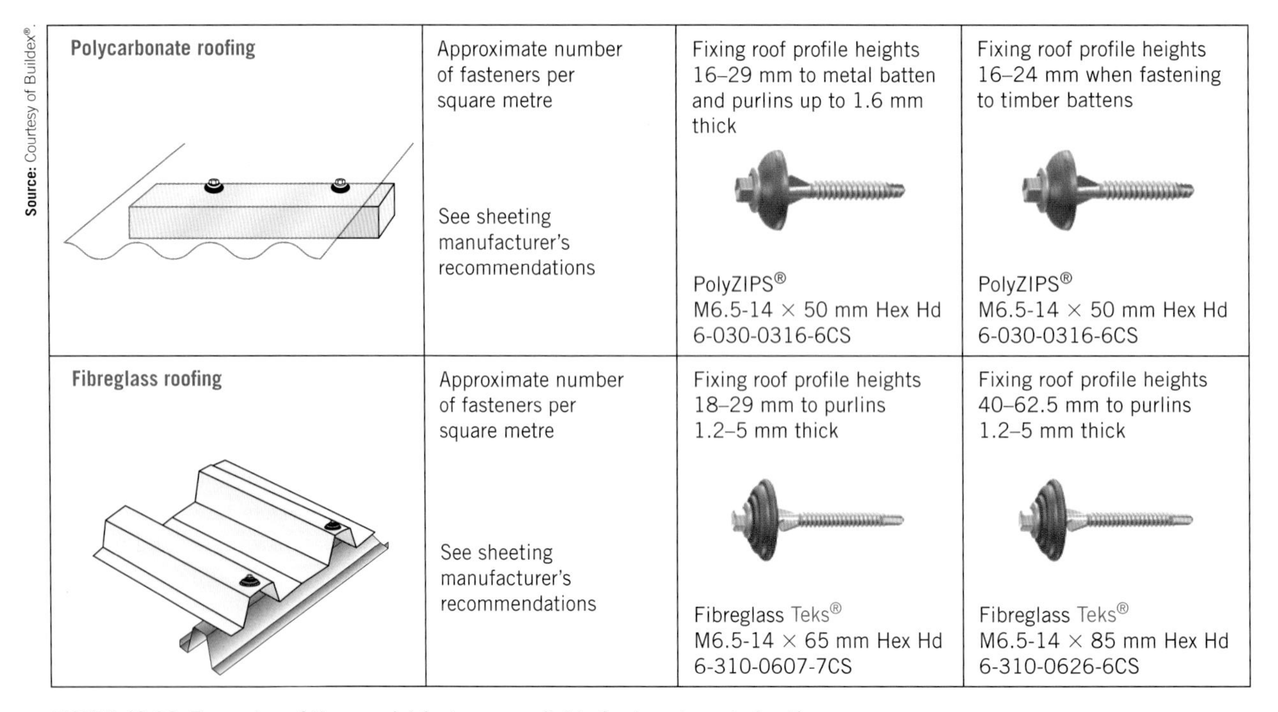

Polycarbonate roofing	Approximate number of fasteners per square metre See sheeting manufacturer's recommendations	Fixing roof profile heights 16–29 mm to metal batten and purlins up to 1.6 mm thick PolyZIPS® M6.5-14 × 50 mm Hex Hd 6-030-0316-6CS	Fixing roof profile heights 16–24 mm when fastening to timber battens PolyZIPS® M6.5-14 × 50 mm Hex Hd 6-030-0316-6CS
Fibreglass roofing	Approximate number of fasteners per square metre See sheeting manufacturer's recommendations	Fixing roof profile heights 18–29 mm to purlins 1.2–5 mm thick Fibreglass Teks® M6.5-14 × 65 mm Hex Hd 6-310-0607-7CS	Fixing roof profile heights 40–62.5 mm to purlins 1.2–5 mm thick Fibreglass Teks® M6.5-14 × 85 mm Hex Hd 6-310-0626-6CS

Source: Courtesy of Buildex®.

FIGURE 12.36 Examples of the special fasteners available for translucent sheeting

Installation requirements for translucent sheets

Some key points for installing translucent sheets are as follows:

- Translucent sheets should overlap steel sheets on both sides of the sheet (see Figure 12.37).
- Translucent sheets are not designed to underlap a steel sheet, and therefore it is standard practice to run a sheet the full length of roof pitch. Where this is not possible, the sheet should be installed at the top of the roof run so that the top flashing can overlay it, while the bottom of the sheet can easily overlap any lower steel sheet.

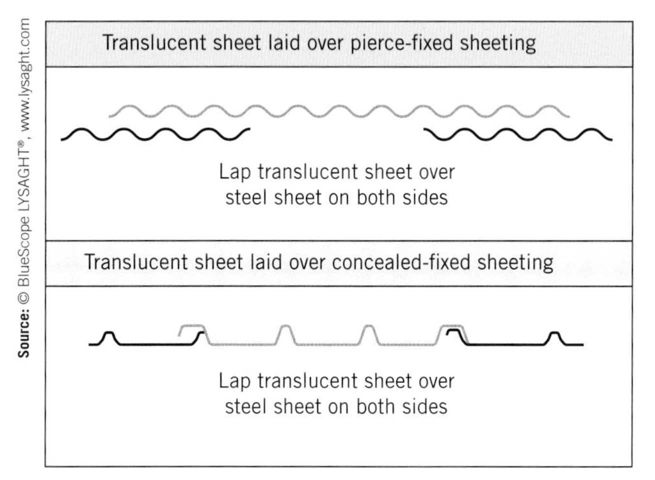

FIGURE 12.37 Translucent sheets normally overlap steel sheets on both sides

- Translucent sheets cannot be weathered in the same manner as steel, so you must use closed-cell foam strips at the top of the sheets to prevent the ingress of water.
- It is standard practice to lay the steel sheeting first, leaving a correctly spaced gap for the translucent sheet to be overlapped onto each side of the steel sheets.
- When you are screwing down translucent sheeting over a concealed-fixed roof system, ensure that you do not pierce the steel deck.

LEARNING TASK 12.6

1 Which section of HB39:2015 references translucent roof materials?
2 What is the reason for fasteners to include special wide washers when used with translucent sheeting?

Clean up

Cleaning up is as much a part of the job as the preparation and installation. The job is not over until all waste is disposed of correctly and tools stored and refurbished according to company policy. Review Chapter 14 for more detail.

SUMMARY

In this chapter you have reviewed the basic requirements related to the installation of roof sheeting with a particular emphasis on concealed-fixed sheet profiles. The importance of detailed planning and product selection was covered with additional reminders relating to considerations for re-roofing and properties with BAL ratings.

Installation considerations included:

- insulation and safety wire where required
- correct sheet set-out
- correct fastener application
- wall cladding procedures
- wall cladding flashings
- translucent sheeting.

COMPLETE WORKSHEET 2

WORKSHEET 1

To be completed by teachers

Satisfactory ☐

Not satisfactory ☐

Student name: ____________________

Enrolment year: ____________________

Class code: ____________________

Competency name/Number: ____________________

Task: Working with your teacher/supervisor, refer to this text, your HB39:2015 and any other relevant resource to answer the following questions.

1 General steel profile roofing can be broken into two main categories. What are they?

____

____

____

2 How are concealed-fixed sheets secured to the roof?

____

____

____

3 Using this text and/or other product information, list at least eight different roof and walling profiles and their respective effective cover measurements.

____

____

____

____

4 Can standard corrugated steel roofing be laid on a 4° roof pitch?

____

____

____

5 What can happen if roof sheets are laid below the specified minimum roof pitch?

____

____

____

6 What details would you be looking for on the plan view of the approved construction plans?

7 List at least four reasons why insulation is installed.

8 What are four sources of product information?

9 How many 50-m rolls of safety mesh would you require to cover a roof area of 500 m^2 if each roll will cover 82.5 m^2?

10 A *gable* roof with an overall width of 45 m is to be sheeted in LYSAGHT KLIP-LOK® 700. How many sheets will you need to do the job?

11 How many rolls of foil-backed insulation blanket do you need to cover a 660 m^2 roof if each roll covers 18 m^2?

WORKSHEET 2

To be completed by teachers	
Satisfactory	☐
Not satisfactory	☐

Student name: ______________________

Enrolment year: ______________________

Class code: ______________________

Competency name/Number: ______________________

Task: Working with your teacher/supervisor, refer to this text, your HB39:2015 and any other relevant resource to answer the following questions.

1 What are the five basic stages of installing a concealed-fixed roof?

2 True or false: When laying safety mesh, the longitudinal wires are positioned uppermost.

3 When purchasing safety mesh, with what Australian Standard should you ensure that the product complies?

4 Why should RFL be allowed to sag slightly below the sheet between each purlin?

5 What is the minimum overlap between each run of insulation blanket?

6 Before installing over-flashing, what needs to be done to the upper end of each sheet? Why?

7 What is the minimum distance that a roof sheet should overhang a gutter? Provide a code reference to support your answer.

8 What are at least three methods of preventing damage to sheets when moving around the roof area?

9 Where roof sheets are laid at less than 10°, what needs to happen to the end of each sheet?

10 Where safety mesh is sandwiched between a purlin and a translucent sheet, what do you need to do to prevent abrasion of the sheet?

11 What safety mesh lap requirements apply where purlin spacing exceeds 2200 mm? Circle the correct answer.

i 450 mm

ii 150 mm

iii 300 mm

12 When installing pierce-fixed wall cladding, do you screw through the crest of the profile or through the tray of the sheet?

____

____

____

13 What is the purpose of an insulation spacer?

____

____

____

13 FABRICATE AND INSTALL EXTERNAL FLASHINGS

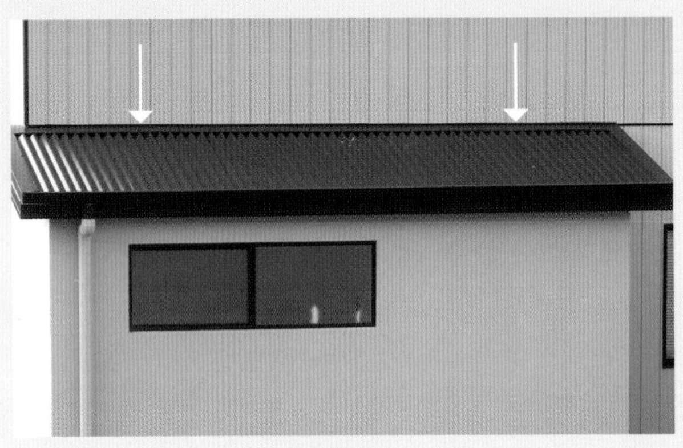

FIGURE 13.1 The intersection of this lean-to roof with a wall is protected from water damage by a transverse apron flashing

Flashings are roofing components that are installed at the intersection of the building structure with roof sheeting and other roofing products, with the primary purpose of keeping the weather and wind out at critical joints and junctions (see Figure 13.1). The majority of roof and cladding leaks occur around poorly fabricated and installed flashings. Getting your flashings right is fundamental to your success as a roof plumber.

What should I read first?

Before proceeding with this chapter it would be advisable to revise Part A of this text with a particular focus on Chapter 7 'Capillary action'. This chapter should also be read in conjunction and cross-referenced with Chapter 12 'Select and install roof sheeting and wall cladding', Chapter 14 'Install roof sheets, wall cladding and complex flashings' and Chapter 15 'Flash penetrations through roofs and walls'.

Overview

In this chapter you will be introduced to the following subjects related to the fabrication and installation of flashings:

- materials
- tools used
- transverse wall flashings
- sloping wall flashings
- pressure flashings
- cappings.

Through each of these subjects, regular reference will be made to good trade practice and the HB39:2015 code requirements. The author also suggests that this chapter be cross-referenced as need be with the latest version of the 'LYSAGHT Flashing Guide – Architectural Detailing Manual' which can be found at this link: http://www.professionals.lysaght.com > resources > user guides and manuals > roofing and walling

Flashing names

As with many other aspects of the plumbing industry, a wide range of names for the same products, components and processes is found right across Australia, and it is not possible to cover them all here. The author has attempted to align names of flashing items as closely as possible to those used within codes, standards and manufacturer product documents, but even from these sources considerable variation and inconsistency often exists.

FROM EXPERIENCE

If you live in an area where a particular item is known by an entirely different name, simply cross out these text references and insert your own. The most important thing is that you, your boss and your trade teacher are talking about the same thing.

Materials suitable for flashings

The materials used for flashings are the same as or similar to those used in the broader roofing industry. The selection of which material is used for a given application is often based on three desirable qualities that are sought in flashings. These are:

- *Compatibility.* The flashing material must be compatible with other materials with which it is in contact or that are downstream of it. Consideration must always be given to the potential for electrolytic corrosion, drip-spot corrosion and inert catchment. For example, the use of lead flashing over ZINCALUME® steel would be unacceptable as they are incompatible when placed together. Sheet zinc would be a better option. Review Chapter 8 'Corrosion' to revise this subject.
- *Malleability.* Malleability refers to the degree to which a metal can be formed into a particular position or shape. Lead is a very malleable metal and is still used extensively for flashings over complex shapes such as roof tiles. Sheet zinc is another malleable metal.
- *Durability.* Durability describes the strength of a material and its ability to resist the elements and last. In most instances, you would want your flashing to last as long as the materials it is designed to cover and protect. Metals such as lead and copper are very durable. ZINCALUME® steel is more durable than galvanised sheet steel. However, remember that durability is subject to correct application. The incorrect use of a durable material can reduce its working life.

The materials used as flashings include those shown in Table 13.1.

See Chapter 5 'Material types' for a more in-depth coverage of the roofing materials that are available.

TABLE 13.1 Materials suitable for flashing and capping

Material	Qualities and applications
Lead	• Very malleable • Very durable when used correctly • Used for strip and collar flashings around complex shapes • Not to be used on roof catchments collecting water for drinking
Butyl-based flexible strip flashing	• High malleability • Self-adhesive • Lead free • Used in all applications where strip lead flashing is used
Zinc sheet	• Malleable • Durable • Often used as a substitute for lead in the same applications
Galvanised steel	• Can be used in all forms of sheet metal flashing • Caution must be exercised in relation to compatibility • Generally more malleable than ZINCALUME® steel but less durable
ZINCALUME® steel	• Very durable • Low malleability • Can be used in all sheet flashing applications
COLORBOND® steel	• As for ZINCALUME® steel but with the added benefit of colour matching and protection
Copper	• Extremely durable • More malleable than steel but less so than zinc or lead
Aluminium	• Durable • Malleable to a degree • High coefficient of expansion requires custom application

A note about aluminium

The HB39:2015 warns that the standard installation practices relevant to the use of steel flashing should not be applied to aluminium flashings. Aluminium has a coefficient of expansion that is approximately twice that of steel (see Chapter 10 'Fabricate and install roof drainage systems' to revise expansion and contraction). You must therefore refer to and follow the manufacturer's specifications when using aluminium products.

A note about lead

Lead is a very good flashing material and it is still used, particularly over roof tiles. However, with domestic water tanks increasingly being retro-fitted to older dwellings, you have a responsibility to inspect the roof to ensure that lead products do not form part of the roof catchment area.

Lead is poisonous and water running over bare lead flashings will wash lead ions into the drinking water supply.

HB39:2015 SECTION 8.5

LEARNING TASK 13.1

1. Which of the suitable flashing materials mentioned has the highest coefficient of expansion?
2. Which two suitable flashing materials mentioned have the least malleability?

Minimum material thickness

Having selected a particular material for your flashings and capping, you must also ensure that the thickness of the material meets the minimum requirements as set out in the Code of Practice. Review this now.

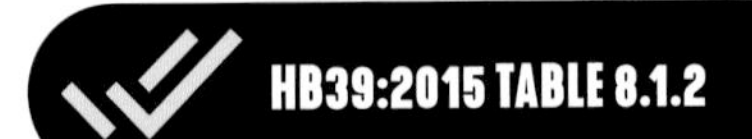

HB39:2015 TABLE 8.1.2

You will note that for each type of material listed, requirements are detailed relating to: the coating class (where applicable); the use as a flashing, capping or both; and the minimum thickness permitted. For example:

- For ZINCALUME® steel:
  - coating class: AZ150
  - usage: flashings and capping
  - minimum thickness (BMT): 0.55 mm.
- For lead:
  - coating class: n/a
  - usage: flashings only
  - minimum thickness: 1.7 mm.

Whenever you are going to design a flashing and have it fabricated by your sheet-metal worker (or fold it up yourself), you need to ensure that the correct material thickness is specified.

Tools for flashings

Marking out and fabricating flashings requires a set of tools similar to those used in most other roof work. You will generally require a selection of the following:

- hacksaw
- straight snips
- aviation-style left- and right-hand snips
- square
- rivet gun
- drill and correctly sized bits
- cordless driver with range of hex and Phillips bits
- self-locking clamps
- multigrips
- bevel
- hammer
- sealant gun
- hand folder
- rags and spatula
- fine marker pen
- relevant personal safety equipment.

In particular, you will also find yourself using these additional items:

- raking chisel – used for raking out the mortar between blocks and bricks so that you can insert your counter flashing
- lead-working tools (including dressers, bossing stick and mallet) – these tools allow you to work lead flashings correctly, enabling you to shape and form the lead while reducing the risk of tearing
- diamond blade cutting tools – angle grinders and small circular saws fitted with diamond blades allow accurate and fast chasing of rebates for raked hanging flashings and for clearing the mortar out of brick courses. Examples can be seen in Figure 13.2.

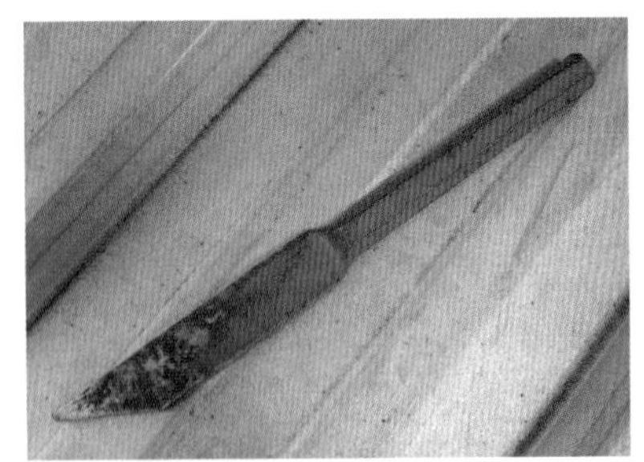

FIGURE 13.2 Tools for flashings (a) Raking chisel (b) Lead-working tools (c) Diamond blade cutting tools

If you work for a business that has a workshop set up to do your own sheet-metal work, you may also find yourself using equipment such as:

- guillotine
- press brake
- folders
- lock seamer.

Review Chapter 9 'Roof plumbing tools' for a broader review of general roofing tools.

Capillary action

Capillary action describes the phenomenon where water that is trapped between two close-fitting sheets of metal may be drawn upwards and leak into a building. The fabrication of flashings requires that you be ever mindful of the potential for capillary action to occur and to prevent it through the correct design of flashing components.

Frequent reference to capillary action will be made in this chapter and, if necessary, you should revise Chapter 7 'Capillary action' before proceeding further.

Minimum covers for flashings and capping

The shape of a flashing or capping is determined by its application and position. However, the actual dimensions of the product need to be checked against the requirements set out in the code.

In determining flashing dimensions, you need to be familiar with the application of two terms:

- *transverse* – where there is a reference in the code to a transverse flashing, it means a flashing that is installed horizontally at 90° to the sheet run
- *sloping* – as the name suggests, this is a flashing that follows the slope of the roof. Also known as a longitudinal flashing.

Follow the boxes in Figure 13.3 in sequence to see how to determine flashing basic dimensions.

Looking at the example in Figure 13.3, you can see that the code tells you that where a transverse flashing is used over a pierce-fixed roof, it requires a minimum 150-mm cover and the end of this cover must be turned down a minimum 20 mm at 30°.

The same transverse flashing over a concealed-fixed roof would require the same 150-mm cover but a minimum 40-mm turn-down at 90°.

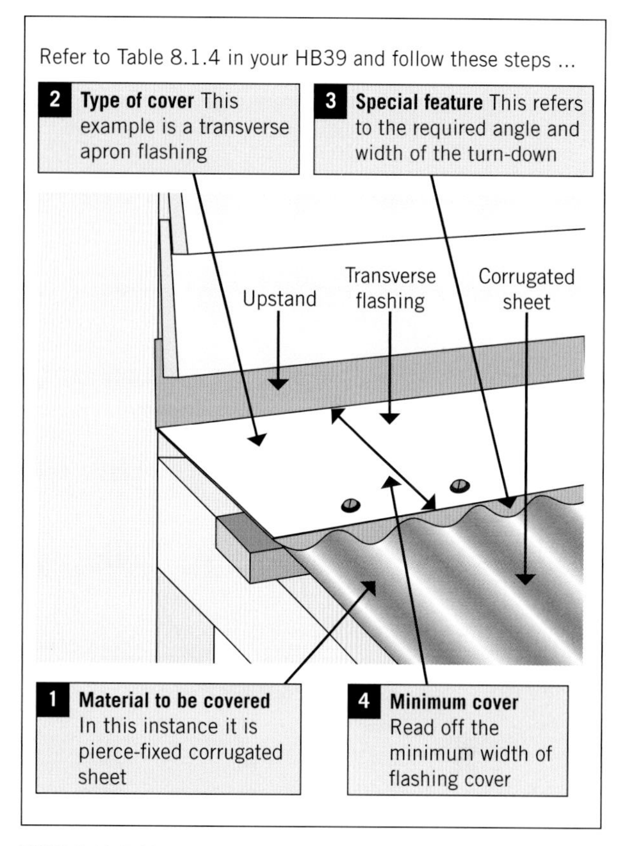

FIGURE 13.3 How to determine flashing and capping dimensions

Refer to this table in the code whenever you are working out the minimum size of your flashings.

Basic wall flashings

Wall flashings are used when a roofline intersects with an adjacent transverse or sloping wall. Flashings are normally installed from the bottom up, with each succeeding layer overlapping the one below. Water is then able to run down the wall over the flashing and then on to the roof. Depending upon the wall structure and cladding material, both transverse and sloping wall flashings are normally made up of two main components:

- a lower apron flashing
- an overhanging counter flashing – sometimes not required where wall sheeting overlaps the apron flashing

Transverse wall flashing

A transverse wall flashing features an upstand that abutts a wall with its lower cover surface projecting over the roof sheeting (see Figure 13.4).

The key feature of a transverse apron flashing is that it must be used with some form of counter flashing, weatherboards or sheet cladding that overlaps the apron upstand to make it completely watertight (see Figure 13.5).

The lower edge covers the roofing profile and is often notched or scribed to take its shape. The lower edge must always have a minimum width of 150 mm, with an additional turn-down sized to suit sheet profiles as required.

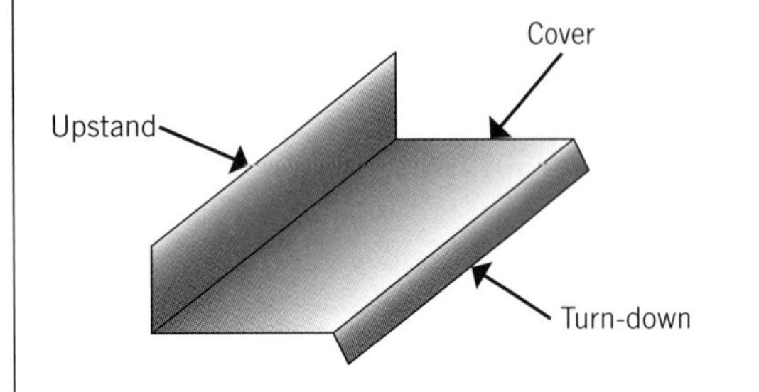

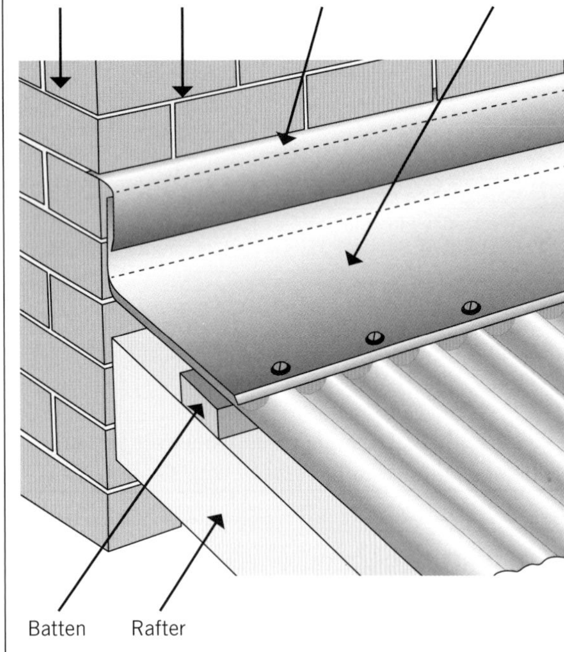

FIGURE 13.4 A transverse apron and chased counter flashing

FIGURE 13.5 Vertical sheet-metal cladding overlapping the upstand of a sloping apron flashing

Notching of transverse flashing

A transverse apron flashing installed over wide-pan roof sheets must be notched into the profile. This is the process of cutting or notching the turn-down of the flashing to enable it to sit down fully into the sheet profile as seen in Figure 13.6. This is important to prevent wind-driven rain and debris getting in under the flashing.

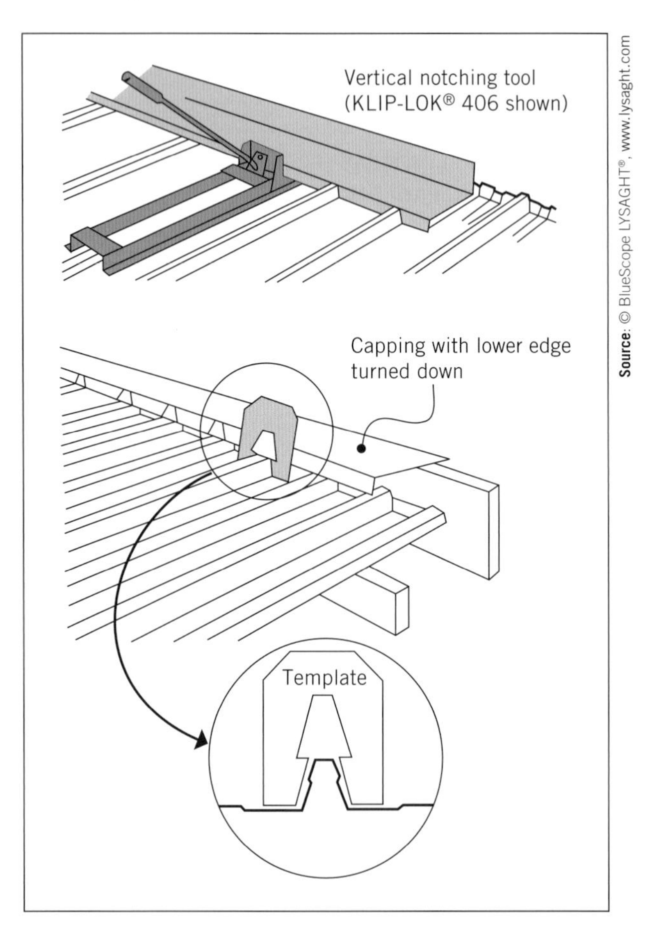

Source: © BlueScope LYSAGHT®, www.lysaght.com

FIGURE 13.6 All flashing over wide-pan sheets must be notched into the profile

Notching can be achieved with a special tool matching the roof sheet profile or by cutting out each notch with tin snips.

In most tropical areas, the turn-down of flashings above corrugated roof sheets are also cut into the profile. This is known as 'scribing' or 'fashioning'. Scribing takes some skill to master but is normally always done by hand, using aviation-style snips. In southern regions the flashing turn-down is normally quite small and scribing is not carried out.

Transverse counter flashing

A counter flashing is one that is normally inserted into brickwork above an apron or soaker flashing, with a hanging section that overlaps the upstand of the apron by at least 50 mm (see Figure 13.7).

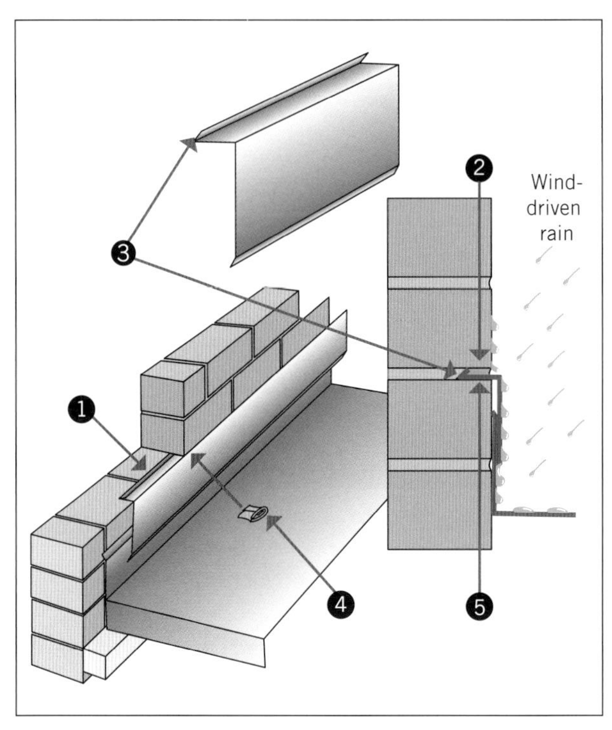

FIGURE 13.7 Use of a chased counter flashing over a transverse apron flashing

Follow the number sequence in Figure 13.7 to see how a hanging flashing is used:

1. Select a mortar course at a suitable height above the apron flashing.
2. Use a diamond blade or raking chisel to remove the brick mortar to a depth of 25 mm.
3. The top of the flashing penetrates the wall by at least 22 mm and incorporates a 10-mm return weathering fold that prevents water from seeping further into the joint.
4. Use compatible metallic wedges at 500-mm intervals to hold the flashing in place. You cannot fasten the flashing directly to other roof components!
5. Once the flashing is inserted and wedged, fill the joint with mortar or silicone to complete the joint.

A critical part of the counter flashing is the lower anti-capillary break. This must be 10 mm wide and be formed at a 30° angle to prevent water from making its way up between the apron upstand and the counter flashing overlap.

HB39:2015 SECTION 8.3(A)

Figure 13.8 shows some flashing detail from a heritage building in Darwin. You can see how a counter flashing has been lapped and the two pieces secured with a holding cut. Remember that no flashing is to be fastened directly to the underlying flashing. Note also how the transverse apron flashing has been scribed to fit into each corrugation. This is standard practice in higher rainfall and wind speed areas of Australia, but it is not normally done in southern regions.

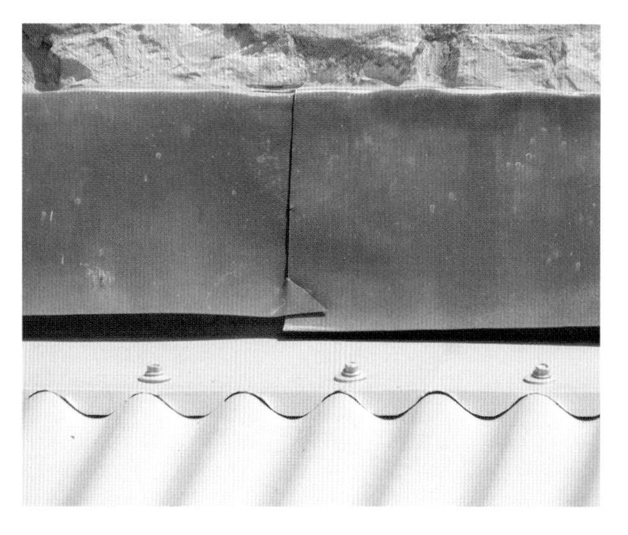

FIGURE 13.8 Counter and transverse apron flashing detail on a heritage building in Darwin

Sloping wall flashing

Sloping wall flashings are simply variations on transverse wall flashings. A lower apron flashing runs parallel with the roof sheet and is overlapped by the wall sheeting or counter flashing. There are two main forms of sloping counter flashing that you need to be aware of. These are:

- stepped flashing
- sloping counter flashing.

Stepped flashing

Stepped flashing is found where a sloping roof abutts a higher brick wall. The intersection of the wall with the roof is covered firstly by an apron flashing and then individual flashing sections are stepped down the slope of the roof, with each one overlapping the next and the apron upstand.

Before the advent of power tools, diamond blades and abrasive discs, the only flashing option available to a roof plumber was to use a hammer and raking chisel to remove the mortar from each brick course and install a lead stepped flashing over the apron upstand.

Stepped flashings are commonly seen on older chimney work, parapets and decorative brick turrets. Some examples can be seen in Figure 13.9.

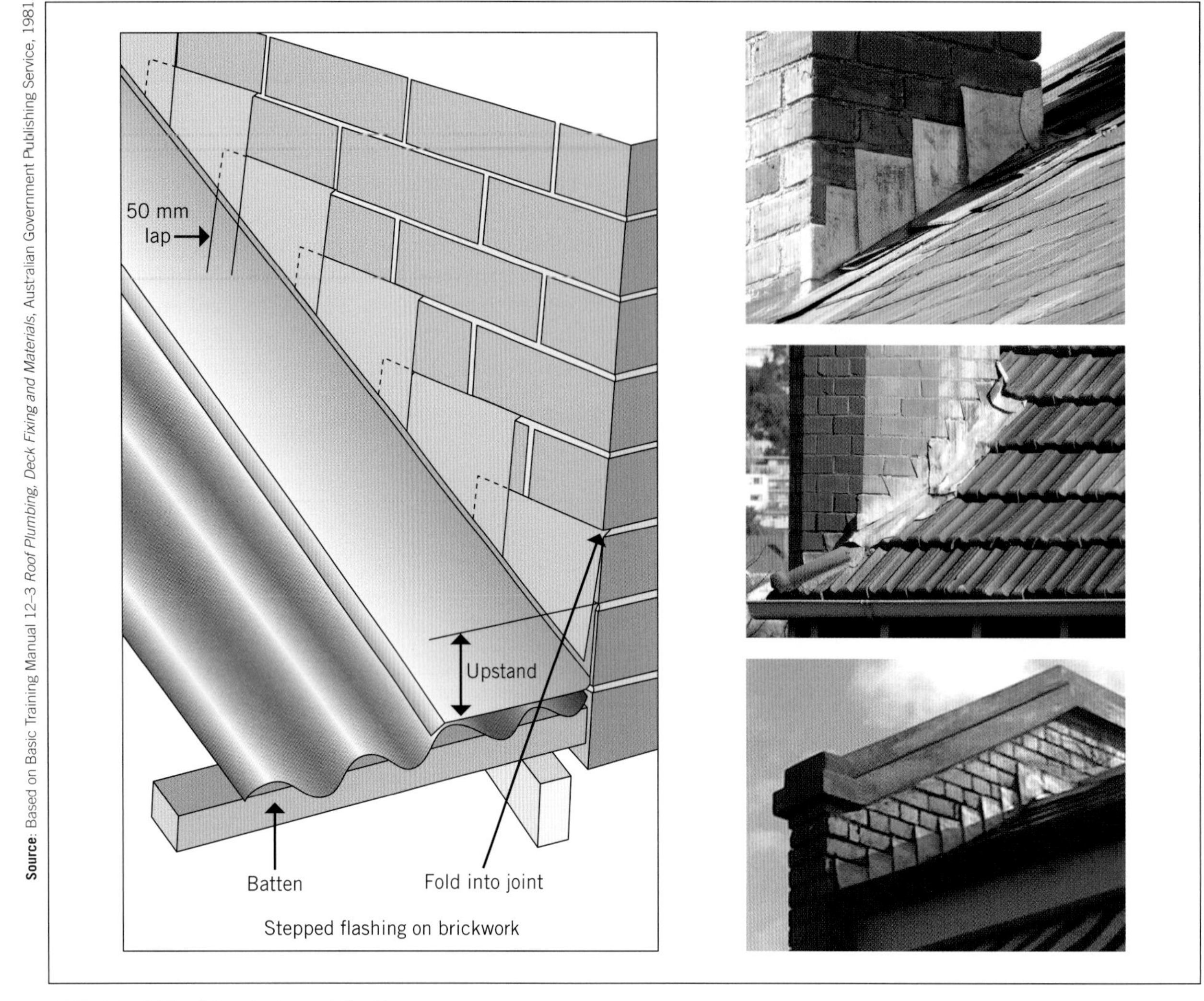

FIGURE 13.9 Traditional stepped flashing

- Each stepped flashing section must overlap the next one down the roof in the direction of flow by at least 50 mm.
- A holding cut should be used to secure the lower edge of each flashing piece.
- The cut of each stepped flashing section should neatly match the pitch of the roof and overlap the apron by at least 50 mm. A clearance of at least 25 mm should be maintained between the bottom of each stepped flashing piece and the apron cover to prevent any build-up of debris.
- The flashing weathering fold should be at least 22 mm wide with a 10-mm return and be inserted into a 25-mm chase in the brick course. Hold it in place with compatible metallic wedges, and then seal the joint with mortar or silicone.

The restoration of heritage buildings is one example where stepped flashing should be used wherever possible, in order to maintain the traditional look of the building. Lead is the most malleable material for the job, though where any rainwater run-off is being collected for drinking purposes you must use an alternative material suitable for drinking water catchment areas.

In most modern applications the stepped flashing process has been superseded by a chased sloping counter flashing, which can be installed more quickly.

Sloping counter flashing

A sloping counter flashing shares all the features of a transverse counter flashing but is installed on a masonry wall above an apron flashing that runs parallel with the pitch of the roof. Rather than being stepped into horizontal brick courses, a sloping counter flashing is inserted into a cut chased into the brickwork above the apron (see Figure 13.10), as follows:

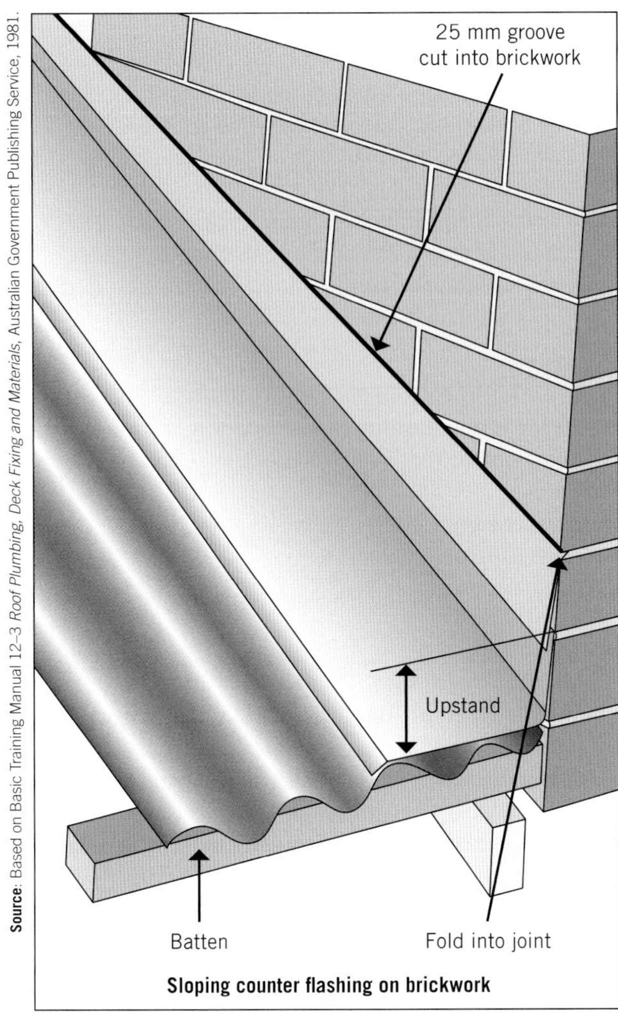

Source: Based on Basic Training Manual 12–3 *Roof Plumbing, Deck Fixing and Materials*, Australian Government Publishing Service, 1981.

FIGURE 13.10 Diagram showing how a sloping counter flashing is chased into brickwork

1. Use a diamond blade on a circular saw to make a 25-mm chase in the brickwork, following the same angle of the roof.
2. The top of the flashing itself must be 22 mm wide incorporating a 10-mm return weathering fold that is inserted into the cut.
3. The flashing is secured with compatible metallic wedges and then sealed with mortar or silicone.

Pressure flashings

Also known as a 'K' flashing, this is a form of purpose-made counter flashing that can be installed above an apron or soaker without having to chase into the masonry wall itself and is applicable to both transverse and sloping wall flashing situations.

Importantly, you need to understand that this flashing solution can only be used where the wall is smooth-finished and any mortar courses are flush-pointed with the outer brick surface itself. The flashing relies upon very specific dimensions, silicone sealant and closely spaced fasteners to ensure a watertight seal (see Figure 13.11).

This form of counter flashing option has the potential to save considerable installation time but must only be fabricated and fitted in accordance with the requirements of the HB39:2015.

Additional details can be found on the Victorian Building Authority website.

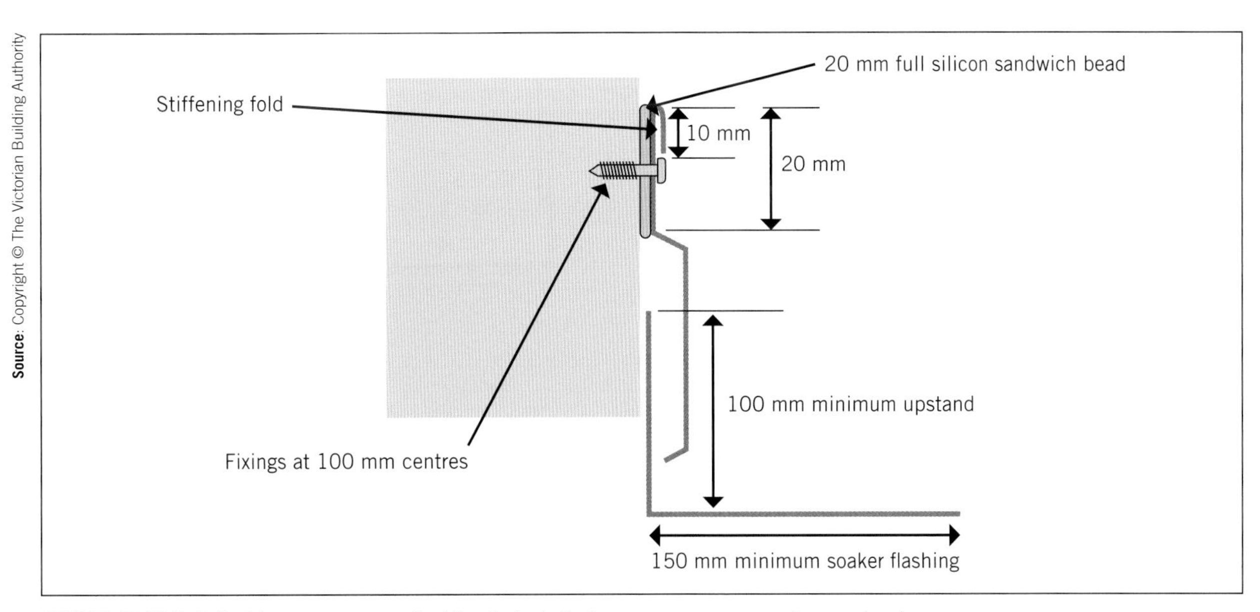

FIGURE 13.11 Detail of how a pressure flashing is installed over an apron or soaker upstand

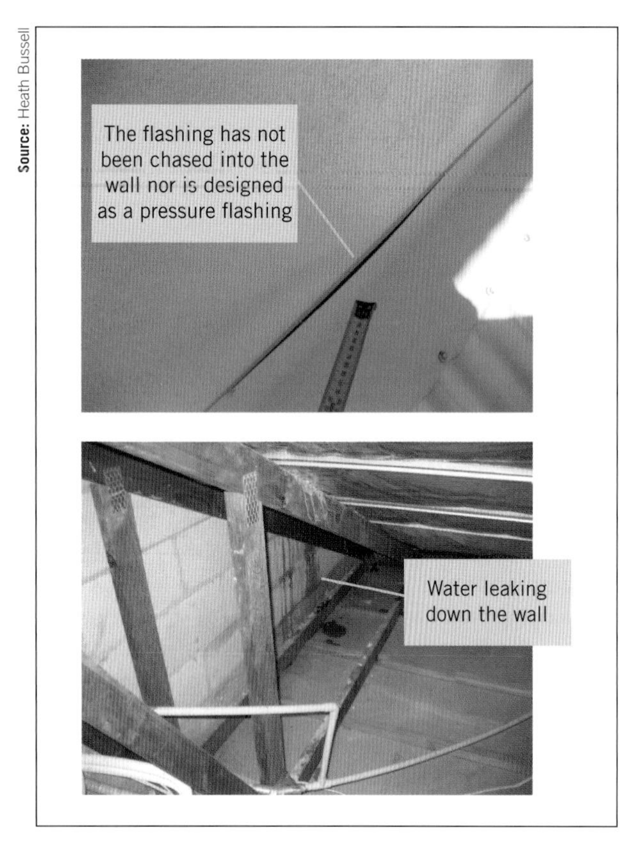

FIGURE 13.12 This poorly designed flashing leaks straight into the house.

LEARNING TASK 13.2

1. Transverse and sloping flashings are normally made up of two main components. What are they?
2. What are the two main forms of sloping counter flashing?
3. Which hand tool is used for raking out the mortar between blocks and bricks so that you can insert your counter flashing?

Cappings

A capping is any flashing that sits on top of a lower flashing, wall or roof sheet intersection. Cappings include:

- parapet capping
- ridge capping
- barge capping.

Parapet capping

A parapet is a wall that extends above and past the roofline. The parapet therefore needs to be protected from the ingress of water. A parapet capping is designed to protect the top of the wall and can be combined with a counter flashing and/or apron on the roof side. The front and rear of the parapet capping must incorporate an anti-capillary break.

Parapet cappings must protect the wall and shed water in the right direction. Follow the number sequence in Figure 13.13 to see the main features of a parapet capping:

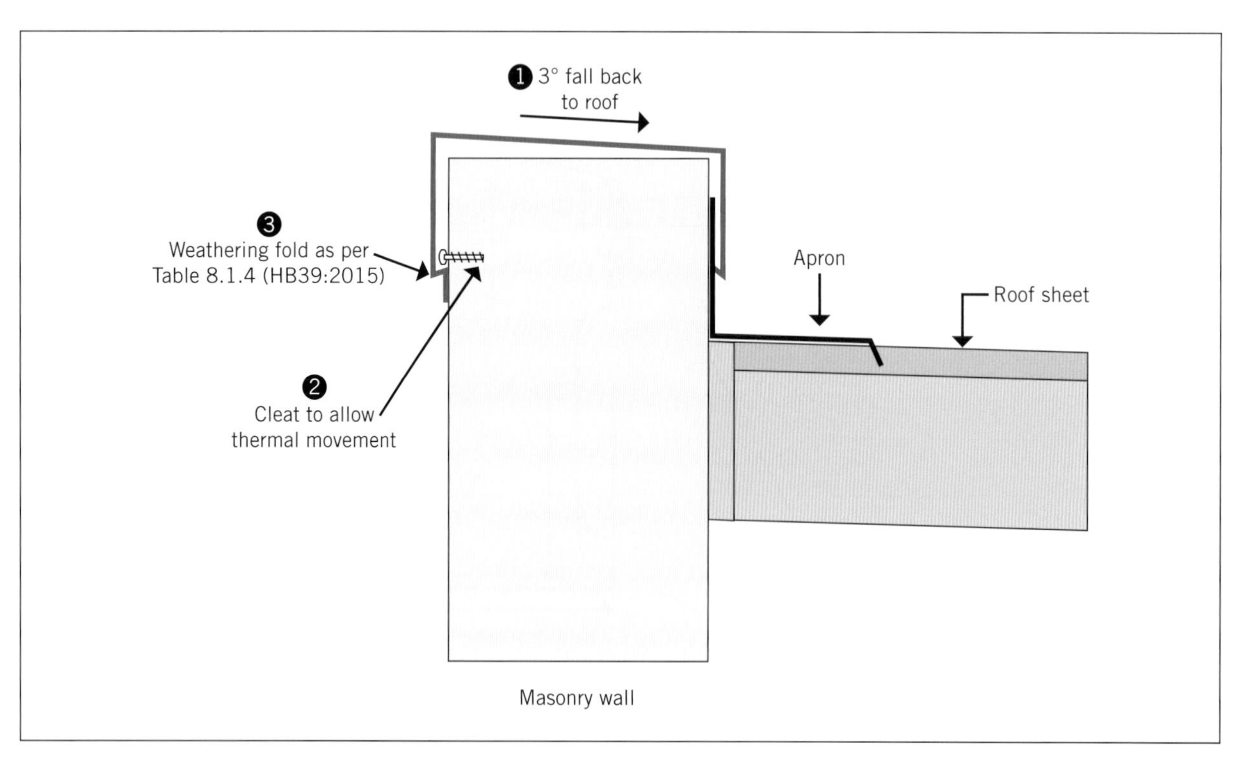

FIGURE 13.13 The main features of parapet capping

1. The parapet capping must have a minimum 3° fall to the back of the flashing so that water does not cascade down the front of the wall, potentially causing a nuisance and eventual staining.
2. The capping must be secured by a series of cleats at 500-mm intervals so as to allow for expansion and contraction.
3. Anti-capillary breaks must be built into the front and rear overlaps.

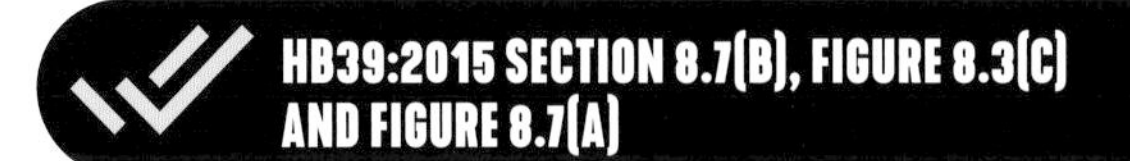

Barge capping

A barge board is normally found at the intersection of a wall top with gable or skillion roof sheeting. The barge capping covers this area and not only keeps water out but protects the edge of the roof sheets from wind lift and damage.

Barge capping comes in a wide range of profiles to suit the style and age of the building. Many gutter profiles also have matching barge capping and barge roll profiles to suit. Custom profiles are often designed and ordered to suit non-standard solutions.

Figure 13.14 shows examples of custom barge capping to suit two very different roofs.

- Barge capping should be secured into the crest of the roof sheets at 500-mm centres.
- There must be a minimum 50-mm cover over the side of the barge board and it must include an anti-capillary break as specified in the code.
- The laps of joins should be a minimum 25 mm, with fasteners every 40 mm.

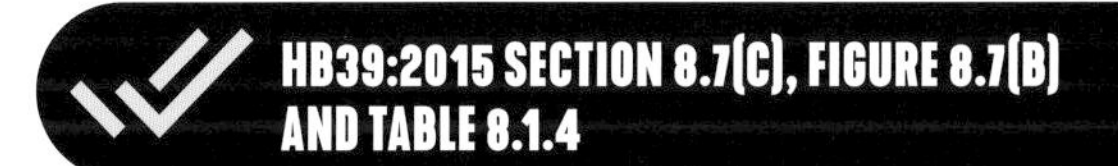

Ridge capping

Wherever two roof planes meet at a ridge or hip a capping is required to prevent water ingress at the intersection. This is known as ridge capping, and extends an equal distance on either side of the ridge centreline. Ridge capping is available around Australia in a wide range of profiles and sizes; however, the main forms are:

- roll-top ridge – the centre of the capping is formed in a half-round shape to provide rigidity and a strong line to the product. It is often matched to the use of corrugated sheeting.
- three-break ridge – a simple three-fold product that is often used for a wider range of roof sheet profiles and is principally used where wide pan sheets are specified.

Table 13.2 shows the LYSAGHT® range of ridge capping and demonstrates the variability of the product between each state and territory.

Points to remember when working with ridge capping include:

- Sheets must be weathered under ridge capping (see Figure 13.15).
- The lower ends of ridge capping must be bird-proof.
- Ridge capping must be fastened at the following minimum intervals:
  - every 4th crest on corrugated (non-cyclonic)
  - every 2nd crest on corrugated (cyclonic; see Figure 13.15)
  - 300 mm maximum on other profiles.

Where the installation is subject to a BAL rating, then ridge capping must comply with the relevant level.

Notching

Where barge or ridge capping sits transversely above wide-pan roof sheets, then the turn-down must be notched into the profile. In tropical areas the same applies to flashing placed above corrugated sheets.

FIGURE 13.14 Two examples of barge capping

TABLE 13.2 Variations in the size and profile of LYSAGHT ridge capping around Australia

Qld	Tas	NSW
Roll-top ridge capping 0.4 mm × 310 mm Custom cut to length (Nth Qld have 450 mm girth roll top ridge and matching three break) 50 120 120 25 25 **Ridge capping – three break** CUSTOM ORB® & TRIMDEK® Standard 20 pitch Custom cut up to 8000 mm 0.4 mm × 400 0.55 × 400 170 170 30 120 30	**Ridge capping** CUSTOM ORB® V-cap ridge capping 175 175 75 310 V-top ridge capping 145 145 27 27 305 KLIP-LOK® V-top ridge capping 145 145 42 310 42 CUSTOM ORB® & TRIMDEK® Roll-top ridge capping extra scribing break 50 185 185 310	**Ridge capping** *Nominate roof pitch CUSTOM ORB® & BLUE ORB® 180 180 30 * 120° 30 TRIMDEK® & SPANDEK® 170 170 30 * 120° 30 KLIP-LOK® 160 160 40 * 120° 40 CUSTOM ORB® & BLUE ORB® 50 127 * 127 19 300 19
Qld	**Tas**	**NSW**
Ridge capping SPANDEK® & TRIMDEK® KLIP-LOK® **Roll top ridge capping** (Scribing break extra on nett price) 300 mm girth 350 mm girth 65 115 115 25 270 25	**Ridge capping** 50 120 120 25 25 310 V-ridge 180 180 75 310 KLIP-LOK® Code: BK 145 145 42 42 310 TRIMDEK® & SPANDEK® Code: ST 145 145 27 27 305	**Ridge capping** All custom cut 30 170 170 30 120 * 120 TRIMDEK® Code: RC2 20 162 162 20 120 * 120 CUSTOM ORB® 395 mm girth Roll to suit the hip cap Suits up to 25 pitch. 395 mm girth available up to 8000 long. Other widths (girths) available with max. length 4000. Roll ridge to suit fibreglass & plastic curving sections Material 0.55 BMT

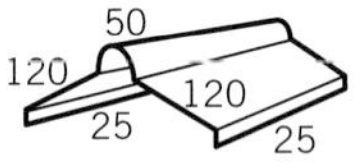

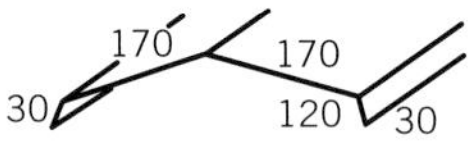

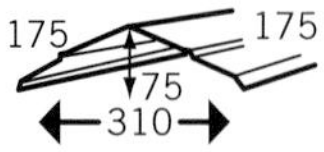

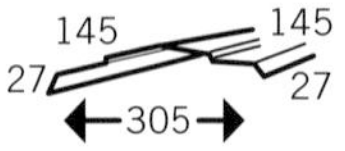

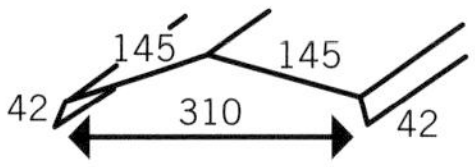

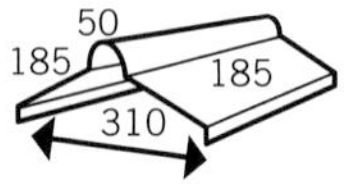

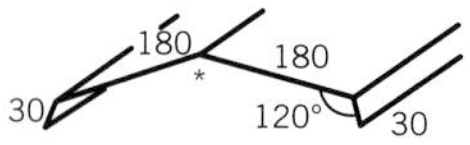

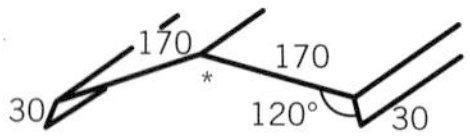

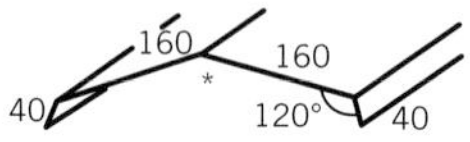

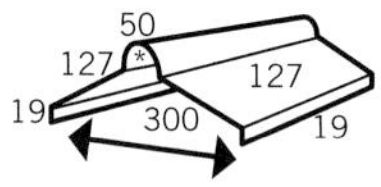

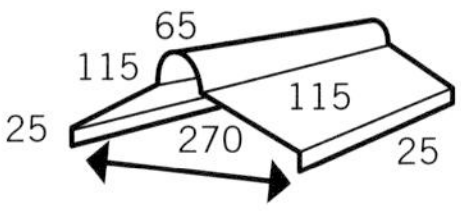

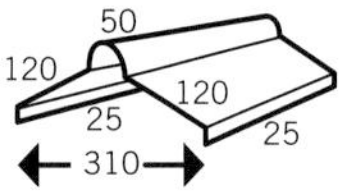

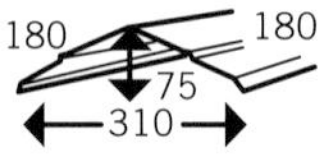

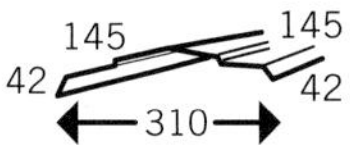

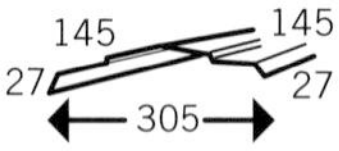

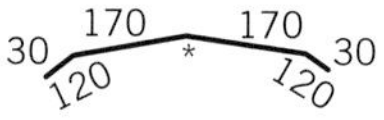

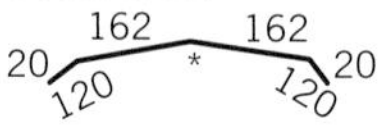

Source: © BlueScope LYSAGHT®, www.lysaght.com

FIGURE 13.15 Weathering the ends of sheets before fitting flashing

Similar details apply here as described in the section on transverse flashing earlier in the chapter (see Figure 13.17).

HB39:2015 TABLE 8.2(B)

See Figure 13.18 for the method for joining ridge capping, and Figure 13.19 for one way that a roll-top ridge can be finished off at the hip-ridge intersection.

Other ridge capping products

Not all ridge capping is made from sheet metal. Other products include:

- lead strip
- synthetic rubber strip flashing
- fibreglass.

These products are particularly useful when completing the ridge capping on curved roofs (see Figure 13.20). This will be covered in more detail in Chapter 17 'Install roof coverings to curved roof structures'.

LEARNING TASK 13.3

1. Name three types of cappings.
2. On which roof types might you expect to use barge capping?
3. In non-cyclonic regions, what is the minimum interval for fastening ridge capping to a corrugated steel sheet roof?

Clean up

Cleaning up as you go and at the completion of your work is an important aspect of site workplace health and safety. Poor site waste control can make the workplace dangerous. The fabrication and installation of flashings creates considerable offcuts and refuse.

Points to consider include the following:

- Remove all fabrication swarf and debris by washing and/or brushing down the job regularly. Ensure that the gutters are flushed as well!
- Dispose of recyclable materials in the appropriate site bin or refuse point.
- All plastic protective film must be collected and disposed of according to policy.
- Any offcuts or other materials that present a trip hazard should be disposed of or stored in the approved location.
- Flashing installation often requires the installation of hanging flashing in brickwork. Mortar droppings and masonry cutting residue can cause corrosion and must not be left on the roof.

HB39:2015 SECTION 8.4(H)

Tools and equipment must be checked for serviceability before being returned to the truck or store. If the tool is not working correctly, is damaged or its tag is coming up for its scheduled test, make sure that you let your supervisor know. Don't just throw it back on the truck and forget about it.

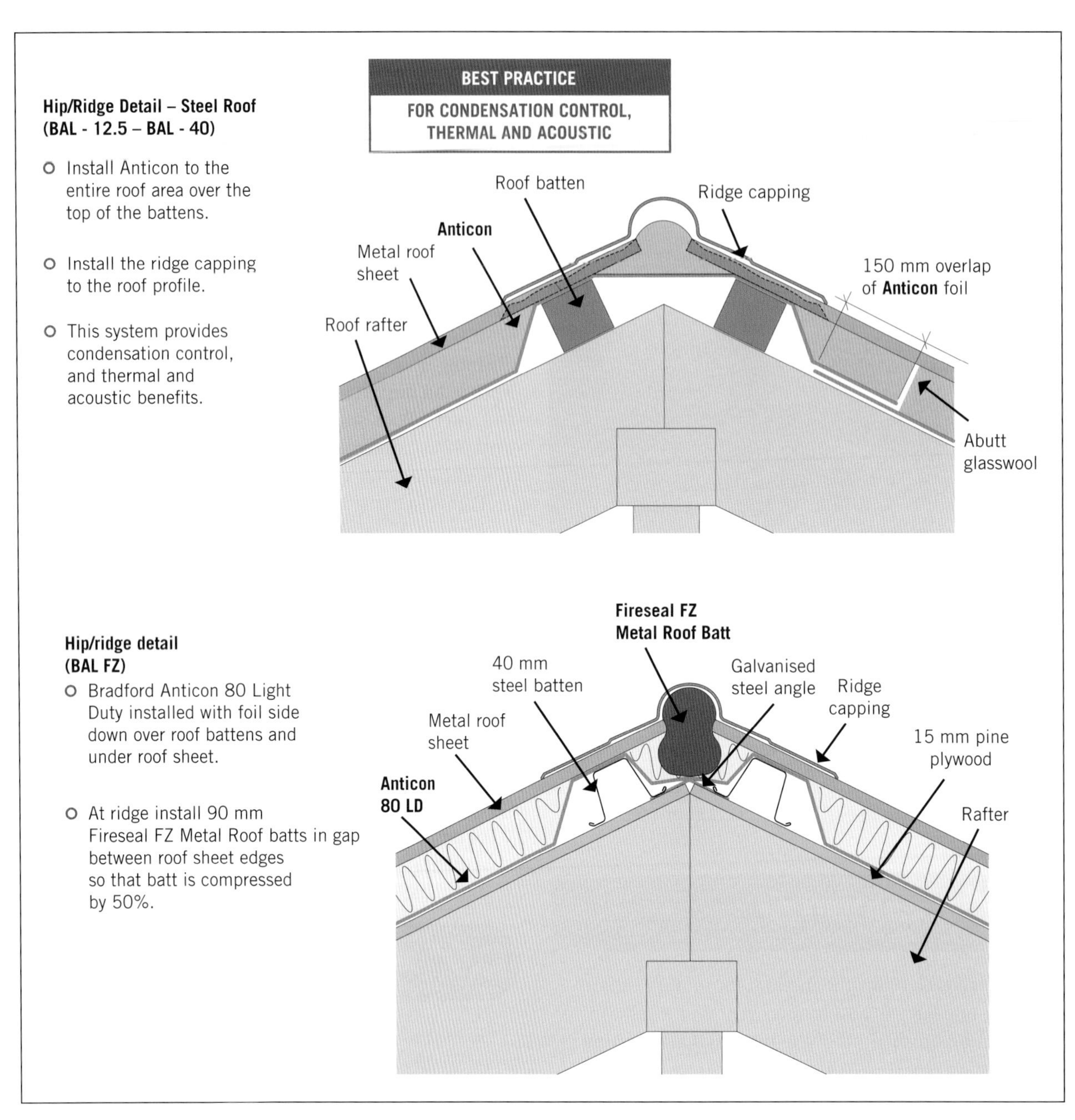

FIGURE 13.16 Ridge capping fitted in bushfire areas must match the specified BAL rating

FIGURE 13.17 Ridge capping scribed into corrugated sheet in a cyclone zone

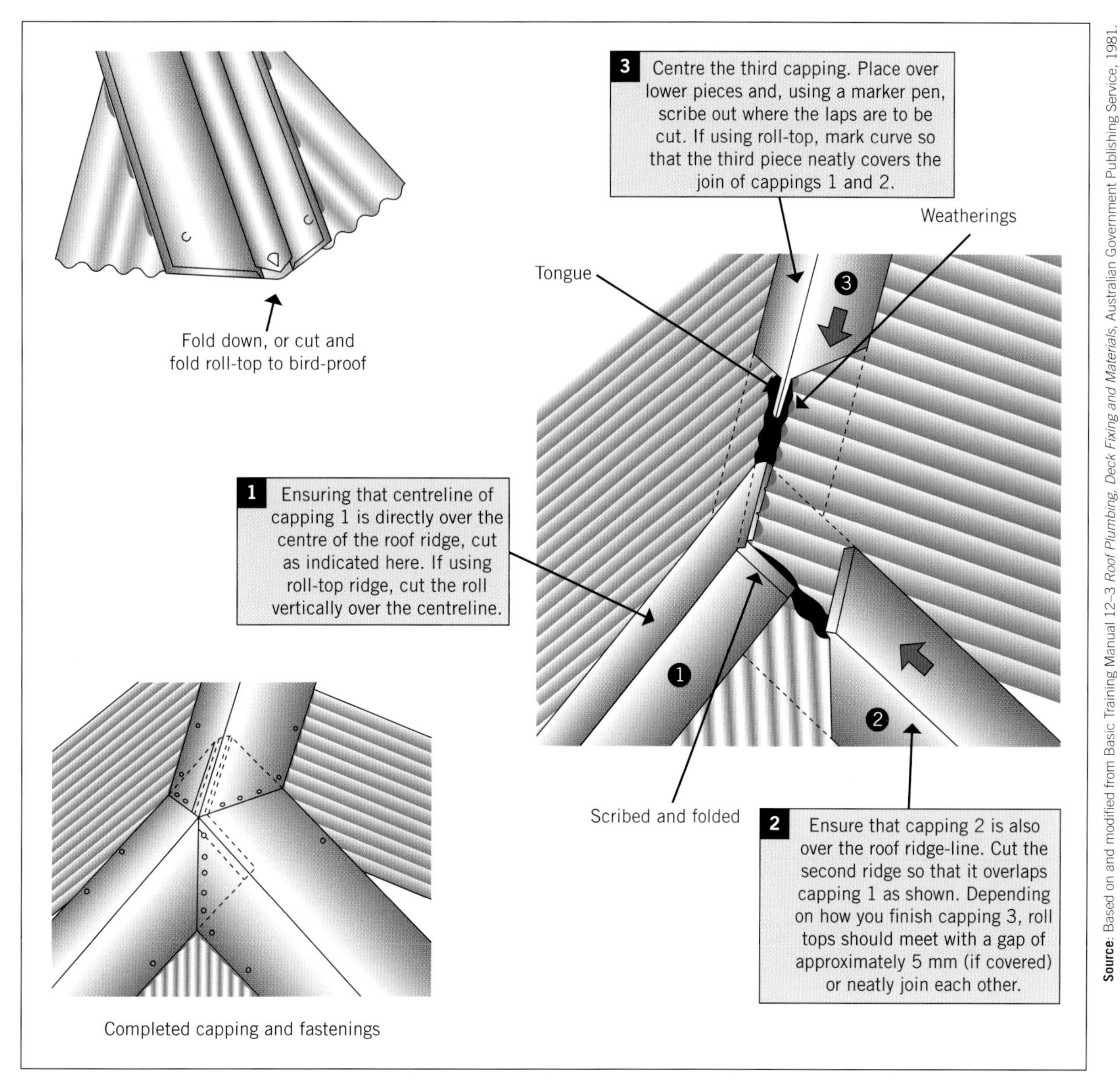

Source: Based on and modified from Basic Training Manual 12–3 *Roof Plumbing, Deck Fixing and Materials*, Australian Government Publishing Service, 1981.

FIGURE 13.18 Three steps for joining ridge capping

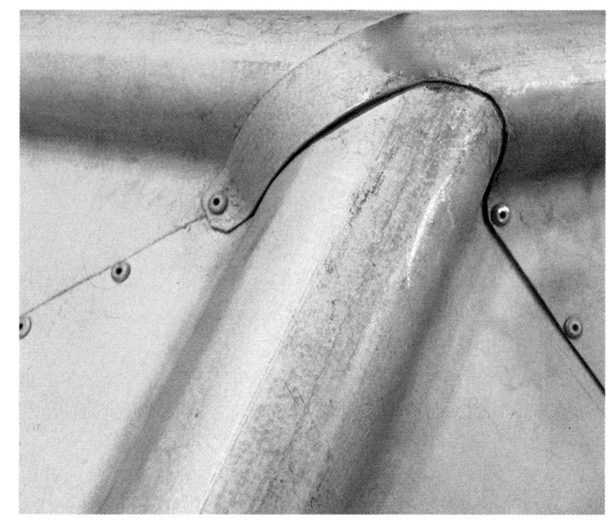

FIGURE 13.19 One method of finishing off a roll-top ridge intersection

FIGURE 13.20 Coated lead strip flashing neatly worked to take the shape of a reverse-curved ridge

SUMMARY

In this chapter you have reviewed the basic requirements related to the installation of commonly applied external flashings. The importance of detailed planning and product selection was covered with additional reminders relating to considerations for properties with BAL ratings.

Installation considerations included:

- material selection ensuring compatibility
- types of flashings
- wall flashings
- cappings
- examples of BAL installation requirements.

COMPLETE WORKSHEET 2

WORKSHEET 1

To be completed by teachers	
Satisfactory	☐
Not satisfactory	☐

Student name: ____________________

Enrolment year: ____________________

Class code: ____________________

Competency name/Number: ____________________

Task: Working with your teacher/supervisor, refer to this text, your HB39:2015 and any other relevant resource to answer the following questions.

1 What are three desirable qualities that you would look for in a flashing?

____

____

____

2 Which material is more malleable: lead or galvanised steel sheet?

____

____

____

3 What may happen if a ZINCALUME® steel flashing is installed above a galvanised corrugated roof sheet?

____

____

____

4 In your own words, describe the meaning of the term 'capillary action'.

____

____

____

5 Why can't bare lead flashing be used where roof water is collected for drinking?

____

____

____

6 You wish to use a ZINCALUME® steel soaker around a large roof penetration. What base metal thickness (BMT) will you specify for your sheet-metal worker to use in making the flashing? Provide a code reference to support your answer.

7 What type of silicone is recommended for roof plumbing?

8 What is the primary purpose of a flashing?

WORKSHEET 2

To be completed by teachers	
Satisfactory	☐
Not satisfactory	☐

Student name: ______________________________

Enrolment year: ______________________________

Class code: ______________________________

Competency name/Number: ______________________________

WORKSHEETS 13

Task: Working with your teacher/supervisor, refer to this text, your HB39:2015 and any other relevant resource to answer the following questions.

1 Is a transverse flashing positioned across or with the slope in the roof?

____
____
____

2 Where flashings are lapped, what is the minimum lap size and maximum interval of lap fasteners? Provide a code reference to support your answer.

____
____
____

3 What is the minimum cover that a transverse apron flashing must provide over a pierce-fixed roof? Provide a code reference to support your answer.

____
____
____

4 Draw a shape in cross-section for each of the following flashing/capping profiles:

Apron Pressure Ridge Parapet

5 By what minimum amount should any capping overlap the upstand of an apron flashing? Provide a code reference to support your answer.

____
____
____

6 You are about to secure a roll-top ridge capping to a corrugated steel sheet roof in a non-cyclonic area. What fastener interval should you apply? Provide a code reference to support your answer.

____
____
____

7 What depth should you set the diamond blade on your circular saw before you chase a 10-mm raked saw cut down the side of a wall? Provide a code reference to support your answer.

---

8 What two substances are recommended for sealing the brick course after the insertion of a counter flashing weathering fold? Provide a code reference to support your answer.

---

9 At what interval should you use fasteners on a sloping apron flashing? Provide a code reference to support your answer.

---

10 By how much should each stepped flashing overlap the one below it? Provide a code reference to support your answer.

---

11 In fabricating a transverse counter flashing to go over the flat upstand of an apron, what angle should you make the anti-capillary break and how wide should it be? Provide a code reference to support your answer.

---

12 If you live in a cyclonic wind area in northern Australia, what standard practice should you apply to the turn-down of a ridge capping or apron flashing when fitted over corrugated steel roof sheeting?

---

13 What is the minimum depth that a transverse counter flashing or stepped flashing should be cut into the brick course? Provide a code reference to support your answer.

____

____

____

14 At what maximum interval should barge capping be fastened into the crest of the roof sheet? Provide a code reference to support your answer.

____

____

____

15 What is the minimum width of silicone sealant required when applying a pressure flashing above an apron? Provide a code reference to support your answer.

____

____

____

16 Parapet capping must have a fall back toward the roof of at least ....degrees. Insert the correct answer and provide a code reference to support your answer.

____

____

____

17 What process must you carry out when installing transverse flashing or capping over wide-pan roof sheeting? Provide a code reference to support your answer.

____

____

____

14 INSTALL ROOF SHEETS, WALL CLADDING AND COMPLEX FLASHINGS

The chapter has as its primary focus the use of metallic and non-metallic profile roof sheeting to cover composite roofs. In its simplest form, a composite roof is the structural combination of a hip roof and a gable roof, but may also be a multi-hip construction including broken ridges, Dutch gables and attached lean-to roofs (see Figure 14.1). Essentially, a composite roof is a mix of structural angles, plane and pitch, as compared to a simple low-pitch or gable roof.

Composite style roofs are generally found in residential applications and often require sheet and flashing installation techniques that can be more complex than those found in commercial applications. Therefore, this chapter will primarily look at residential installation under the following areas of focus:

- Roof types
- Sheet profiles and pitch
- Planning, task sequencing and product selection
- Quantity calculations
- Installation of sheets and flashings

FIGURE 14.1 Example of a domestic composite roof

What should I read first?

Before proceeding with this chapter it would be advisable to revise Part A of this text with particular attention to Chapter 7 'Capillary action', as well as pre-reading and cross-referencing with Chapter 12 'Select and install roof sheeting and wall cladding' and Chapter 13 'Fabricate and install external flashings'.

Overview

To avoid unnecessary repetition, this chapter will have as its primary focus the installation of pierce-fixed roof systems, though many aspects will equally apply to concealed-fixed roofing profiles. It should be reviewed in conjunction with Chapter 12 'Select and install roof sheeting and wall cladding', where concealed-fixed roofing is covered in more detail.

Types of roof and cladding systems

Profiled roofing and cladding sheets provide designers and builders with a product that offers important features for a range of construction applications. These include that it:

- is lightweight
- offers ease of handling
- can be installed quickly
- is suitable for curving
- is durable
- is available in a wide range of colours.

The various products available for roof and walling installation can be broadly separated into two specific categories: *pierce-fixed systems* and *concealed-fixed systems*. Each offers advantages and characteristics that provide designers, builders and roofing contractors with the choices to best match systems to the building type, application, location and weather zones.

Pierce-fixed systems

A pierce-fixed roof and cladding system is secured to the building structure with fasteners that pass through the sheet. The selection and application of particular fasteners must be matched to the particular profile being installed.

Roofing sheets are generally pierced through the crests of the sheet profile, while wall cladding sheets are pierced through the pan (see Figure 14.2).

The traditional corrugated roofing and walling products that are commonly seen throughout Australia are pierce-fixed systems with the sheets screwed to either metal purlins or timber battens.

Concealed-fixed systems

A concealed-fixed roof and wall cladding system is secured to the structure with the use of special clips that do not pass through the sheet at all. The clips are located beneath the sheet itself and are screwed

FIGURE 14.2 Corrugated pierce-fixed sheets used on a composite roof

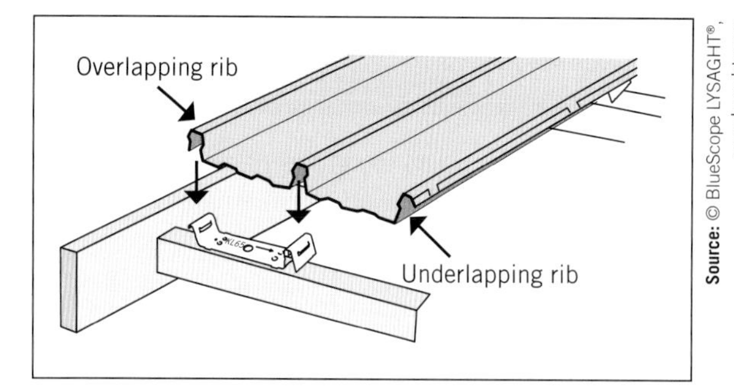

Source: © BlueScope LYSAGHT®, www.lysaght.com

FIGURE 14.3 Securing concealed-fixed roof sheet profiles with clips beneath the sheet

directly to the purlin (see Figure 14.3). There are various systems of holding the sheet in place, but a common method is for the sheet to be snap-locked into position over the specially shaped clips.

With no fasteners piercing the sheets, this roofing and walling system is inherently watertight. The proprietary clips are specifically made for each profile and must be used in strict accordance with the manufacturer's instructions.

Concealed-fixed roofing and cladding products are especially popular for use on commercial and industrial buildings. They are less commonly used on domestic composite roofs.

Profiles, effective cover and minimum pitch

The 'profile' of a sheet is a description of its cross-sectional shape. A wide range of profiles is available from different manufacturers. As well as providing rigidity to a sheet, each profile has been designed to satisfy certain requirements and applications. These may include:

- aesthetic difference
- better holding capability
- different pitch applications
- suitability for curving
- ease and speed of installation.

For many jobs, the building designer will have already chosen a sheet profile to satisfy the needs of the client and the structural requirements. The approved plans and specifications will provide most of these details. In other instances, you will need to make the correct selection based on a site visit and client requirements.

Each manufacturer has a particular name for each of its respective roof sheet and walling products. However, when referencing plans, specifications, codes and standards, such profiles will generally be referred to with a generic description, requiring you to become used to matching these with product names. These profile categories include:

- corrugated
- trapezoidal
- close-pitched trapezoidal
- wide rib
- narrow rib.

HB39:2015 SECTION 7.1 'TYPICAL PROFILES'

Table 14.1 shows some examples of corresponding LYSAGHT® product and generic profile names.

Effective cover

Another feature of the various roof sheet profiles is the difference between the actual width of the sheet and what is known as 'effective cover' (or 'cover width'). *Actual width* is the measurement from one side of the sheet to the other, whereas *effective cover* is the measurement of the amount of cover provided on the roof less the lap on one side.

The effective cover of any sheet is a vital measurement for every roof plumber to understand as it allows you to calculate roof sheet quantities and assists in roof setting out. A selection of roof sheet and wall cladding profiles comparing the difference between their respective actual widths and effective cover is shown in Table 14.1.

Minimum pitch

Roof sheet profiles are designed and approved for installations that meet minimum pitch requirements. You must always check that the roof sheet profile you select is approved for use on the roof pitch specified in the plans or measured on the job.

Table 14.2 shows some of the variations in minimum pitch between different profiles.

For instance, the rib or crest height of some profiles is much lower than others. If such a profile is installed

TABLE 14.1 Comparison between actual sheet width and effective cover of various roof sheet profiles

Profile	Cross-section	Actual width(mm)	Effective cover(mm)
CUSTOM ORB® (corrugated)		838	762
CUSTOM BLUE ORB® (corrugated)		838	762
TRIMDEK® (trapezoidal)		816	762
SPANDEK® (close-pitched trapezoidal)		754	700
KLIP-LOK® 406 (wide rib)		432	406

Source: © BlueScope LYSAGHT®, www.lysaght.com

TABLE 14.2 Minimum roof pitch of different profiles

Profile	Profile cross-section	Minimum roof pitch
KLIP-LOK®406 (wide rib)		1°
TRIMDEK® (trapezoidal)		2°
SPANDEK® (close-pitched trapezoidal)		3°
CUSTOM ORB® (corrugated)		5°

Source: © BlueScope LYSAGHT®, www.lysaght.com

Source: Heath Bussell

FIGURE 14.4 Corrugated sheets must be laid at a minimum of 5° pitch.

below the recommended minimum pitch, it is quite possible that the pans or tray of the sheet will flood and allow water to overflow the lap between sheets. Avoid this by checking the roof frame with a pitch gauge and ensure that your selected profile is suitable.

Translucent sheeting

In addition to standard metallic roof and walling profiles, a number of translucent products are available that have been manufactured to match many standard metallic roof sheet profiles. Translucent sheets are commonly specified to provide natural lighting for domestic, commercial and industrial workspaces and living areas, and are available in a range of tints to suit different applications.

To ensure that plastic roof sheeting has adequate strength, you must check that the products you install have been manufactured according to AS 1562.3 'Design and installation of sheet roof and wall cladding – Plastic'.

Some specific installation requirements relate to the use of translucent sheeting, and these are covered in more detail later in this chapter.

LEARNING TASK 14.1

1 In reference to Figure 14.4, what pierce-fixed profile would you use to replace the unsuitable corrugated sheets?

2 Which of the following is the effective cover of CUSTOM ORB corrugated sheet? Circle the correct answer.

800 mm   700 mm   762 mm   754 mm

General planning and preparation

A successful installation depends on the efficient research and collation of information, plans, specifications, authorities' documentation and confirmation through site visits. Some of these points are covered below.

Scope of work

Immediately after being informed that you are to carry out an installation, you need to establish the scope of work so that you can start to plan tasks, determine product requirements and sequence job stages. Sources of this primary information include:

- your supervisor (for a general outline)
- company quality assurance protocols
- the approved plans
- the specifications
- manufacturers' product information
- the weather bureau.

Re-roof requirements

When determining the scope of work, a special mention needs to be made of the renewal of roofs on older buildings. Many roofing contractors will undertake the regular replacement of roofing for older domestic dwellings. This is fine, and is normally done by the roof plumber as the principal contractor, without a builder or project manager.

However, in this instance you may need to allow for a professional inspection of the roof frame at the quotation stage to ensure that it is structurally suitable for a new roof. This is because new roofs are not always compatible with older building frames. A new roof and its underlying insulation create a very effective sealed 'cap' over the roof structure. During high winds, the pressure differential between that in the roof space and that above the house can create significant uplift forces on the roof. Although a modern house frame is built to resist these forces through a system of tie-downs from the roof structure to the rest of the frame, older dwellings may not have adequate tie-down systems and components in place. As a result, new roof cladding may not be compatible with an old frame without structural improvements. See Table 14.3 for a comparison of the characteristics of old and new roof installations.

An older roof may survive for decades with minimal trouble from high winds because it effectively 'leaks' pressure, thus reducing the uplift forces on the roof sheets. However, when you replace this with a new 'sealed' roof using modern materials and fastening methods, you may be fitting a complete cap to the building that is stronger than the frame itself. During high winds, this can sometimes lead to the entire roof, including its battens, being torn from the frame in one piece. Older roof frames may therefore require upgrading before being renewed (see Figure 14.5).

TABLE 14.3 The general characteristics of old and new roof installations

Older dwellings	New dwellings
Short sheets Sheets nailed to timber battens Battens skew-nailed to rafters Minimal tie-down between roof members and house frame Gaps around sheets and flashings No reflective foil laminate (RFL) or insulation	Continuous-length sheets Sheets screwed to battens/purlins Battens often screwed to trusses Comprehensive tie-down of roof trusses to house frame Minimal gaps into roof space RFL or insulation, creating a thermally sealed roof space
Result: an 'unsealed' roof space with limited tie-down but lower uplift forces.	**Result: a 'sealed' roof space with greater uplift forces but stronger tie-down.**

FIGURE 14.5 An older roof frame in need of upgrading

When planning to install a new roof on an older building frame you need to consider the following:

- Firstly you MUST check local state and territory licensing requirements. In some areas only an accredited builder is permitted to install roof battens and purlins.
- Old timber battens may be split lengthways from previously used spring-head nails and will need replacement. Never re-use split battens!
- Timber battens that have previously been skew-nailed into rafters will need to be fastened more securely with 'batten screws'.
- Many old roof frames were originally constructed of 'green' hardwood that has since shrunk considerably, leaving ill-fitting joints and loose fasteners.
- The rafters may require better tie-down to other frame members.
- Timber roof members suffering from 'dry-rot' will need replacement.

While roof contractors will often replace and screw down new battens, any structural changes and new tie-down requirements will need to be determined and carried out by an accredited builder. Think about this before your next re-roof.

Roofing requirements in areas subject to bushfires

In certain areas land and houses must be evaluated against the risk of bushfire damage and then assigned a Bushfire Attack Level (BAL) rating. BAL ratings are as follows:

- BAL – Low
- BAL – 12.5
- BAL – 19
- BAL – 29
- BAL – 40
- BAL – FZ (Flame Zone)

The BAL rating influences your component purchases and subsequent installation. If necessary, revise the section on BAL ratings in Chapter 12 'Select and install roof sheets and wall cladding'.

Plans and specifications

The plans and specifications for a job are essential for the preparation process. The approved plans will include:

- Plan view – a bird's-eye view of the building showing a dimensioned outline of all roof sections, locations of gutters, downpipes, penetrations, sumps and rainheads
- Elevations and sections – dimensioned side and sectional views of the building also showing height above natural ground level, roof pitch, shape and/or curve, direction of flow, special flashing cross-sections and relevant rainwater goods locations
- Site plan – a plan of the site showing its slope, contour intervals, access points, approved points of stormwater discharge, boundaries and other services.

The specifications may be provided as an integral part of the plans, or may be a separate document. In relation to roof and wall cladding, specifications may include:

- covering material – profile type, material type, thickness, coating and colours
- rainwater goods – gutter profile and effective cross-sectional area, material type, thickness, coating, colours and overflow types/sizes
- safety mesh requirements
- property BAL rating
- insulation requirements
- fasteners – type, gauge, length and coating
- standards and codes – the specifications may list relevant manufacture and installation codes in reference to which all work must be conducted.

Plans and specifications are evaluated together as part of the local authority approval process, and you are therefore required to adhere to the content of the specifications.

Manufacturers' product information

All roofing and walling products are backed by information produced by the manufacturer that is designed to assist you in the selection and installation of its products. Important product information may include:

- details of Australian Standards compliance
- BAL rating
- performance parameters
- assembly instructions
- coating details
- installation and use instructions
- any product care and maintenance recommendations.

As an example, this chapter will make regular reference to the following three sources of manufacturer information:

- LYSAGHT® product range, available at http://www.lysaght.com (> products > roofing and walling)
- LYSAGHT® *Roofing & Walling Installation Manual*, available at http://www.lysaght.com (> installation and maintenance > roofing and walling)

Bradford™ *Bushfire Roofing Systems Design Guide* is available on the Bradford website, and similar publications are available for free download from the relevant company websites.

Insulation requirements

All new buildings must now meet thermal performance requirements that are more stringent than they were in the past. A significant part of this performance is determined by the correct installation of roof and wall insulation products. Almost all roof and wall cladding jobs now require the roof plumber to also install some type of insulation membrane, batt or blanket.

More general detail on insulation products is found in Chapter 12 'Select and install roof sheeting and wall cladding'.

GREEN TIP

In order to meet the energy efficiency provisions set out by the National Construction Code, designers and builders select very specific insulation solutions. These products will only work to standard if they are installed correctly. As such, the modern roof plumber must know as much about roofing insulation materials and installation requirements as how to install these products.

Site inspection

While the approved plans detail the original intent of the project, the site inspection is all about confirmation of what is actually there. Variations occur in all construction jobs, and you always need to account for these with a site visit before and often during your activities on the site. Examples of what you will be doing during a site visit include:

- consulting with the site foreman, WHS representative, builder and/or owner
- checking the site access
- checking the storage areas
- checking for overhead electrical hazards
- checking the roof access requirements
- confirming all measurements
- confirming the intended direction of sheet laying.

Safety

All aspects of the job must be planned for and carried out with constant observance of all safety considerations. Identification of hazards is vital on every job (see Figure 14.6 for an example). This is also the time where you may need to ensure all personal fall arrest equipment is in complete working condition. Review Chapter 3 'Basic roof safety' for more details.

FIGURE 14.6 Power line 'tiger tails' visually identifying an electrical hazard near the work area

Task sequencing

In its simplest form, the task sequencing for a standard roof sheet installation might include the following stages:

- Receive the materials onsite.
- Lay any required safety mesh and tie it off.
- Lay the insulation blanket or RFL and secure the sheets.
- Install the flashings.
- Clean up.

Of course, there is much more to the job than this, particularly where you must work in with the builder's schedule and the activities of other trades. You need to examine each part of your own tasks and determine how your work might affect others around you and how their activities will impact upon your own. The questions you need to ask include:

- What stage and completion deadlines do you need to be aware of?
- What safety systems do you need to put in place and how will this affect other trades?
- What is the lead time required between ordering materials and their arrival onsite?
- How will you sequence delivery and plan for materials storage onsite?
- Which trades do you need to work with during the installation?

It is important that you review and assess your task sequencing on almost every day of the job in order to accommodate delays due to bad weather, delivery problems and the activities of other trades.

Tools and equipment

The tools and equipment required for roof sheeting and wall cladding include all of those items listed in Chapter 9 'Roof plumbing tools'.

LEARNING TASK 14.2

1. What is the main reason older roofs survive for decades with minimal trouble from high winds?
2. Who should be consulted if structural changes are required to an older timber frame?
3. What is the name of the safety covering applied to live electrical hazard near the work area?

Product selection

While the approved plans and specifications will detail general requirements for particular aspects of the job, the selection of individual products to meet specification and performance requirements will often be left to the individual contractor.

FROM EXPERIENCE

Each job has its own requirements, and you should be wary of falling into the habit of simply selecting the products that you are familiar with or just because that is '... what everyone else uses'.

Product selection has never been simpler, with a range of information sources available that include:

- knowledgeable suppliers' representatives
- manufacturers' sales and product representatives
- hard-copy product information
- internet searches for installation and product guides.

In this section you will see some examples of how to select certain installation products.

Selection of roof sheets

Although there are many roof sheet profiles available, not all are suitable for the different forms of roof installation. Variations in fixing requirements, compatible fasteners, roof pitch and support spans mean that you need to interpret the requirements of the specifications and match these to suitable products. If the profile is not specified in the plans, you may need to research and confirm a suitable profile yourself. For example, information to support a pierce-fixed roof installation is available via simple internet searches, as shown in Figure 14.7.

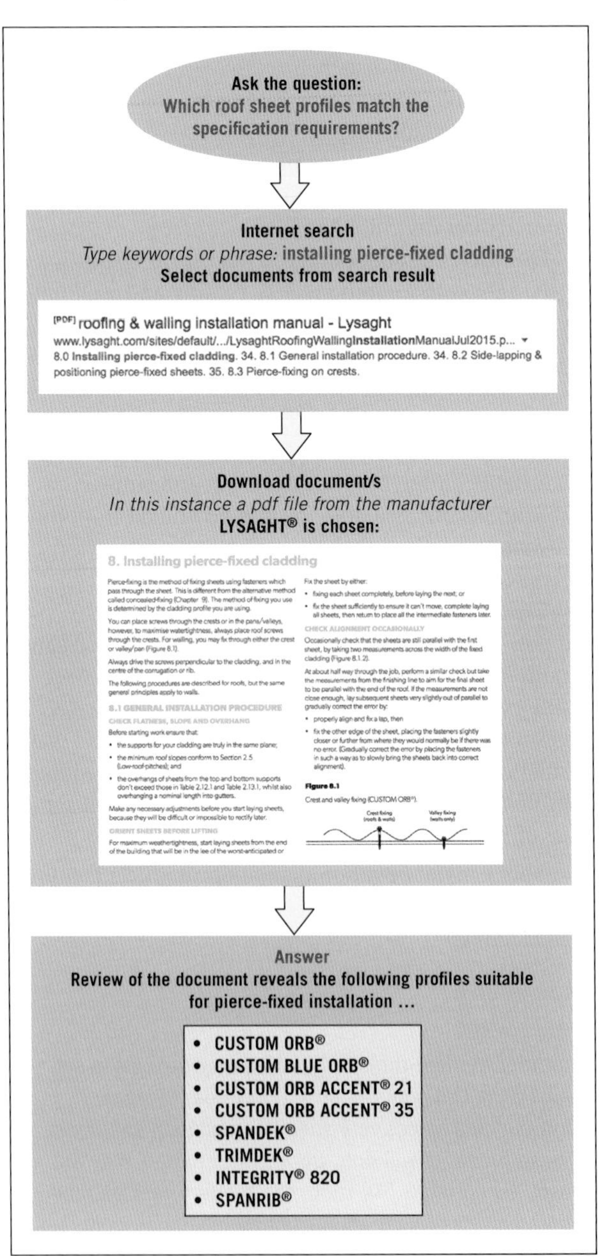

FIGURE 14.7 Using an internet search to find information for pierced-fixed roof installations

Source: © BlueScope LYSAGHT®, www.lysaght.com.

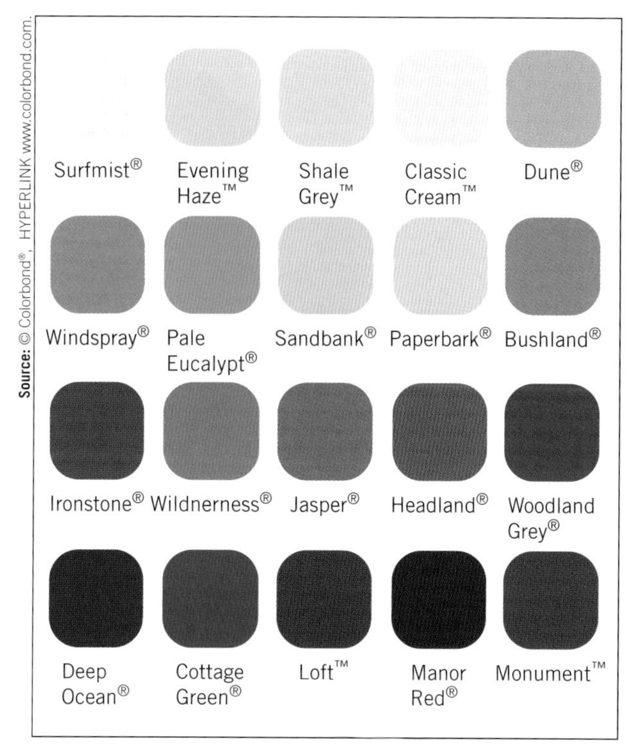

FIGURE 14.8 Selecting roof colour for thermal performance

Colour selection

Product colour selection may be detailed in the plan specifications or you may need to determine this through negotiation with the builder or owner. Often, your advice will be sought on the range of colours on offer. Not all colours are available in all areas, and you will need to check with your suppliers on what colour choices are on offer for various products in your region. Refer to relevant websites for specific information, but don't forget to have hard copy colour charts on hand to show and discuss with clients.

Colours and sustainable building practices

Colour selection is an important aesthetic choice made by the owner and designer for any building project (see Figure 14.8). A point often missed, however, is the impact that colour selection can have on building heating and cooling running costs and its accumulative effect on the environment.

GREEN TIP

Heating and cooling of houses is a significant ongoing cost for most building owners. The correct selection of colours can assist considerably in reducing the energy consumption of a building.

Dark roof colours will absorb substantially more heat than lighter colours or plain ZINCALUME® steel. This is not to say that dark colours should always be excluded, but they should be selected in reference to the following:

- regional climate zone
- roof type
- ceiling type
- colour and material thermal absorbance
- insulation combination.

The full colour/insulation selection process to meet National Construction Code requirements is beyond the scope of this text, but where you need to deal with this question you should refer to the insulation supplier, roofing product supplier and/or building designer for assistance.

Selection of fasteners

The choice of fasteners will depend on two sources of information:

- the roof sheet manufacturer's recommendation for size, gauge and coating
- the fastener manufacturer's recommendation of products that meet these size, gauge and coating requirements.

You will need to choose fasteners for a wide range of applications and materials, including:

- crest fastening
- pan/valley fastening
- ridge fastening
- fastening of concealed fixings
- flashing fasteners
- translucent sheet fastening.

In relation to pierce-fixed roofing requirements, you will need to select fasteners that are suitable for securing the profile to metal purlins or timber battens, subject to the requirements of your job. Cyclonic zones and insulation thickness will also influence selection. To find a solution, you need to determine the type of fastener recommended by the roof/wall product manufacturer and then find a matching product from the fastener manufacturer.

The example in Figure 14.9 demonstrates some steps in the process of fastener selection.

LEARNING TASK 14.3

1 Which aspects of the proposed installation might not make the roof sheet profile suitable?
2 What are two sources of information that can help with the selection of product?
3 The choice of fasteners will depend on two sources of information. What are they?

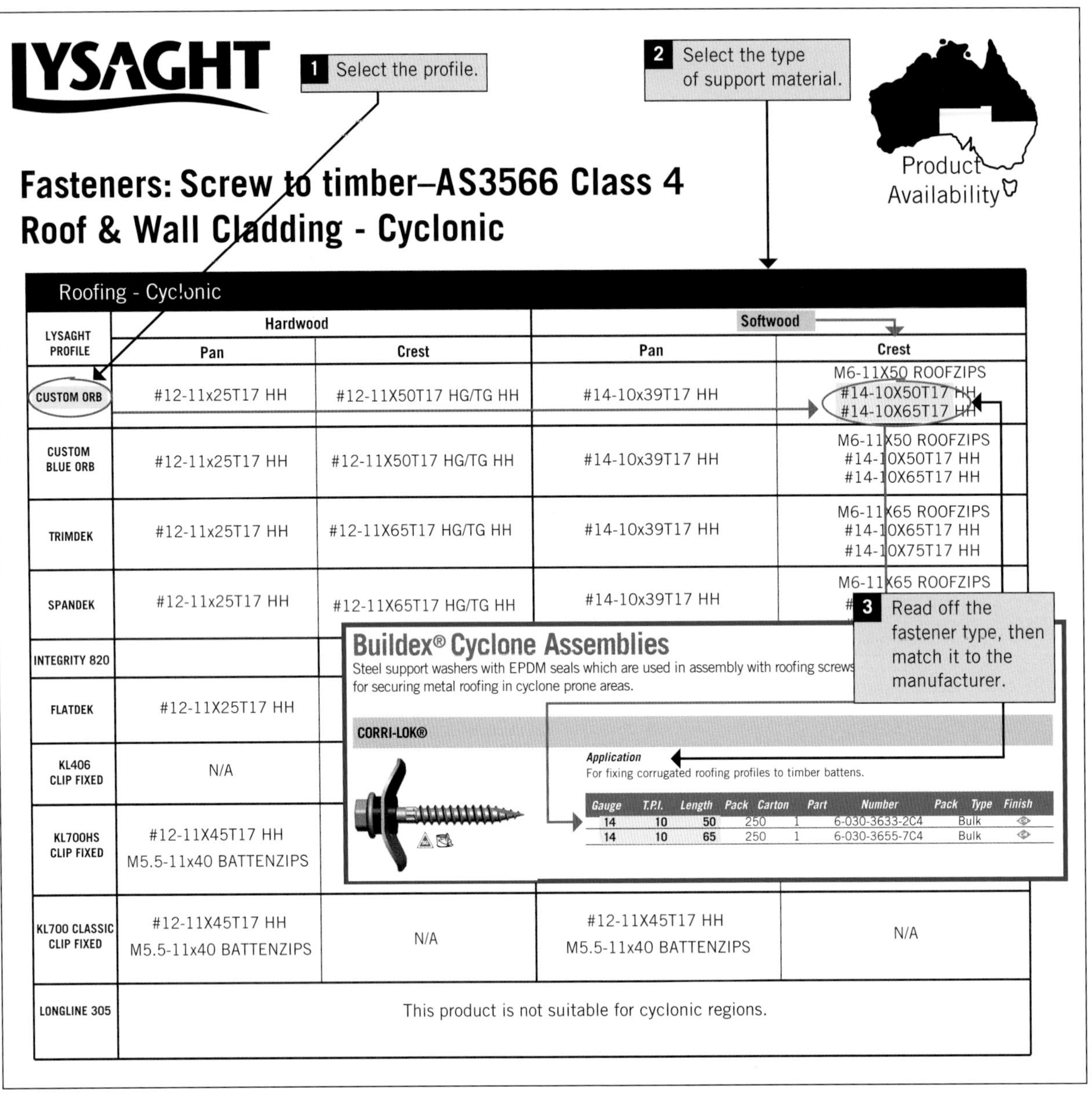

LYSAGHT PROFILE	Hardwood Pan	Hardwood Crest	Softwood Pan	Softwood Crest
CUSTOM ORB	#12-11x25T17 HH	#12-11X50T17 HG/TG HH	#14-10x39T17 HH	M6-11X50 ROOFZIPS #14-10X50T17 HH #14-10X65T17 HH
CUSTOM BLUE ORB	#12-11x25T17 HH	#12-11X50T17 HG/TG HH	#14-10x39T17 HH	M6-11X50 ROOFZIPS #14-10X50T17 HH #14-10X65T17 HH
TRIMDEK	#12-11x25T17 HH	#12-11X65T17 HG/TG HH	#14-10x39T17 HH	M6-11X65 ROOFZIPS #14-10X65T17 HH #14-10X75T17 HH
SPANDEK	#12-11x25T17 HH	#12-11X65T17 HG/TG HH	#14-10x39T17 HH	M6-11X65 ROOFZIPS
INTEGRITY 820				
FLATDEK	#12-11X25T17 HH			
KL406 CLIP FIXED	N/A			
KL700HS CLIP FIXED	#12-11X45T17 HH M5.5-11x40 BATTENZIPS			
KL700 CLASSIC CLIP FIXED	#12-11X45T17 HH M5.5-11x40 BATTENZIPS	N/A	#12-11X45T17 HH M5.5-11x40 BATTENZIPS	N/A
LONGLINE 305	This product is not suitable for cyclonic regions.			

Gauge	T.P.I.	Length	Pack	Carton	Part Number	Pack Type	Finish
14	10	50	250	1	6-030-3633-2C4	Bulk	
14	10	65	250	1	6-030-3655-7C4	Bulk	

FIGURE 14.9 The process of choosing fasteners that match the sheet manufacturer's requirements

Source: © BlueScope LYSAGHT®, www.lysaght.com and courtesy of Buildex®.

Quantity calculations

The determination of material quantities is based on some simple calculations that relate the specified product requirements to various relevant dimensions on the job. These may be lineal measurements, square metres or multiples of an object or product.

The examples examined in this section are based upon a small, equal-hip roof frame measuring 20 m long and 10 m wide to be covered in LYSAGHT TRIMDEK® roof sheets.

Roof area

Knowing the roof area is useful when ordering materials such as safety mesh and RFL. For items such as these, it is generally sufficient to determine dimensions directly from the plan view and elevations of the approved plans (length of sheets should always be measured onsite).

An equal-hip roof has the same pitch on all four sides. The method for calculating the area of an equal-hip roof is essentially the same as calculating the area of a gable roof of the same pitch. Remember to always add 50 mm on either side for overhang into the gutter.

Referring to the plan, use the procedure shown in Figure 14.10.

Applying this to our example roof frame:

Roof area = (Common rafter length + 50 mm) × (Plan length + 100 mm) × 2

= (5.78 + 0.05) × (20 + 0.1) × 2

= 5.83 × 20.1 × 2

= 234.36 m^2

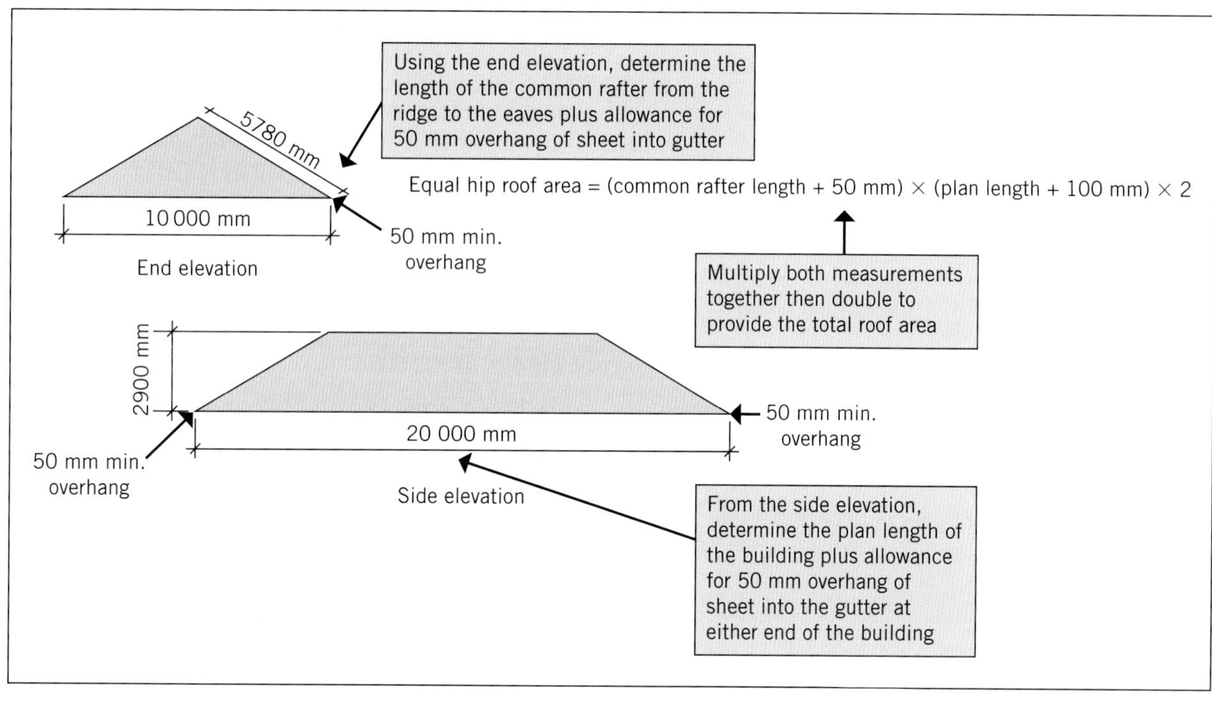

FIGURE 14.10 Calculating the area of a roof

EXAMPLE 14.1

Determine the length of the common rafter

Where you are unable to determine the length of the common rafter you will need to calculate this using Pythagoras' theorem:

$$H^2 = A^2 + B^2$$

Where:

H = length of the common rafter (hypotenuse)
A = vertical height from ceiling to ridge
B = plan width from fascia to vertically below ridge

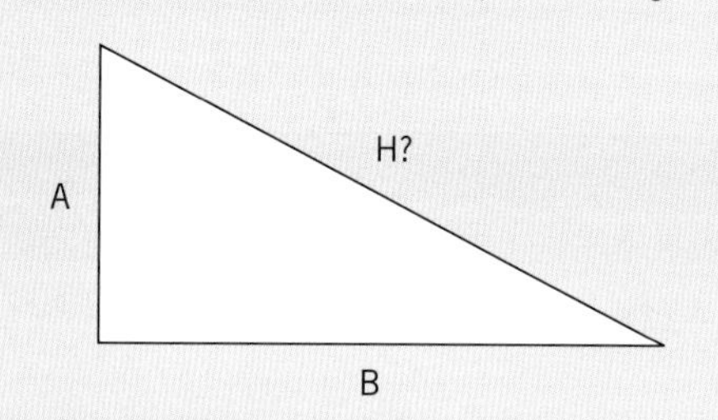

Following the example:

$H^2 = A^2 + B^2$
$= 2.9^2 + 5^2$
$= 8.41 + 25$
$H^2 = 33.41$
$H = \sqrt{33.41}$
$H = 5.78$ m

Insulation quantity

The specifications for this job also require the installation of an R1.3 foil-backed insulation blanket. A local insulation sales representative tells you that this product is available with a nominal coverage of 18 m² per roll. How many rolls do you need to cover this equal-hip roof?

EXAMPLE 14.2

Calculate number of insulation rolls required

$$\text{Number of insulation rolls} = \frac{\text{Roof in square metres}}{\text{Square metres per roll}}$$
$$= \frac{234.36}{18}$$
$$= 13.02$$
= 14 (answer can be rounded up to the next highest)

Number of roof sheets

Once you know the effective cover measurement of the specified profile, you can calculate the number of sheets required for the roof. Remember that you need to double your answer to ensure that you account for the other side of the ridge.

The number of sheets is determined by dividing the total width of the roof by the effective cover of the particular sheet profile, then multiplying by 2:

$$\text{Number of sheets required} = \frac{\text{Roof total width}}{\text{Effective cover}} \times 2$$

EXAMPLE 14.3

Calculate number of roof sheets required

The hip roof in our example has an overall width of 20 m. The specifications require the use of LYSAGHT TRIMDEK® sheets for the job. After referring to the manufacturer's information for the effective cover, you calculate the number as follows:

$$\text{Number of sheets required} = \frac{\text{Roof total width}}{\text{Effective cover}} \times 2$$
$$= \frac{20}{0.762}$$
$$= 26.247 \times 2$$
$$= 52.49$$

Number of sheets required = 53 (rounded up to whole sheet)

You will notice that in this calculation the final answer has been rounded up to the next whole sheet. Always round up your answer to ensure that you have full coverage of the roof.

Manufacturers also provide tables for each profile that have been pre-calculated to match simple roof widths to numbers of sheets (see Figure 14.11). You simply round up to the next full metre and then read off the number of sheets. In this example, you can see that a total roof width of 40 m requires 53 sheets – the same as in the previous calculation. Of course, if your roof is wider than indicated or precise measurements are required, you will need to work it out manually.

Fastener quantity

Each roof sheet profile has specific fastener needs that are detailed by the manufacturer. For pierce-fixed roofing products, a certain number of fasteners will be specified for both end and intermediate supports. Subject to building design, structures in cyclonic zones may also need additional fasteners to provide added protection from high winds. Table 14.4 shows some examples of different profiles and fastener needs.

Fastener quantity for pierce-fixed roof sheets

The formula for determining the number of fasteners for pierce-fixed sheeting is as follows:

Number of fasteners = Number of fasteners per support × Number of sheets × Number of supports

LYSAGHT TRIMDEK® sheet coverage																				
Width of roof (m)	3	4	5	6	7	8	9	10	11	12	13	14	15	16	17	18	19	20	30	40	50
Number of sheets	4	6	7	8	10	11	12	14	15	16	18	19	20	21	23	24	25	27	40	53	66

FIGURE 14.11 Example of a manufacturer's table for calculating sheet coverage

Source: © BlueScope LYSAGHT®, www.lysaght.com

TABLE 14.4 Examples of different profiles and their fastener requirements

Profile fastener location and number per support			
Profile	Support location	Number*	Fastener location
CUSTOM ORB® (corrugated)	Intermediate	3	
	Sheet ends	5	
TRIMDEK® (trapezoidal)	All supports	4	
SPANDEK® (close-pitched trapezoidal)	Intermediate	3	
	Sheet ends	4	

* Lap fastener is only counted once per sheet/support.

Source: © BlueScope LYSAGHT®, www.lysaght.com

EXAMPLE 14.4

Determine the number of fasteners for a sheet of TRIMDECK®

In the example of our 20 m × 10 m equal hip roof, there are a total of 7 purlin runs spaced at 900 mm centres on either side of the ridge across the width of the building. This means a total of 14 purlin runs.

A review of the manufacturer website reveals that a sheet of LYSAGHT TRIMDEK® requires 4 fasteners per support. You have previously worked out that there are 53 sheets required for this job. Therefore, the number of fasteners is determined as follows:

Number of fasteners = No. fasteners per support × No. of sheets × No. of supports
= $4 \times 53 \times 14$
= 2968

LYSAGHT TRIMDEK® is a profile that needs every crest over every support to be screwed down. You will notice that other profiles have different fastener requirements for the intermediate and sheet end supports. If the specifications for this job had required the use of a corrugated profile, you would need to break the calculation into two parts.

EXAMPLE 14.5

Determine the number of fasteners for a sheet of LYSAGHT CUSTOM ORB®

In the example of our 20 m × 10 m equal hip roof, there are a total of 7 purlin runs spaced at 900 mm centres on either side of the ridge across the width of the building. This means a total of 14 purlin runs. Of these, 12 purlin runs would be intermediate and 4 purlin runs would be positioned at the top and bottom of the sheets.

A review of the manufacturer's website reveals that a sheet of LYSAGHT CUSTOM ORB® requires 3 fasteners for each intermediate support and 5 fasteners for the sheet end supports. You have previously worked out that there are 53 sheets required for this job. Therefore, the number of fasteners is determined as follows:

Number of fasteners = No. fasteners per support × No. of sheets × No. of supports for intermediate supports
= $3 \times 53 \times 12$
= 1908

Number of fasteners = No. fasteners per support × No. of sheets × No. of supports for end supports
= $5 \times 53 \times 4$
= 1060

Total fasteners required = intermediate + end supports
= 2968 m

Some suppliers will provide fastener-reckoning charts that will enable you to quickly determine quantities by matching profiles with square metres. Speak to your supplier for any recommendations to assist you in this process.

COMPLETE WORKSHEET 1

Installation of a pierce-fixed roof

In this section you will be introduced to the basic procedures required to lay a standard pierce-fixed roof. Of course, jobs will vary considerably, but a sound knowledge of fundamental installation practice and code requirements will enable you to adapt your approach for more complex applications.

Once all the preparatory work has been completed, the installation can be broken up into a number of stages. These are:

1 Install rainwater goods.
2 Lay safety mesh (where required).
3 Lay insulation.
4 Fix sheets.
5 Install flashings.

HB39:2015 SECTION 7.9 'PIERCE-FASTENED DECKS'

Install rainwater goods

Some rainwater goods need to be installed before the roof sheets. These include valley gutters, eaves and box gutters. You may also need to prepare some large penetrations before installing the sheets. More details dealing with the installation of these parts of the roof system are covered in Chapter 10 'Fabricate and install roof drainage systems'.

Lay safety mesh

Many roofs will require the installation of correctly specified and installed safety mesh. This is vital for the safety of all roof workers during the installation and also, at a later date, for maintenance workers where translucent sheeting is used. You are permitted to use only approved safety mesh that is manufactured in compliance with AS/NZS 4389. One commonly encountered product is manufactured in 2-mm gauge wire with a rectangular 300 mm × 150 mm mesh.

HB39:2015 SECTION 4.4 'SAFETY MESH'

The longitudinal wires (down the roof pitch) are 150 mm apart and the transverse wires (across width of roof) are 300 mm apart.

More details on the use and installation of safety mesh are included in Chapter 12 'Select and install roof sheeting and wall cladding'.

Lay insulation

While the details relating to insulation are provided separately here, be aware that insulation is in fact laid at the same time as the roof sheets. The following points relate to the use of insulation blanket and reflective foil laminate (RFL).

Insulation blanket

Insulation blanket is commonly made from glasswool or rockwool fibre that is adhered to a grade of reflective foil laminate and performs as anti-condensation control and thermal insulation under the metal sheets. Insulation blanket acts as a major fire retardant component in BAL rated properties. Figure 14.12 below shows some of the differences between BAL 12.5–40 and the higher rated BAL FZ.

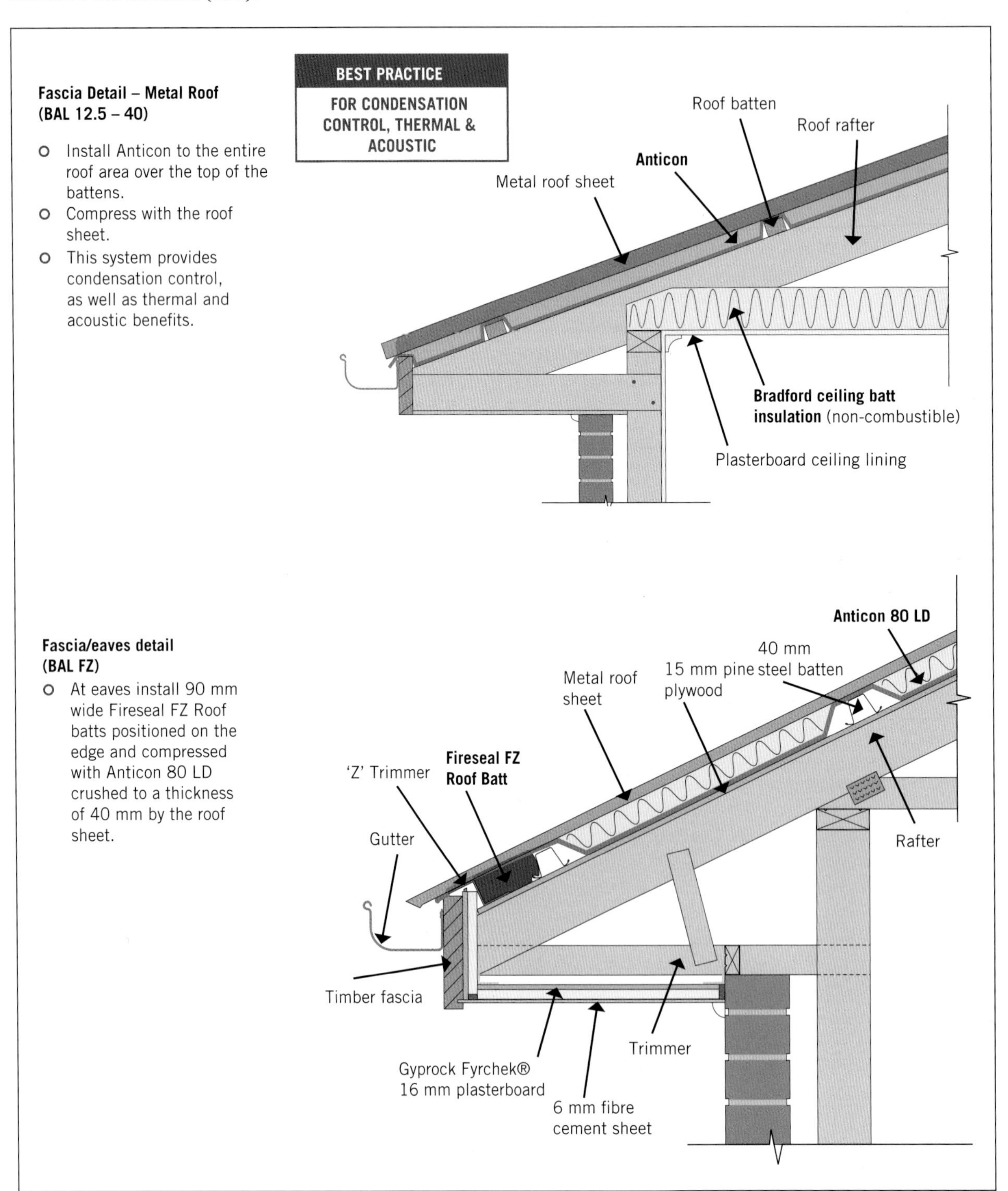

FIGURE 14.12 Ensure that you install insulation and associated components in accordance with the BAL rating

FIGURE 14.13 Insulation blanket being laid

Basic installation considerations are as follows:

- Using a stringline, check that all purlins are in plane so that all sheets can be fixed with an even grade. Once the insulation is on you will not be able to do this.
- Lay the blanket directly on top of the safety mesh, with RFL facing downwards (see Figure 14.13). (Note that the blanket used in tropical areas may need to be installed with the RFL facing up or, alternatively, installed in a conventional manner but with a second layer of RFL laid on top. Check with the manufacturer and local building requirements.)
- Lay the blanket down the pitch of the roof and overlap the preceding roll with the 150-mm side lap.
- With concealed-fixed profiles, fit spacers where necessary.
- Lay only as much insulation as you can soon cover with sheets.
- Prevent exposed insulation from becoming wet.
- Prevent exposed insulation from flogging in a high wind.

Reflective foil laminate

Reflective foil laminate (RFL) should be laid in a similar manner to insulation blanket, and the following points should be considered:

- Ensure that RFL is sagged between each batten (see Figure 14.14). The air gap between the membrane and the roof sheet is necessary for effective condensation control. Do not pull the RFL tight.
- Lap each length by at least 150 mm.
- Seal each lap and end-join with the adhesive tape recommended by the manufacturer.
- Do not allow excessive RFL to hang into the gutter.
- Prevent RFL from flogging in the wind.

Fix sheets

The following steps relate to the fixing of the roof sheets to the structure. It is always recommended that roof sheets be installed so that the laps face away from the prevailing weather, lessening the chance of wind-driven rain being forced past the sheet lap.

FIGURE 14.14 RFL should be laid with a slight sag between each batten

This means that if the wind generally comes from a westerly direction, you should start setting out and laying sheets from the eastern side of the roof so that each overlying lap will face away from the wind direction.

Mark out roof for sheet alignment

The first sheet that you secure on any roof must be correctly aligned and square with the frame. Failure to check and mark out your sheet positions can result in all remaining sheets being out of square to the fascia or gable ends. At times, this is so bad that it is visible from the ground, and it is evidence of extremely poor installation.

It is possible to correct misaligned pierce-fixed sheets gradually by 'creeping' them till the measurement is correct. However, this is more difficult and sometimes impossible with concealed-fixed sheets. It is better to spend more time in getting the set-out correct in the first place than to try and do it halfway into the installation.

On a *gable and skillion roof* you need to check that the end rafters are square with the fascia.

On a *hip roof* the first sheet is normally laid to intersect with the highest point of the hip. In this instance you need to use the fascia as a base line and establish a square set-out to the top of the hip.

When doing small fabrication tasks, you would normally use a square to check that something like a gutter angle is set at 90°. However, this is not a practical option when you need to establish a large-scale square line on a building frame. The simplest way to do this is to triangulate the area.

The 3:4:5 rule

The 3:4:5 rule (also known as the 6:8:10 rule – it's the same thing) is based on the relationship between each side of a right-angle triangle. If each side of a triangle is equal to multiples of the units 3:4:5, then the angle must be square. You simply use a tape measure to mark out 3 units in one direction and

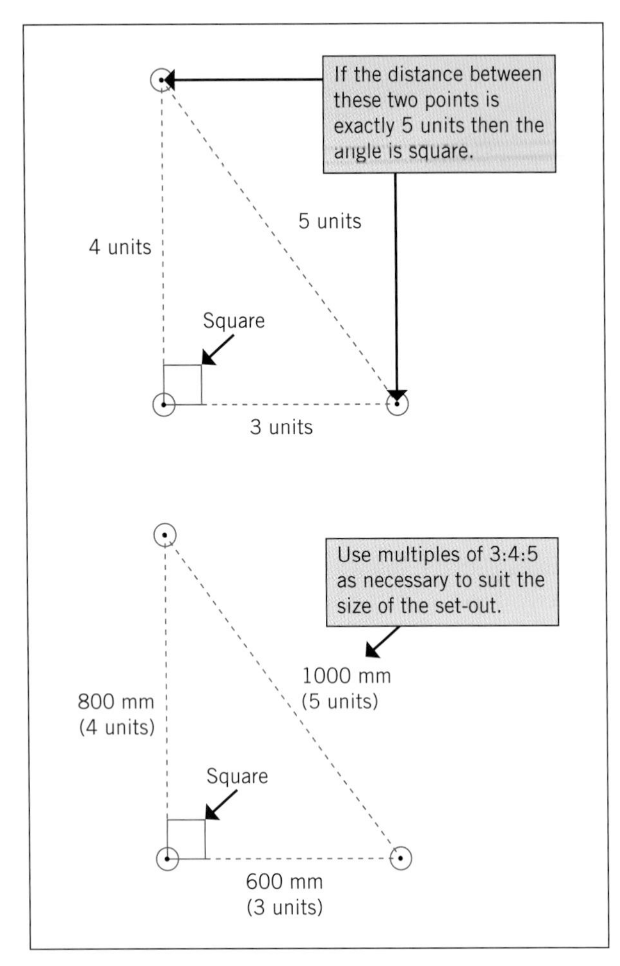

FIGURE 14.15 The 3:4:5 rule

4 units in the other. If the distance between each measurement is 5 units then you know that the shape is square. Figure 14.15 shows examples of how this relationship works.

Setting the first sheet (pierce-fixed)

There are some minor variations between different profiles and products, but the following procedure is a general guide:

1 The first sheet laid on a hip roof is normally positioned at the point where the ridgeline intersects with the hip (see Figure 14.16).
2 If you intend to fix the first sheet into position immediately, lay the first run of any RFL or insulation blanket.
3 Use the 3:4:5 rule to confirm that the fascia is square with the line of the first sheet and a similar point at the other end of the roof.

   The end of run may be at a gable, a wall or the highest point of another hip. If the line is not square, determine the amount of error so that the first sheet can be positioned to accommodate the inaccuracy. This is not normally a problem on a new construction, but older buildings will often have considerable variation in the frame.

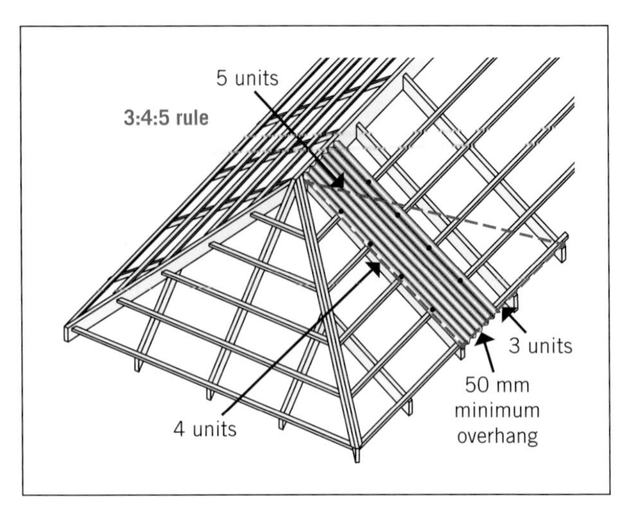

FIGURE 14.16 Positioning the first sheet on a hip roof

Source: Based on Basic Training Manual 12–3 *Roof Plumbing, Deck Fixing and Materials*, Australian Government Publishing Service, 1981.

4 Clearly mark the line square to the fascia.
5 Fit the first sheet into position, ensuring that it projects into the gutter by a minimum of 50 mm.

Turn up trays/valleys

The highest end of any roof sheet *must* be weathered to prevent wind-driven rain from flowing back past the end of the sheet into the building. In many cases you will be fitting sheets where the highest end will meet at a ridge or abut a parapet wall, so you will need to turn up the trays/valleys of each sheet before fixing as it is not normally possible to use the folder effectively once the sheet is in place. Turn up all trays in the manner recommended by the sheet manufacturer.

Installation of corrugated steel on a standard trussed hip roof will normally allow you to weather sheets with a pair of multigrips or a shifter after fixing (see Figure 14.17). Other profiles will require a large folder to do the job, so the sheets may require folding before fitting. Proceed as appropriate.

Correct use of fasteners

The performance of a pierce-fixed roof depends on the correct use of the selected fasteners. This involves checking that your screw-gun over-ride adjustment is set so that the screws are not over-driven or under-driven (see Figure 14.18). It is also important to drive the screws in at a perpendicular angle to the roof pitch.

Be aware that impact drivers have no depth adjustment and will almost always result in inconsistent fastener application. They may damage screw anti-corrosion coatings so check with the fastener manufacturer before using. The only way to ensure correct screw setting is by using a Tek gun with depth setting.

Begin laying sheets

Once the first sheet is fixed in position, you are ready to begin laying subsequent sheets in the manner

FIGURE 14.17 Weathering the valley of a corrugated sheet

described by the manufacturer. However, before doing so it is a good idea to position some check marks to ensure that the sheets are not drifting out of alignment as you progress. While this is more of a problem with pierce-fixed sheeting, it is always wise to regularly check concealed-fixed decks as well.

Regular alignment checks

You will have identified any inaccuracies between the fascia line and the end-wall/barge board when you were setting out the alignment of the first sheet. Your next goal is to ensure that when you have finished sheeting that particular section, the sheets are still parallel with the end-wall/barge board and perpendicular to the fascia line. The following method is one way of ensuring that this is achieved (see Figure 14.19):

1. Once the first sheet is in position and parallel to the far end-wall/barge board, measure back from the completion point and mark off a series of regular checkpoints on the top and bottom purlins.
2. The distance between each mark should be short enough that one person can make a quick check while the sheets are being laid without having to waste time measuring all the way to the end of the run. A 1–2-m interval is usually fine.
3. On a regular basis, simply measure between the edge of the last sheet and the next purlin mark to check for any misalignment.

Can't I just measure the sheet overhang past the fascia?

Some roof workers will measure each side of the overhang of the sheet past the fascia in an attempt to maintain sheet alignment. This method can't be recommended, particularly when long sheets are being installed. A 1-mm discrepancy at the bottom may translate into a 10-mm misalignment at the top

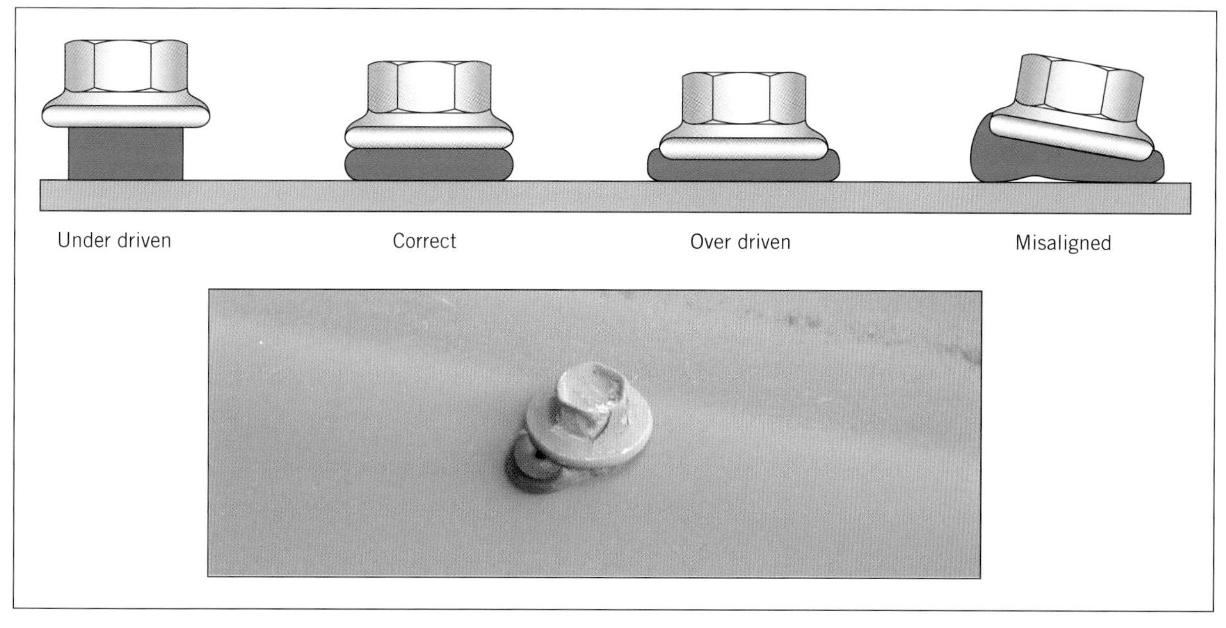

FIGURE 14.18 The correct and incorrect use of fasteners

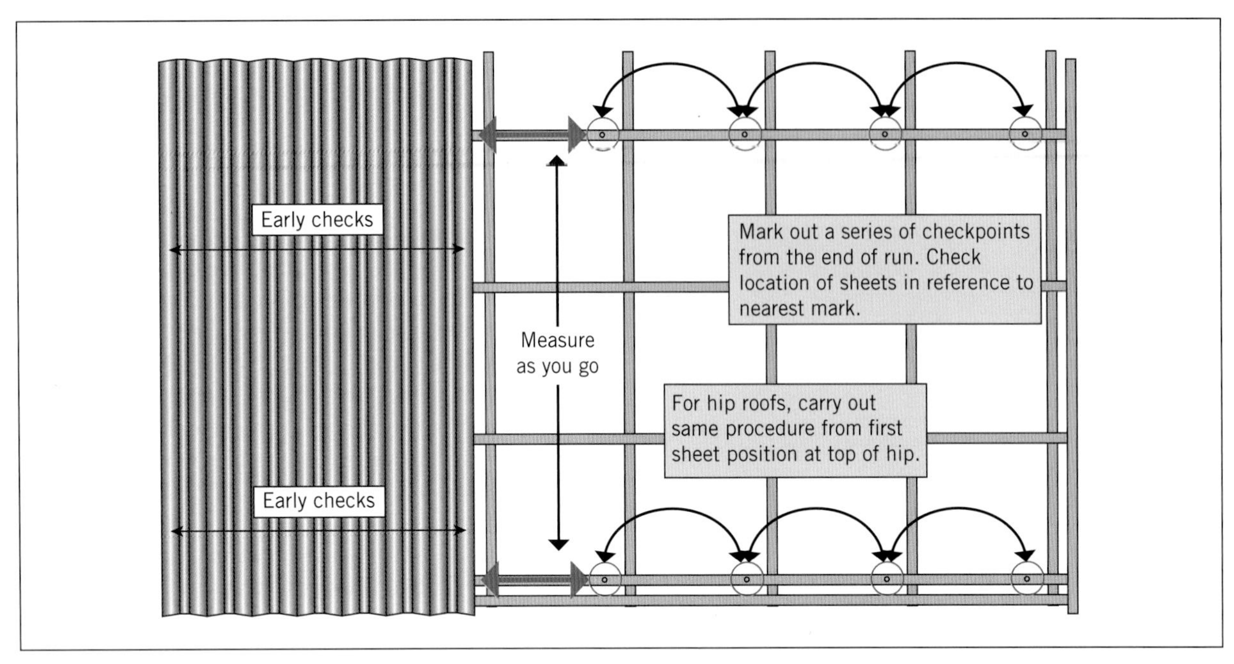

FIGURE 14.19 Simple checks to keep your sheets running true

of the sheet. You can only maintain true alignment by *regularly checking both the bottom and top of the sheet against a perpendicular base line from the fascia.*

Correction of misaligned sheets

When you identify that the sheets are beginning to drift out of alignment, you need to take immediate steps to correct the problem. Pierce-fixed sheets can be gradually stretched or compressed back into position, as shown in the following example.

EXAMPLE 14.6

Imagine that the top leading edge of the sheet has drifted 10 mm closer to the next check mark than the bottom leading edge of the sheet. To correct this, screw down the lap side of the sheet top and bottom; then, using your foot or the leverage of a wood chisel, compress the top end of the sheet back approximately 5 mm and screw it into position while holding it. The profile will absorb this adjustment much like a spring. Make up the remaining 5 mm in the next sheet. Gradual adjustments will not be visible.

Where concealed-fixed sheets have drifted out of alignment, the method you use to correct this problem will be subject to the type of clipping system employed. Consult the manufacturer's installation instructions to find the recommended procedure. However, regardless of the method used, it is most important to ensure that the correction is gradual and made over a number of sheets.

What to do where the frame is badly out of square

You will occasionally encounter an older building frame that is significantly out of square. Where it is not practicable to have structural work done to remedy the problem, you will need to determine how you will accommodate the misalignment in the lay of the sheets. In such a case, the following should be considered:

- Regardless of the solution you choose, ensure that your work still remains watertight and compliant.
- Try to 'take up' the misalignment by allowing the sheets to 'saw-tooth' into the gutter where they can be later trimmed, or at the very least kept from view. This should always be your last option, but at times it may be necessary.
- Ensure that the intersection between flashings and roof sheets is visually even wherever possible, so that the job still looks aesthetically sound.

Misaligned building structures will certainly test your skills but, by thinking ahead and applying basic principles, you can generally achieve a satisfactory result.

Fastening sheets

Fastening of sheets normally includes the following steps:

1. Place the sheet into position with the required lap.
2. Check that the sheet is extending into the gutter by at least 50 mm.
3. Use a tape measure to check that the distance between the top leading edge and the bottom leading edge to the next check marks is the same. Adjust as needed.
4. If necessary, and particularly on higher pitch roofs, use a pair of self-locking grips at the top of the sheet to hold it in position (see Figure 14.20).

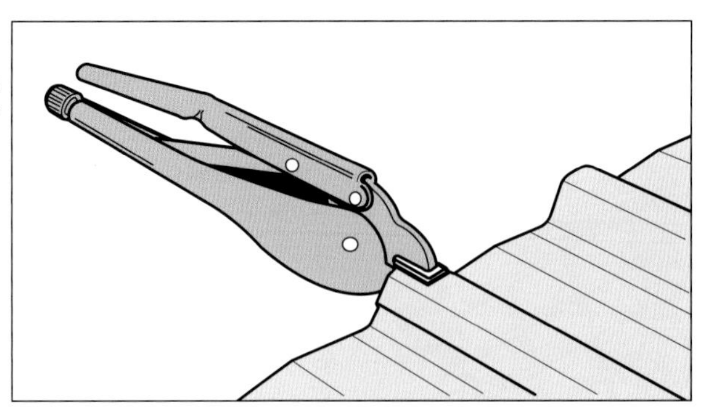

Source: © BlueScope LYSAGHT®, www.lysaght.com

FIGURE 14.20 Using self-locking grips to stop a sheet from sliding out of position

5 Screw the sheets through the lap rib to secure them.
6 Ensure that the screws are driven directly into the middle of the purlin/batten.
7 At this stage, use only as many screws to secure the sheet as necessary. It is common practice just to 'pin' the sheets until the end of the run is reached.
8 Once the required section of roof is covered, spring a chalk-line over each support to show the run of screws (see Figure 14.21). Ensure that you use washable blue or white chalk. Red chalk is often indelible and will not come off – not a good look on a white roof!
9 Screw down all the required ribs over all support positions.

Laying procedure at the hip

Having normally completed the run of full sheets first, you now need to sheet those areas that intersect at the hips of the roof.

As previously stated, an equal-hip roof can be measured up and sheets ordered as full lengths just as if it were a simple gable roof. Because the pitch and angles on an equal-hip roof are all the same, you are able to use the offcuts from one side of a hip on the opposite corner. In other words, offcuts from a left-side hip can be swung around to another left-side hip. Offcuts from a right-side hip can be swung around to another right-side hip. This means that there is essentially no wastage of material, as all cut sections are used somewhere else on the roof.

An example of how this can be applied is shown in Figure 14.22.

1 Position as many sheets as possible in place until the overhang across the hip is too great. This will be about halfway down the hip.
2 Spring a chalk-line across the sheets and remove the offcuts with snips.
3 For the remaining sheets, measure from the gutter overhang up to the hip, then cut a piece of long length before fitting. (Alternatively, these sections can wait until the corresponding offcut is available from one of the other corners; see Figure 14.22.)
4 Place all offcuts in a location where they will be ready for use on an opposite corner.
5 Mark them if necessary to make identification easier and ensure that they are weighted down so that they do not blow around the site.

Laying procedure at valleys

Cutting sheets to fit between the valley gutter and the ridge is a little more involved than a hip roof cut because the sheet cannot simply be laid in position and cut. The following steps show you the basic requirements for laying sheets at valley intersections.

FIGURE 14.21 Use a chalk-line to achieve straight runs of screws

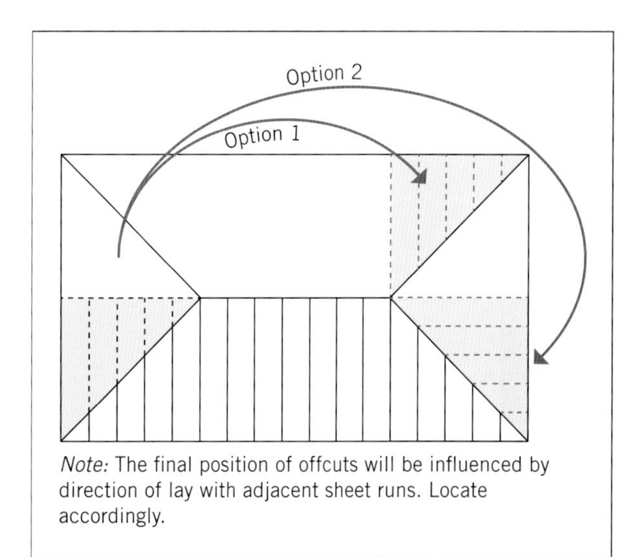

FIGURE 14.22 Example of how the offcuts from one hip can be used on the opposite side

HOW TO

LAY SHEETS AT VALLEY INTERSECTIONS

Step 1	a Sheets must overlap the valley gutter by at least 100 mm. Spring two chalk-lines up the full length of the valley. b Slide sheet 1 down into position (with 1½ corrugation overlap) until the leading edge corner intersects with the chalk-line. c Tack the sheet in place if necessary.	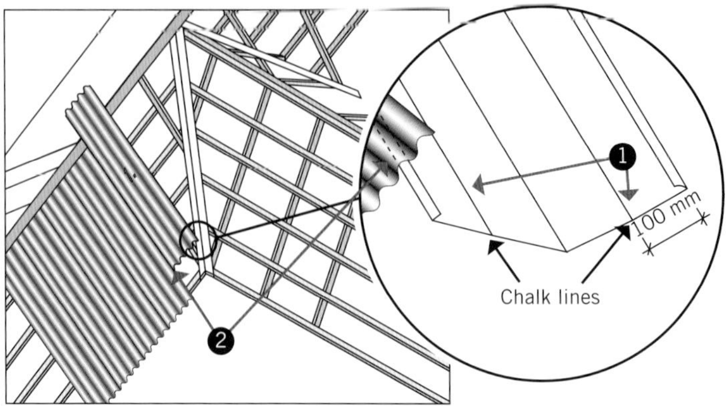
Step 2	a Measure distance A from the point at which the bottom corner of sheet 1 will overlap into the gutter to the chalk-line. b Transfer distance A up to the bottom of sheet 1 and mark it clearly as point X.	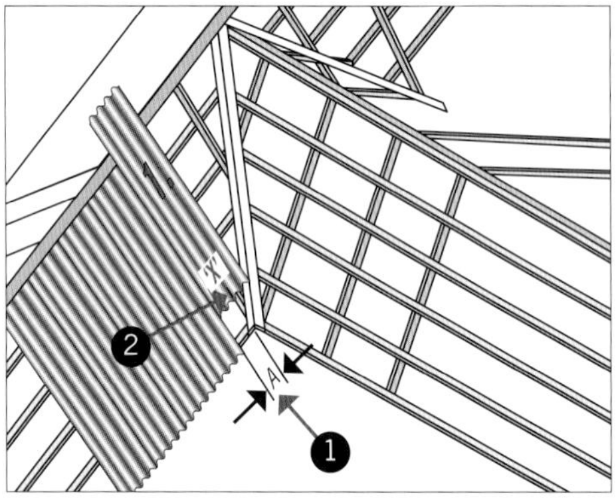
Step 3	a Measure distance B from the gutter overlap position up to the bottom of sheet 1. b Transfer distance B, starting from the corner of sheet 1 in the valley gutter right up the leading edge, and mark it as point Y. c Use a straight edge to line up point X and point Y and mark a line. d Cut with snips and keep the offcut for use on opposite side of valley. e Fasten sheet 1 in position.	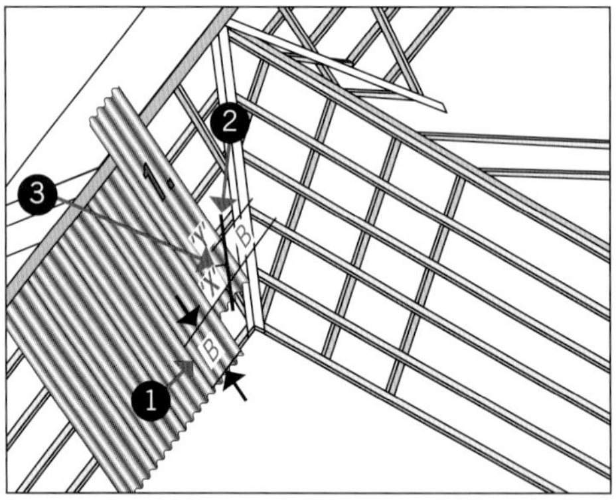
Step 4	a Slide sheet 2 down into position (with 1½ corrugation overlap) until the leading edge corner intersects with the chalk-line. b Measure distance A from where the over-corrugation of sheet 2 will intersect with the chalk-line up to the ridge. c Transfer distance A to sheet 2, measuring down from the top trailing edge of the sheet to point X.	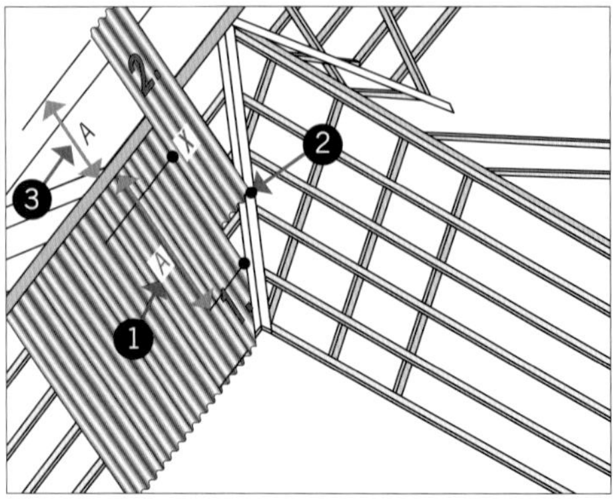

>>

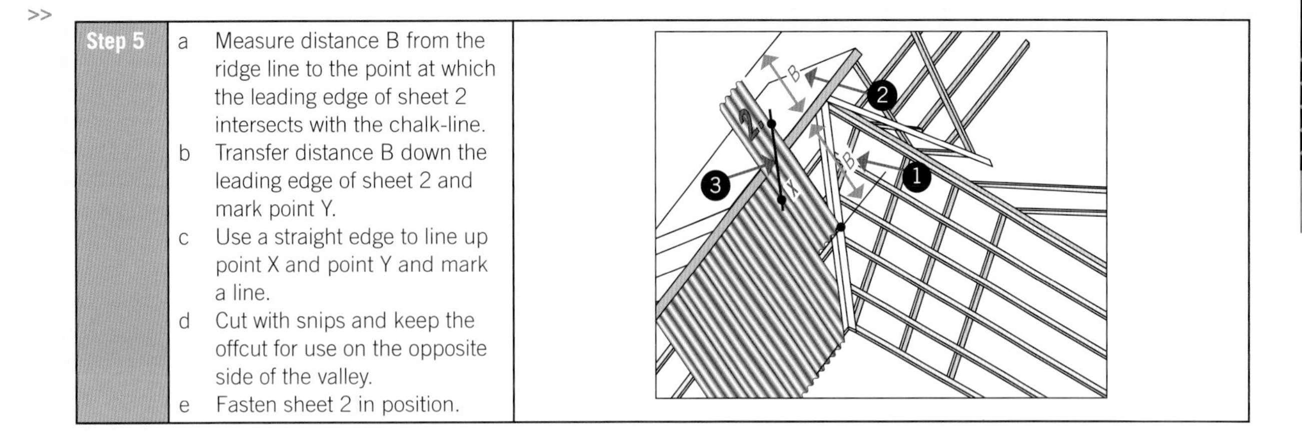

Step 5

a. Measure distance B from the ridge line to the point at which the leading edge of sheet 2 intersects with the chalk-line.
b. Transfer distance B down the leading edge of sheet 2 and mark point Y.
c. Use a straight edge to line up point X and point Y and mark a line.
d. Cut with snips and keep the offcut for use on the opposite side of the valley.
e. Fasten sheet 2 in position.

Figure 14.23 shows the suggested order for laying roof sheets. Additional considerations for laying sheets at valley intersections are as follows:

- This procedure can be continued for all subsequent sheets up the valley. Alternatively, a cut sheet can be used as a template for subsequent sheets, requiring only each trailing-edge measurement from the chalk-line to the ridge to position.
- Remember to save offcuts so that they can be used on the opposite valley if necessary.
- Valley gutters are a structural straight line that attracts the eye of the observer, so poorly cut sheets will stand out more than normal. You must therefore pay close attention to the quality of your cuts so that the sheet ends on both sides of the valley are not only neat but parallel to each other.

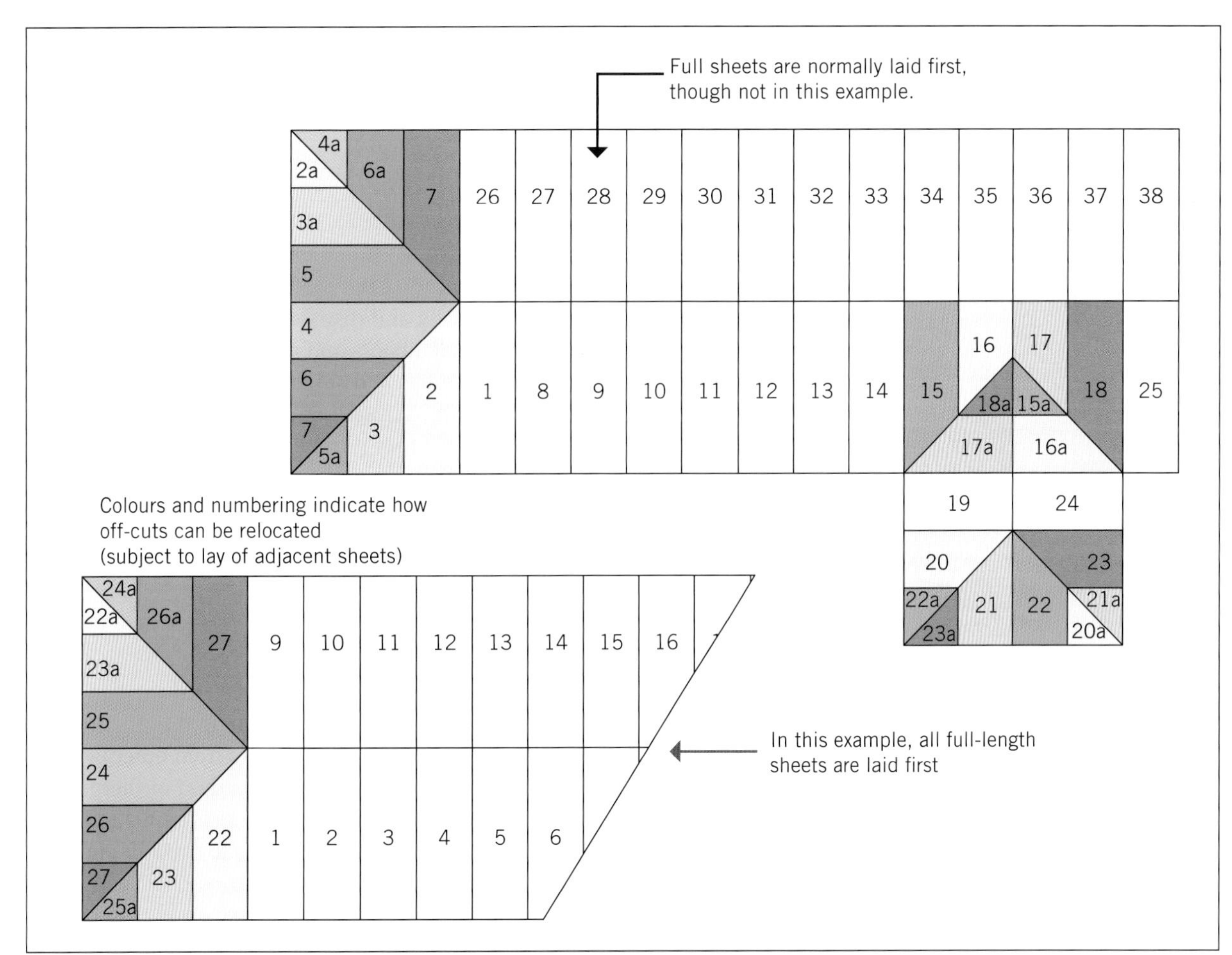

FIGURE 14.23 Suggested sheet laying order for a hip and valley roof

Source: Based on and modified from Basic Training Manual 12-3 *Roof Plumbing, Deck Fixing and Materials*, Australian Government Publishing Service, 1981.

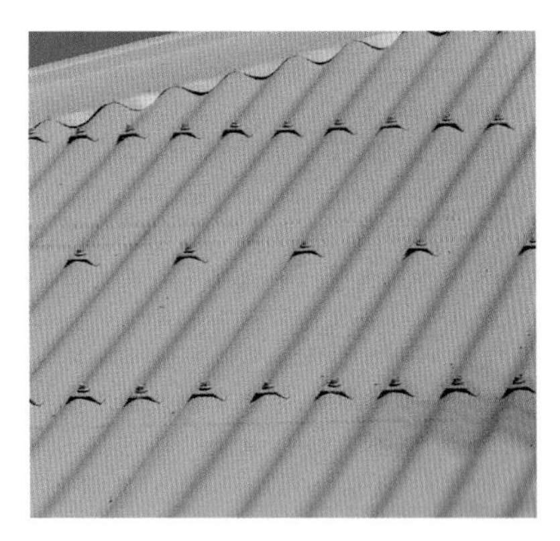

FIGURE 14.24 Additional fasteners on roofs in cyclone regions

- The final row of screws for the sheets on either side of the valley *must not pierce the gutter itself*. Position the run of screws just outside the valley gutter and spring a chalk-line from top to bottom to provide a straight line of screws.

Fastening requirements for cyclone regions

In declared cyclone regions of Australia, pierce-fixed roof sheets must be secured to supports with fasteners that are fitted with broad washers. These washers increase the bearing area and holding power of the fastener. In addition, corrugated sheet profiles will require additional fasteners to be used (see Figure 14.24). Intermediate supports will normally require a screw at every second rib and sheet-end supports will require a screw on every rib.

Specific details relating to support spans, fasteners and screw interval will normally be provided in the specifications sheets.

Access the LYSAGHT Cyclonic Area Manual, found on the LYSAGHT website, to determine roof fastening requirements in cyclonic areas.

FROM EXPERIENCE

If you come from a non-cyclonic zone and intend to work in tropical coastal areas, it is vitally important that you become familiar with the significant differences in product installation requirements in cyclone areas.

Take care when walking on roof sheets

Roof workers and other trades unnecessarily damage many roofs through careless movement around the roof. Kinks and buckles in a low-pitch roof profile can create low points where water can pool and lead to premature corrosion (see Figure 14.25). At the very least, buckles in the roof profile are unsightly and they are not a good recommendation for your business.

FIGURE 14.25 Damage caused to corrugated steel by careless movement

The following work practices should be observed when moving around the roof:

- Tread carefully and distribute your weight evenly.
- Wear soft-soled boots and regularly check for trapped swarf, trimmings and stones caught in the tread.
- When moving across the roof, always try to walk on or close to the roof purlins.
- When moving up and down the roof on close-ribbed and corrugated roofing, walk on at least two ribs at once.
- When moving up and down on pan-type roofing, walk directly in the pans.

Rib closure

Rib closure is the process of ensuring that the end of each rib is sealed. This is often done to prevent access by birds, to prevent bushfire embers from entering the roof space and to act as a further method of thermal control. In BAL rated areas specific requirements relate to the use of mineral wool products to prevent embers getting under each rib.

Rib stop-ends can be fitted during or after the installation of roof sheets, depending upon product variations, and are available in both metallic stops, mineral wool and closed-cell foam filler strips (see Figure 14.26). Foam filler strips for corrugated profiles are normally laid with each sheet.

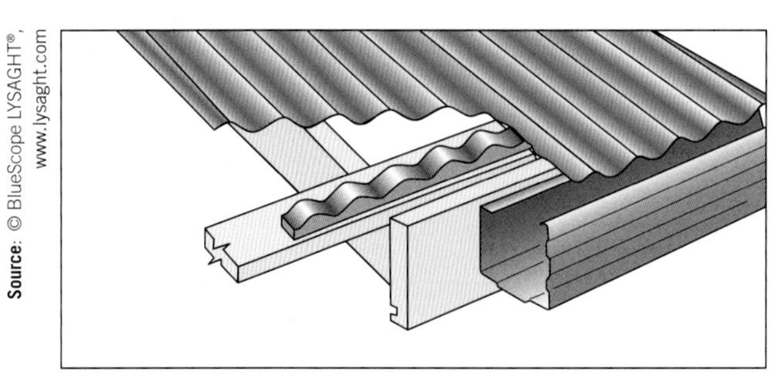

Source: © BlueScope LYSAGHT®, www.lysaght.com

FIGURE 14.26 Use of foam filler strips to seal the end of corrugations

HB39:2015 SECTION 7.13 'RIB END STOPS' AND SECTION 7.14 'FILLER STRIPS'

Turn down sheet ends into the gutter

Where any wide tray roof sheeting is laid at a pitch of less than 10°, you must turn down the pans/valleys of the sheets into the gutter. The turn-down is necessary to prevent surface tension, capillary action or wind from causing water to run back underneath the sheet into the building. Also, the low pitch and lack of turn-down does not permit the sheet to wash clean, allowing the build-up of dust, salts and contaminants on top of the sheet end (see Figure 14.27). This can lead to premature corrosion. The pans must be turned down by approximately 20° using a folder or tool recommended by the manufacturer. Be careful not to tear the sheet when doing so. For roof sheet installation of 5° or less, you may also need to trim off the corner of the underlying lapped sheet to prevent capillary action from drawing water up into the lap. (Source: LYSAGHT *Roofing and Walling Installation Manual*, 2017 Edition. Website: LYSAGHT® *Roofing and Walling Installation Manual*, available at http://www.lysaght.com > installation and maintenance > roofing and walling.)

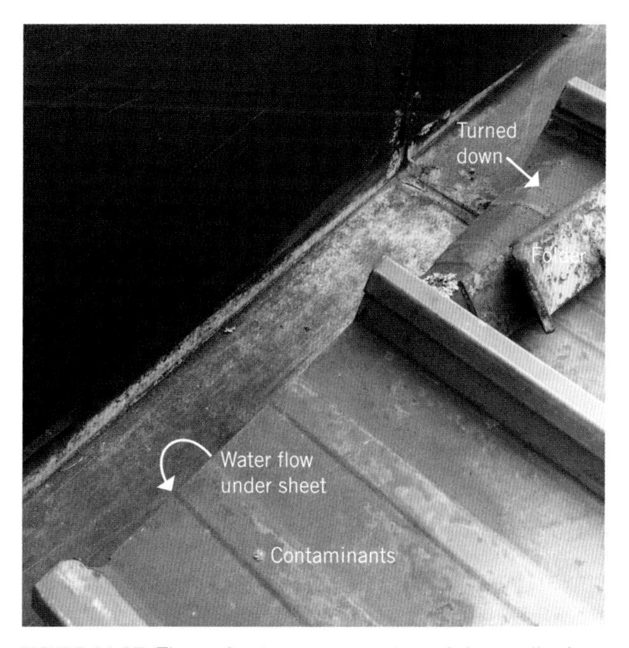

FIGURE 14.27 These sheets were never turned down, allowing surface tension and capillary action to cause considerable internal water damage. Also note the build-up of contaminants at the sheet end

Install flashings

Once the roof sheets have been installed, you are ready to start installing all over-flashings to make the roof watertight. General flashing types and installation requirements are covered in more detail in Chapter 13 'Fabricate and install external flashings' and Chapter 15 'Flash penetrations through roofs and walls'.

However, one important type of flashing that will be discussed here is that which is required when changing the pitch of a roof within a run of sheets.

Change of pitch flashing

A change of pitch flashing (also known as a transition flashing) is required where the pitch of the roof changes along the run of sheets. For example, this situation is commonly found where a steep pitched roof of perhaps 30° changes to a veranda roof of 5°. The end of the steep pitched sheets must transition to the lower pitch over a transverse flashing or a gap must be left between the higher and lower sheets (subject to BAL rating). This can be seen in Figure 14.28.

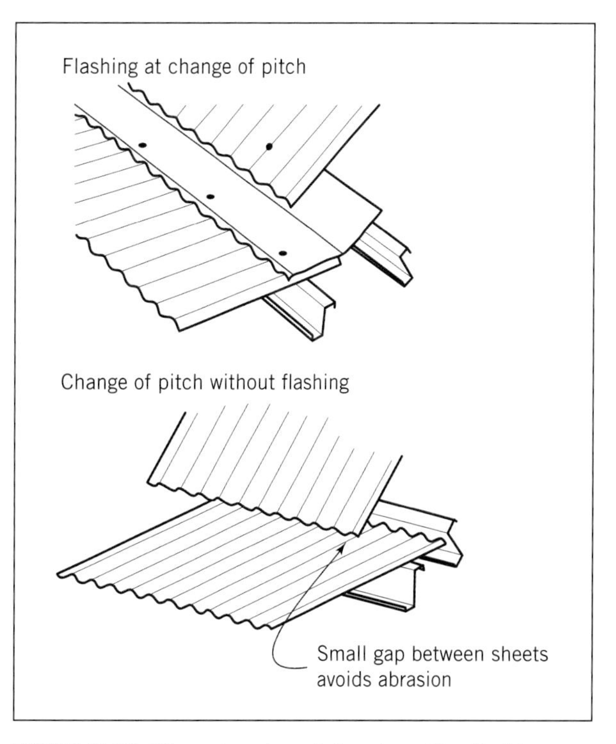

FIGURE 14.28 When changing pitch, a transition flashing or small gap is required between sheets

Do not simply fit the upper sheet hard against the lower sheet. Due to often significant daily fluctuations in temperature, roof sheets continually expand and contract longitudinally (lengthways). Where the ends of the upper sheets are installed to sit hard against the lower pitched sheets, this small but constant movement will cause premature and serious corrosion. Sheets will perforate in a very short space of time.

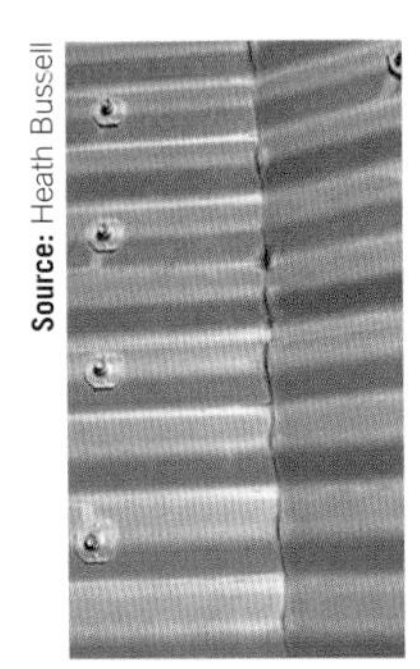

Source: Heath Bussell

FIGURE 14.29 Premature corrosion seen in both ZINCALUME® and COLORBOND® sheets where no transition flashing is used

This can only be avoided by the use of one of the two methods described above. In both instances, the installation of such a flashing must be in accordance with any require BAL rating that may apply to the property.

Touch-ups

Scratches in COLORBOND® steel roofing and rainwater goods products should be touched up only if it is visually important to do so.

Only use a fine brush or special pen and just cover the scratch itself. *Never use spray paint products to cover scratches!* Spray paint will fade at a different rate to the base colour, and will become unsightly within a short time (see Figure 14.30).

Where a scratch is not in a visible position, it is generally best to leave it alone. Remember that pre-painted roofing products are actually based on ZINCALUME® steel sheet and will self-sacrifice if scratched.

Laying short sheets

Today, the clear majority of roofing jobs can be achieved with continuous, cut-to-measure sheeting. However, before the 1970s, roofing sheets were often available only in short (normally 2400 mm and 1800 mm), pre-cut lengths and had to be overlaid in rows up the full length of the roof pitch. One of the few instances where you might need to do this today is on heritage properties where galvanised sheets need to be installed in the same way that they were originally laid, in order to maintain a degree of period authenticity (see Figure 14.31).

FIGURE 14.30 Spray can touch-up paint always fades differently to the base sheet colour

FIGURE 14.31 A new galvanised short-sheeted roof being installed on a heritage building

The sequence of laying a series of short sheets can be seen in Figure 14.32.

The particular requirements for sheet end-lapping depend on the pitch of the roof. These dimensions apply not only to the application of short sheets on older properties. Large commercial, industrial and curved design installations are often impossible to complete in one sheet run and, in these instances,

Source: Based on and modified from Basic Training Manual 12-3 *Roof Plumbing, Deck Fixing and Materials*, Australian Government Publishing Service, 1981.

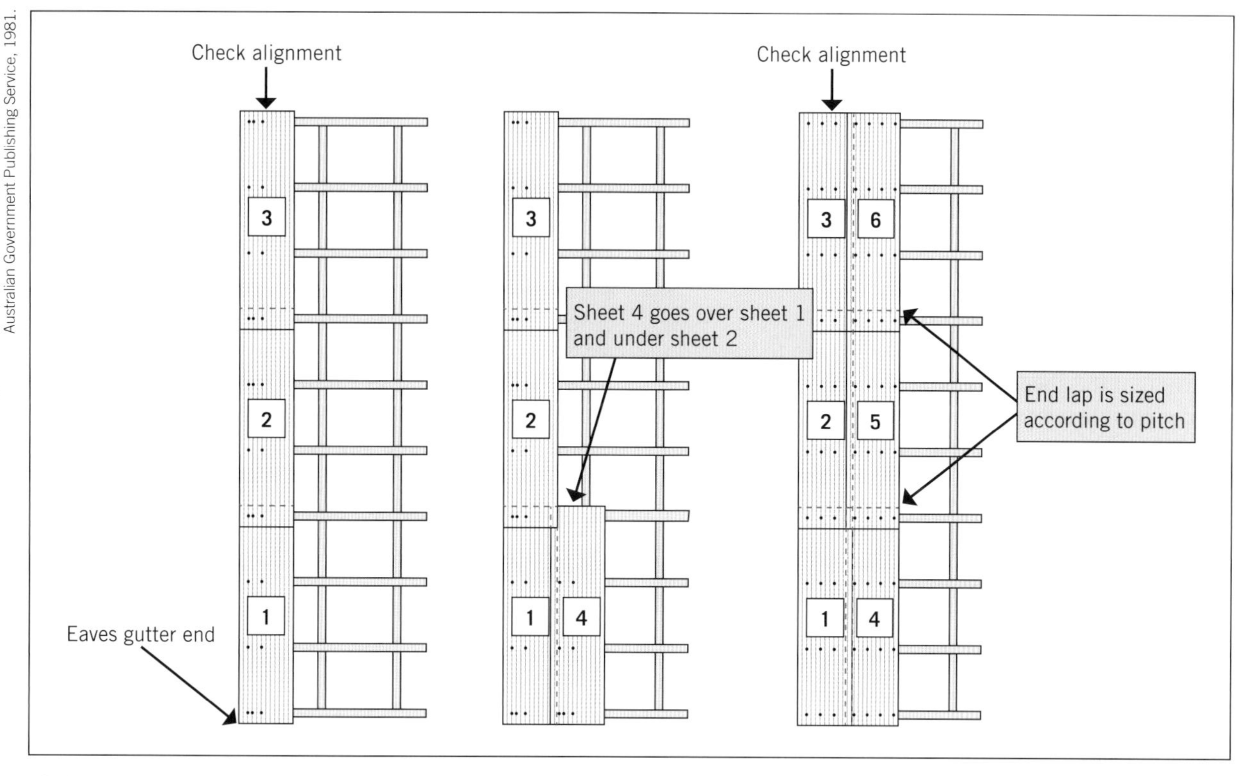

FIGURE 14.32 The recommended laying order when installing short sheets

extra-long lengths or curved profiles must be joined to other sheets. Table 14.5 details these dimensions.

TABLE 14.5 Requirements for the width of roof sheet end laps

Roof pitch	Minimum end lap (mm)	Maximum end lap (mm)
Less than 15°*	200	300
Greater than 15°	150	250

* End laps laid at less than 15° must be sealed with silicone in a manner approved by the manufacturer.

Be aware that sheet laying sequence can be subject to the requirements of the manufacturer. You will note this difference in Figure 14.33.

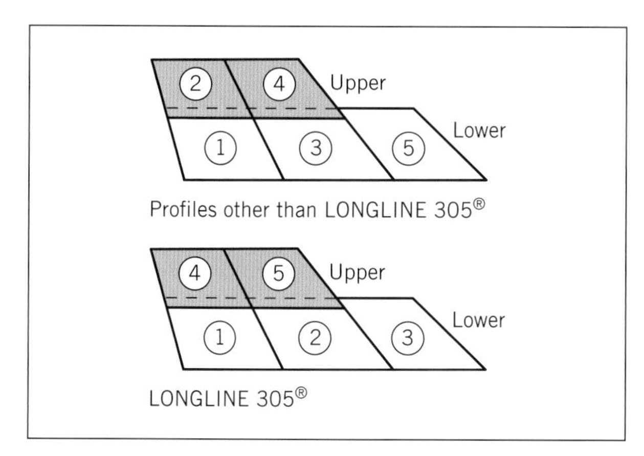

FIGURE 14.33 LYSAGHT recommend a different laying sequence when using different profiles

You should always check manufacturer product information to establish any particular requirements. In this respect, the reference in the HB39 would only refer to the installation of the LYSAGHT Longline 305 profile.

LEARNING TASK 14.4

1 At which angle should you drive the roof screws in, with relation to the roof pitch?
2 What is the minimum end lap required where sheets must be joined on a roof pitch of 15°?
3 When wide tray roof sheeting is laid at a pitch of less than 10°, you must turn down the pans/valleys of the sheets into the gutter by approximately what degree?

Installation of translucent sheeting

Fibreglass and polycarbonate products are available in a range of profile shapes to match complementary metal roof sheets. Translucent sheets must be used in reference to specific code and installation requirements to ensure that their design performance is maintained.

Safety mesh

In almost all cases, safety mesh must be installed beneath translucent sheets. In years past, injuries and deaths have occurred where roof workers have fallen through old sheeting that has become brittle and sometimes painted over. Always check the manufacturer's information to determine if the product can be installed without the use of safety mesh.

Protection of sheets over purlins

You must provide some form of protection to prevent abrasion of the sheet where the safety mesh is sandwiched between the purlin and the sheet. This may be in the form of a foam or plywood strip secured over each purlin.

Fastener selection for translucent sheets

All translucent sheeting must be pierce-fixed, including those profiles normally used for concealed-fixed applications. Fasteners suitable for use with translucent sheeting include special wide washers to improve holding power and flexibility.

Due to the greater thermal expansion and contraction of translucent sheeting, the installation code requires that fastener holes be larger than the screw itself. These holes can be pre-drilled; however, purpose-made fasteners also include special wings that create a hole wider than the diameter of the screw in the sheet, so that the sheet is free to move (see Figure 14.34).

Installation requirements for translucent sheets

Some key installation points are as follows:

- Translucent sheets should overlap steel sheets on both sides of the sheet (see Figure 14.35).
- Translucent sheets are not designed to underlap a steel sheet, and therefore it is normal practice to run a sheet the full length of roof pitch. Where this is not possible, the sheet should be installed at the top of the roof run so that the top flashing can overlay it, while the bottom of the sheet can easily overlap any lower steel sheet.
- Translucent sheets cannot be weathered in the normal manner, so you must use closed-cell foam strips at the top of the sheets to prevent the ingress of water.
- It is standard practice to lay the steel sheeting first, leaving a correctly spaced gap for the translucent sheet to be overlapped onto each side of the steel sheets.
- When you are screwing down translucent sheeting over a concealed-fixed roof system, ensure that you do not pierce the steel deck.

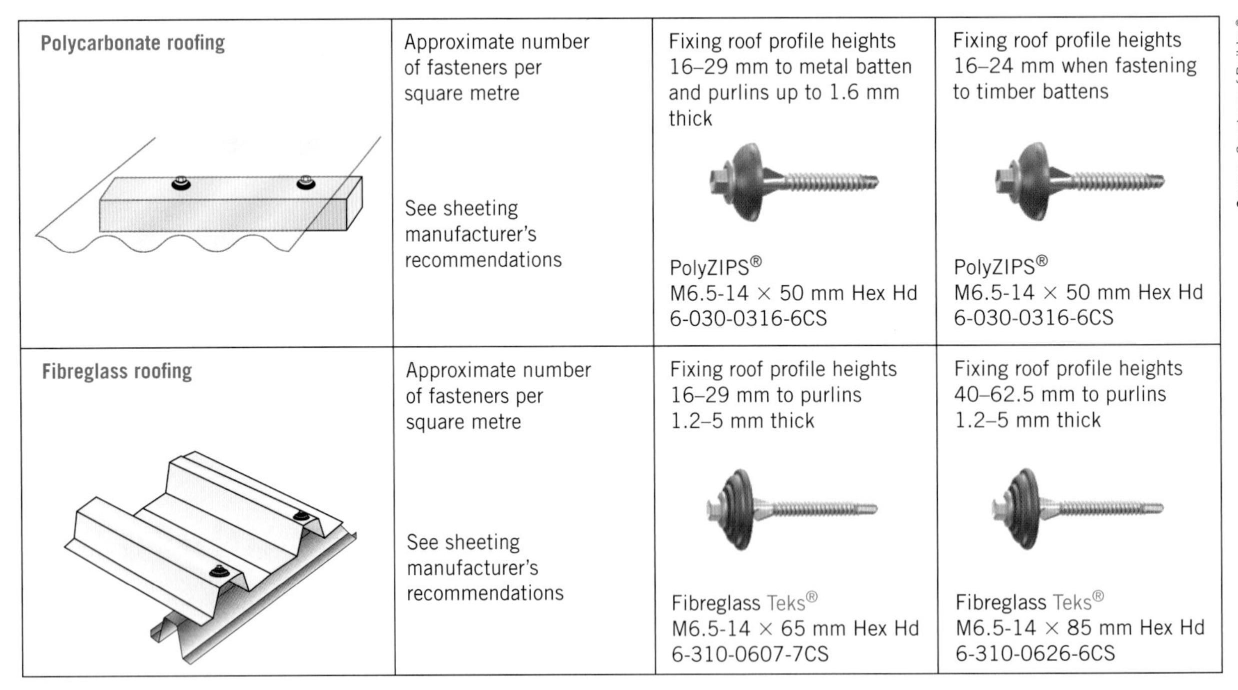

Polycarbonate roofing	Approximate number of fasteners per square metre See sheeting manufacturer's recommendations	Fixing roof profile heights 16–29 mm to metal batten and purlins up to 1.6 mm thick PolyZIPS® M6.5-14 × 50 mm Hex Hd 6-030-0316-6CS	Fixing roof profile heights 16–24 mm when fastening to timber battens PolyZIPS® M6.5-14 × 50 mm Hex Hd 6-030-0316-6CS
Fibreglass roofing	Approximate number of fasteners per square metre See sheeting manufacturer's recommendations	Fixing roof profile heights 16–29 mm to purlins 1.2–5 mm thick Fibreglass Teks® M6.5-14 × 65 mm Hex Hd 6-310-0607-7CS	Fixing roof profile heights 40–62.5 mm to purlins 1.2–5 mm thick Fibreglass Teks® M6.5-14 × 85 mm Hex Hd 6-310-0626-6CS

Source: Courtesy of Buildex®.

FIGURE 14.34 Special fasteners available for translucent sheeting

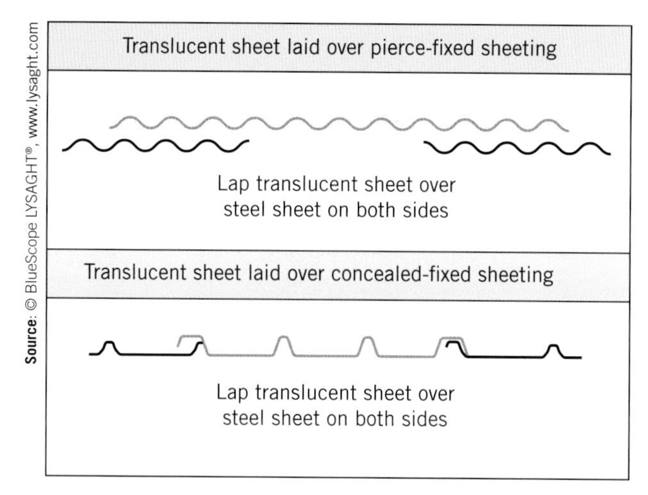

Source: © BlueScope LYSAGHT®, www.lysaght.com

FIGURE 14.35 Overlapping of translucent sheets on both sides of steel sheets

HB39:2015 SECTION 9.2 'INSTALLATION PROCEDURE'

LEARNING TASK 14.5

1 Where safety mesh is sandwiched between a purlin and a translucent sheet, what do you need to do to prevent abrasion of the sheet?

Clean up

Cleaning up is as much a part of the job as the preparation and installation. The job is not over until all waste is disposed of correctly and tools are stored and refurbished according to company policy. The following points will apply to many jobs:

- Disposal areas must be identified at the beginning of the job so that waste can be deposited both during the job and at the end.
- Wherever possible, you should clean up after every day's work so that no build-up of hazardous material occurs.
- Offcuts of insulation blanket, metal, RFL and strippable plastic coatings are prone to being blown off the job. Lightweight refuse such as this should always be weighted down around the site until it can be removed and disposed of in the appropriate bin.
- Cutting and shaping sheet metal creates considerable swarf and small offcuts. Ensure that the job is swept and/or washed down as much as possible, including the gutters. It only takes one night's condensation to permanently stain and damage an otherwise good job.

GREEN TIP

From a sustainability perspective, larger offcuts of material should always be considered for re-use wherever possible. Where identified as such, store such materials in accordance with your company policy. Material that can't be re-used should be allocated to the appropriate bin for recycling or disposal.

Where old and rusted material has been removed from a roof, ensure that it is stored in a safe manner before it is removed from the site. This is especially important around domestic premises where the movement and activities of occupants is often unpredictable. Ensure that the storage area is well barricaded.

Tools and equipment

You are responsible not only for the correct use and care of tools but also to check for damage and faults during and at the end of work.

- Tools showing any form of damage must be reported to your supervisor immediately.
- Where you find electric tool test tags to be out of date, take the tool out of service and return it to your supervisor.
- Hire equipment must be cleaned, refurbished and returned to the hire company in the condition required by the contract.
- Check that your fall arrest gear and other personal protective equipment remain in a serviceable condition and replace items as necessary.

LEARNING TASK 14.6

1 Which quality assurance processes might apply to company equipment and hired equipment?

SUMMARY

In this chapter you have reviewed the basic requirements related to the installation of roof sheeting with a particular emphasis on pierce-fixed sheet profiles. The importance of detailed planning and product selection was covered with additional reminders relating to considerations for re-roofing and properties with BAL ratings.

Installation considerations included:

- Insulation and safety wire where required
- Correct sheet set-out
- Correct fastener application
- Sheet offcut use
- Valley gutter cuts
- Change of pitch flashings
- Short sheet laying

COMPLETE WORKSHEET 2

WORKSHEET 1

To be completed by teachers

Satisfactory ☐

Not satisfactory ☐

Student name: ____________________

Enrolment year: ____________________

Class code: ____________________

Competency name/Number: ____________________

Task: Working with your teacher/supervisor, refer to this text, your HB39:2015 and any other relevant resource to answer the following questions.

1 General steel profile roofing can be broken into two main categories. What are they?

____

____

____

2 In your own words, what is meant by the term 'composite roof'?

____

____

____

3 Using this text and/or other product information, list at least four different pierce-fixed roofing profiles and their respective effective cover measurements.

____

____

____

4 Can standard corrugated steel roofing be laid on a 6° roof pitch?

____

____

____

5 What is the effective cover of standard corrugated roof sheeting?

____

____

6 Using manufacturer's brochures or the weblink below, find out what the minimum pitch and effective cover is for the following corrugated profiles:

Link – **http://www.lysaght.com** > roofing and walling

CUSTOM ORB Accent®21

- Effective cover – ______________________
- Minimum pitch – ______________________

CUSTOM ORB Accent®35

- Effective cover – ______________________
- Minimum pitch – ______________________

7 On what minimum pitch of roof could a trapezoidal pierce-fixed profile be used?

____

____

____

8 List at least four reasons why insulation is installed.

____

____

____

____

9 When planning to re-roof an older house, what should you be particularly aware of in relation to the building roof structure? Why?

____

____

____

10 How many 50-m rolls of safety mesh would you require to cover a roof area of 425 m^2 if each roll will cover 82.5 m^2?

____

____

____

11 An equal-hip roof with an overall width of 45 m is to be sheeted in LYSAGHT CUSTOM ORB®. How many sheets will you need to do the job?

____

____

____

12 How many rolls of foil-backed insulation blanket do you need to cover a 350 m^2 roof if each roll covers 18 m^2?

____

____

____

13 What does roof sheet colour selection have to do with sustainable construction practices?

______

______

______

14 Determine the roof area of an equal-hip roof measuring 25 m long and 13 m wide, with a common rafter length of 7.5 m.

______

______

______

______

______

15 The end elevation of a building shows a roof frame with a vertical height of 2.5 m from the ceiling to the ridge and a width of 5.5 m from the fascia to a point vertically below the ridge. What would be the length of the common rafter?

______

______

______

______

______

16 You are going to replace the roof sheets on a small, equal-hip building. The building measures 15 m long and 9 m wide and is to be covered in corrugated sheets. There is a total of 6 purlin runs on either side of the building. Determine the number of fasteners required. (Calculate for a cyclonic or non-cyclonic zone as appropriate for your area.)

______

______

______

______

______

WORKSHEET 2

To be completed by teachers	
Satisfactory	☐
Not satisfactory	☐

Student name: ______________________

Enrolment year: ______________________

Class code: ______________________

Competency name/Number: ______________________

Task: Working with your teacher/supervisor, refer to this text, your HB39:2015 and any other relevant resource to answer the following questions.

1 In standard use, should the reflective foil side of insulation blanket face up or down when it is being laid?

____

____

____

2 What is the lap required between each run of insulation blanket?

____

____

____

3 Where the prevailing weather is generally from the north, from what direction should you begin laying the roof sheets?

____

____

____

4 Why should RFL be deliberately allowed to sag between each purlin?

____

____

____

5 You are setting up a stringline to determine your first roof sheet position. What method would you use to confirm that the stringline is perpendicular to the fascia?

____

____

____

6 What do you need to do to the high end of a sheet before fixing it into position?

7 What is the minimum distance that each sheet should extend into the gutter?

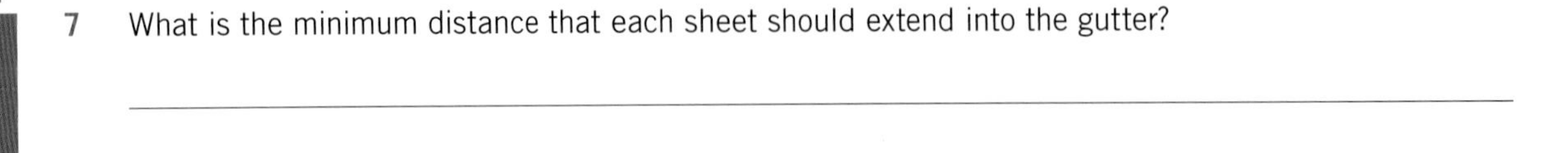

8 When walking around a roof area, where should you place your feet when walking up and down the roof sheets?

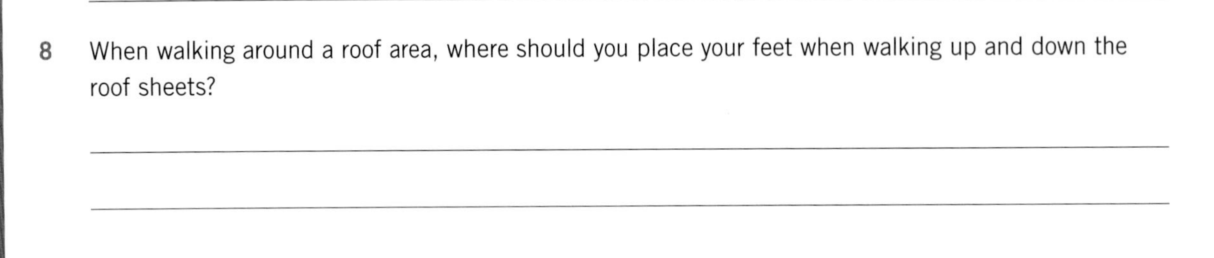

9 Which of the following screws has been driven correctly? Circle your answer.

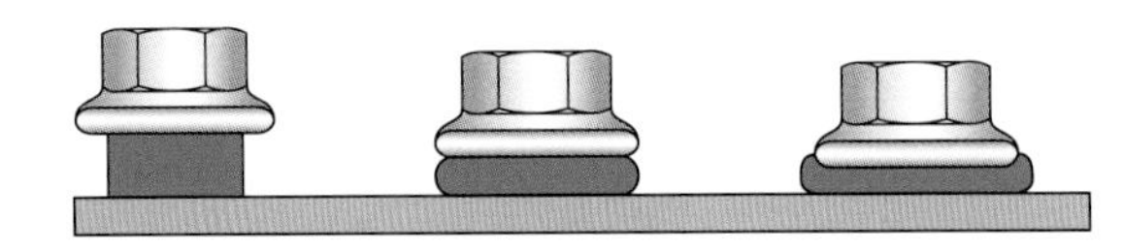

10 A wide tray sheet profile is installed on a roof with an 8° pitch. What do you need to do to the trays at the end of all sheets?

11 On the adjacent roof plan, where could the offcuts (shown in yellow) be installed? Circle your answer/s.

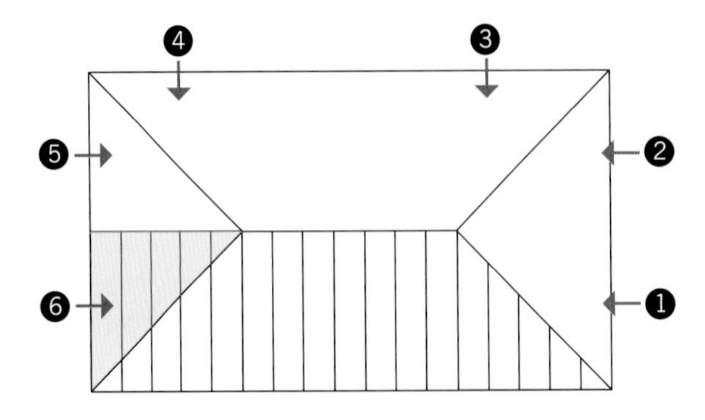

12 During installation of the hip roofing sheets, what should you do with any offcuts while you are working?

____

____

____

13 What is the minimum end lap required where sheets must be joined on a roof pitch of 30°?

____

____

____

14 You need to install a translucent sheet measuring 7.5 m in length. What is the diameter of the oversized fastener hole required in a sheet this long? Provide a code reference to support your answer.

____

____

____

15 What is the name of the flashing that should be used where a roof changes mid run from a steep pitch to a flatter pitch?

____

____

____

16 What is the key issue related to side laps when laying translucent sheeting?

____

____

____

17 When laying sheets over a valley gutter, what is the minimum discharge width to be left in the centre of the valley? Provide a code reference to support your answer.

____

____

____

18 On this short-sheet roof plan and in reference to the LYSAGHT *Roofing and Walling Manual*, indicate with numbers the preferred sheet laying sequence for a corrugated profile. Insert numbers where indicated.

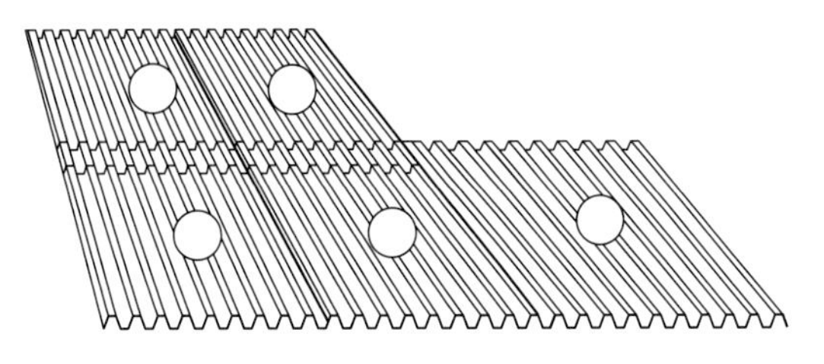

19 What key housekeeping task are you required to do as often as possible to prevent staining and corrosion of the roof?

____

____

____

20 When setting out a square sheet run, what triangular rule do you apply to the first sheet?

____

____

____

FLASH PENETRATIONS THROUGH ROOFS AND WALLS

15

Whenever a pipe, duct or structural component penetrates the wall or roof on a building, it must be flashed so that the building remains watertight. Flashings are the components that a roof plumber uses to create this watertight seal.

The flashing of a penetration must be done in such a way that it allows for movement, expansion and contraction between the building and the penetration without being detrimental to the drainage capacity of the roof.

Before proceeding with this chapter it would be advisable to revise Part A of this text with particular attention to Chapter 7 'Capillary action'. Other sections will be referred to where relevant.

Overview

A roof plumber may be required to perform the flashing of penetrations in the following circumstances:

- as part of a new project, generally after roof sheets have been laid
- as part of a roof replacement, so that sheets are laid around existing penetrations and flashed by the roof plumber as work progresses
- as a part of general plumbing installation of services to existing buildings (e.g. new gas heater flue, boiler flow and return pipes)
- to repair and replace leaking, corroded and poorly installed existing flashings.

Figure 15.1 shows a small commercial roof space crowded with various penetrations and flashing solutions.

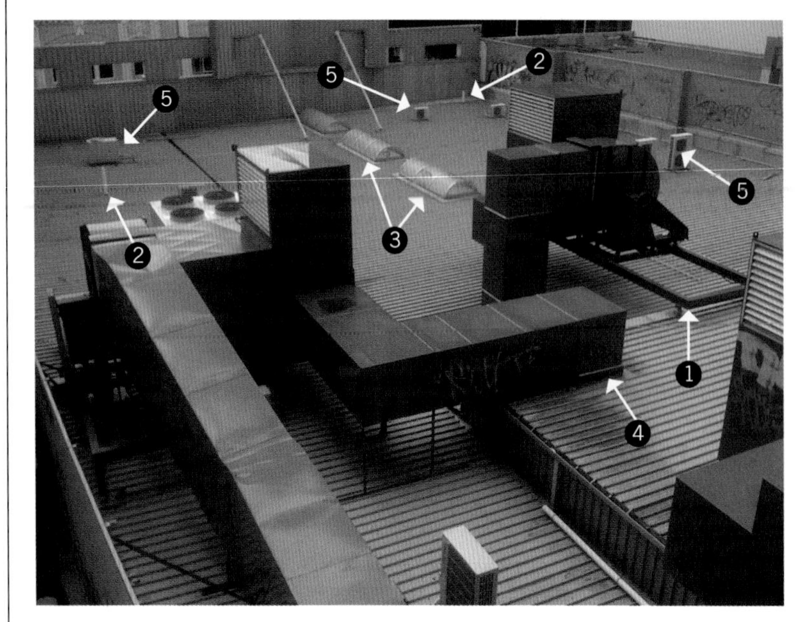

❶ Structural penetration
❷ Vent pipe penetration
❸ Skylight penetration
❹ Heating and cooling duct penetration
❺ Power conduit and refrigeration gas pipe penetration

FIGURE 15.1 Multiple roof penetrations for building services

Penetrations

Penetrations of roofs and walls are made with the intent of creating access for building services or structural members between the internal and external parts of the building. Examples of these include:

- exhaust ducts
- sanitary vent pipes
- gas flues
- heating and cooling ducts
- power
- water and gas pipes
- skylights
- structural penetrations for platforms or other building components.

The various types of roof and wall penetrations require a range of flashing solutions that are available in diverse sizes and shapes.

Identification of requirements

A successful penetration of a roof or wall depends on efficient planning and preparation. Penetration and flashing solutions vary considerably and you must be able to identify the individual requirements of each penetration.

Approved plans and specifications

For new work or major refit projects, details of flashing requirements are normally found on the approved plans and detailed within the specifications documents. You must review these in the planning stage. Ensure that you identify any required Bushfire Attack Level (BAL) rating so that matching components are purchased and flashings comply with minimum requirements. The following ratings can apply to different properties:

- BAL – Low
- BAL – 12.5
- BAL – 19
- BAL – 29
- BAL – 40
- BAL – FZ (Fire Zone)

Where plans have any rating between BAL 12.5 and FZ you must comply with specific installation requirements related to roofing and penetrations. These will be detailed within the plans or your supervisor will need to cross-reference to the AS 3959 as and when required.

Site visit

Regardless of the type of job and whether approved plans and specifications are available or not, you must confirm all details through a site visit. During the site visit you will need to do some or all of the following:

- check the requirements for safe access and fall prevention
- confirm where the penetration is to be located
- check the location of other services in proximity to the penetration point
- identify the material to be penetrated
- determine the tools and equipment that will be needed

- check the structural details and impediments to the proposed penetration point
- confirm the site quality control and safety requirements (review Chapter 3 'Basic roof safety').

Type and application of penetration

Sometimes the application of a particular penetration may dictate the type of flashing selected. One example is where the installation of a heating and cooling duct requires that your flashing allow for vibration and not transfer this vibration to the building structure. Gas flues are another example: some gas flues operate at a temperature that would be too high for standard flashing collars and another material may need to be specified. If you are not sure and do not hold a gas licence yourself, you and/or the builder must arrange for a gas fitter to confirm compliance requirements. Never re-flash existing gas flues without a compliance check.

Structural issues

As part of new construction, penetrations are normally well planned in terms of location, so that structural problems are avoided. However, where penetrations are planned for existing buildings, you must check for some of the following issues:

- Structural members in the way of the planned penetration position. Any alteration of structural members must be approved and certified. Don't cut anything unless you have approval to do so and a remedial solution is already in place.
- Dislocation of thermal insulation. Movement of insulation may affect its performance.
- Cutting of the underlying safety wire. Cuts to safety wire may be carried out only when there is a solution in place for the wire to be re-affixed to the structure in an approved manner.
- Roof sheet support. The roof sheets surrounding the penetration must be provided with additional transverse and longitudinal support as part of the installation process. A lack of sheet support will lead to the collapse of the sheet near the penetration, allowing water to pond in it.

Materials selection and compatibility

When selecting your flashing material or product, you must ensure that it will be compatible with the proposed or existing roof material. Incorrect selection may lead to premature corrosion of the flashing or the roof itself. To revise issues associated with material compatibility, see Chapter 8 'Corrosion'.

Common materials used for penetration flashings include:

- lead – collars for penetrations through tile roofs
- zinc – collars and counter flashings where lead is incompatible
- ZINCALUME® steel, COLORBOND® steel and galvanised sheet steel – wherever compatible with surrounding products
- copper – used where matching roof materials permit
- aluminium – where compatible
- synthetic rubber flashings:
  - EPDM (ethylene propylene diene monomer) rubber (the standard flashing for vents, twin-wall gas flues, pipes and conduit)
  - silicone (a high-temperature flashing material suitable for single-wall gas flues).

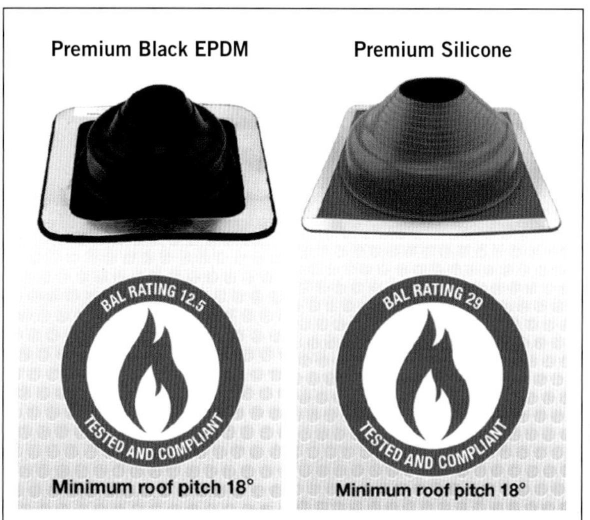

FIGURE 15.2 Two 50–70 mm polymer flashings from the same company, but with very different operating temperatures and BAL ratings

If necessary, revise Chapter 5 'Material types'.

Tools

Your standard set of roofing tools may all be required for the fabrication and installation of penetrations through roofs and walls. As a reminder, these include:

- hacksaw
- straight snips
- aviation-style left- and right-hand snips
- square
- rivet gun
- drill and correctly sized bits
- cordless driver with a range of hex and Phillips bits
- multigrips
- bevel
- hammer
- sealant gun
- hand folder
- rags and spatula
- fine marker pen
- relevant personal safety equipment.

In addition, you will also find yourself using the following items:

- Plumb bob. This is particularly useful in ceiling spaces where you need to transfer the position of a ceiling penetration up to the roof, or vice versa. It is the best way of ensuring that your centres line up.
- Lead-working tools. These include dressers, bossing stick and mallet. These tools allow you to work lead flashings correctly, enabling you to form the lead while reducing the risk of tearing.
- Nibbler. Cutting penetrations through existing roof sheets can be difficult with snips alone. A nibbler allows you to cut across and with the run of the sheet profile.
- Pilot drill bits. A small exploratory pilot hole can save you a lot of trouble when locating the final penetration position. Small-diameter drill bits on long extensions are handy for this.
- Flexible camera probes. Concealed ceiling spaces may hold electrical cables and other services that you are not aware of. Flexible camera probes allow you to view these cavities for potential safety hazards before making larger penetrations.
- Power cutters. Angle grinders and small circular saws fitted with appropriate sheet steel cutting blades can save a lot of time when cutting out large penetrations. However, remember that carborundum discs should never be used on roofing products, as they burn the cut edge, preventing sacrificial protection of the cut steel.

Review Chapter 9 'Roof plumbing tools' if required.

Capillary action

Where two flat surfaces are in close contact, it is possible for water to be drawn upwards through the process of capillary action. The fabrication of penetration flashings requires that you are always mindful of the potential for capillary action to occur, particularly in relation to the penetration directly through decking laps and contact between counter flashings and apron or soaker upstands.

If necessary, revise Chapter 7 'Capillary action' before proceeding further.

LEARNING TASK 15.1

1. Which tool can be used to transfer a vertical measurement in a ceiling space to ensure your centres line up?
2. How might a flexible camera probe be used as a safety device when preparing to make a penetration?

Penetrations and safety

There are many standard safety issues that you need to observe when carrying out penetrations of roofs and walls. Review Chapter 3 'Basic roof safety' to revise these. However, there is one point that needs to be made here that you must never forget:

Never cut a penetration through roofs and walls without checking for other services.
Especially power and gas lines.

Cutting through a power or gas line may be the last thing you ever do. Older buildings and concealed spaces are particularly risky. Damage to other services such as water may end up flooding rooms or spaces below and causing thousands of dollars worth of damage. Therefore, spend some time to do the following checks:

- Consult with the building owners/managers in relation to service locations.
- Use voltage sensors to detect power cables behind wall linings and concealed ceiling spaces.
- Access ceiling spaces and confirm that exactly where you're going to cut is clear of hazards. Never touch aluminium sheet foil insulation that has been placed directly upon ceiling joists and wiring.
- Be extra cautious around raked ceilings with exposed rafters, or concealed ceiling spaces below low-pitch roofs. Lift roof sheets if necessary and look for electrical cables, as these will be run in many areas to power lights and fans. Use a portable inspection camera if available.

Precautionary investigations at this stage can save job profitability and, most importantly, your life and the lives of others on the job.

Flashing of small penetrations

In this section, the flashing of small penetrations relates to collar flashing of penetrations up to approximately 300 mm. This can be achieved using either a fabricated flashing or prefabricated collar flashings.

Use of prefabricated collars

The most commonly used prefabricated flashing is the synthetic rubber collar. These are actually made from either ethylene propylene diene monomer (EPDM) or, for higher temperature applications, silicone. Follow the steps in the following 'How to' box to see how a synthetic rubber flashing is installed.

A key requirement that you must observe when installing any flashing is the need to ensure that it will still permit total drainage of the area above the penetration.

Figure 15.3 shows two installation options, depending upon upstream drainage requirements.

Where a soaker flashing is required

Some small penetrations can be simply flashed using a prefabricated collar directly attached to the roof profile when there is little risk in restricting the flow of water down the roof. These factors include:

- roof pitch – a higher pitch will generally allow water to flow more easily around the collar
- roof sheet profile – a wide pan roof profile allows more space for water to flow past the collar
- catchment area – a small catchment area above the penetration may minimise the risk of water build-up
- rainfall intensity – an area of low rainfall intensity means that the risk of restricting water flow is reduced.

However, when one or more of these factors cannot be satisfied, you may need to install a soaker flashing around the penetration so that water flowing past the collar is not impeded or allowed to back up and potentially flood back past the end of the sheet into the building.

Figure 15.3 shows how the use of a soaker flashing at the base of a gas flue allows free flow of water around the penetration. Compare this to the situation in Figure 15.4 where multiple cable penetrations have been put through the same low-pitch corrugation with no thought of water flow. A careful examination of the photo reveals how the sheet is beginning to corrode because of water ponding upstream of the last penetration.

Key points to consider when installing a small soaker include:

- Mark out your cut a minimum of 100–150 mm upstream and downstream of the penetration.
- Cut the sheet and the soaker itself so that the soaker is lapped in the direction of flow.

This installation shows a gas flue penetrating a deck directly through the pan of the roof sheet. The distance to the upstream ridge is not far and it was deemed that there was no chance of significant ponding being caused by the flue.

However, where the catchment area of the sheet is much greater and there is a risk of water backing up and flooding of ridge and laps, then the solution below would be a better option. Here the rib has been cut away above and below the penetration and a soaker tray fitted to allow better water flow across two pans in the sheet.

FIGURE 15.3 Two options to ensure complete drainage above a penetration

- Install support around the cut so that the sheet does not sag and cause ponding of water.
- Fit the soaker and ensure that if a sheet ridge lap is penetrated, make an anti-capillary cut immediately before the rib stop-end.

HOW TO

PROCEDURE FOR INSTALLING A SYNTHETIC RUBBER COLLAR FLASHING

A synthetic rubber collar flashing can be installed by following these steps.

Step 1	a The collar is graduated with increasing diameter markings. b Measure the diameter of your penetration pipe. c Trim the collar with a pair of snips to the diameter that will provide a tight fit (not too tight).
Step 2	a Ensure that you have provided new sheet support beneath the cut and that at least 20 mm is left all round between the pipe and sides of the cut. b If necessary, lubricate the pipe with some water and slide the collar down to the roof. c *If the penetration is passing through a sheet lap, note where the collar sits and cut an anti-capillary slot in the rib just upstream from the collar.* See Figure 7.4 if required.
Step 3	a Press the aluminium band of the collar base around and over any ribs in the sheet until the collar is sitting as it will when secured. b Once the base is shaped, lift the collar base up and apply silicone sealant.
Step 4	a Press the collar back into position. b Ensure that the collar is not too stretched so that there is allowance for pipe movement.
Step 5	a Screw down base with appropriate fasteners, ensuring a tight fit around the base of rib profiles where the collar will tend to lift. b Visually check that no gaps are left around the base. c Where the pipe has a grooved seam, ensure that you run a bead of silicone up the full length of the exposed groove and around the top of the synthetic collar.

Source: www.deks.com.au

FIGURE 15.4 A soaker flashing should have been used around these penetrations to provide effective drainage.

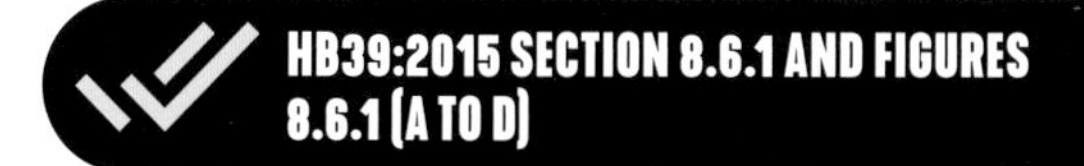

Ensure support around the cut

An important aspect of cutting in any collar flashing is to provide support to the ends of the cut sheet. Once the sheet has been cut it will lose much of its normal rigidity. Any pressure applied near the area of the penetration will cause deformation of the sheet and allow water to pond in it (see Figure 15.4). You will need to fit metal or timber bracing upstream and downstream of the hole so that no deflection is possible.

Use of fabricated collars

The flexibility, convenience and speed associated with the use of synthetic rubber pre-fabricated collars means that it would be very unusual nowadays for you to fabricate your own collar. However, it may still be necessary to do so in some circumstances, such as when a repair warrants maintenance of the existing system or where a BAL 40 or FZ rating may dictate the use of a non-combustible metallic flashing around a metal vent pipe. All sealants used must also match the same rating.

The features of a fabricated penetration flashing are shown in Figure 15.5.

LEARNING TASK 15.2

1. What considerations might suggest a soaker flashing should be chosen over a prefabricated collar?
2. How would you transfer the pipe diameter to the prefabricated collar to ensure the cut is accurate?

Flashing of large penetrations

Large penetrations such as skylights (see Figure 15.6) require additional attention be applied to the effective drainage of water around the obstruction, as well as particular care with structural support around the penetration hole.

Confirm location of penetration

The plans and specification or job briefing will show the location of the proposed penetration. However, you need to confirm these details on the job so that unexpected obstacles can be identified before you begin the job.

Therefore, you should:

- check the location of sheet laps in relation to the penetration – if you can avoid a lap by slightly shifting the penetration, and the plans allow for it, save some work by doing so
- check for the location of structural members beneath the sheets that may be in the way of the penetration – again, check if you have any flexibility in location
- measure the upstream drainage requirements, as this will influence the size of the penetration hole.

Determine upstream drainage requirements

A large penetration will almost always obstruct the normal flow of water down the pans of the roof sheets. Your penetration must include a suitably sized soaker flashing that will efficiently divert water around the penetration without retarding and backing-up the flow.

To achieve correct flow around the penetration, the width of the soaker is determined by the amount of catchment area directly above the penetration.

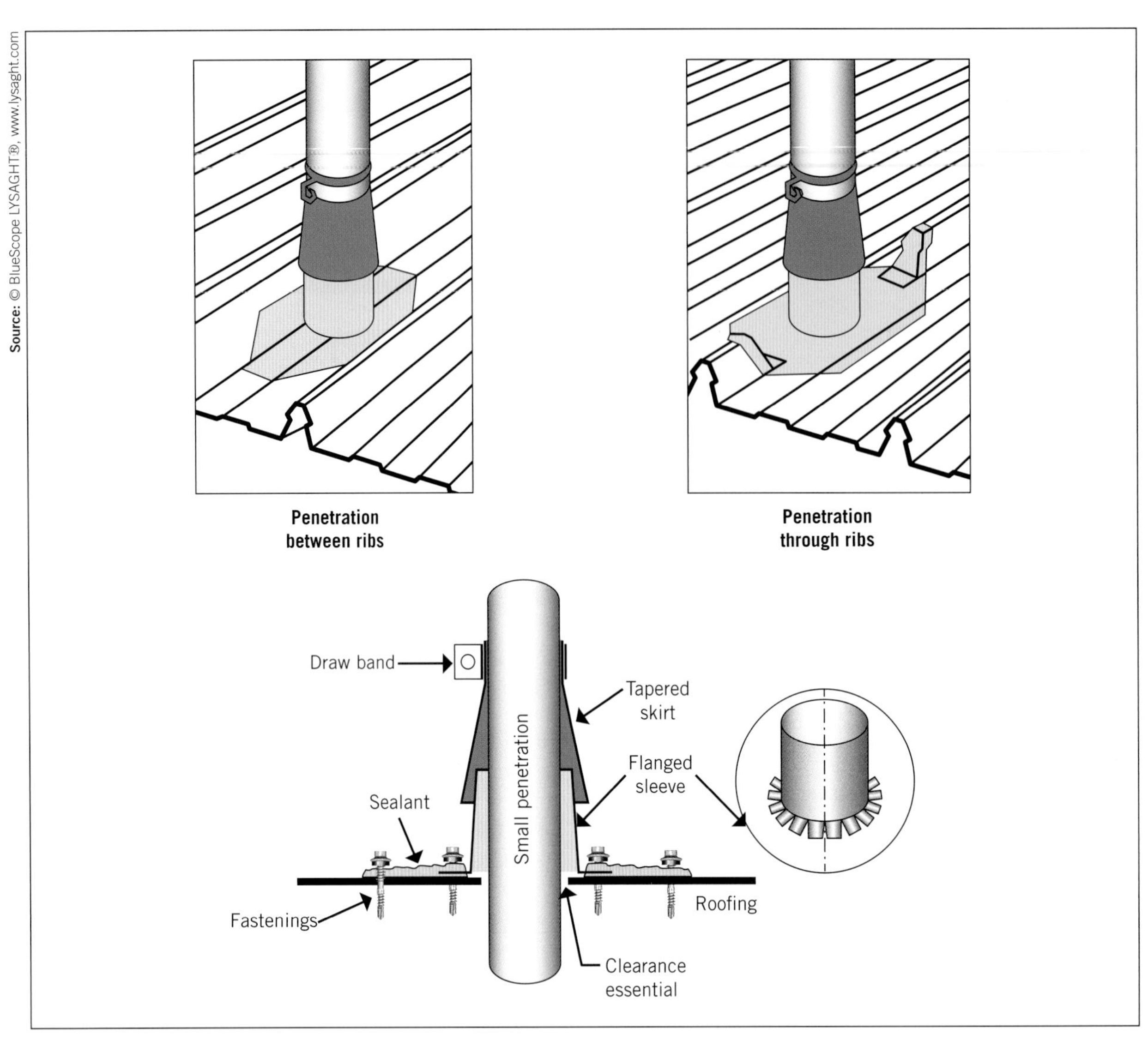

FIGURE 15.5 A fabricated metallic collar flashing

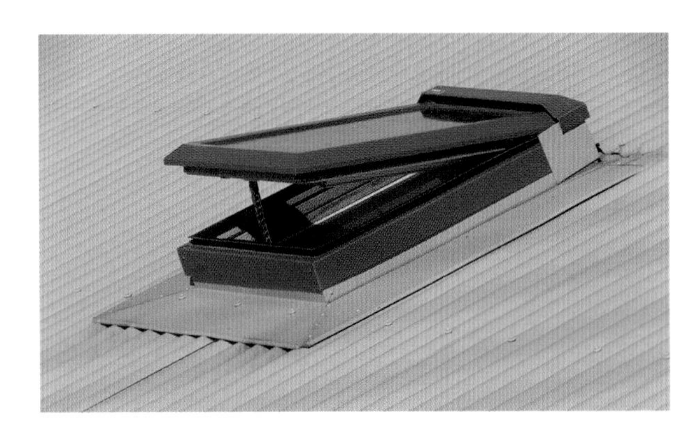

FIGURE 15.6 Apron flashing around a skylight

FIGURE 15.7 Prior planning should have prevented this situation

HB39:2015 SECTION 5.5 AND FIGURE 5.5(A)

- Where a catchment area directly above a penetration is up to 40 m², an equivalent of two tray/pan widths must be allowed for in soaker width on each side of the penetration.
- For a catchment area directly above a penetration that is between 40 m² and 80 m², an equivalent of three tray/pan widths must be allowed for in soaker width on each side of the penetration.

Figure 15.8 shows how this is determined.

Cut out and install structural support

Having determined the width of your penetration soaker, you can now finalise any structural and/or support modifications necessary around the penetration hole, including the following points:

- If your penetration affects in any way the load-bearing or tie-down characteristics of the building, the work may need to be certified before proceeding.
- Depending upon ceiling space access, the supports may be installed before or after the hole is cut in. If weather is a concern, try to have the supports in and ready to go before opening the roof.
- Transverse supports should be correctly sized and secured to rafters in an approved manner. The sheet on each side of the hole must be supported.
- To allow for thermal and normal building movement, the roof sheet cut-out must allow a minimum of 20 mm between the nearest part of the sheet and the side of the penetration.

Flash the penetration

The following points apply to the flashing of the penetration:

- Use a power cutter, hacksaw, snips and nibbler as required to cut through the ribs and pans of the sheets.
- Using compatible material, fabricate and install the soaker flashing. The top of the soaker should be lapped under the upstream sheets in the direction of flow. The lower end of the soaker should be lapped over the downstream sheets in the direction of flow (where a full cut-out is done). Seal all laps with non-acetic roofing silicone sealant.

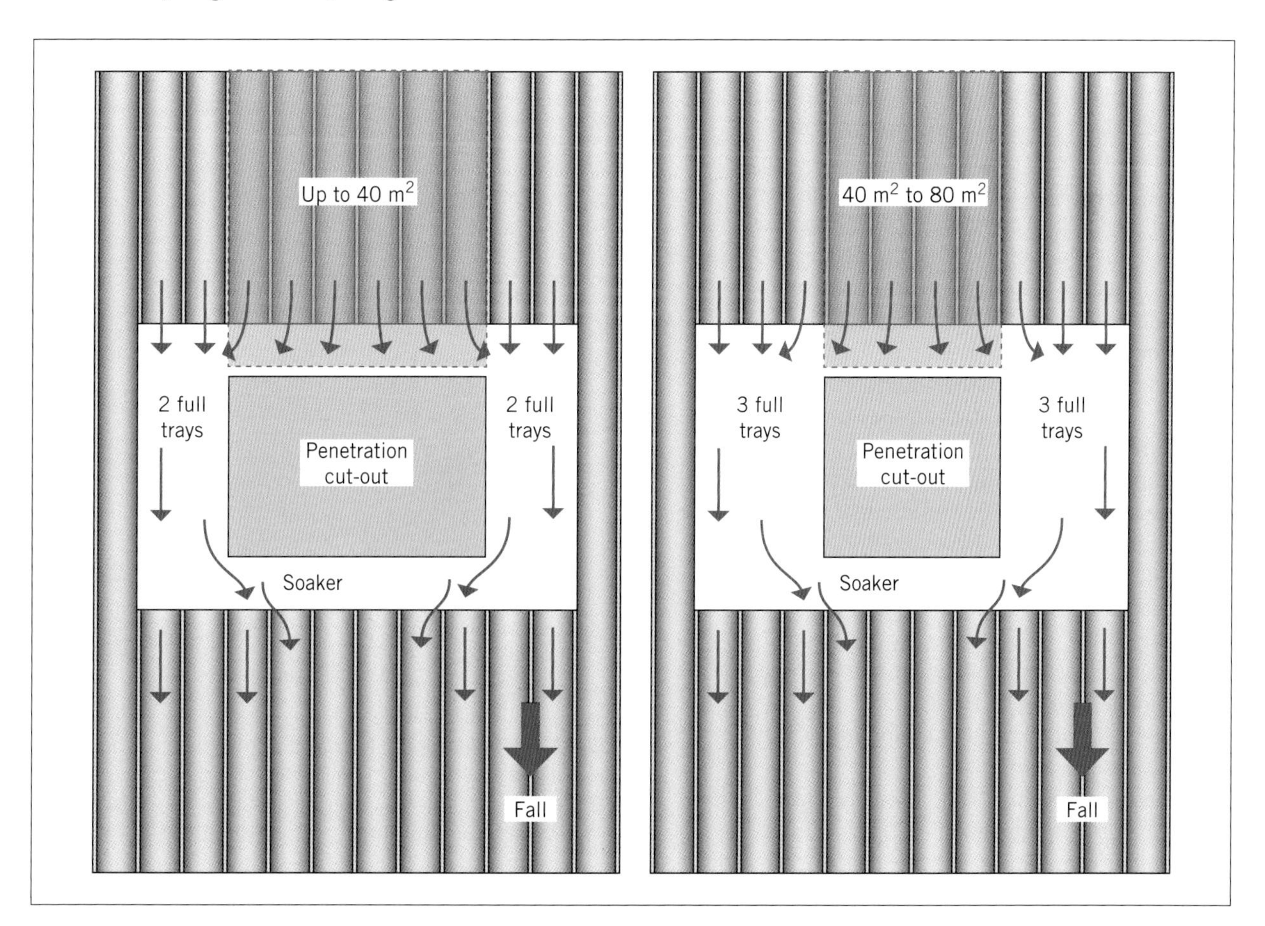

FIGURE 15.8 Determining the width of a soaker: (a) in a catchment area directly above a penetration that is up to 40 m² and (b) in a catchment area directly above a penetration that is 40–80 m²

- Where an apron flashing is used to simply over-flash the downstream sheets, ensure that the pans of each sheet are fully turned up and weathered. The apron turn-down must be fully notched around the ribs.
- Upstream ribs should be closed either by using proprietary rib-caps or by cutting and folding the ribs down to seal them.
- Install all over-flashing according to the needs of the penetration and with compatible material.

LEARNING TASK 15.3

1. What two additional considerations should you consider when installing a large penetration through a roof?
2. To determine the correct width of the soaker, what do you need to calculate?

Chimney flashing

Chimney flashing incorporates a selection of the following flashing components:

- chimney gutter – also known as the V-flashing or back-gutter; may form part of the full soaker or flow on to an apron flashing
- soaker or apron flashing – used at both sides and the downstream side of the penetration
- counter flashing – either stepped or sloping.

Various methods are employed for chimney flashing, depending upon the type of roof covering (e.g. slate tiles, masonry/terracotta tiles or profile steel sheeting). As always, there is more than one way to do a job and each has advantages and disadvantages. The following procedure is one way of flashing a chimney penetrating a corrugated steel roof:

- Leave a gap around the chimney to allow for building and thermal movement (see Figure 15.9).
- Determine the minimum width of the apron flashing on either side and the rear cut-out for the chimney gutter (see Figure 15.10).
- Fit the front and side aprons (see Figure 15.11).
- Fit the chimney gutter into position (see Figure 15.12).
- Fit the stepped or sloping counter flashing to complete (see Figure 15.13).

Some key points relating to apron flashings for large penetrations are as follows:

- Sheets must be fully supported around the full perimeter of the penetration hole.
- Maintain a minimum annular clearance around the penetration of 20 mm.
- The apron must have a minimum upstand height of 100 mm.

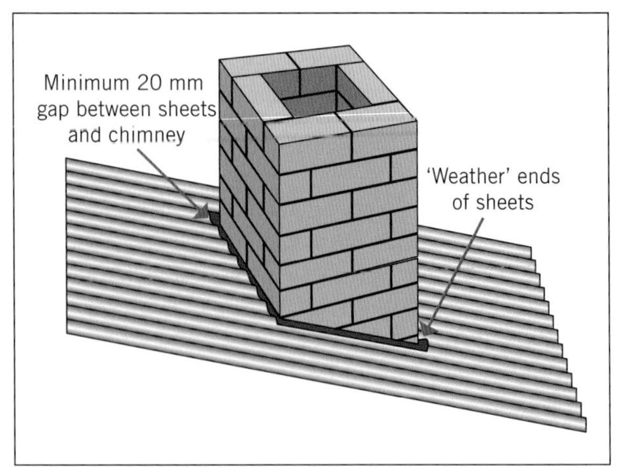

FIGURE 15.9 A minimum annular clearance must be left around a chimney.

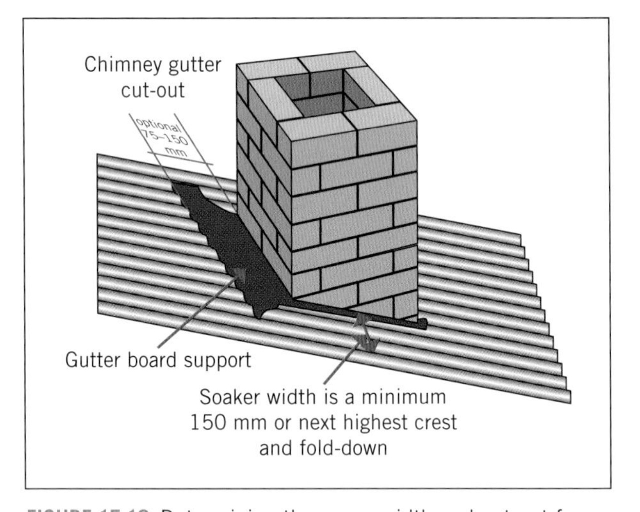

FIGURE 15.10 Determining the apron width and cut-out for a chimney gutter

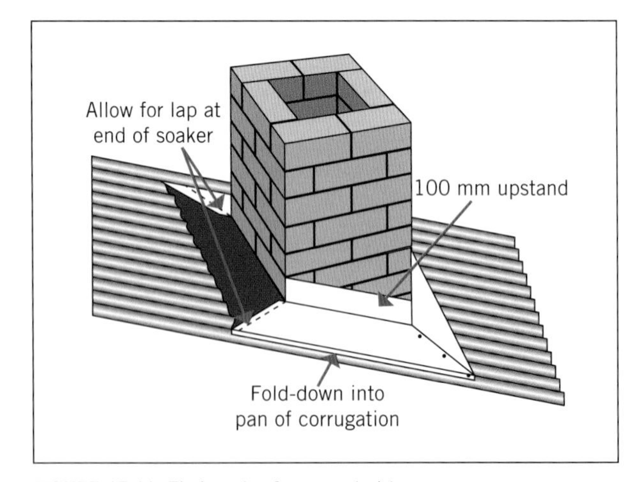

FIGURE 15.11 Fitting the front and side aprons

- Counter flashing must overlap the upstand by 50 mm and remain 25 mm clear of the apron cover to ensure free drainage.
- Where rib sealing on the upstream side of the chimney gutter is carried out on a roof sheet

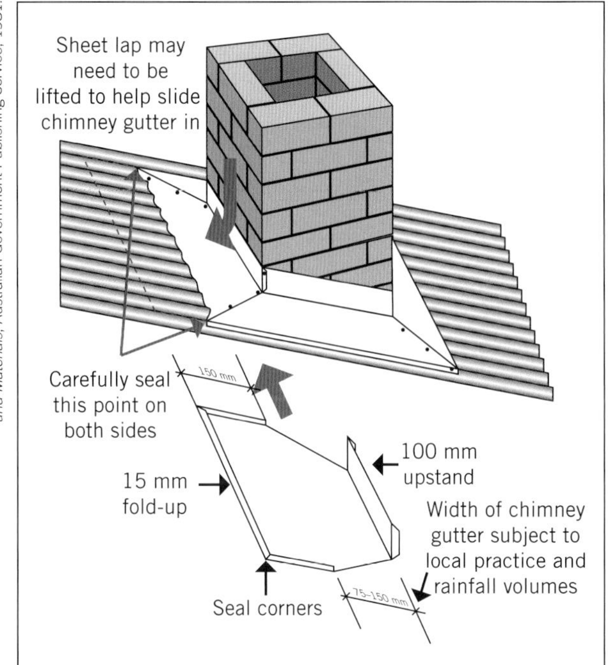

Source: Figures 15.9 to 15.12 all based on Basic Training Manual 12-3 *Roof Plumbing, Deck Fixing and Materials*, Australian Government Publishing Service, 1981.

FIGURE 15.12 Fitting the chimney gutter into position

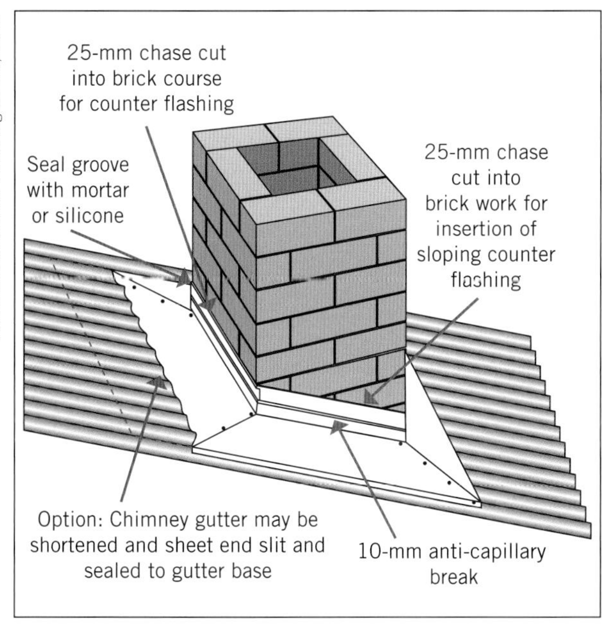

Source: Based on Basic Training Manual 12-3 *Roof Plumbing, Deck Fixing and Materials*, Australian Government Publishing Service, 1981.

FIGURE 15.13 Fitting a stepped or sloping counter flashing

lap, ensure that an anti-capillary cut is made immediately before the closed rib.

LEARNING TASK 15.4

1 What names might a chimney gutter also be referred to as?
2 What is the correct sequence of fitting the chimney flashings?

Dry-pan flashing

Where your penetration is located close to an upstream ridge or wall flashing, one option that you may consider is to install a type of over-flashing called a 'dry-pan' flashing (also known as a 'flat-tray' flashing). This sheet is inserted underneath the wall flashing or ridge capping and extends down to and around the penetration in one piece (see Figure 15.14).

The advantages of a dry-pan flashing are that:

- it may reduce installation time
- the penetration hole is normally smaller and the structural support requirements are reduced.

The disadvantages of a dry-pan flashing are that if the flashing sheet is too large or long:

- a consequence can be premature corrosion through 'sweating' between the roof sheets and the over-flashing. With little or no ventilation between the sheets, water vapour will condense and remain trapped there, accelerating the corrosion process
- an excessive drumming noise from heavy rain and 'bangs' caused by thermal expansion and contraction of the sheet may become a nuisance to the building's occupants.

Done correctly, a dry-pan flashing can be a legitimate flashing option that will save you time when installing some of your penetrations (see Figure 15.15). Unfortunately, it is often misused by services installers as an easy way of getting around the problem of correct penetration installation (see Figure 15.16).

Clean up

Creating both small and large penetrations will leave a considerable amount of debris and waste material. You must ensure that the entire area is cleaned up and checked before leaving the job. The following points should be considered:

- All swarf, rivet shanks, screws and metal trimmings must be swept and/or washed from not only the roof, but the gutters as well! The smallest amounts will cause corrosion damage to the sheet coatings (see Figure 15.17).
- Ensure that all offcuts are disposed of in accordance with the waste requirements for the site.
- Check that all excess silicone sealant has been removed.
- Unused materials should be returned to the workshop or store.
- Ensure that all tools are returned to their respective store area in a serviceable condition. If any are damaged or faulty, report this to the appropriate person.

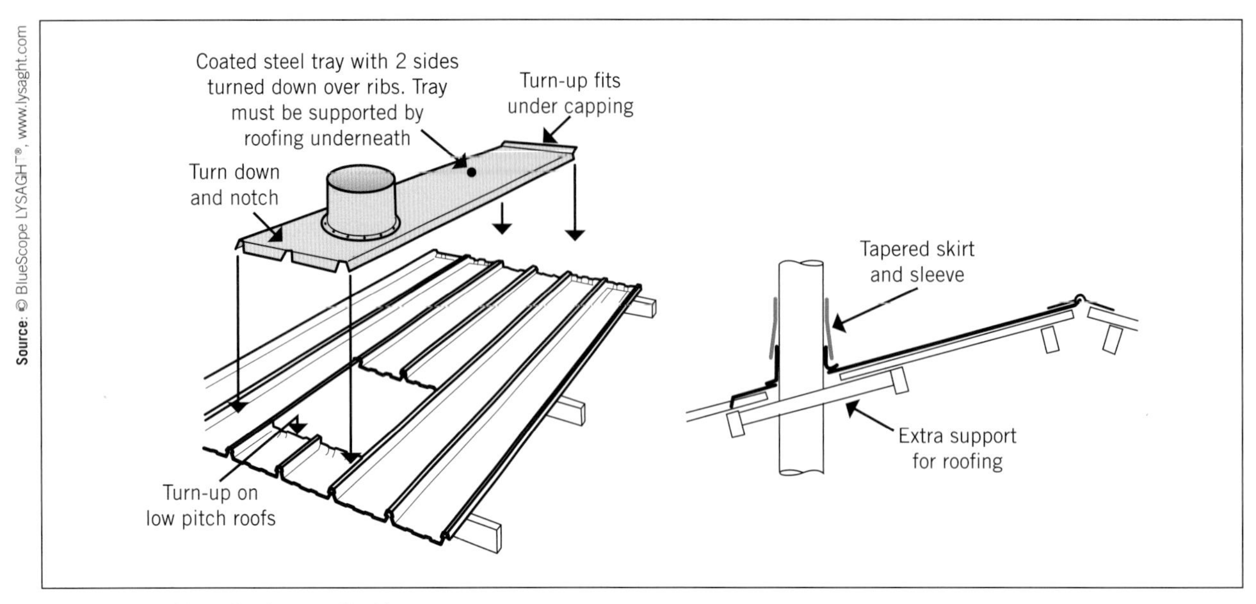

FIGURE 15.14 Use of a dry-pan flashing

FIGURE 15.15 Effective use of a dry-pan flashing (arrows indicate direction of water flow)

FIGURE 15.16 A flashing that is arguably too long, too large and unsightly

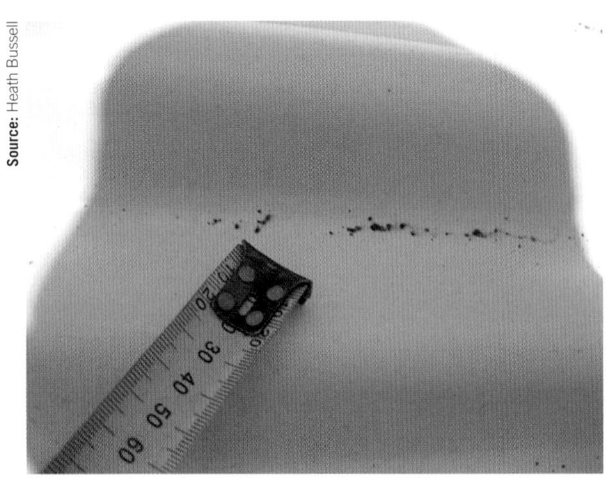

FIGURE 15.17 Swarf corrosion to a COLORBOND® roof sheet. Entirely unnecessary.

SUMMARY

In this chapter you have reviewed the basic requirements related to the preparation and installation of both small and large services penetrations through roofs. These included the use of polymer and metallic flashings as dictated by the type of job. Importantly the influence of BAL ratings was discussed and how these different ratings will direct your choice of flashing material and seals.

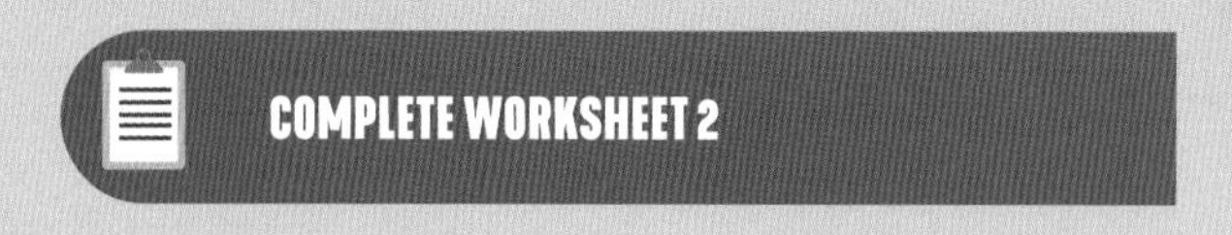

WORKSHEET 1

To be completed by teachers	
Satisfactory	☐
Not satisfactory	☐

Student name: ______________________

Enrolment year: ______________________

Class code: ______________________

Competency name/Number: ______________________

Task: Working with your teacher/supervisor, refer to this text, your HB39:2015 and any other relevant resource to answer the following questions.

1 Name at least six types of roof penetration.

____

____

____

____

____

____

2 Would a hanging flashing made of lead be compatible if used over a ZINCALUME® steel roof?

____

____

____

3 What would be at least four items that you wish to confirm during a site visit?

____

____

____

____

4 What is the difference between EPDM and silicone synthetic rubber collar flashings?

____

____

____

5 Referring to Figure 15.2 earlier in this chapter, could the high temperature red silicone flashing be used on a new construction where the plans indicate a property BAL 40 rating?

____

____

____

6 How would you prevent water being drawn up between a soaker upstand and a counter flashing?

____

____

____

WORKSHEET 2

To be completed by teachers	
Satisfactory	☐
Not satisfactory	☐

Student name: ______________________

Enrolment year: ______________________

Class code: ______________________

Competency name/Number: ______________________

Task: Working with your teacher/supervisor, refer to this text, your HB39:2015 and any other relevant resource to answer the following questions.

1. Your roof penetration has gone through the roof sheet lap. What needs to be done to this lap just upstream of the collar? Provide a code reference to support your answer.

   ____

   ____

   ____

2. What minimum annular clearance must you maintain all around the penetration and the roof sheets? Provide a code reference to support your answer.

   ____

3. True or false: Where the catchment area above a penetration and soaker gutter exceeds 40 m^2, at least two trays on either side of the penetration must be available for water to be diverted into.

   ____

4. A counter flashing must overlap an apron or soaker upstand by at least .... mm. Provide a code reference to support your answer.

   ____

5. Having completed the fitting of a synthetic rubber collar around a flue pipe, what must be done to the grooved seam running up the side of the pipe?

   ____

   ____

   ____

6. Which code reference directs you to ensure that, '... where a section of the roof has been removed for a penetration, the perimeter of the opening should be restored to the original strength of the roof'?

   ____

   ____

   ____

7 Where the downstream transverse capping of the penetration covers the roof sheets, what must be done to each of the roof sheets before securing the flashing? Provide a code reference to support your answer.

______

______

______

8 What is the minimum size lap and distance between fastenings required when installing flashings and cappings? Provide a code reference to support your answer.

______

______

______

9 What is the minimum width of the apron flashings on either side of a chimney penetration? Provide a code reference to support your answer.

______

______

______

10 Circle the correct answer. The minimum height of the upstand for a chimney apron flashing is .... mm.

i 300 mm

ii 100 mm

iii 75 mm

iv 25 mm

11 What is one disadvantage of a dry-pan flashing if it is made too large?

______

______

______

12 What code reference directs you to ensure that all debris is cleaned from the roof?

______

INSTALL ROOF COMPONENTS

16

The primary focus of this chapter is the installation of roof ventilation components. As a roof plumber, your task would be to deal with the penetration and flashing of the ventilator through the roof. While many roof ventilator systems are delivered onsite as a pre-assembled unit, you may also have to assemble ventilator components in accordance with the manufacturer's specifications and instructions. There are clear requirements within the National Construction Code (NCC) relating to the control of condensation and roof plumbers have a key role in satisfying this aspect of construction in both residential and commercial applications.

What should I read first?

Before proceeding with this chapter it would be advisable to revise Part A of this text and Chapter 15 'Flash penetrations through roofs and walls'. Other sections will be referred to where relevant.

Overview

Roof ventilator installations may range from the simple installation of a single, small domestic turbine on a dwelling or shed, right through to multiple, large-scale installations on industrial sheds and factories. The key focus in all your work must be to ensure that the ventilator is installed according to manufacturer's instructions and that your flashing solution is entirely weatherproof. Importantly, you must also ensure that both the component itself and the flashing solution you employ are compliant with any property Bushfire Attack Level (BAL) rating that may apply.

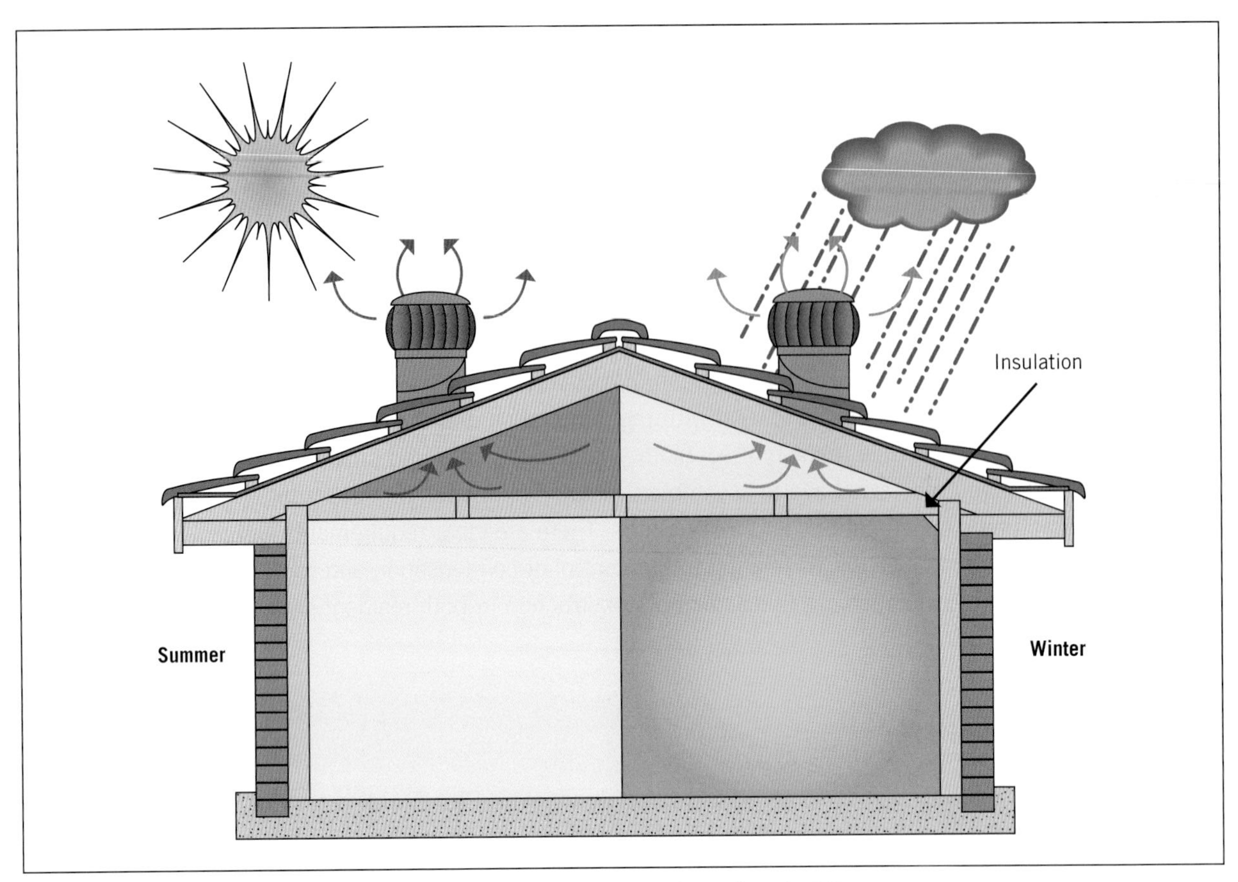

FIGURE 16.1 How roof ventilators work to remove excess heat and moisture from buildings

The function of roof ventilators

Roof spaces need some form of ventilation to remove excess heat and prevent a build-up of water vapour and the resulting condensation. Figure 16.1 illustrates how roof ventilators remove excess heat and moisture from buildings.

Excess heat

The sun heats the surface of roofs each day, causing the temperature within the roof space to rise too. Without any means of escape, the heat in such a confined space builds up and radiates heat into the rooms below. In some areas of Australia, it is not uncommon for the temperature within domestic roof spaces to reach 70°C! Fashionable dark-coloured roof sheets and tiles only make things worse if not adequately insulated.

GREEN TIP

Uncontrolled heat in a roof space makes the rooms below uncomfortable and overloads air-conditioning systems, making them less efficient, more expensive to run and environmentally unsustainable.

Water vapour

The build-up of water vapour within a roof space can occur in both hot and cooler climates. In colder areas, where night-time temperatures can fall close to zero, the temperature differential between the heated rooms of the building and the unheated roof space can be so large that water vapour can condense inside the building. You have probably seen condensation form on the outside surface of a cold drink on a hot day; it's a similar process.

Buildings in areas of high humidity may also require the removal of roof space air to avoid the formation of mould and mildew.

Roof space condensation can have the following serious outcomes:

- Insulation batts can become wet, reducing their R-value.
- Puddles can form, causing damage to plasterboard.
- Timber can suffer from 'dry rot'.
- Steel fixings can corrode.
- Mould and mildew can form and represent a serious health hazard for the building's occupants.

It is for these reasons that condensation is required to be controlled in accordance with the provisions within the NCC.

Air displacement

In addition to the primary roof ventilators that are normally mounted at the highest point of the roof structure, there normally needs to be some form of air intake vent system. For any roof ventilator to remove a certain quantity of air, the same amount of air must come into the roof space. This is called displacement air and it is normally drawn in via intake vents installed at low points in the roof space area. The arrows in Figure 16.2 indicate how air is displaced from the roof space.

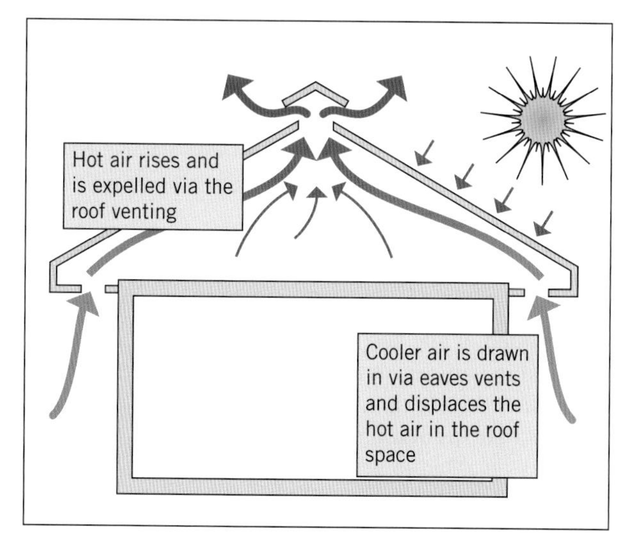

FIGURE 16.2 The process of air displacement from a roof space

For domestic houses, such vents ideally would be fitted into the eaves. These vents are usually sized to match the exhaust capacity of the roof ventilators themselves; however, you must be guided by the manufacturer's recommendations in all cases.

If buildings were only fitted with ventilators at the highest point of the roof, airflow would tend to 'short-circuit', and despite the assistance of thermal convection would be quite inefficient (see Figure 16.3).

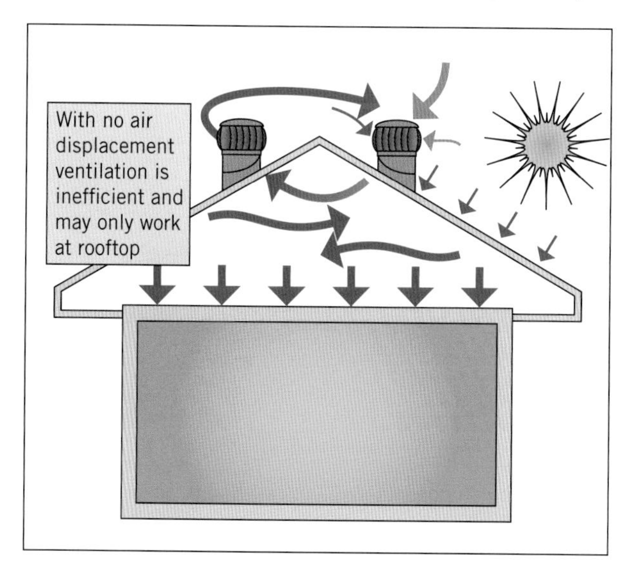

FIGURE 16.3 How air flow 'short-circuits' when there is no allowance for air displacement

A greater volume of air is displaced if the correct allowance is made for displacement air.

Ventilation for BAL-rated properties

All forms of ventilation into a building potentially represent a dangerous pathway for ember attack during a bushfire emergency. It is vital that you purchase and install components in reference to the BAL rating assigned to the property. If necessary, revise the overview of BAL system in Chapter 12 'Select and install roof sheeting and wall cladding'.

LEARNING TASK 16.1

1. Why is it important to have air intake vents when installing primary roof ventilators?
2. What are the possible serious conditions that can occur in areas of high humidity if the removal of roof space air is not considered?

Types of ventilators

Roof ventilators are available in a broad range of sizes, shapes and capacities. Building designers will evaluate and specify one of these systems based on some of the following considerations:

- the purpose of the building
- type of building materials used
- local climatic conditions
- property BAL rating
- strength and wind resistance
- aesthetic impact
- corrosion resistance
- maintenance needs.

The choice of roof ventilator system includes the following examples:

- ridge/slope ventilators
- fixed and adjustable louvre systems:
  - cupola
  - gable
  - box
- rotary turbine ventilators
- gravity intake and relief ventilators
- powered ventilators

Ridge/slope ventilator

A ridge/slope ventilator (also known as a 'continuous' ventilator) is a prefabricated unit that is installed along the ridge of a building or along the slope of a roof. These systems may run the full length of the building or they may be installed at certain intervals (see Figure 16.4).

FIGURE 16.4 Examples of ridge/slope ventilators

This system relies on the natural thermal convection movement of warm air rising to the highest point of the structure and then being vented to atmosphere (see Figure 16.5).

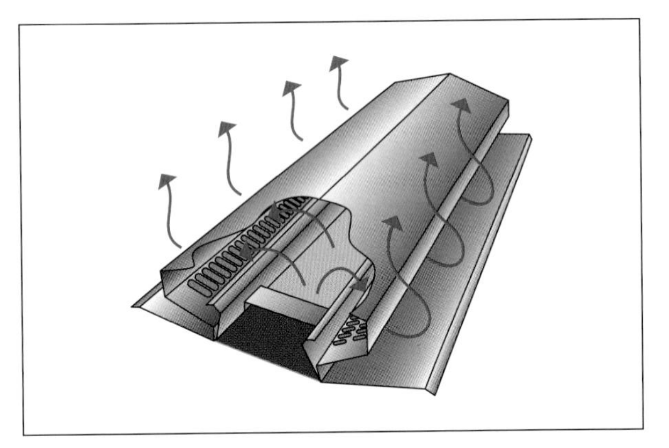

FIGURE 16.5 Cross-section of a typical ridge/slope ventilator showing path of air movement

The design of the ridge/slope ventilator allows air to escape while still being weatherproof and bird-proof. Some units may also incorporate adjustable dampers that allow the user to regulate the amount of air movement through the device.

Local authority advice should be sought in respect to the installation of any ventilators on existing houses as mandatory BAL compliance may apply to any work undertaken on such buildings in high-risk bushfire areas.

Source: Smoothline ventilator for corrugated roofs. Manufacturer: Universal Roof Ventilators

FIGURE 16.6 A BAL 40 rated ridge ventilator showing ember guard mesh

Fixed and adjustable louvre systems

A standard louvre vent features a series of horizontal, overlapping slats that are angled so that air movement is permitted, while keeping rain and wind out. Louvres are used in various configurations to provide roof ventilation. In addition to the conventional wall louvre systems that will normally be installed by the builder, you may be asked to work on the following:

- Box louvre. Box-framed louvre assemblies are mounted on roofs to provide a large but protected opening for natural draught ventilation. Since all four sides of the box louvre are slatted, ventilation will occur regardless of which direction the wind is coming from.
- Cupola. These are effectively the same as the box louvre, though they are often more decorative in appearance and are commonly seen on larger colonial- and Federation-era buildings (see Figure 16.8).
- Gable louvre. The gable louvre is installed vertically in the gable wall of a building (see Figure 16.9). As a result, its optimum performance can be subject to changes in wind direction. However, its efficiency is increased if it is balanced with other soffit and/or gable vents (see Figure 16.10).

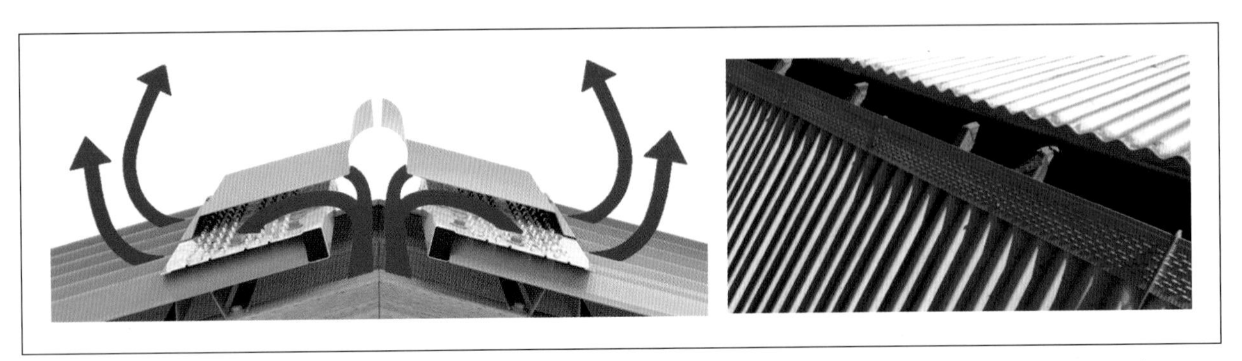

FIGURE 16.7 Full length Lysaght Vent-A-Roof® BAL 40 rated ridge ventilation components being retro-fitted to an existing house

FIGURE 16.8 A traditional cupola ventilator

Louvre slats may either be fixed in position or they can be adjusted by the user to increase or decrease air movement. Adjustment may be affected by:

- electric servo motors
- pneumatic controls (compressed air)
- direct and flexible mechanical winders and levers.

Similar ventilators are available as BAL-rated products fitted with the maximum 2-mm mesh or slotted perforations.

FIGURE 16.9 A fixed gable louvre

Rotary turbine ventilators

Rotary turbine ventilators are a style of ventilator commonly found on domestic, commercial and industrial buildings (see Figures 16.11 and 16.12).

FIGURE 16.11 A commercial multiple turbine ventilator installation

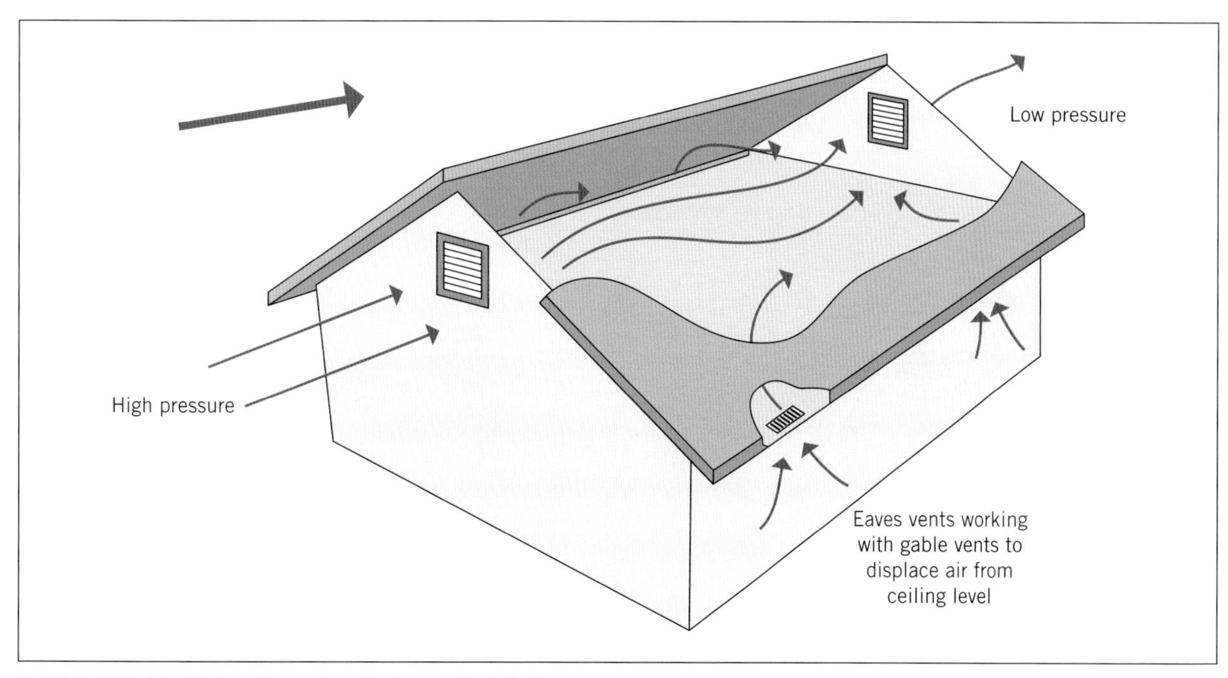

FIGURE 16.10 Ventilation through gable louvre installations

FIGURE 16.12 A domestic-sized turbine ventilator

The unit consists of a bearing-mounted rotating turbine made from overlapping vanes that sits on top of a tube or turret and flashing. Domestic units are available in a range of COLORBOND® colours and can be flashed for tile or steel profile roofing.

Many smaller units incorporate a variable pitch adjustment in the turret riser. An important installation requirement is that this turret be adjusted to ensure that the turbine is level. Figure 16.13 provides a generic guide to how this is done.

LEARNING TASK 16.2

1. In which style of buildings would you see a cupola ventilator installed?
2. Which commonly used domestic ventilator requires the installer to level the turret in each direction prior to fitting the ventilator?

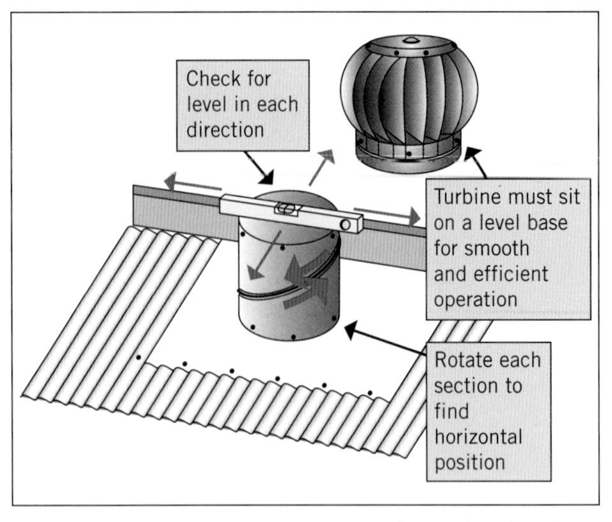

FIGURE 16.13 How to ensure that the variable-pitch flashing is level

Gravity intake and relief ventilators

Gravity intake and relief ventilators are non-powered, passive devices that operate through both thermal convection and the pressure differential that exists between the inside and outside of buildings (see Figure 16.14).

Ventilator installation

The performance of any roof ventilator is dependent on the correct installation procedures being followed. The building designer's specifications and/or manufacturer's instructions must be adhered to. Attempts to install these products without reference

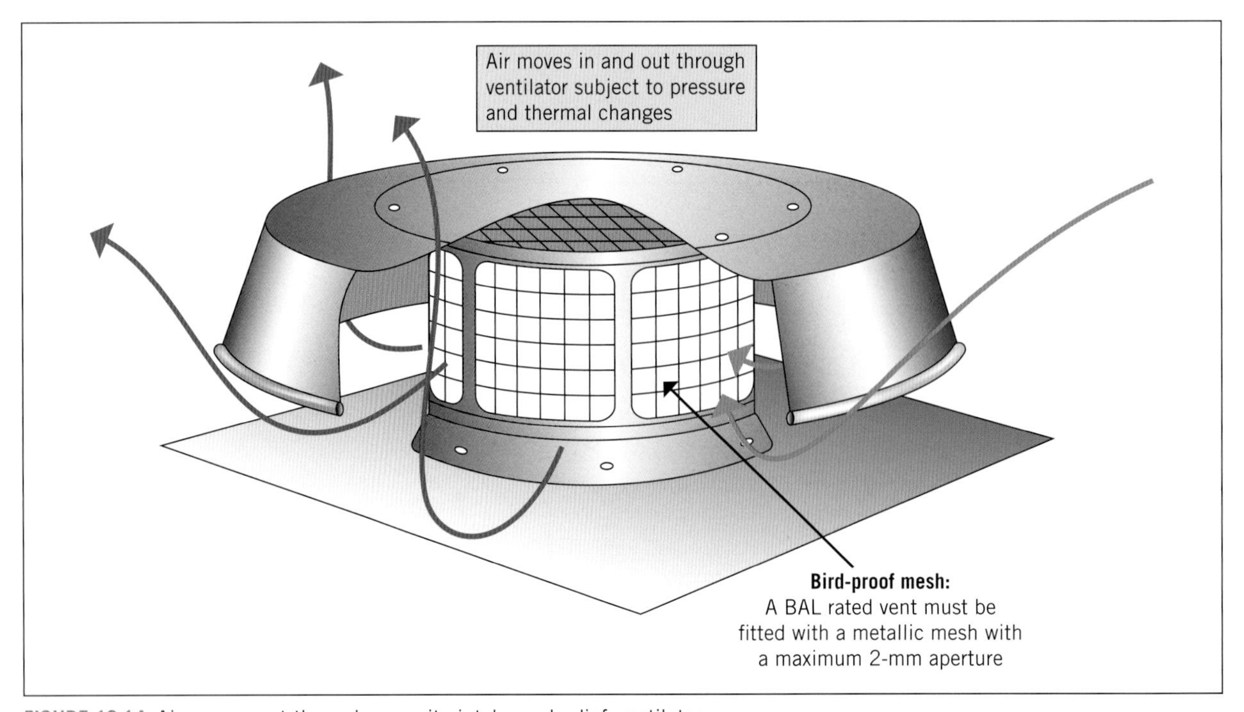

FIGURE 16.14 Air movement through a gravity intake and relief ventilator

to such documents may void warranty provisions and adversely affect the performance of the product.

Preparation

As a central part of your preparation, you must access the plan specifications and/or the manufacturer's instructions and confirm some of the details outlined below.

Determine property BAL rating

You must determine the BAL rating of the property and confirm any local authority requirements relating to existing buildings. Not only does this affect the type of product you purchase but also informs how you install the device and ensure the flashing is sealed in accordance with the rating level.

Flashing options

Different ventilators have a range of flashing requirements. Some may come with soakers pre-flashed to suit certain metallic roof profiles. Others may come with no flashing or, alternatively, complete with a proprietary flashing system that you must use. Depending upon the BAL rating and product characteristics, you may need to seal the flashings with metallic mesh, mineral wool or fire-rated sealant in accordance with AS 3959.

FROM EXPERIENCE

Research your products before final selection. Commonly encountered faults are flashing systems that are incompatible with the roof catchment system or do not meet the BAL rating construction requirements. Get in the habit of researching the requirements of all jobs prior to commencement.

Support requirements

While some units may be entirely supported by an internal frame mounted with the flashing itself, other larger units may require the installation of a support system. Look for these details and check what other materials you may need to complete the job.

Tie-down requirements

The tie-down requirements are those components that you must use, or procedures that you must follow, to ensure that the ventilator is securely fixed to the roof and will not be dislodged by high winds. These components may include straps, brackets and rods that need to be installed as part of the assembly process.

Handling requirements

While a small domestic rotary turbine is easily carried up a ladder, many large commercial and industrial ventilators are so large that some form of mechanical assistance will be required. This might include:

- forklift
- scissor lift
- boom lift
- material hoist
- cranes.

Mark-out and flashing

In this section, refer to Chapter 15 'Flash penetrations through roofs and walls' as required.

Where a penetration for a large roof ventilator is to pass through the roof sheets, there are two main forms of soaker and flashing system applied. These are:

- *Method 1: Full soaker with sheet ribs cut and sealed 100–150 mm from rear and front of penetration.* This method requires that the soaker extend all around the penetration, with the rib ends cut and either sealed or folded to make them watertight and ember proof as required by any relevant BAL rating (see Figure 16.15).

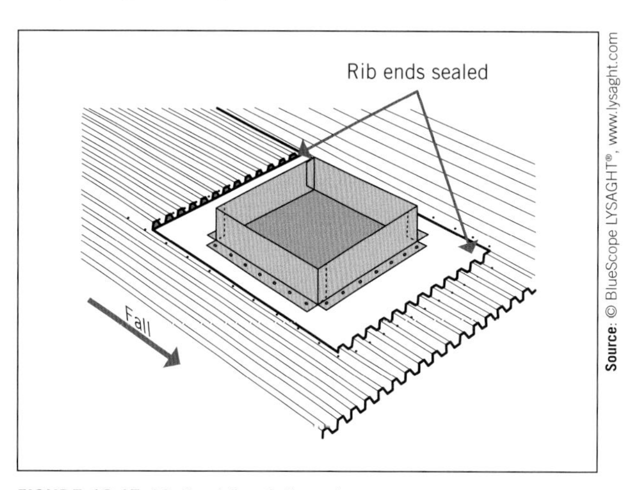

Source: © BlueScope LYSAGHT®, www.lysaght.com

FIGURE 16.15 Method 1 – full soaker

- *Method 2: Three-sided soaker with downstream sheet pans stop-ended.* In this method, the soaker extends only across the back and down the sides of the penetration. The area in front of the penetration is covered entirely with an over-flashing that is turned down and notched into the pans of the sheets (see Figure 16.16).

  Your decision on which method to adopt will be influenced by the type of penetration, the manufacturer's recommendations and personal preference.

  The penetration mark-out for each method is explained in the following procedures.

Ventilator supports

Regardless of what method you use to flash the penetration, ensure wherever possible that ventilator and roof supports are in place before opening the roof, reducing the time that the roof is open and at risk of water damage should it rain.

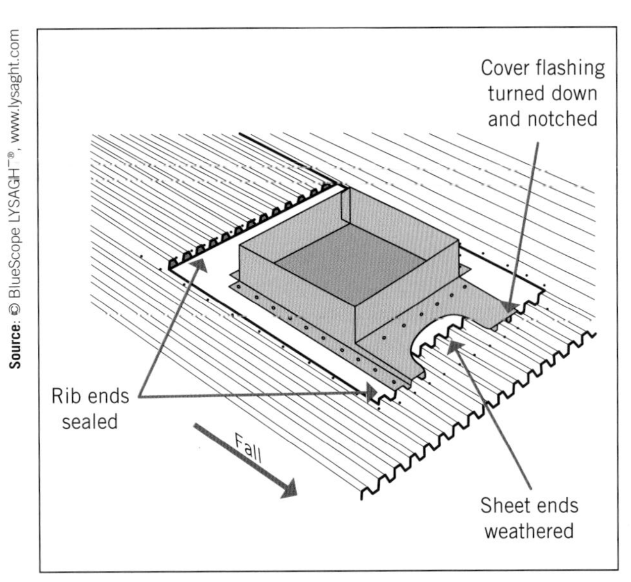

FIGURE 16.16 Method 2 – three-sided soaker

FROM EXPERIENCE

An attribute valued by employers is the ability to anticipate problems before they happen and put in place effective solutions. Develop the habit of checking the job first before getting stuck into it.

Where this can't be done, make sure that you have all materials ready so that once you make the breach in the roof you can get the support system in place as soon as possible.

Remember that the full perimeter of your penetration must be supported to the same roof strength as it was before the hole was cut.

HB39:2015 SECTION 5.5(A)

Soaker and flashing – Method 1 (full soaker)

The following illustrations and text describe the basic requirements when installing a full soaker flashing and upstand.

Mark out penetration size

Using the manufacturer's instructions and your own check measurements from the ventilator itself, mark out the size of the penetration on the roof. To this measurement you must add a minimum of 20 mm around the full penetration to allow for movement (see Figure 16.17).

HB39:2015 SECTION 5.5(E) AND FIGURE 5.5(C)

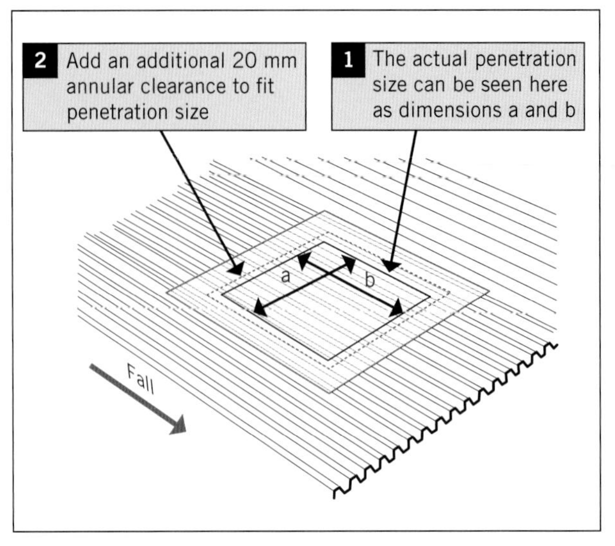

FIGURE 16.17 Penetration and annular clearance mark-out

Mark out the soaker dimensions

Once you have marked out the penetration dimensions plus the clearance allowance, you need to add to this the soaker mark-out. Ribs must be removed to allow water flow around the penetration (see Figure 16.18).

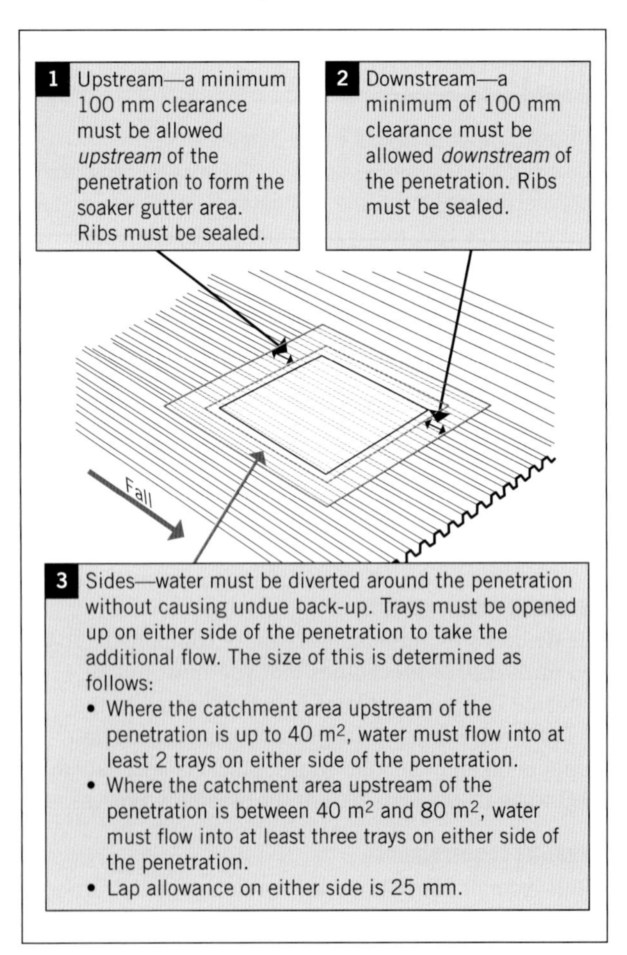

FIGURE 16.18 Penetration and full soaker mark-out

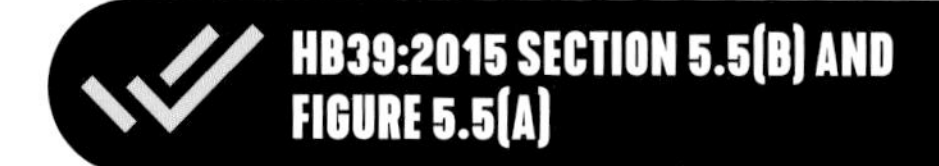

HB39:2015 SECTION 5.5(B) AND FIGURE 5.5(A)

Do not use a standard pencil to do your mark-out, as this will cause corrosion. A marker pen is the best option.

FROM EXPERIENCE

It is surprisingly easy to start cutting in the wrong place, particularly when you are under time pressure. Make sure that you use a marker pen to cross-hatch all waste material in your mark-out. This helps to ensure that your cut lines stand out from your fold lines and laps.

Cut out the hole

The tools and procedure you use to cut out your penetration hole will depend to some extent on the type of roof sheet rib profile. The following is a guide only:

- Use a hacksaw or appropriate power tool to cut through each of the ribs. It is not normally possible to get right to the bottom of the rib. Finish off the cut down to the tray with a hammer and an old wood chisel.
- Use the appropriate snips or a nibbler to cut out the flat material of the roof sheet; the side of an old chisel or half of an old pair of Gilbows are good tools to hit with a hammer and quickly cut out sections.
- Never use carborundum disc power cutters, as these will burn the sacrificial edge of the cut and the sheet will rust as a result.

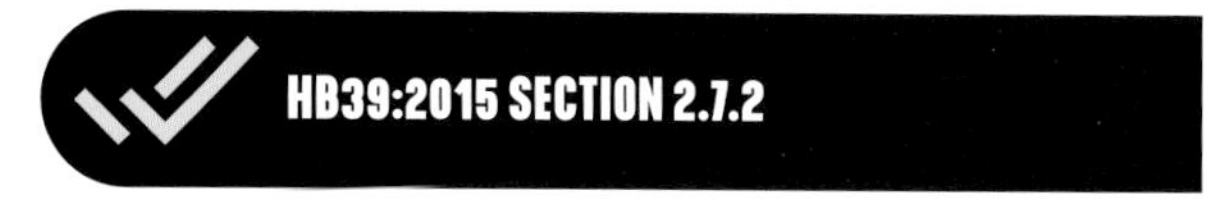

HB39:2015 SECTION 2.7.2

Fabricate the soaker

Depending on its overall size, the soaker plate is normally made up from three separate sections:

- the soaker base plate (including two folded upstands)
- two separate side upstands and corner laps.

The soaker upstand should be at least 100 mm high, and all fasteners should be no more than 40 mm apart (see Figure 16.19).

Fit the soaker

Your soaker should be inserted as follows (see Figure 16.20):

1. Insert the upstream end a minimum of 25 mm beneath the roof sheet so that it is lapped in the direction of flow.
2. The sides of the soaker will overlap the sides of the roof sheet by a minimum of 25 mm as per the mark-out.
3. The downstream end of the soaker will sit on top of the roof sheet pan with notches taken out in front of the ribs.
4. Seal and secure soaker with appropriate fasteners at 40-mm intervals.
5. Remove excess silicone with a spatula.

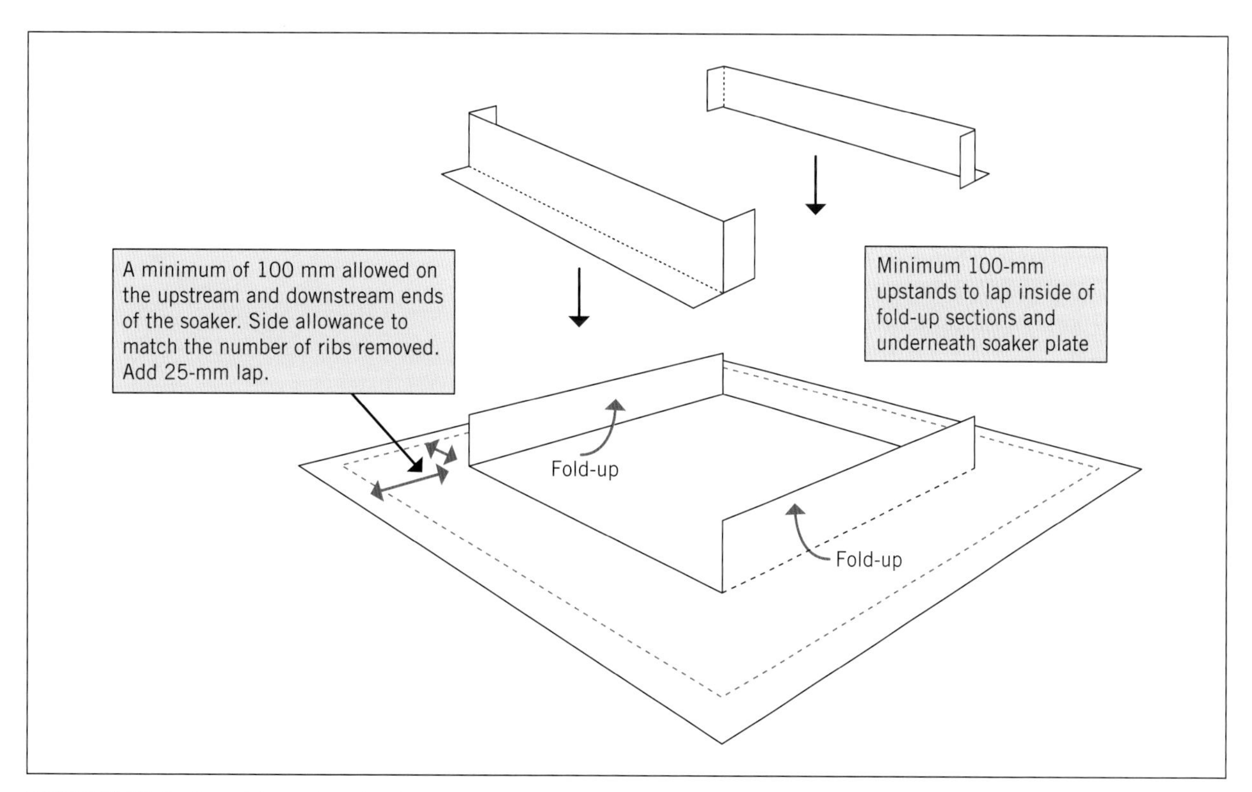

FIGURE 16.19 Soaker plate and upstands

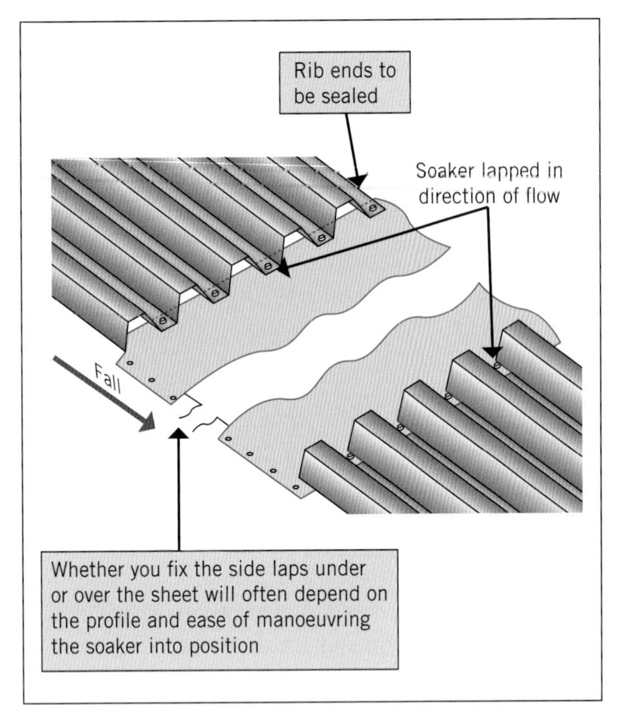

FIGURE 16.20 Soaker lap detail

The ribs should be closed using a method that is suitable for the roof sheet profile and the material being worked on. More detail on rib closure is provided at the end of this section.

Fabricate and fit the flashing

Some ventilator units come with an integral flashing kit, but in most cases you will need to put a flashing solution in place yourself. For this installation method a simple four-sided flashing is the most effective. The flashing sections are sealed and secured to the side of the penetration body and have 15-mm laps at each corner.

The actual measurements will be subject to individual circumstances, but the drawings and measurements in Figure 16.21 should act as a basic guide.

When fitting the flashing, ensure that you maintain at least 10-mm clearance between the top edge of the soaker upstand and the over-flashing. This provides an allowance for movement between the roof and the penetration.

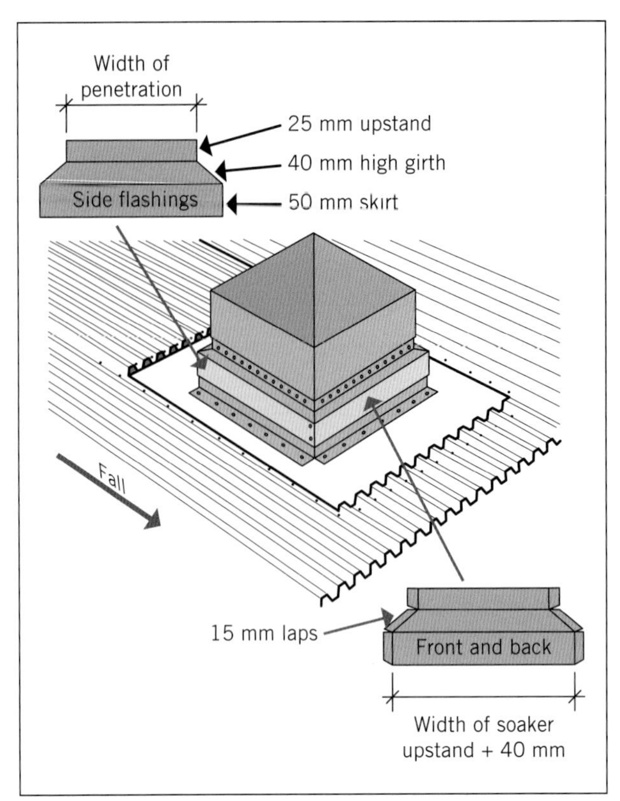

FIGURE 16.21 Flashing detail for Method 1

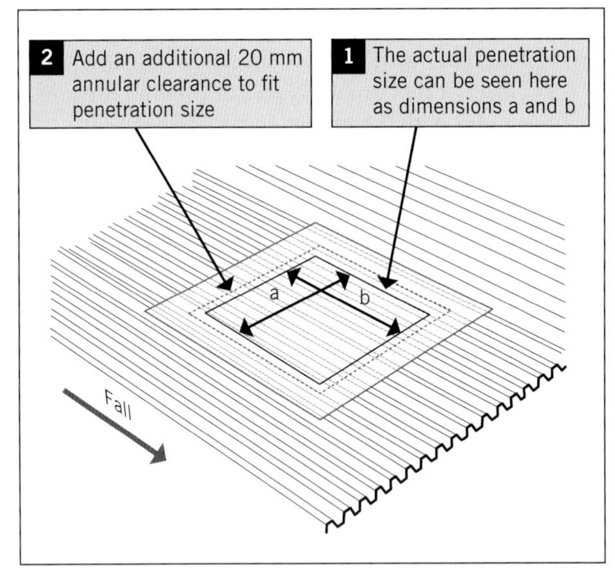

FIGURE 16.22 Penetration and annular clearance mark-out

Soaker and flashing – Method 2 (three-sided soaker)

While less common than a full soaker installation, some components may dictate the use of a three-sided soaker and front over-flashing. Follow the steps detailed below.

Mark out the penetration size

Using the manufacturer's instructions and your own check measurements from the ventilator itself, mark out the size of the penetration on the roof. To this measurement you must add a minimum of 20 mm around the full penetration to allow for movement (see Figure 16.22).

Mark out the soaker

Once you have marked out the penetration dimensions plus the clearance allowance, you need to add to this the soaker mark-out. Ribs must be removed to allow water flow around the penetration (see Figure 16.23).

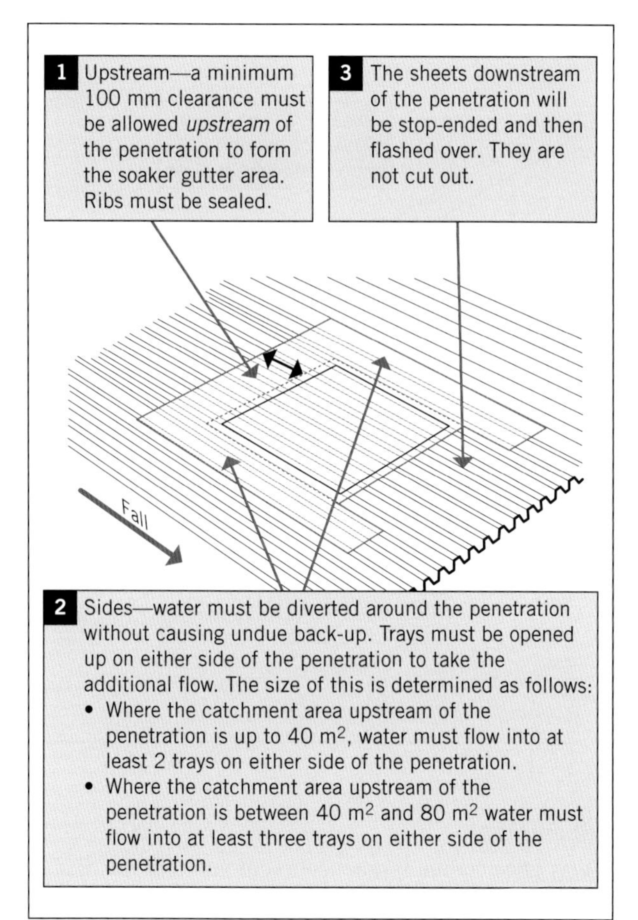

FIGURE 16.23 Penetration and three-sided soaker mark-out

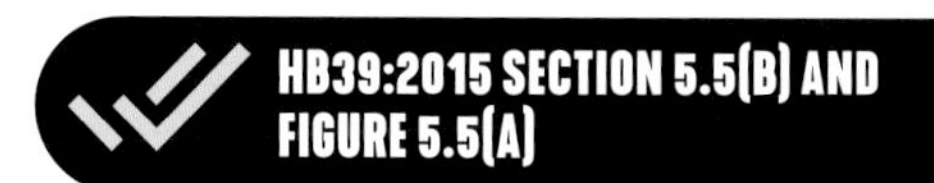

Do not use a standard pencil to do your mark-out as this will cause corrosion. A marker pen is the best option.

Cut out the hole

The tools and procedure you use to cut out your penetration hole will depend to some extent on the type of roof sheet rib profile. The following is a guide only:

- Use a hacksaw or appropriate power tool to cut through each of the ribs. It is not normally possible to get right to the bottom. Finish off the cut down to the tray with a hammer and old wood chisel.
- Use the appropriate snips or a nibbler to cut out the flat material of the roof sheet; the side of an old chisel or half of an old pair of Gilbows are good tools to hit with a hammer and quickly cut out sections.
- Do not use carborundum disc power cutters, as these will burn the sacrificial edge of the cut and the sheet will rust as a result.

Stop-end the sheets

Despite being covered in an over-flashing, the sheets directly downstream of the penetration must be turned up to prevent wind-driven rain from flowing past them and into the building. Figure 16.24 shows an example of how this may be done. Use the method and tools recommended by the roof sheet manufacturer.

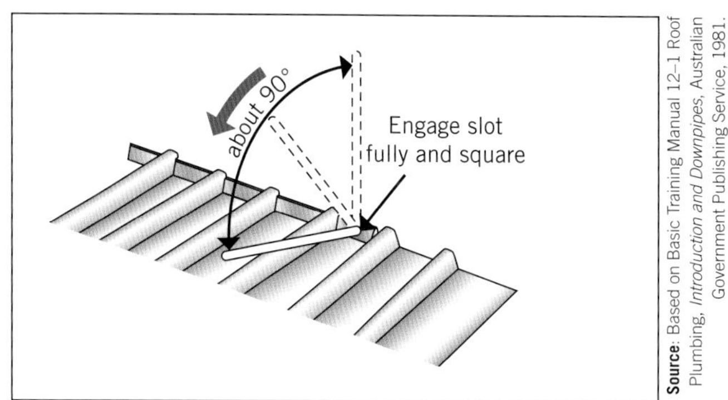

Source: Based on Basic Training Manual 12-1 Roof Plumbing, *Introduction and Downpipes*, Australian Government Publishing Service, 1981.

FIGURE 16.24 Turning up sheet ends

Fabricate the soaker

Depending on its overall size, the soaker plate is normally made up from four separate sections:

- the soaker base plate (including one folded upstand at the upstream end)
- two separate side upstands and corner laps
- the combined cover flashing and front upstand.

The soaker upstand should be at least 100 mm high, and all fasteners should be no more than 40 mm apart (see Figure 16.25).

The ribs should be closed using a method that is suitable for the roof sheet profile and the material being worked upon. More detail on rib closure is provided at the end of this section.

Fabricate and fit the flashing

Some ventilator units come with an integral flashing kit, but in most cases you will need to put in place a flashing solution yourself. For this installation method the back and sides of the flashing are very similar to the style shown in Method 1; however, the front flashing must be longer to provide sufficient cover of the sheets. The front flashing must extend a minimum 100 mm over the downstream sheets and be turned fully down into the pan for a length suitable for the particular profile. The turn-down must be notched to provide a weatherproof fit.

The flashing sections are sealed and secured to the side of the penetration body and should have 15-mm laps at each corner.

The actual measurements will be subject to individual circumstances, but the drawings and measurements in Figure 16.26 should act as a basic guide.

When fitting the flashing, ensure that you maintain at least a 10-mm clearance between the top edge of the

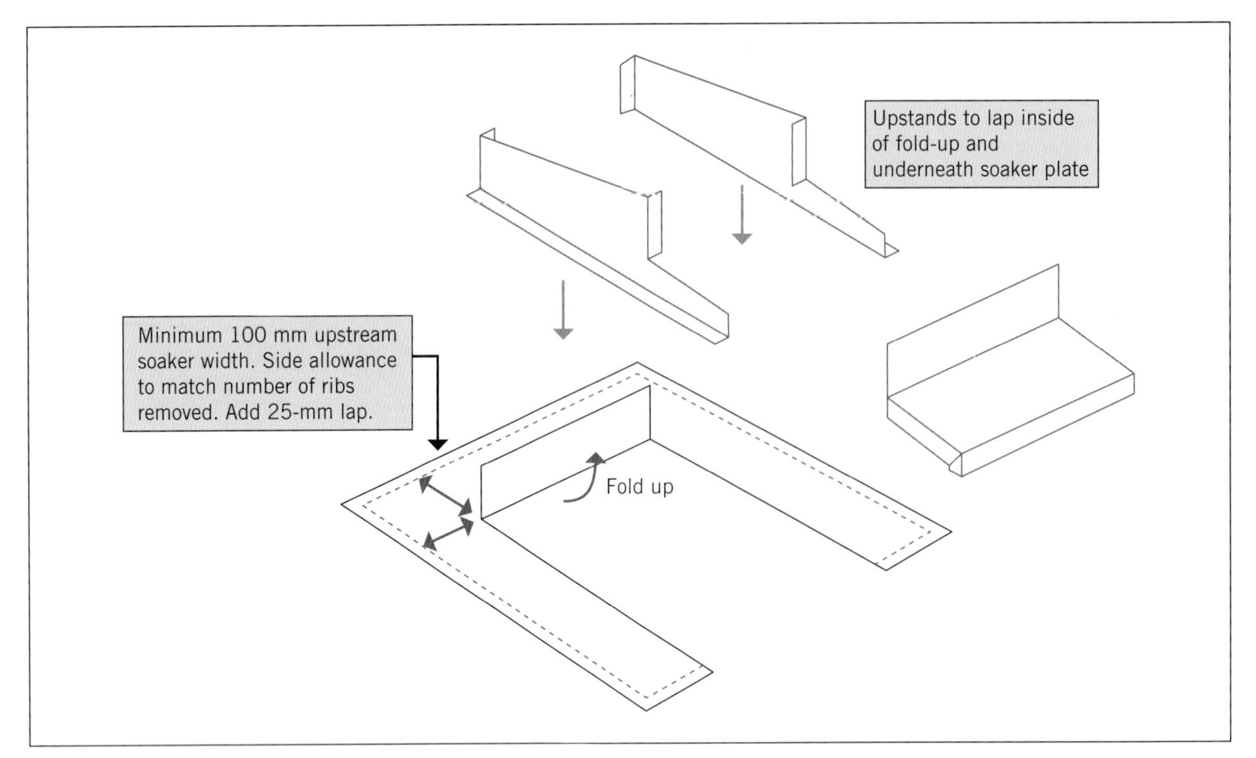

FIGURE 16.25 Soaker plate and upstand for Method 2

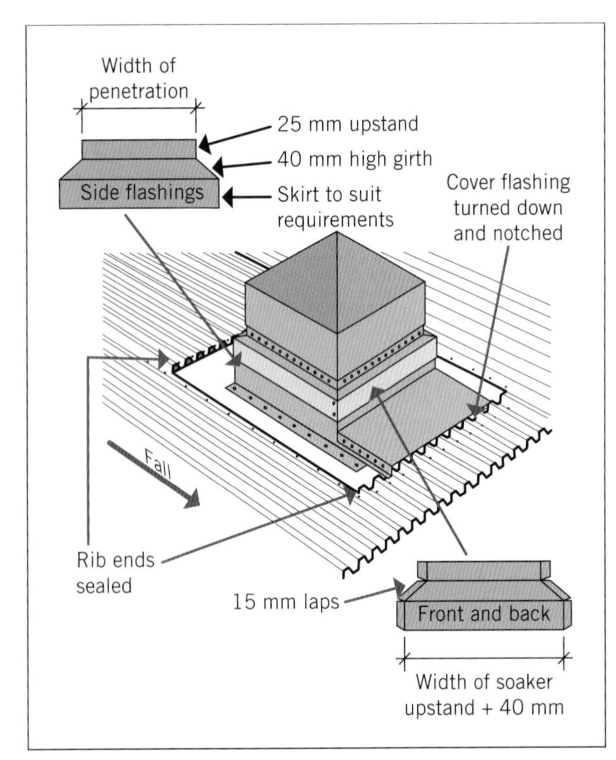

FIGURE 16.26 Flashing detail for Method 2

soaker upstand and the over-flashing. This provides an allowance for movement between the roof and the penetration.

Rib closing

It was once common practice to cut and fold the ribs down to seal them. Older roof sheets were marginally thicker and slightly more malleable, making this a relatively easy process. This method is still acceptable today; however, modern high-tensile ZINCALUME® steel roofing sheets are more difficult to work and close.

It comes down to personal preference, but you might find that the use of simple, vertically cut ribs with prefabricated end closers is just as fast and often provides a neater finish when closing sheet ribs.

Figure 16.27 shows some options for closing ribs.

Use of a dry-pan flashing

If your penetration is situated close to the ridgeline of the roof, you may wish to install a dry-pan flashing from under the ridge capping down to and around the penetration (see Figure 16.28). This may save considerable time if it is suitable. More detail on dry-pan flashings is included in Chapter 15 'Flash penetrations through roofs and walls'.

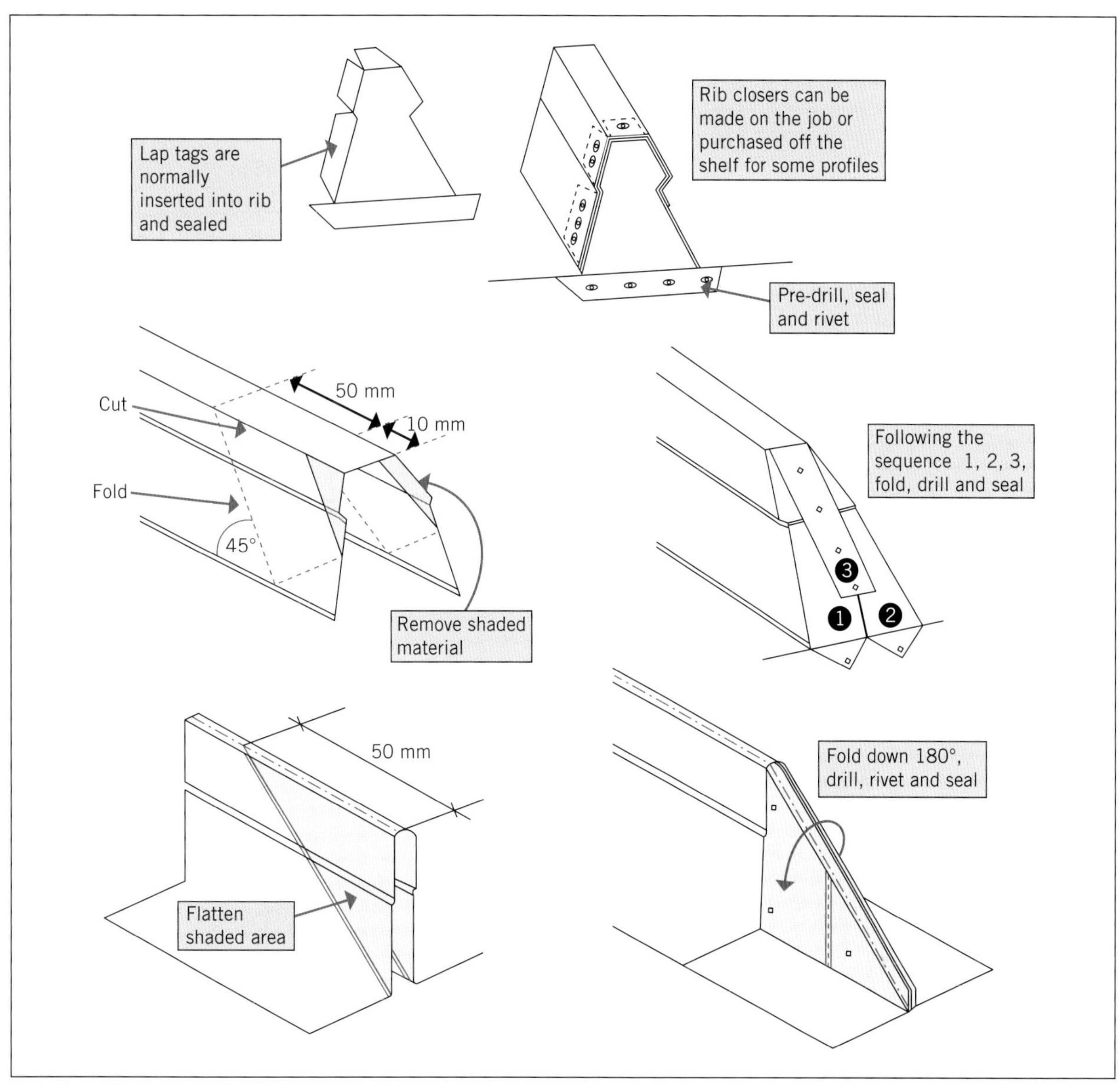

FIGURE 16.27 Rib closing options

FIGURE 16.28 Dry-pan flashing used between multiple rotary turbine ventilators

LEARNING TASK 16.3

1 What three factors will influence your decision whether to use a full soaker or three-sided soaker?
2 Which other type of flashing might be considered for a penetration close to the ridgeline?

Clean up

The cutting in of roof penetrations and installation of roof components will generate considerable waste that must be removed as soon as possible from the job. The following points should be considered:

- Clean up as you go to lessen the amount of work to be done at the end of the job. Good worksite waste control is also important from a health and safety perspective.

- The use of nibblers, drills and snips will create a lot of swarf and small offcuts that must be swept and/or washed from the roof every day whenever possible. Swarf can stain a roof very quickly and small trimmings can get caught in your boots, scratching painted roof surfaces.
- Lightweight offcuts of insulation, vapour membrane and strippable protective film can easily blow around and off the roof, creating potential slip hazards and a general litter problem. Store such refuse under a weight while you are working and bring it down for disposal at the end of the day.
- Larger offcuts should be checked for re-use or recycling as appropriate. Larger sites will often have designated refuse bins. Dispose of offcuts accordingly.
- Check that your tools are in good working order and report any faults to your supervisor immediately. Don't simply return to the store and leave it for someone else. In certain circumstances such negligence may lead to your workmates being injured, or worse.

SUMMARY

This chapter has provided an introduction to the selection and installation of roof components with particular emphasis on various roof ventilators including:

- ridge/slope ventilators
- fixed and adjustable louvre systems
- rotary turbine ventilators
- gravity intake and relief ventilators
- powered ventilators.

Large penetrations require particular attention to be placed upon sheet support around the penetration and correct fabrication and assembly of soaker and relevant over-flashings. All roof component installations now must be also planned and carried out in reference to designated BAL ratings for both product selection and subsequent installation.

COMPLETE WORKSHEET 1

WORKSHEET 1

To be completed by teachers	
Satisfactory	☐
Not satisfactory	☐

Student name: ______________________

Enrolment year: ______________________

Class code: ______________________

Competency name/Number: ______________________

Task: Working with your teacher/supervisor, refer to this text, your HB39:2015 and any other relevant resource to answer the following questions.

1 There is approximately 45 m^2 of catchment area above your planned ventilator penetration. How many trays should be allocated for drainage on each side of the penetration to ensure adequate water flow? Provide a code reference to support your answer.

____

____

____

2 Your ventilator penetration measures exactly 600 mm × 600 mm. What will be the actual hole size within your soaker? Provide a code reference to support your answer.

____

____

____

3 What parts of the roof penetration should be supported? Circle the correct response below and provide a code reference to support your answer.

i each side only

ii upstream and downstream end

iii the complete perimeter of the penetration

____

____

____

4 On an older commercial building, you plan to re-flash a large box-louvre ventilator. The existing concealed-fixed roof is in good condition and made from unpainted galvanised steel. What material will you specify for your new soaker and over-flashings? Why?

____

____

____

5 What should be the minimum height of the soaker upstand?

6 Section 3.6 of AS 3959:2018 lists certain general requirements related to all BAL-rated vents. What is the maximum permitted mesh or perforation size required to be fitted to all vent openings?

7 What does the term 'weathering' mean in relation to roof sheets?

8 Referring to this text, what minimum clearance should you leave between the top edge of the soaker upstand and the over-flashing?

9 List at least four types of roof ventilator.

10 Referring to this text, what would be the minimum width of your soaker flashing upstream of the penetration?

11 Is it fine to use a carborundum disc as long as you immediately clean up after cutting? Explain your answer.

17

INSTALL ROOF COVERINGS TO CURVED ROOF STRUCTURES

This chapter focuses on the skills and knowledge required to confidently undertake the installation of various forms of curved roofs. Curved roofs such as the one shown in Figure 17.1 are used in many modern construction designs, and are incorporated into building design for both functional and aesthetic reasons. The constant improvements in building materials and installation techniques allow designers to be more adventurous in how the building is constructed and appears.

What should I read first?

Before proceeding with this chapter it would be advisable to revise Part A of this text and, as a minimum, work through Chapter 12 'Select and install roof sheeting and wall cladding'. Other chapters will be referred to where relevant.

Overview

The installation of curved roof products shares many similarities with the use of conventional roof and cladding. This chapter will look more specifically at the extra needs and variations that you will find with curved roof work.

FIGURE 17.1 A modern curved roof

Types of curved roof installations

Curved roofing falls into two basic categories, based upon how the curve itself is made: sprung roofs and roll-formed (pre-curved) roofs.

Sprung roofs

Standard corrugated and trapezoidal sheets are made from high-tensile steel with a base metal thickness (BMT) of 0.42 mm or 0.48 mm. These sheets are suitable for sprung curves.

A spring-arched roof covering is created when standard flat roofing sheets are gradually fixed to a purpose-designed roof featuring a gentle, free-form curve from eave to eave.

A spring-curved roof can be made across the ridge of the roof with the remainder of the sheets remaining flat.

Roll-formed roof sheets

Some curves are too acute and if you tried to spring-curve a standard high-tensile steel sheet around such a shape, the metal would buckle.

In this instance, the roof plumber must take the exact dimensions of the roof from the approved plans and have the flat sheets run through a special roll-forming machine to make roll-formed roof sheets.

The sheets used for roll-formed curving are more ductile than high-tensile steel and they have a BMT of 0.6 mm. With this grade of roof sheet, acute curves can be accurately created to suit individual building requirements. Figure 17.2 shows roll-formed roof sheets used on a commercial building.

FIGURE 17.2 Use of roll-formed roof sheets on a commercial building

Curved roof shapes

A name is given to each of the curved roof shapes or profiles used for different effects and purposes. Of course, profile names will vary between contractors, regions and manufacturers. As you can see in Figure 17.3, there are many types of curved roof profiles. Some can be sprung and others must be roll-formed. The curved roof shapes reviewed here include:

- spring-curved ridge
- spring-arched (convex)
- spring-curved (concave)
- 180° barrel vault
- bullnose
- Old Gothic (cyma recta)
- reverse curve (cyma reversa)

If agreed by your teacher and supervisor, change any of the names used here to one in more common use from your own region.

Some of these profiles are reviewed in more detail in the next section.

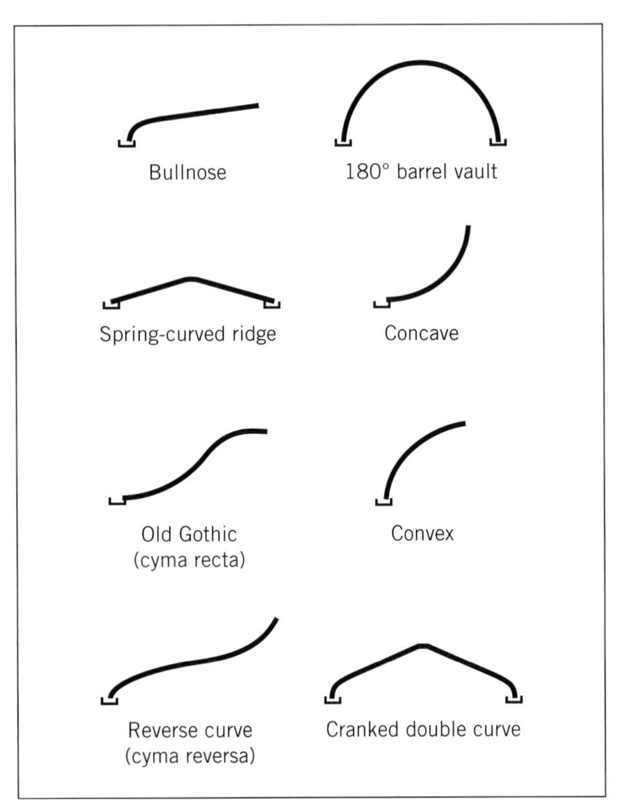

FIGURE 17.3 Basic curved roof profiles

Spring-curved ridge

The spring-curved ridge roof is created when flat sheets are fixed and curved over the ridge of a gable roof. The ridge is the only part of the sheet that is curved, while the rest of the sheet length remains flat.

Spring-arched (convex)

While the spring-curved ridge roof is only curved over the ridge area itself, the spring-arched (convex) roof is laid in a single radius curve from one eave to the other (see Figure 17.4).

FIGURE 17.4 A spring-arched roof

Spring-curved (concave)

A spring-curved (concave) roof is the opposite of a convex roof. The roof sheets are spring-curved inwards onto specially located purlins. As with a convex roof, there is a minimum roof radius onto which the sheets are fixed (see Figure 17.5).

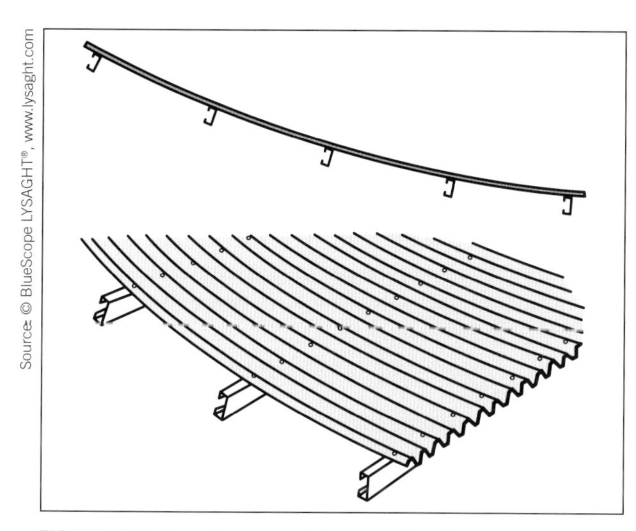

Source: © BlueScope LYSAGHT®, www.lysaght.com

FIGURE 17.5 A spring-curved (concave) roof

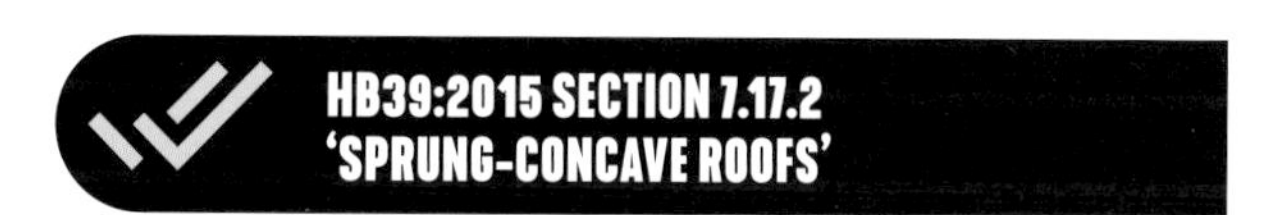

Which is it – concave or convex?

It is easy to get confused over which curve is concave or convex. One curves inwards and one curves outwards.

One way to remember is to associate the word concave with the word 'cave'; a concave roof curves inwards just like a 'cave' in a hillside.

180° barrel vault

The sheets of a 180° barrel vault roof form a complete 180° arc from one side of the roof to the other (see Figure 17.6). In most instances this will entail joining additional sheets to create the full curve. The vault design provides a cost-effective roof space area through the use of lightweight roof frame construction.

Ensure that silicone is used in all laps at the flattest part of the curve. Continue this until the roof sheet is falling at or above the minimum pitch requirements.

FIGURE 17.6 180° barrel vault roof

Bullnose

The bullnose roof is a pre-curved roof profile that has been popular in Australia since the colonial era (see Figure 17.7). Most often used on verandas, the bullnose roof can play a valuable role in the design of modern energy-conscious and sustainable buildings in hotter climate zones. The radius curve of the sheet into the gutter protects the building walls and windows against the sun's rays while still maintaining adequate roof height under the veranda.

FIGURE 17.7 A bullnose roof veranda

More information on the installation of bullnose roofs is provided later in this chapter.

Old Gothic (Ogee or cyma recta)

Old Gothic is a pre-curved profile featuring a convex curve above a concave curve. This creates a flowing

S-curve that is used by designers to provide more visual interest and shape to a roof design (see Figure 17.8). The terms 'Ogee' (OG – Old Gothic) or cyma recta are also used to describe this shape.

Like standard spring-arched concave roofs, the lower end of the sheets should not fall below the minimum pitch specified for the sheet profile.

Source: © BlueScope Steel Limited.

FIGURE 17.8 Applications of Old Gothic roofs in both modern and colonial designs

Reverse curve (cyma reversa)

As the name suggests, the reverse curve roof profile is essentially the opposite of the Old Gothic design, having a concave curve above a convex curve. This profile is sometimes used to act as a smooth transition between the main roof pitch and the veranda (see Figure 17.9).

FIGURE 17.9 Colonial-era building featuring a reverse curve roof

LEARNING TASK 17.1

1. What name is given to the form of sprung roof that curves inwards in a gentle curve from one end of the sheet to the other?
2. What is the BMT of steel used for a bullnose roof?

General planning and preparation

Every job has its own variations, but in most instances planning and preparation for your curved roof installation requires at least the following tasks:

- gathering information
- product selection
- calculating requirements.

Some points particularly relevant to curved roof installation are covered in more detail in the following section.

Gathering information

A successful curved roof installation depends on the efficient research and collation of information, plans, specifications and authorities' documentation, as well as confirmation through a site visit. Some of these points are covered below.

Scope of work

You need to determine the scope of work so that you can plan all tasks, establish product requirements for the job and accurately sequence each step of the job. You can source scope of work information from some of the following sources:

- your supervisor – for a general outline
- company quality assurance protocols
- approved plans – must be stamped as approved
- specifications – normally included on the plans or as an accompanying document
- manufacturers' product information
- the weather bureau – planning around inclement weather is vital to ensure that the job runs efficiently, profitably and safely. Re-roof work in particular must be well planned in order to avoid water and wind damage to existing buildings.

Re-roof considerations

As part of determining the scope of work, when carrying out the replacement of roof sheets on an older building you must ensure that the existing structure will be suitable for the installation of new products. Older building frames may lack the necessary tie-down requirements and the existing roof frame components may have deteriorated over time. As an example, where older bullnose roofs must be replaced, you will often find that the existing battens are undersized or have been split by spring-head nails. Some veranda frames will also reach over-span and additional timber work will need to be fitted.

Roof replacements can be lucrative work but you need to be cautious and know exactly what you are taking on. Some additional considerations relating to roof replacements are included in Chapter 12 of this text.

Plans and specifications

The plans and specifications are vital documents that provide most of the core information that you need to undertake the job. Review Chapter 12 for more detail on what the plans and specifications contain.

Manufacturers' product information

All curved roofing products and ancillary materials are backed by information produced by the manufacturer that is designed to assist you in the selection and installation of their products.

In addition to the information provided in Chapter 12, this chapter makes regular reference to the following sources of manufacturer information:

- LYSAGHT® product range, available at http://www.lysaght.com > products > roofing and walling
- LYSAGHT® *Roofing & Walling Installation Manual*, available at http://www.lysaght.com > installation and maintenance > roofing and walling
- Buildex® technical data sheets, available at http://www.buildex.com.au > technical advice

Almost all companies will provide similar information on their respective websites.

Insulation for curved roofs

The correct selection and installation of insulation products is a central requirement of most roof installations. This applies also to curved roof installations, though simple veranda coverings will often not require insulation.

The approved plans and specifications will detail the minimum R-value required for insulation in walls and ceilings. It will be your responsibility to source a product that will meet these minimum requirements and also be compatible with the roofing system that you are using.

Increased thermal performance requirements for buildings have led to higher R-values and therefore thicker insulation material. This will have an impact on the selection and use of roof system fasteners:

- *Pierce-fixed roofs*. Where insulation blanket is specified for use beneath the roof sheets, you will need to accommodate the thickness of this extra material with the selection of longer screws.
- *Concealed-fixed roofs*. The compression of insulation blanket between roof sheet, fasteners and purlins will reduce the overall thermal rating of the installation. This may cause the building to fail to meet minimum performance measures. In this case, manufacturers can provide proprietary spacers that sit between the purlins and roof sheet and provide a gap that prevents insulation compression.

More detail on the selection and use of insulation is included in the installation section of Chapter 13, 'Select and install roof sheeting and wall cladding'.

Site inspection

Never plan for a job without conducting an initial and ongoing regular site visits. There are almost always variations from the original plans and you need to accommodate these changes for your installation. For curved roof jobs it is particularly important to confirm that support spacing and specified radii comply with manufacturer and code recommendations.

Review Chapter 14 'Install roof sheets, wall cladding and complex flashings' for more details relating to site visits.

Safety

All aspects of the job must be planned for and carried out with constant observance of all safety considerations. Chapter 3 'Basic roof safety' covers this issue in more detail.

Product selection

For some work, the scale or type of project may dictate that almost all material items will have been selected during the planning and approvals stage, leaving you only the task of installation. In other instances and particularly for small domestic jobs, you may have the responsibility of selecting a wide range of products to suit the job. As long as such products satisfy the performance criteria within the plans and specifications, you may have considerable flexibility to choose profiles, material types and colours. For many of these jobs you may be the sole contractor dealing directly with the client.

Treat every job as an individual case and don't fall into the trap of simply selecting products with which you are familiar or that are used all the time by other contractors. Do your research and choose the best solutions for each application.

Sources of product selection information include:

- the codes and standards
- local government recommendations and requirements
- materials suppliers' representatives
- manufacturers' sales and technical representatives
- hard copy product information
- internet searches for installation and product guides.

Some examples relating to product selection are given below.

Selection of roof sheets

There are many roof sheet profiles available for general roofing, but not all are suitable for the different forms of curved roof installation. If the profile is not specified in the plans, you may need to research and confirm a suitable profile yourself. Most information to support your installations is available via simple internet searches, as shown in Figure 17.10.

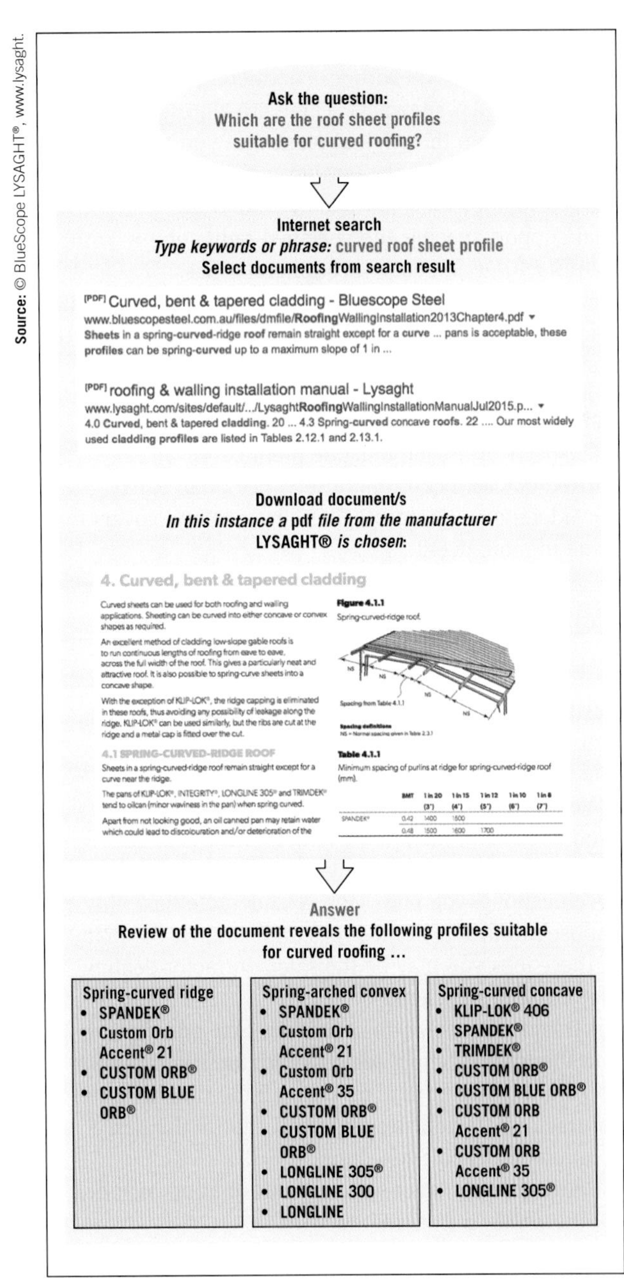

Source: © BlueScope LYSAGHT®, www.lysaght.

FIGURE 17.10 Using an internet search to find information for your installation

Selection of flashings

Flashing materials used for curved roofs include:

- ZINCALUME® steel
- COLORBOND® steel
- lead – uncoated or uncoated, however, not where drinking water is being collected
- fibreglass
- polyethylene
- zinc
- synthetic rubber.

The application of curved roofing is very broad, and there is a considerable variation in the sizes, arcs and pitch used. Therefore, while some flashing products will be available off the shelf directly from the manufacturer, others will need to be fabricated in accordance with the job requirements.

Off the shelf

Products such as lead, zinc and synthetic rubber can be formed and installed to flash curves and compound shapes. Other off-the-shelf products are made to match curves with a specified radius. These include fibreglass and polyethylene capping, valleys and barge (see Figure 17.11).

Customised flashing products

Sometimes the individual shape and dimension of a curved roof job may dictate that you arrange for the flashings and capping to be prefabricated. For this you will need to provide your supplier or sheet-metal contractor with all plans and specifications to ensure accuracy. Curved barge capping and wall flashings are often fabricated in two sections and run through a small roll-former to create what is known as a Pittsburgh lock seam. The two halves are then joined together as a complete curved flashing (see Figure 17.12).

Selection of fasteners

Fasteners must be selected carefully for each individual job and match both the sheet manufacturer's and the fastener manufacturer's recommendations. Review Chapter 12 for more details.

The example shown in Figure 17.13 demonstrates some steps in the process of fastener selection.

Selection of sealants

For the installation of curved roof sheets, a neutral-cure silicone sealant is normally recommended where you need to seal side laps over a cranked ridge, sheet end-laps laid at less than 15°, or where a section is below pitch. Always check with the manufacturer's product literature to ensure that this is the case. A more detailed review of sealants is included in Chapter 6 'Fasteners and sealants'.

Curved roof calculations

The preparation and planning for a curved roof job will require you to carry out some fairly simple calculations to determine quantities and materials lengths. Even

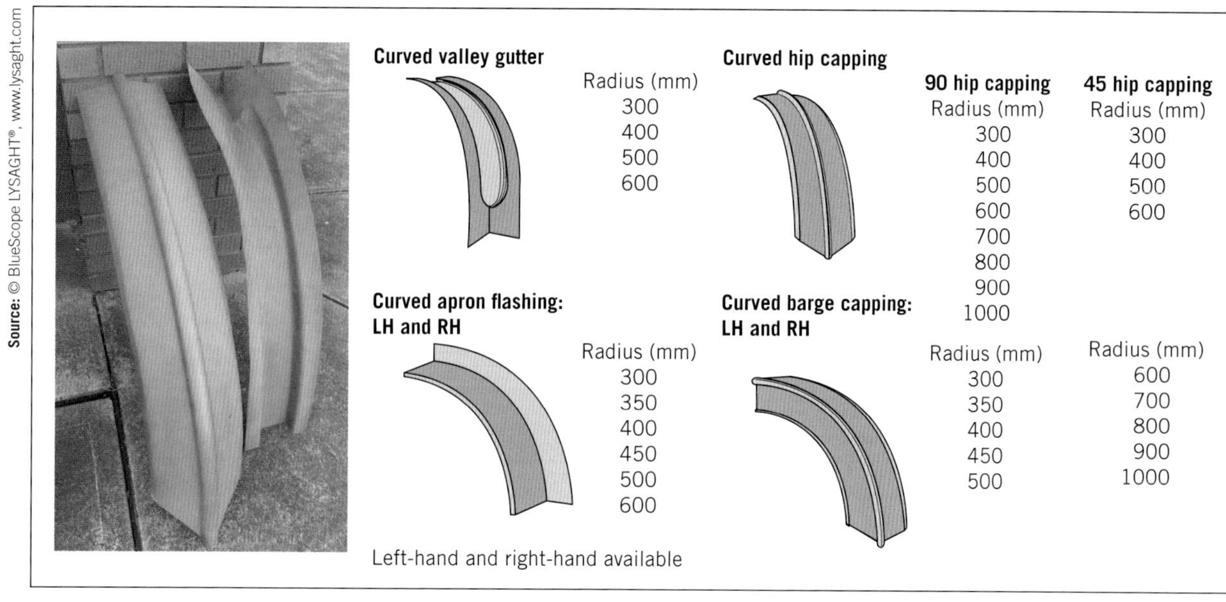

FIGURE 17.11 Curved polymer flashings and specified radius

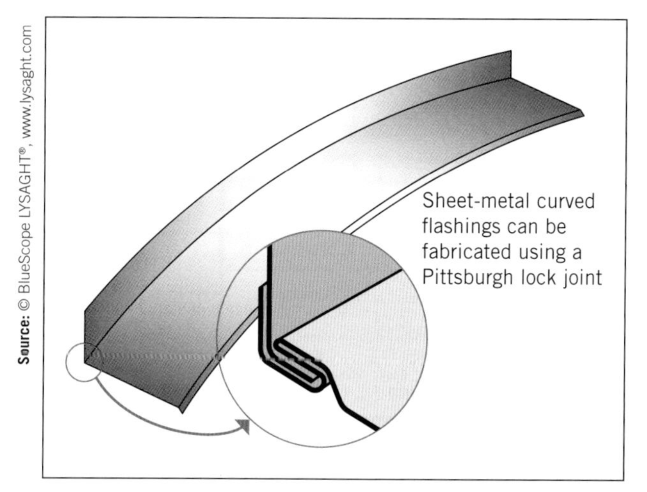

FIGURE 17.12 Use of the Pittsburgh lock seam to create a curved apron flashing

if much of this information is found on the approved plans and specifications, you will still need to confirm all details through a site visit. It is not uncommon for variations to occur between the original plans and the actual job. Therefore, you need to be able to confirm the job requirements.

Number of sheets

Roof sheets suitable for curving are available in a range of profiles and some of these profiles have different widths. The width measurement most used by roof plumbers is that known as 'effective cover', or sometimes 'width of cover'. This measurement is the amount of cover provided on the roof less the lap on one side.

The effective cover of any profile is provided by the manufacturer, and before ordering material and commencing any job you need to access the relevant documents to confirm the effective cover of the specified profile. The differences between the actual sheet width and its effective cover can be seen in Table 17.1.

Once you know the effective cover measurement, you can calculate the number of sheets required for the job. This is determined by dividing the total width of the roof by the effective cover of the particular sheet profile:

$$\text{Number of sheets required} = \frac{\text{Roof total width}}{\text{Effective cover}}$$

EXAMPLE 17.1

Calculate the number of sheets for a curved roof

You need to sheet a concave spring-curved gable roof that has an overall width of 30 m. The specifications require the use of LYSAGHT TRIMDEK® sheets for the job. How many sheets do you require?

$$\begin{aligned}\text{Number of sheets required} &= \frac{\text{Roof total width}}{\text{Effective cover}} \\ &= \frac{30}{0.762} \\ &= 39.37 \\ &= 40 \text{ (rounded up to whole sheet)}\end{aligned}$$

You will notice that in this calculation the final answer has been rounded up to the next whole sheet. Always round up your answer to ensure that you have full coverage of the roof.

Manufacturers also provide tables for each profile that have been pre-calculated to match simple roof widths to numbers of sheets (see Figure 17.14). To

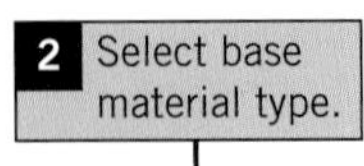

Fasteners: Screw to steel–AS3566 Class 4
Roof & Wall cladding - non-cyclonic

Roofing - non-cyclonic

LYSAGHT PROFILE	CREST			PAN		
	Single & Total Lapped ≥0.55 up to 1.0mm BMT	Single ≥1.0mm up to 3.0mm BMT	Lapped ≥1.0mm up to 1.9mm BMT	Single & Total Lapped ≥ 0.55 up to 1.0mm BMT	Single ≥1.0mm up to 3.0mm BMT	Lapped ≥1.0mm up to 1.9mm BMT
CUSTOM ORB	M6-11X50 ROOFZIPS	#12-14X35 HH HG/TG M5.5-14X39 AUTOTEKS	#12-14X35 HH HG/TG M5.5-14X39 AUTOTEKS	#10-16X16 HH M5-16X25 DH M6-11x25 ROOFZIPS	#10-16X16 HH M5-16X25 DH	#10-16X16 HH
CUSTOM BLUE ORB	M6-11X50 ROOFZIPS	#12-14X35 HH HG/TG M5.5-14X39 AUTOTEKS	#12-14X35 HH HG/TG M5.5-14X39 AUTOTEKS	#10-16X16 HH M5-16X25 DH M6-11x25 ROOFZIPS	#10-16X16 HH M5-16X25 DH	#10-16X16 HH
TRIMDEK	M6-11X50 ROOFZIPS	#12-14X45 HH HG/TG #12-14X55 HH HG/TG M5.5-14X50 AUTOTEKS	#12-14X45 HH HG/TG #12-14X55 HH HG/TG M5.5-14X50 AUTOTEKS	#10-16X16 HH M5-16X25 DH M6-11x25 ROOFZIPS	#10-16X16 HH M5-16X25 DH	#10-16X16 HH
SPA						5X16 HH
INTEG						
FLA						5X16 HH 4x20 HH
KL CLIP						5X16 WH 5X22 WH
KL7 CLIP						4x20 HH 4x30 HH
KL700 CLIP						4x20 HH 4x30 HH
LONGL						5X16 WH 5X22 WH

Corrugated roofing profiles

3 Read off the fastener type, then match to the manufacturer.

	Fixing in non-cylonic conditions: Valley Fixing	Fixing in non-cylonic conditions: Crest Fixing	Fixing in cylonic conditions: Valley Fixing	Fixing in cylonic conditions: Crest Fixing	Fixing in cylonic conditions: Ridge Cap Fixing
Fixing to timber supports	RoofZIPS® M6-11x25 Hd/Seal 6-030-3308-3C4	RoofZIPS® M6-11x50Hd/Seal 6-030-3306-1C4	14-10x25mm Hex Hd SL Type 17 6-030-3007-4C4	14-10x50mm Hex HdCori-Lok® Type 17 6-030-3633-2C4	14-10x50mm 25mm Ridge Cap Assy Type 17 6-030-3093-2C4
Fixing to metal 0.42–1-5 mm	RoofZIPS® M6-11x25 Hd/Seal 6-030-3308-3C4	RoofZIPS® M6-11x50Hd/Seal 6-030-3306-1C4	RoofZIPS® M6-11x25 Hd/Seal 6-030-3308-3C4	RoofZIPS® M6-11x50 Cori-Lok® Cyl Assy 6-030-3312-2C4	RoofZIPS® M6-11x50 25mm Ridge Cap Assy 6-030-3314-4C4

Buildex® Product Catalogue

FIGURE 17.13 Choosing fasteners to match the sheet manufacturer's requirements

TABLE 17.1 Comparison between actual sheet width and effective cover

Profile	Cross-section	Actual width (mm)	Effective cover (mm)
CUSTOM ORB® (corrugated)		832	762
CUSTOM BLUE ORB (corrugated)		838	762
TRIMDEK® (trapezoidal)		816	762
SPANDEK® (close-pitched trapezoidal)		756	700
KLIP-LOK® 406 (wide rib)		432	406

LYSAGHT TRIMDEK® sheet coverage																					
Width of roof (m)	3	4	5	6	7	8	9	10	11	12	13	14	15	16	17	18	19	20	30	40	50
Number of sheets	4	6	7	8	10	11	12	14	15	16	18	19	20	21	23	24	25	27	40	53	66

FIGURE 17.14 Example of a manufacturer's sheet coverage calculation table

Source: © BlueScope LYSAGHT®, www.lysaght.com

use these tables, you round up the width to the next metre and then read off the number of sheets. Of course, if your roof is wider than indicated or precise measurements are required, you will need to work it out manually.

Length of curved sheets

The length of roof sheets for most jobs is simply determined from the plans and then checked with a straight measurement on the job during a site visit prior to ordering. However, checking the length of curved sheets can be more involved.

Between any two fixed points, a curved line is always longer than a direct line. Look at Figure 17.15 and you can see that when you 'stretch' the curved line out, it is longer than the distance between the two fixed points.

The same principle applies to your curved roof installation. The curved section of any roof can be worked out if you can source the following two pieces of information from the plans or designer and apply the formula:

- the radius of the roof sheet curve
- the degrees between each end of the curve measured from the centre point of the arc (see Figure 17.16).

When calculating the length of roof sheets, always remember to include the overhang of sheet into the gutter. If you omit this allowance your sheets will be too short and the curve inaccurate.

The formula used in Figure 17.16 can also be used to determine the curved length of bullnose sheeting. To this answer, you would need to add the length of sheet overhang into the gutter and the straight roof section.

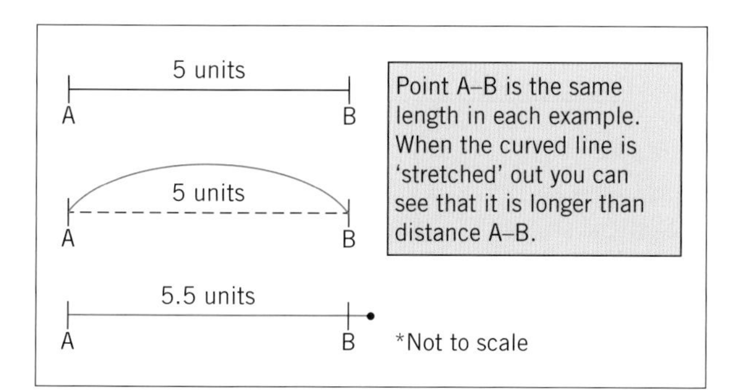

FIGURE 17.15 Comparison of a straight and a curved line between two fixed points

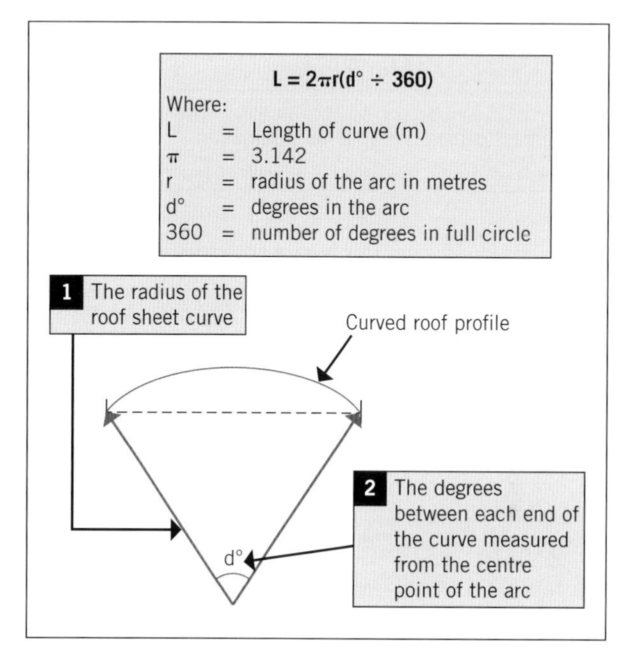

FIGURE 17.16 Calculating the length of a curve

EXAMPLE 17.2

Calculate the length of the curve for a spring-arched roof

Prior to ordering your sheets for a spring-arched roof, you go and discuss the job with the building designer. The job specifies the use of LYSAGHT CUSTOM ORB® corrugated sheets with a BMT of 0.48 mm. The roof curve has a radius of 10 m and an arc of 30°. Allowing an additional 50-mm overhang into the gutter on either side of the building, you use the following formula to work out the sheet length:

$$\begin{aligned}\text{Length of curve (m)} &= 2\,\pi\,r\,(d° \div 360)\\ &= 2 \times 3.142 \times 10\,(30° \div 360)\\ &= 62.84 \times 0.0833\\ &= 5.235 + 0.1 \text{ (gutter overhang)}\\ &= 5.335 \text{ m}\end{aligned}$$

However, on many smaller jobs where detailed specifications sheets may not be available, it may be difficult to determine the degrees of arc for the curve. Fortunately, some manufacturers have made this process simpler by providing special bullnose order forms that will assist you in getting the right result. An example of this is shown in Figure 17.17. Such information is easily obtainable during a site visit and

LYSAGHT

Bullnose order form

1 CUSTOMER

Customer

Mr. Bloggs

Contact person (First name Surname)

Order number

1230

Delivery address

10 Knobbies Lane

2 DATA

Number of sheets required

12

Topside colour (or finish)

Z/A

Direction of laying (see FRONT VIEW in right-hand column)

Left to right ✓ Overlap

Right to left ☐ Overlap

Does the bullnosed sheet end-lap onto a straight sheet?

Yes ☐

No ✓

Have you supplied a template? (Preferred method)

Yes ☐

No ✓

3 DIMENSIONS

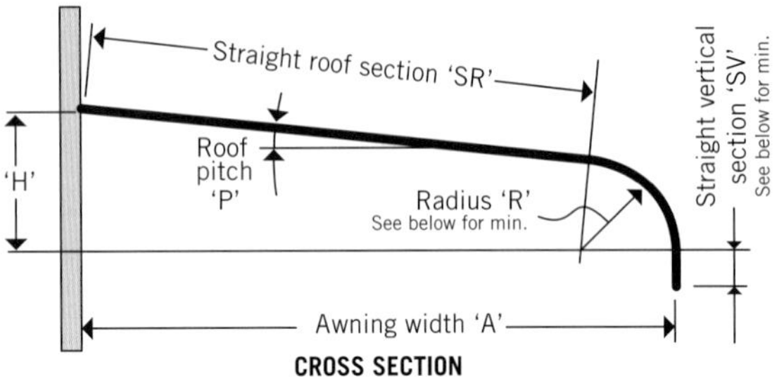

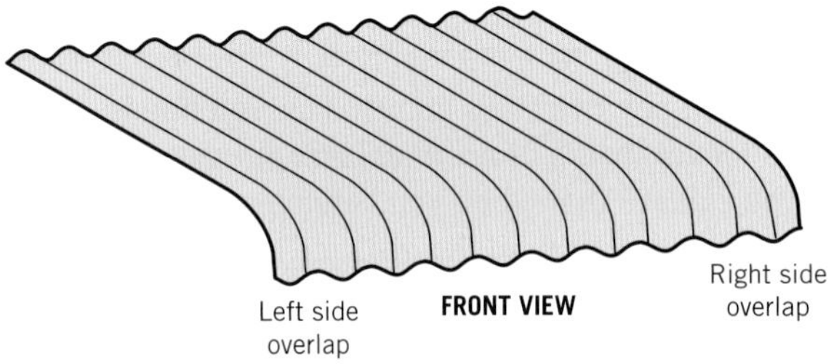

Straight roof section 'SR'

2900 mm (in WA 75 mm)

Straight vertical section 'SV'mm (100mm min., 80mm in Vic.)

Radius 'R'

400 mm

(300 min., 400 minimum in Qld & 400 minimum in W.A.)

Roof pitch 'P'

5° degrees (5° min.)

Awning width 'A'

3250 mm

Height less straight vertical 'H'

650 mm

BlueScope Lysaght use only

Curved length ____ mm

Straight roof section 'SR' ____ mm

Straight vertical section 'SV' ____ mm

Total length of sheet for manufacture ____ mm

Pass length ____ Nip depth ____ Pass reduction ____

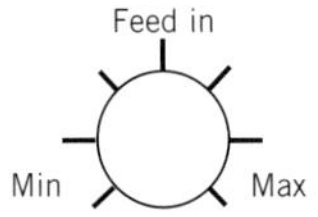

Feed in

Min Max

FIGURE 17.17 Information required before ordering your bullnose roof sheets

TABLE 17.2 Examples of fastener requirements for different profiles

Profile fastener location and number per support			
Profile	**Support location**	**Number***	**Fastener location**
CUSTOM ORB® (corrugated)	Intermediatc	3	
	Sheet ends	5	
TRIMDEK® (trapezoidal)	All supports	4	
SPANDEK® (close-pitched trapezoidal)	Intermediate	3	
	Sheet ends	4	

*Lap fastener is only counted once per sheet/support.

Source: © BlueScope LYSAGHT®, www.lysaght.com

also from the building plans. Once you have all of the details, your supplier is able to fabricate your curved sheets.

Number of fasteners

The manufacturer of the roof sheets provides details on how each particular product is fastened to the roof. Fastener applications for pierce-fixed roof sheet depend on the profile and any differences between the intermediate and end supports. Concealed-fixed roofs will require brackets/clips to be fastened onto each support. Subject to the design of the building, structures in cyclonic zones may also need additional fasteners to provide added protection from high winds. Table 17.2 shows some examples of different profiles and fastener needs.

The formula for determining the number of fasteners for pierce-fixed sheeting is as follows:

Number of fasteners = Number of fasteners/support × Number of sheets × Number of supports

EXAMPLE 17.3

Calculate number of fasteners required

For a simple bullnose roof measuring 17 m wide and a width of 3 m roof, there are a total of 4 purlin runs between the main building wall and the veranda posts. There would be in this instance 2 intermediate supports and 2 sheet-end supports.

A review of the manufacturer website reveals that a sheet of LYSAGHT CUSTOM ORB® in a non-cyclonic area requires 3 fasteners for each intermediate support and 5 fasteners for the sheet-end supports. Referring to the manufacturer's website once again, you see that to cover a roof 17 m wide, you would need 23 sheets for this job. Therefore, the number of fasteners is determined as follows:

Number of fasteners for intermediate supports = No. fasteners/support × No. of sheets × No. of supports
= 3 × 23 × 2
= 138

Number of fasteners for end supports = No. fasteners/support × No. of sheets × No. of supports
= 5 × 23 × 2
= 230

Total fasteners required = Intermediate + End supports
= 368

To simplify this process, ask your supplier if they have any fastener 'ready-reckoner' tables that may assist in quickly determining total quantities.

COMPLETE WORKSHEET 1

LEARNING TASK 17.2

1 How many sheets of corrugated CUSTOM ORB® would you require to cover a skillion roof that measures 15 m wide? Round up as required.
2 List at least 4 material types used for curved flashing in your local area.

Curved roof installation – general considerations

In the following section you will be introduced to the basic procedures required to lay a standard pierce-fixed roof. Of course, jobs will vary considerably, but a sound knowledge of fundamental installation practice and code requirements will enable you to adapt your approach for more complex applications.

Sheet handling

Sheets should be stored close to where they are going to be used and, if possible, spaced out at appropriate distances to reduce handling time. Point loading on the frame should be considered where suitable, ensuring of course that the frame is capable of handling the weight of each pack (see Chapter 11 'Receive roofing materials').

It is common practice to start laying sheets at the furthest point away from the prevailing weather so that each lap faces away from the wind. Align your sheets so that they are stored and handled with each lap facing the correct way. Turning or relocating sheets on the roof is unsafe and should always be avoided where possible.

Safety mesh

The use of safety mesh will depend on the height of the job and provision of other fall protection systems. Also, the use of any translucent sheeting may necessitate the use of mesh to prevent falls through plastic sheets during installation and maintenance.

The mesh must be laid and secured according to the code and the manufacturer's instructions before the installation of insulation and sheeting. More detail on safety mesh installation can be found in Chapter 12 'Select and install roof sheeting and wall cladding'.

Thermal insulation

When installing curved roofs, the requirements for thermal insulation are largely identical to those for other forms of pierce-fixed and concealed-fixed jobs. Refer to Chapter 14 'Install roof sheets, wall cladding and complex flashings' for more details relating to insulation.

Spring-curved ridge installation

As described earlier in this chapter, a spring-curved ridge roof is installed with standard flat sheets that are curved across the ridge zone of a gable roof. The rest of the sheet remains flat. The following points relate to these installations:

- Each sheet must be fixed to one side only, then pulled down and fixed on the other side, leaving the curve over the ridge.
- *Important:* The side laps of each sheet must be sealed with silicone for the full length of the curve. The curving process will tend to flatten out the roof sheet's inbuilt capillary break, and therefore the use of silicone sealant is essential to prevent capillary action from drawing water past the laps.
- To minimise sheet 'creep' and laying errors, it is recommended that you start laying each alternate sheet on opposite sides of the roof.
- The roof sheet profiles shown in Figure 17.18 are the only types recommended for spring curving. The spacings referred to in this table refer to the distance between the two top purlins adjacent to the roof ridge area. Any top of any ridge board must be finished below the line of the roof sheets so that it does not interfere with the line of the curve.

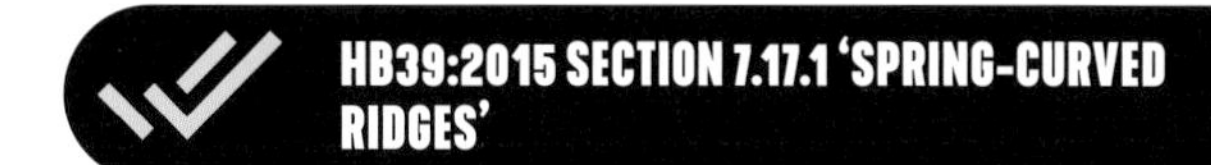

Spring-arched (convex) installation

A spring-arched (convex) roof is installed with standard flat sheets fastened onto a gently curving roof frame from one side of the building to the other. The following points relate to these types of installation:

- Each sheet must be fixed to one side only then pulled down and fixed on the other side.
- To minimise sheet 'creep' and laying errors, it is recommended that you start laying each alternate sheet on opposite sides of the roof.
- The frame and purlin spacing must be designed to suit a spring-arched roof. The ability to curve the sheet depends on the relationship between the following dimensions:
  - the roof width
  - the roof rise
  - the radius of the roof curve.

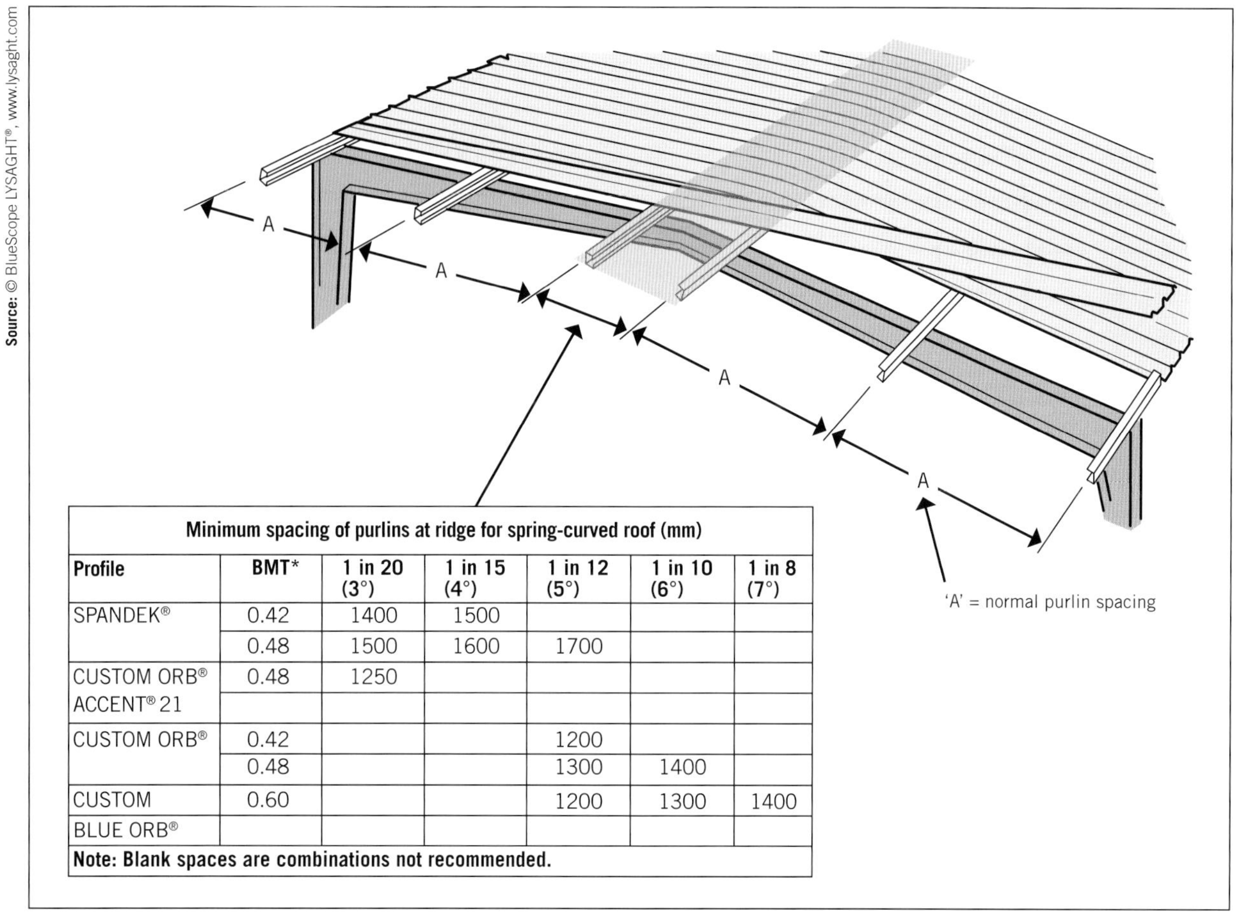

Minimum spacing of purlins at ridge for spring-curved roof (mm)						
Profile	**BMT***	**1 in 20 (3°)**	**1 in 15 (4°)**	**1 in 12 (5°)**	**1 in 10 (6°)**	**1 in 8 (7°)**
SPANDEK®	0.42	1400	1500			
	0.48	1500	1600	1700		
CUSTOM ORB® ACCENT® 21	0.48	1250				
CUSTOM ORB®	0.42			1200		
	0.48			1300	1400	
CUSTOM BLUE ORB®	0.60			1200	1300	1400
Note: Blank spaces are combinations not recommended.						

Source: © BlueScope LYSAGHT®, www.lysaght.com

FIGURE 17.18 Requirements for spring-curved ridge roofs

The recommended sheet profiles and an explanation of the formula used to determine minimum radius for each of these is shown in Figure 17.19.

The crest area of a spring-arched roof is essentially flat and below the minimum pitch requirements for the sheets. The curving of the sheets will also flatten out the side lap capillary break to a degree. To prevent water ingress within this zone, silicone sealant must be applied between the lapped sheets for the specified distance. This length is determined as follows:

Seal length = 0.035 × Radius × Specified minimum roof pitch

See Figure 17.19 for the application of this formula.

Spring-curved (concave) installation

Spring-curved (concave) roofs are installed with standard flat sheets that are fastened onto a roof frame with a gentle internal curve and specially located purlin spacings. An important aspect of concave roof installation is that the end sections of the sheets must not curve so much as to lay lower than the minimum pitch for each particular product (see Figure 17.20).

LEARNING TASK 17.3

1 What safety measure might be used to prevent falls through translucent sheet during installation and maintenance?
2 What is the formula to identify the seal length?

Installation of a bullnose veranda roof

In the following section, you will be introduced to some of the installation considerations and procedures required when installing a simple bullnose veranda roof, such as the one shown in Figure 17.21. Each section contains some points and diagrams to help explain what needs to be achieved.

Lay safety mesh

It is not usual to lay safety mesh for a standard bullnose roof, but this will be subject to the building design, heights above ground, pitch and material selection. Where required, refer to Chapter 12 'Install roof sheeting and wall cladding' for more detail.

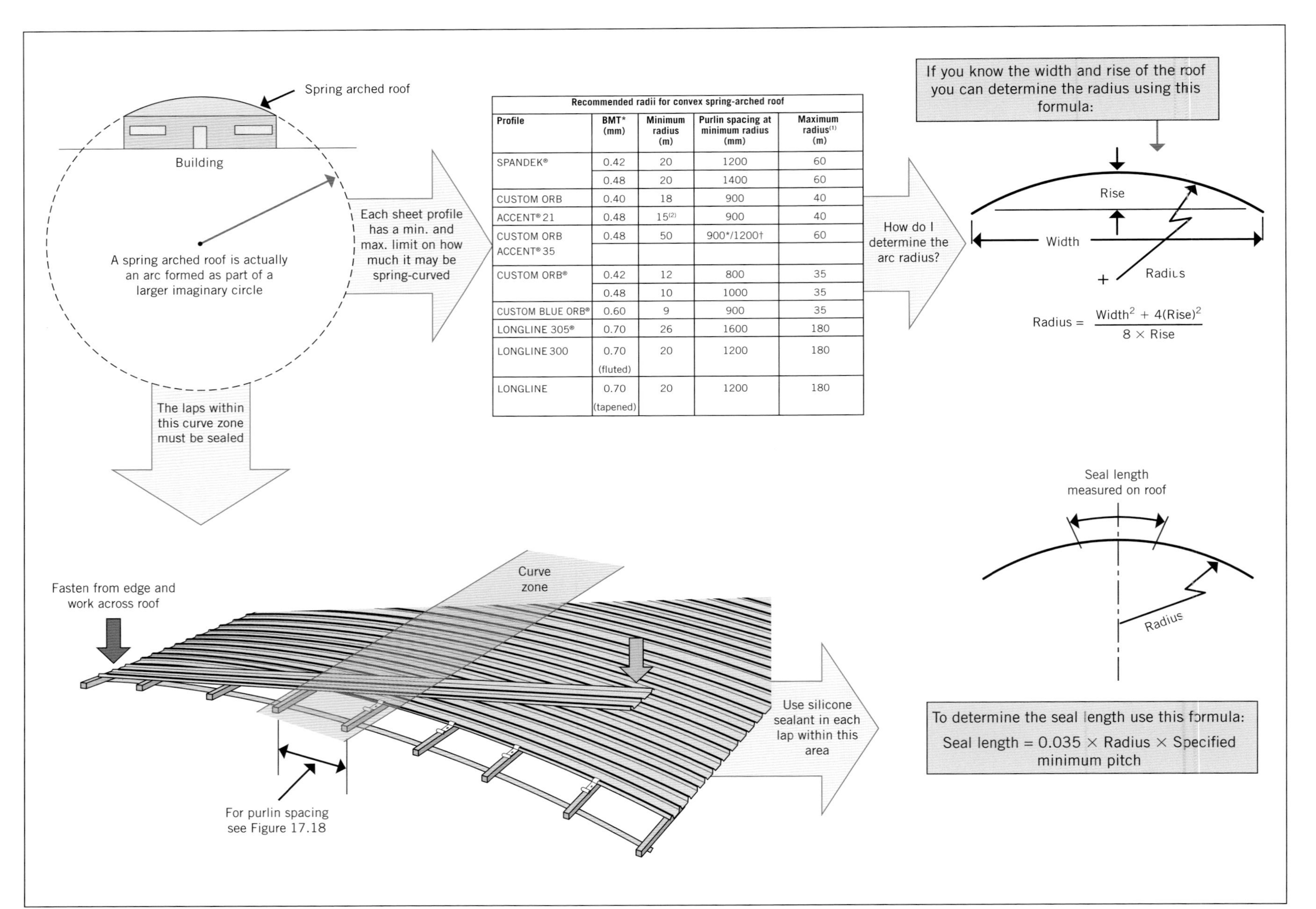

Recommended radii for convex spring-arched roof				
Profile	BMT* (mm)	Minimum radius (m)	Purlin spacing at minimum radius (mm)	Maximum radius(1) (m)
SPANDEK®	0.42	20	1200	60
	0.48	20	1400	60
CUSTOM ORB	0.40	18	900	40
ACCENT® 21	0.48	15(2)	900	40
CUSTOM ORB ACCENT® 35	0.48	50	900*/1200†	60
CUSTOM ORB®	0.42	12	800	35
	0.48	10	1000	35
CUSTOM BLUE ORB®	0.60	9	900	35
LONGLINE 305®	0.70	26	1600	180
LONGLINE 300	0.70 (fluted)	20	1200	180
LONGLINE	0.70 (tapened)	20	1200	180

FIGURE 17.19 Radius requirements for a spring-arched roof

Source: © BlueScope LYSAGHT®, www.lysaght.com

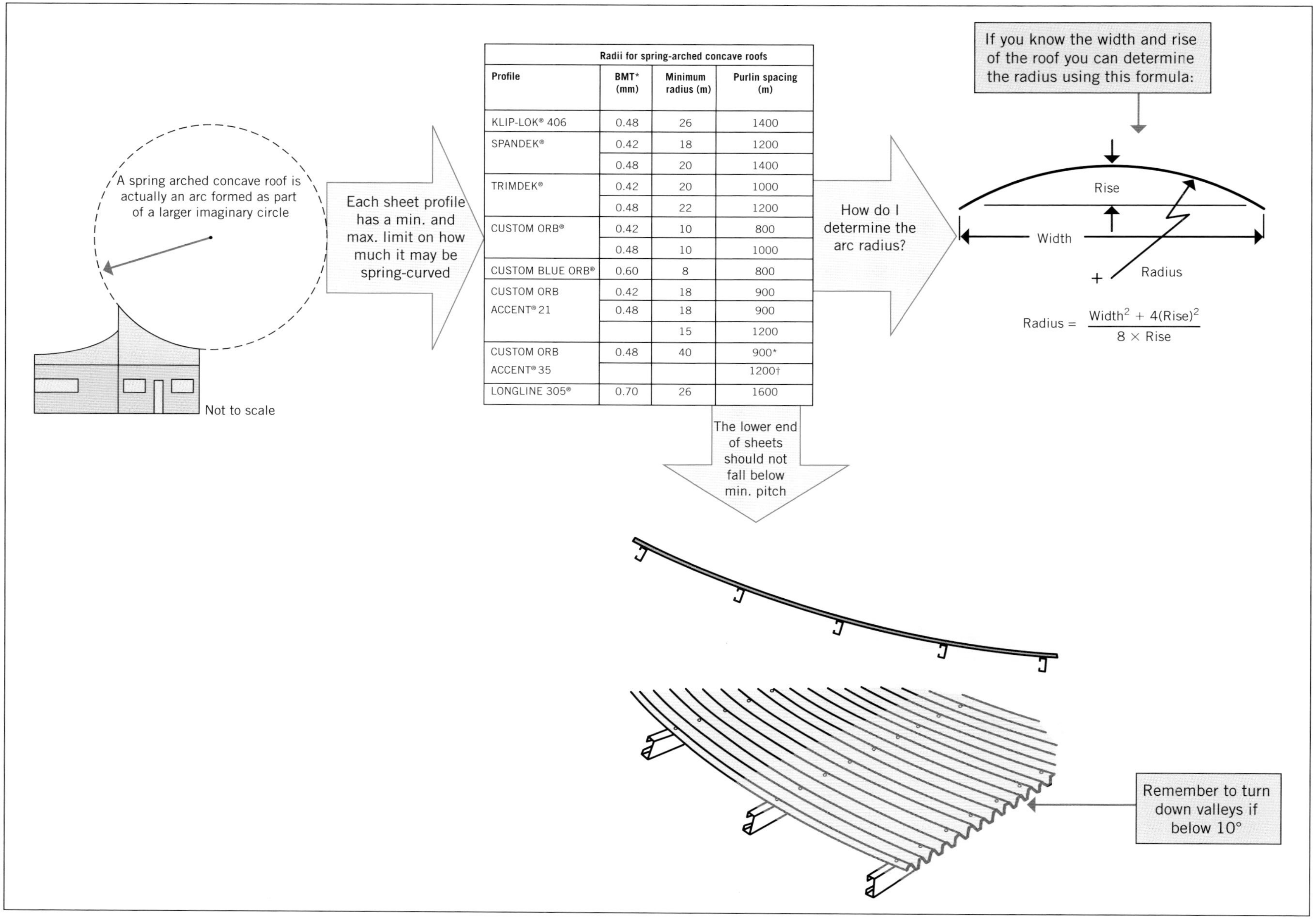

Radii for spring-arched concave roofs			
Profile	BMT* (mm)	Minimum radius (m)	Purlin spacing (m)
KLIP-LOK® 406	0.48	26	1400
SPANDEK®	0.42	18	1200
	0.48	20	1400
TRIMDEK®	0.42	20	1000
	0.48	22	1200
CUSTOM ORB®	0.42	10	800
	0.48	10	1000
CUSTOM BLUE ORB®	0.60	8	800
CUSTOM ORB ACCENT® 21	0.42	18	900
	0.48	18	900
		15	1200
CUSTOM ORB ACCENT® 35	0.48	40	900*
			1200†
LONGLINE 305®	0.70	26	1600

$$\text{Radius} = \frac{\text{Width}^2 + 4(\text{Rise})^2}{8 \times \text{Rise}}$$

FIGURE 17.20 Minimum radii for a concave curved roof

Source: © BlueScope LYSAGHT®, www.lysaght.co

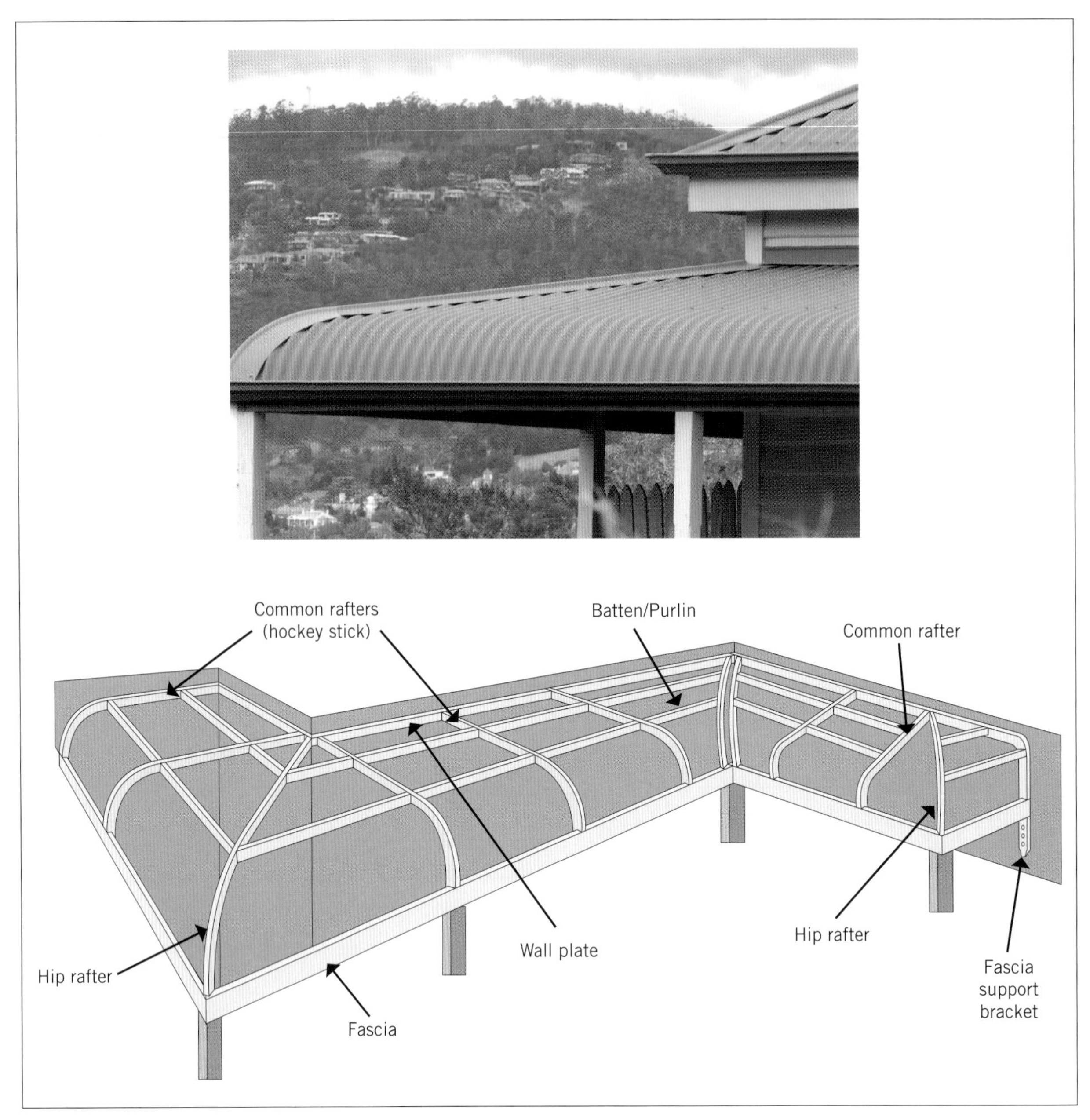

FIGURE 17.21 A typical bullnose roof installation

Install rainwater goods

It is standard practice to fit the gutter into place before the roof sheets are installed. This is particularly so with a bullnose roof, as positioning the gutter brackets can be awkward due to the vertical curve of the sheet-end being in the way. The following procedure should be followed:

1 Fasten the first gutter clip at the proposed high point of the gutter run.
2 Lay a curved sheet into position adjacent to this bracket to confirm the suitability of bracket height. Curved sheets will normally have an 80–100-mm straight section at the end that may need to be trimmed off the sheet. Trimmed sheets should terminate immediately above the brackets (see Figure 17.22).
3 Once the correct bracket height is confirmed, remove the test sheet and fix a stringline along the fascia with the desired fall to the gutter outlet.

Position gutter so that the end of all roof sheets will be hidden from ground view. As each sheet is laid on the one before, the sheet ends can tend to 'saw-tooth'. For bullnose roofs this is common; it is not necessarily a sign of misalignment, but instead is caused by the shape of the curve creating a minor height difference seen at the vertical end of the curve. Nevertheless, any saw-tooth effect is regarded as unsightly and should be kept well below the gutter bead.

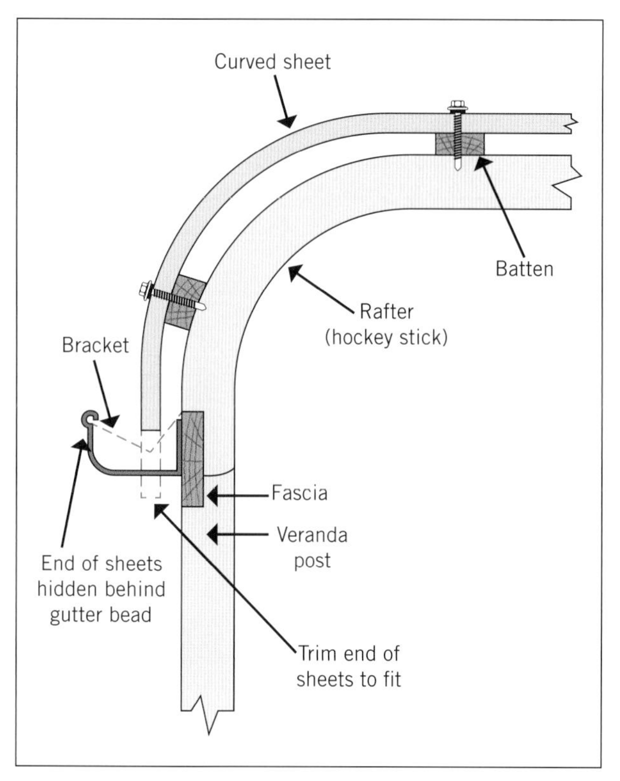

FIGURE 17.22 Trimming the excess length from sheet ends

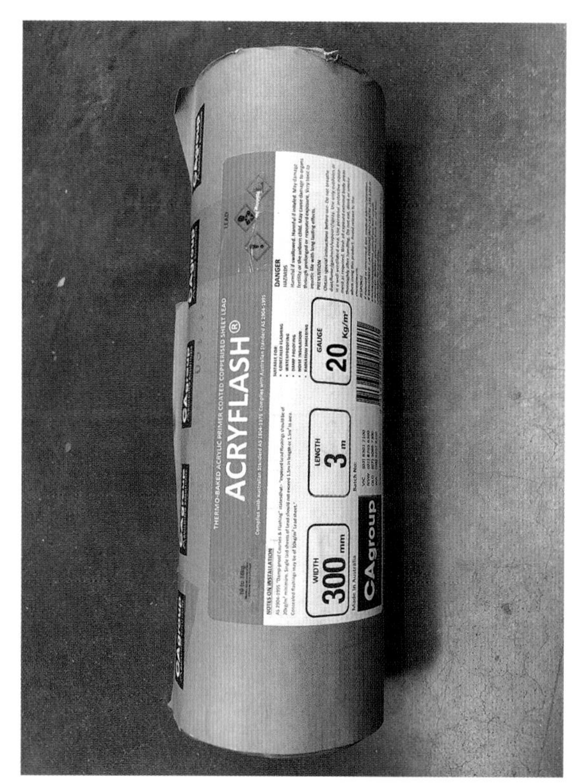

FIGURE 17.23 A roll of 20 kg/m^2 acrylic coated lead strip flashing

Install valley gutter

Where a valley gutter is required, this must be installed before the sheets are laid. A space of approximately 110–130 mm is left between two hip rafters to form a place for the valley to be fitted.

Bullnose roof valley gutters are shaped differently to normal flat roof profiles and almost always service very small catchment areas above the valley. In reflection of this fact, different minimum dimensions traditionally applied to their size. At the time of writing this edition, neither the AS/NZS 3500.3 nor HB39:2015 detailed specific requirements for bullnose roof valley gutters. In the absence of any specific direction, the following measurements are a guide but are subject to local authority ruling or any subsequent interpretation from the AS/NZS 3500.3:

- minimum curved valley central discharge width – 100 mm
- minimum curved valley gutter depth – 50 mm
- minimum curved valley gutter lears (sides) – 100 mm.

Prefabricated valley gutters

Fibreglass, polymer or sheet-metal valley gutters (see Figure 17.11) can be fitted directly on top of the two adjacent hip rafters, ensuring that the end of the valley protrudes correctly into the gutter.

Coated lead valley gutters

Where drinking water may be collected from a bullnose roof, lead flashing materials are not permitted for use.

However, where suitable, the requirements for the use of lead strip in curved roof valley gutters include the following:

- The lead strip should be 300 mm wide.
- The minimum weight is 20 kg per square metre (1.8-mm thickness).
- To allow for expansion and contraction, the maximum length of any strip is 1.5 m. Overlap it with the next strip by a minimum of 150 mm and seal this overlap with a silicone sealant (normally subject to pitch, but almost always for bullnose roofs).

A key aspect of using any lead product is the provision of support. Lead cannot support its own weight, and over time it will sag and can tear if not fully supported. An example of how a lead strip flashing can be used for a bullnose valley gutter can be seen in Figure 17.24.

Lay insulation

Depending on the location and application of the bullnose roof, you may need to lay insulation blanket or RFL. The procedures for this do not vary from the standard practice for flat roof installation. Refer to Chapter 12 for more detail.

Fasten full sheets

With eaves and valley gutters in place, it is standard practice to lay all full sheets first before starting the

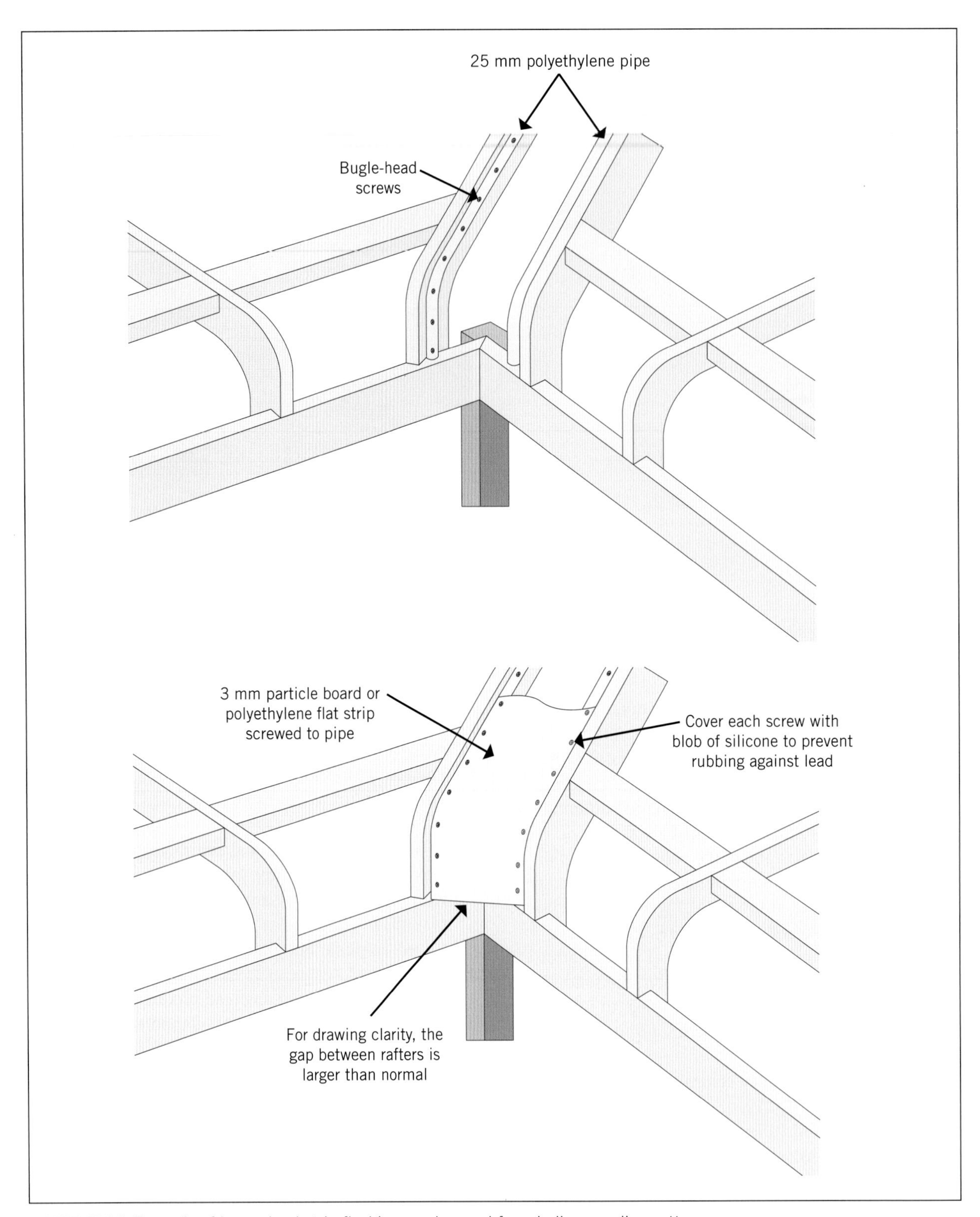

FIGURE 17.24 Example of how a lead-strip flashing can be used for a bullnose valley gutter

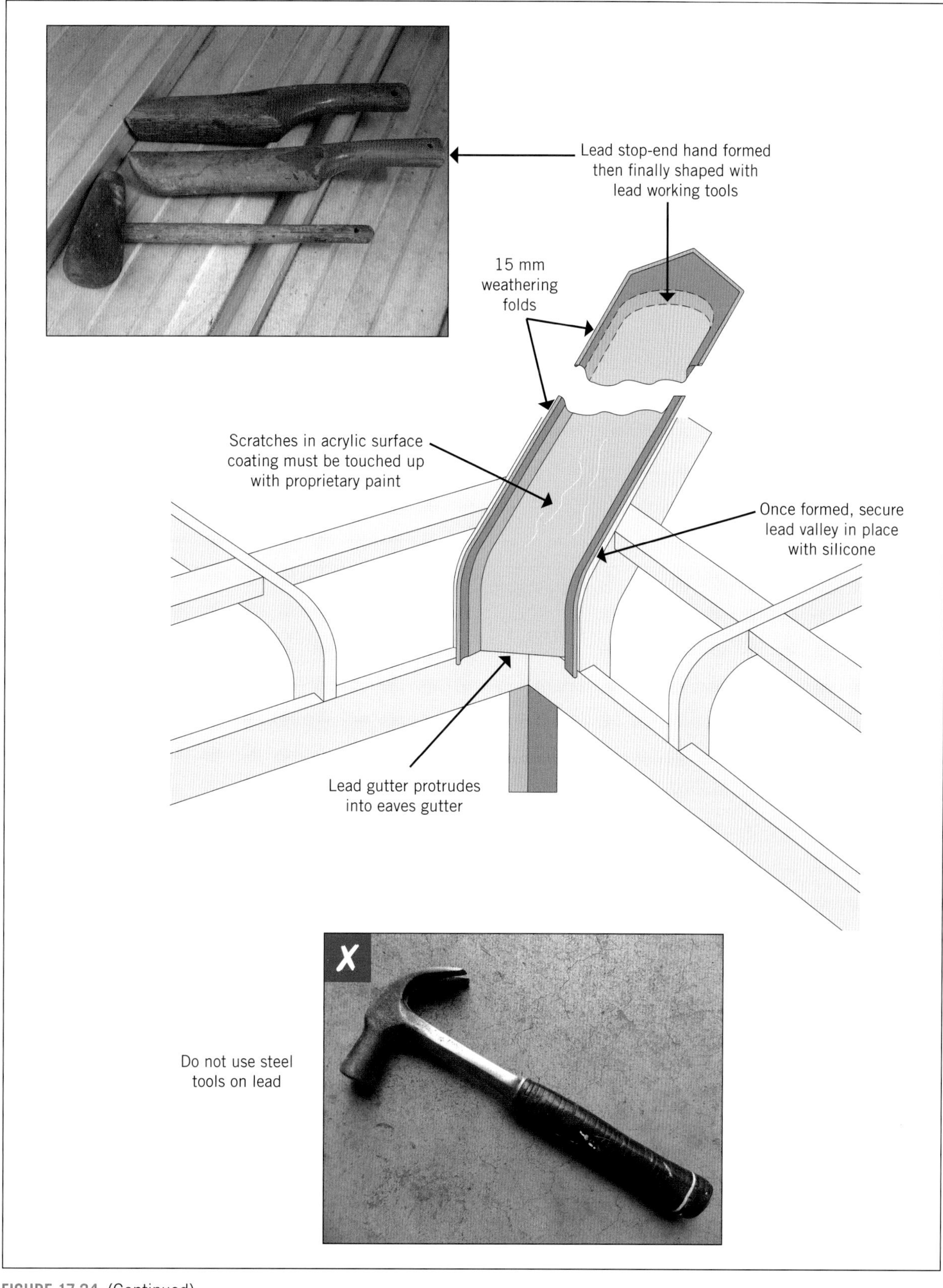

FIGURE 17.24 (Continued)

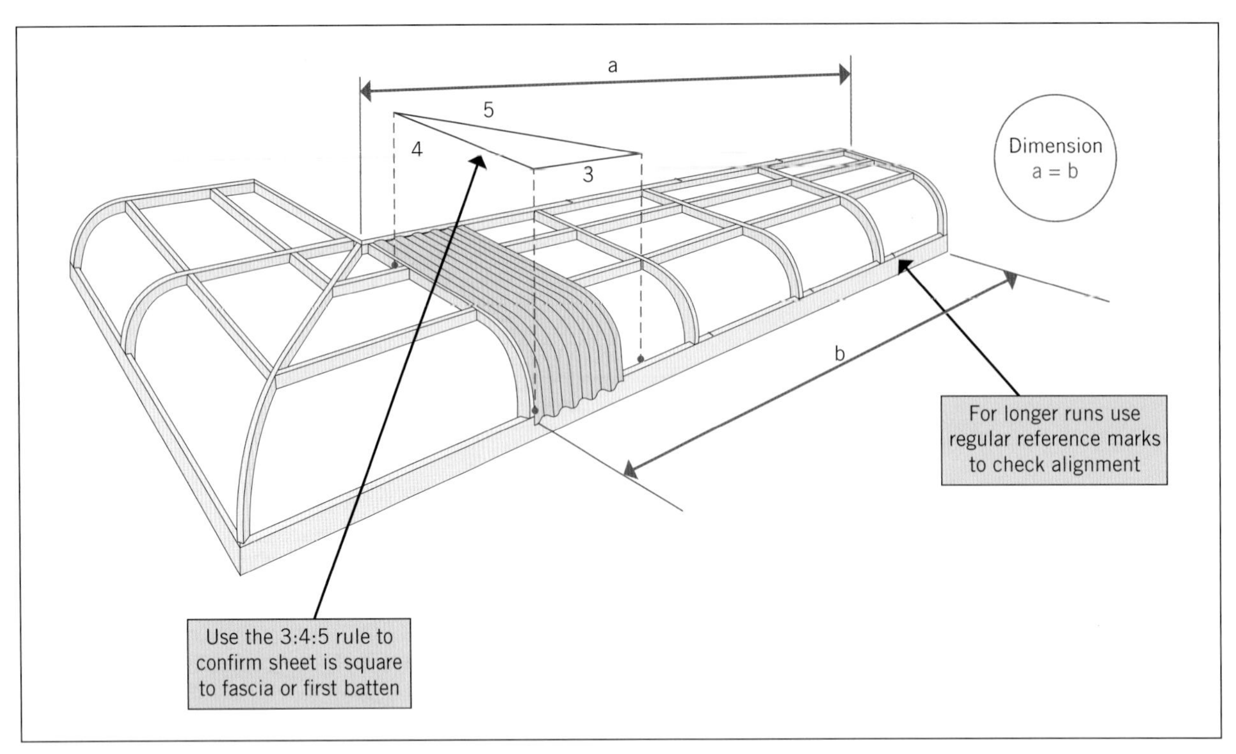

FIGURE 17.25 Ensuring that the first sheet is correctly aligned

valley and hip intersections. This is done as follows (see also Figure 17.25):

1 Confirm the direction of lay if this is important.
2 Weather the top end of each sheet by turning up each valley with an appropriate tool.
3 Lay the first sheet with the overlap intersecting with the top of the hip (this is the normal practice).
4 Use the 3:4:5 rule to confirm that the sheet is set perpendicular to the fascia (see Chapters 12 and 13 if required).
5 On larger jobs, mark out a series of reference points on the top and bottom purlins to allow for regular alignment checks (see Chapters 12 and 13).
6 'Pin' each sheet until the run is completed. Spring a chalk-line to mark the position of subsequent screws.

Mark out and cut sheet intersections

Having fastened all full sheets in place, you now need to mark out, cut and install sheets at hip and valley intersections.

Hip sheet installation

The procedure for installing sheets around an external hip is essentially the same as for a standard flat roof:

1 Position the sheet/s in place over the hip and adjacent sheet with a standard lap and secure this with self-locking grips if necessary.
2 Spring a chalk-line approximately 20 mm short of the hip centreline and remove offcuts with snips or a nibbler.
3 Weather the valleys on each sheet.
4 Secure with fasteners.
5 It may be too awkward to follow this procedure for the final sheet, and this may need to be marked out on the ground (see Figure 17.26).

Valley sheet installation

Unlike the sheets around the valley of a standard flat roof, bullnose roof sheets cannot be marked out in place, as the curve interferes with accurate positioning and the top of the sheets cannot protrude past the adjacent wall. However, it is a simple process to transfer some measurements and mark these out with the aid of a straight-edge on the ground or adjacent roof area (see Figure 17.27).

Unless water flows directly off a higher pitched roof, the run-off from a bullnose roof is generally small. The gap between intersecting sheets over the valley gutter is therefore normally 50–75 mm maximum (subject to local conditions).

Capping over curved hip intersection

Prefabricated curved roof capping

The product and installation quality of the curved hip capping can make or break the job. This section of the roof will stand out and must be completed with care to ensure a professional-looking result. Fibreglass, plastic, polymer or sheet-metal capping (see Figure 17.11) can be fitted directly on top of the two intersecting curved hip roof sheets.

The product should be placed into position and gradually trimmed 20–30 mm at a time till the capping fits tight and neatly against the roof sheets.

Coated lead-strip capping

Lead strip covered with a thermally baked acrylic coating can also be used over the hip intersection.

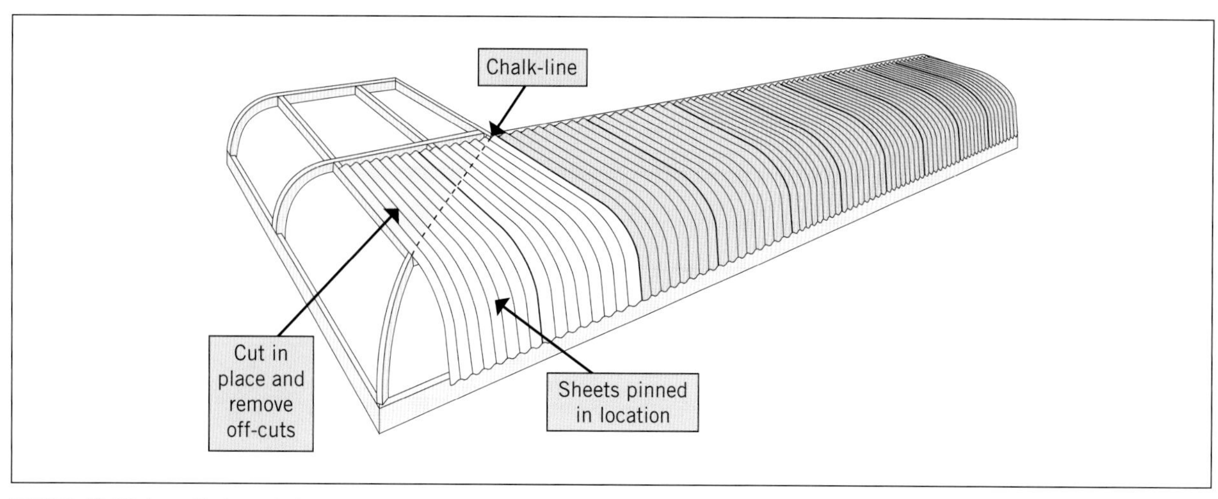

FIGURE 17.26 Installation of hip sheets

Requirements for curved ridge capping are as follows:

- Lead strip should be 400 mm wide.
- It should be a minimum weight of 20 kg per square metre (1.8 mm thickness).
- To allow for expansion and contraction, the maximum length of any strip is 1.5 m. Overlap it with the next strip by a minimum of 150 mm and seal this with silicone sealant (this is usually subject to pitch, but it is almost always done for bullnose roofs).

Some form of support must be placed across the centreline of the sheet intersection to support the lead and also to create a neat roll form. Figure 17.28 shows how this may be achieved.

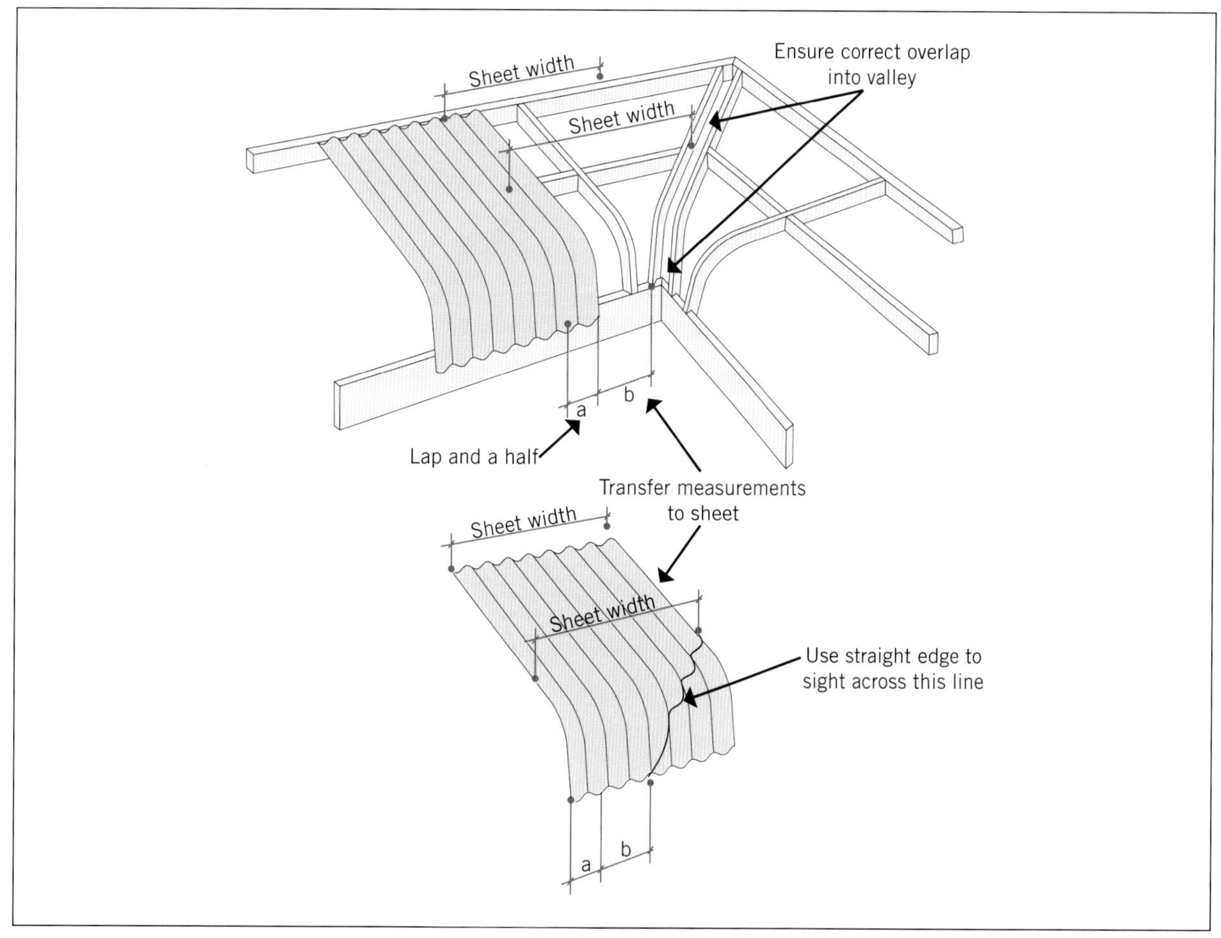

FIGURE 17.27 Mark-out of valley sheets

Clean up

Review Chapter 12 'Select and install roof sheeting and wall cladding' for details on the clean-up of your job site.

LEARNING TASK 17.4

1. When installing bullnose roof sheets, it is common practice to install which other component first?
2. Which type of tools should not be used to dress lead?
3. Which rule can be used to confirm a sheet is square to fascia or first batten?

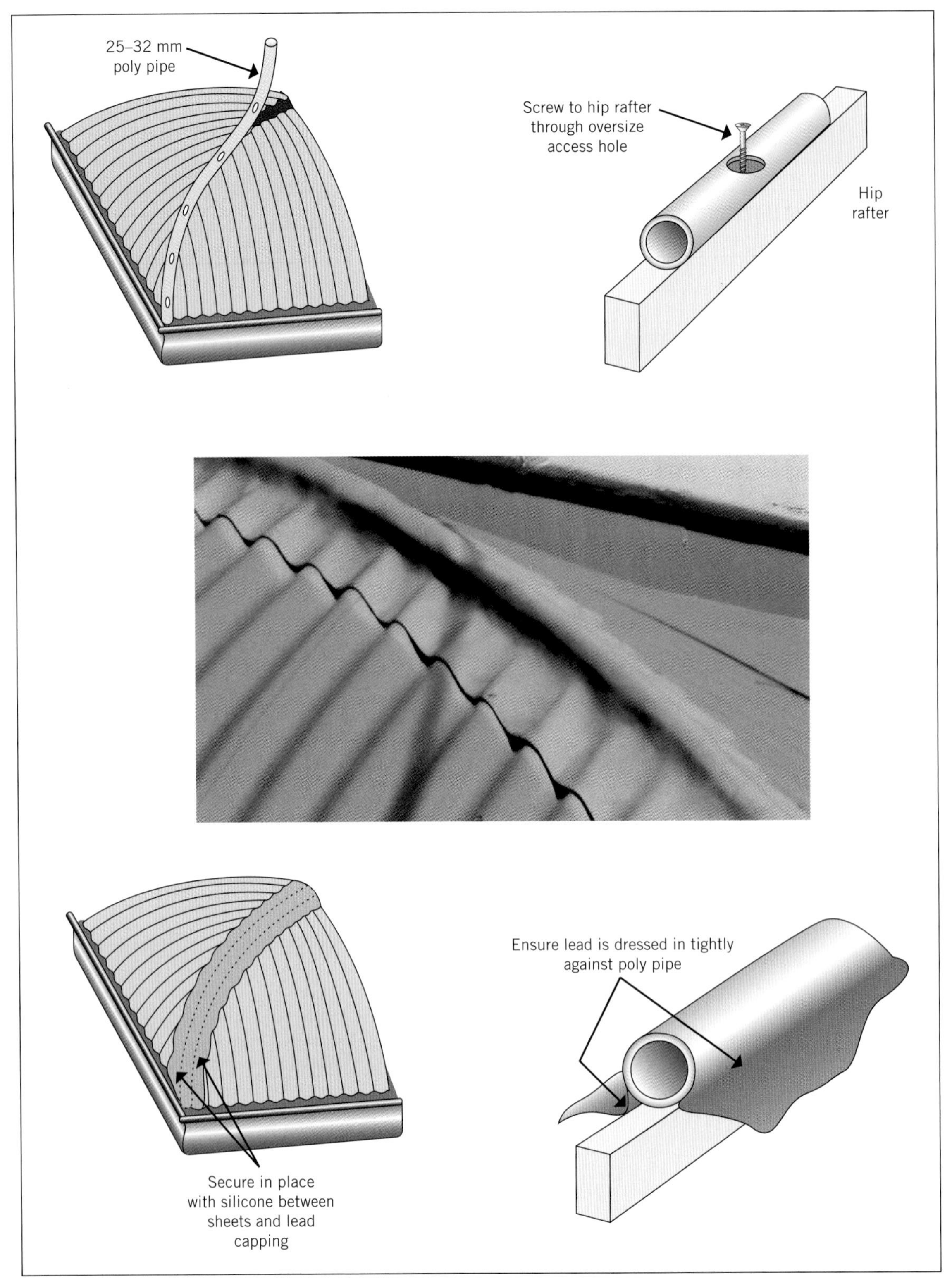

FIGURE 17.28 Use of poly pipe to support lead curved ridge capping

SUMMARY

There is a broad range of curved roof applications used throughout both residential and commercial applications and in this chapter you have been introduced to the following:

- Key forms of categories of curved roof including sprung and pre-curved
  - The common names applied to the various types within these categories
- Curved flashing and relevant material types
- Calculation of roof sheet numbers by knowing the effective cover
- Calculation of curved sheet lengths through use of appropriate formulae
- How to lay a bullnose roof and install both internal and external flashings

COMPLETE WORKSHEET 2

WORKSHEET 1

To be completed by teachers	
Satisfactory	☐
Not satisfactory	☐

Student name: ______________________

Enrolment year: ______________________

Class code: ______________________

Competency name/Number: ______________________

Task: Working with your teacher/supervisor, refer to this text, your HB39:2015 and any other relevant resource to answer the following questions.

1 Curved profile roofing can be broken into two main categories. What are they?

____

____

2 List at least five forms of curved roofing shape.

____

____

____

____

____

3 Is a spring-arched (convex) roof curved over the whole roof from eave to eave or just across the ridge area?

____

____

____

4 Circle the correct answer. What shape defines a reverse curve roof?

i Concave over convex curve

ii Convex curve over concave curve

5 What are at least five considerations to be examined during a site visit?

____

____

____

____

____

6 Referring to the manufacturers' information for products used by your employer, which profiles are suitable for spring-arched convex applications?

7 How many corrugated sheets would you require to cover a bullnose veranda 23 m wide on one side of the building?

8 How many fasteners would you require for the roof in the previous question where five purlin supports are in place? Provide an answer for both non-cyclonic and cyclonic zones.

WORKSHEET 2

To be completed by teachers

Satisfactory ☐

Not satisfactory ☐

Student name: ____________________

Enrolment year: ____________________

Class code: ____________________

Competency name/Number: ____________________

Task: Working with your teacher/supervisor, refer to this text, your HB39:2015 and any other relevant resource to answer the following questions.

1 In normal practice, should insulation blanket be laid with the RFL facing up or down?

2 When laying a spring-curved ridge roof, what must be done to the side laps over the curve zone?

3 You are working with a builder to modify an existing gable-roofed dwelling to have a 6° pitched corrugated spring-curved ridge roof. The builder has modified the ridge so that it will not be in the way, but you need to confirm the minimum spacing between each purlin at the ridge. What should the distance be?

4 By how much should RFL overlap the previous run?

5 When installing a spring-curved roof, what particular laying sequence assists in preventing creeping of the sheets?

6 Subject to local requirements, what is the recommended minimum central discharge width between intersecting sheets over a curved valley gutter?

7 How is a lead curved ridge capping secured to the roof?

8 What is the minimum weight per square metre specified for lead strip flashing?

9 If you are forming up a lead valley gutter, what is the recommended depth?

10 What product is recommended to provide rigidity and a roll form under a lead curved ridge capping? What nominal size should it be?

11 What should be done with all lead strip trimmings?

COLLECT AND STORE ROOF WATER

18

This chapter will focus on the skills and knowledge required of a roof plumber to plan for and install rainwater collection and storage systems. The scope of this chapter includes the connection of the roof drainage system to the storage tank and the connection of the storage tank overflow to an approved point of discharge. It does not detail the connection of rainwater supply to fixtures or other outlets.

In this chapter, regular reference will be made to the HB 230–2008 *Rainwater Tank Design and Installation Handbook*. This Handbook provides a good overview of rainwater tank installation planning and sizing. However, teachers and students should in all instances refer to relevant state and territory legislation to ensure compliance with local requirements.

Overview

Rainwater has always been used in many rural areas to supply water for drinking and ablution needs, and this continues today. Up until the early to mid 1900s, rainwater storage and use was also common in urban areas. However, the steady rollout of town water reticulation networks during last century saw the private use of rainwater storage supplies decline, and in certain areas it was actively discouraged or prohibited.

This is all changing, and domestic and commercial rainwater storage systems are once again becoming common in urban areas. Some reasons for this change include:

- drought and increasingly unpredictable rainfall patterns
- the growing population
- increasingly expensive water treatment and reticulation requirements
- avoiding the construction of large dams, which can sometimes cover valuable farming land
- growing public awareness of the importance and benefits of a sustainable water supply.

Australia is a land that experiences regular and sometimes extended periods of drought, and in recent years the large dam storage supplies of many of our major towns and cities have at times been depleted to critically low levels. With approximately 90% of Australia's growing population living in reticulation areas, it is no longer viable to rely entirely on dam reservoirs as our only source of water.

Government authorities in many areas now require all new homes to be fitted with a rainwater storage system to supply basic laundry, toilet flushing and garden requirements (see Figure 18.1). Many local council authorities will provide rebates for water tank installation, and in some areas even building renovation approvals require the inclusion of a minimum water storage solution. Uses for stored rainwater include:

- laundry water supply
- garden and lawn irrigation

Source: © BlueScope LYSAGHT®, www.lysaght.com

FIGURE 18.1 Tank water supply for a garden and toilet flushing in a new home

- outdoor cleaning of buildings and vehicles
- toilet flushing
- hot water use
- pool top-ups
- commercial cooling towers
- storage for fire fighting
- community storage systems
- drinking water (subject to local health department advice).

Any of the above purposes may also be incorporated into a tank required by a local authority as part of a stormwater detention system. It is beyond the scope of this chapter to discuss stormwater detention and retention, so confirm basic local requirements with your teacher.

GREEN TIP

The broader use of rainwater collection and storage systems will to a degree offset the cost and environmental impact of drought on our water treatment and reticulation systems. Significantly, this places you as a roof plumber at the forefront in providing sustainable water supply solutions to your community.

Tank materials

Rainwater storage tanks are made in a range of materials that feature different advantages and characteristics to suit different installation requirements. Figure 18.2 shows some examples of stainless steel water tanks. The types of materials available for rainwater storage tanks include those listed in Table 18.1.

FIGURE 18.2 Installation of new stainless steel tanks

Materials compatibility

As when installing any other roof and rainwater goods, you need to ensure that the materials you select are compatible with each other. In relation to water tanks, this is no longer the big problem that it was some decades ago. After the broader introduction of ZINCALUME® steel and COLORBOND® steel roofing products during the 1970s and 1980s, many existing galvanised rainwater tanks suffered premature inert catchment corrosion from water flowing down the new roof sheets to the older galvanised gutters and tanks. The majority of materials now used for tanks are unaffected by these problems. However, bare galvanised steel tanks are still manufactured and should be installed with caution. Check all catchment materials above the tank to ensure compatibility. Review Chapter 8 'Corrosion' if required.

SA HB39:2015 TABLE 2.3(B) 'ACCEPTABILITY OF DRAINAGE FROM AN UPPER SURFACE TO A LOWER METAL SURFACE'

Lead

Look out for lead! Always check a roof for the presence of lead flashings where a roof catchment area is going to be used for the collection of rainwater for drinking purposes. Replace any bare lead roofing materials with products compliant with AS/NZS 4020 'Testing of products for use in contact with drinking water'. Lead is prohibited when any part of the roof forms part of a drinking water catchment.

SA HB39:2015 SECTION 8.5 'LEAD FLASHINGS'

Also ensure that where you are ordering a tank to be fabricated, confirm with the manufacturer that where any solder has been used that it is of the lead-free type. Remember, a licensed plumber is legally responsible to ensure that all products are certified and 'fit for purpose'.

Materials compliance with Australian Standards

Materials used in the manufacture of rainwater tanks and materials used to supply water to and away from a

TABLE 18.1 Water tank materials and characteristics

Tank material	Characteristics
Galvanised steel	• Corrugated galvanised steel has been used for water storage in Australia for decades. • Very strong, but does not have the same longevity as more modern products.
AQUAPLATE® steel	• Steel product lined with a food-grade polymer film that is compliant with AS/NZS 4020:2005 'Testing of products for use in contact with drinking water'. • Available in plain corrugated steel, ZINCALUME® or pre-painted COLORBOND® steel colours. • Strong product with very long life.
Polyethylene	• Strong and lightweight UV stabilised product available in wide range of colours. • Corrugated or flat side. • Can be made in wide range of shapes and sizes to suit almost any application. • Can be made for above- and below-ground applications.
Copper	• Expensive but outlasts most other products. • Often used for smaller header tanks.
Stainless steel	• Ideal in aggressively corrosive environments. • More expensive but has very long life. • Commonly used for specialised applications and header tanks.
Concrete	• Very strong and durable. • Available as prefabricated products or can be constructed onsite.
Flexible bladders	• Made from PVC or thermoplastic polyurethane. • Bladders can be ordered to suit a wide range of sizes and are installed in otherwise unused subfloor areas beneath buildings and decks.

rainwater tank must be certified according to a relevant Australian Standard.

As an example, some of the relevant standards are listed below:

- *AS/NZS 4020 'Testing of products for use in contact with drinking water'.* For example, the supply downpipes used on a 'wet system' must comply with this standard if the water is used for drinking. Commonly used plastic downpipes may not be approved for drinking water purposes.
- *AS/NZS 4766 'Polyethylene storage tanks for water and chemicals'.* When purchasing an above-ground poly tank for your customer, ensure that it has been designed and manufactured to this standard.
- *AS 2001.2 'Methods of test for textiles – Physical tests'.* This is the standard to which flexible bladders must be tested. Check before purchase.
- *AS/NZS 3500 'Plumbing and drainage'.* Parts 1 and 3 are particularly applicable in relation to all pipes, outlets and fittings for supplying rainwater.
- *National Construction Code (NCC).* Rainwater tanks are classed as Class 10b structures under the NCC. This requires that certain set-back and siting provisions apply to their installation and once raised above ground on a tank stand you may require a building permit. Your local council must be consulted before proceeding.

It is your responsibility to ensure that the products you purchase for a job meet and are installed to industry and standards requirements.

LEARNING TASK 18.1

1. What is the Australian Standard that applies to all plumbing products that form part of a drinking water catchment?
2. List at least four non-drinking water uses of roof water storage systems.
3. In reference to your HB39, would an installation that drains roof water from a copper gutter system be compatible with a ZINCALUME® storage tank?

Plans and specifications

In areas where new homes are required to accommodate minimum water storage requirements, building designers are required to calculate the requirements of each project. The details of this calculation will be found in the approved plans and attached specification documentation.

In most instances these documents will tell you the size, material, location and connection requirements. The choice of material brand and type is often left up to the contractor, builder and property owner to discuss. All products selected should be 'fit-for-purpose' and comply with the relevant standards as discussed in the previous section.

Confirm details with a site visit

It is very important that you confirm all details with the builder, designer and property owner as applicable through a site visit before ordering materials and commencing the installation. It is not uncommon to find that the actual conditions on a job site have varied considerably from the original plans, and to avoid problems it is best to confirm all details before starting on the job. Items to discuss include the following points:

- Confirm if the storage system is to be used for drinking water purposes. This will inform the following:
  - Compatibility of catchment area materials
  - Compatibility of 'charged' or 'wet downpipe' system. A PVC charged downpipe system supplying a drinking water tank must be compliant with AS/NZS 4020 'Testing of products for use in contact with drinking water'. This is to ensure the pipes contain no lead.
- Ensure that everyone agrees on the tank site and, where applicable, that the support system is compliant.
- Check the tank material for compatibility with the roof catchment area and drainage system.
- Ensure that the owner is satisfied with the colour selection (if applicable).
- Confirm the type of downpipe connection system to be used – gravity drainage or charged downpipe system.
- Check the provision for overflow connection to the approved point of discharge – the local authority will dictate the approved form of connection.
- Where a charged downpipe system is planned, confirm requirements for inspection openings and location of a sediment flush point.
- Check that ground levels are as per the plan.
- Ensure that gutter systems are falling in the right direction and that downpipes are located in positions suitable for tank connection.
- Where excavation of an underground tank is required, check the following:
  - minimum clearance horizontal set-backs from any existing domestic waste water treatment land application area as per AS/NZS 1547
  - access for machinery around the site
  - whether there is sufficient space for the proposed hole
  - that the hole is the specified distance from the footings of the adjacent building
  - whether there is sufficient space for the excavated material to be stored before removal.

- Check the location of all services particularly where any excavation is planned. Use Dial Before You Dig to determine the position of services connected to the property. Importantly, also employ an asset location company to determine services within the property boundary. These include:
  - power
  - telecommunication
  - water
  - gas
  - sewer
  - stormwater.

During the planning stage and site inspection, consideration must also be given to identification and mitigation of any source of rainwater contamination. Items include:

- overhanging vegetation
- run-off from slow combustion heater flues – look for creosote stains
- concentrations of animal droppings
- overflows and drains from roof-mounted appliances such as solar water heaters, evaporative air conditioners and ceiling located central heating furnaces.

HB 230–2008 SECTION 9.2 'MINIMISING CONTAMINATION' AND TABLE 9.1

Check manufacturer's specifications

Once you have confirmed all installation and product requirements, you will need to refer to the manufacturer's specifications to ensure that the products you have chosen will be used and installed in the correct manner. Failure to do this may lead to poor performance or product failure and subsequent loss of warranty, leaving you paying the bill for replacement or repair and possible litigation.

LEARNING TASK 18.2

1 Check your local water utilities web page. Do they require new homes to have a minimum water storage requirement?
2 Which handbook publication should be referenced when collecting and storing roof water?

Local authority requirements

As part of the preparation and planning process for the installation and connection of a rainwater storage system, you must get to know the local authority requirements in your own area.

Australia is a big country with multiple climate zones and variations in rainfall. There is no single answer to local storage needs, and therefore state, territory and local governments have put into place a range of solutions to fit their particular circumstances. To find out what the requirements in your area are, you may need to search the internet for water tank information from one or more of the following sources or entities:

- Plumbing Code of Australia (NCC Vol. 3) including state appendices
- local councils
- local water authorities
- local and state/territory health departments
- the state/territory plumbing licensing regulator
- state/territory government climate authorities
- state/territory water tank monetary rebate systems.

Some authorities produce detailed documents and regulations covering water tank installation, roof drainage connection and the connection of fixtures to water tank supplies. In all cases you must also ensure adherence to the Plumbing Code of Australia and AS/NZS 3500.1 'Water supply'. Remember that strict regulatory requirements apply to all plumbing work, and you must hold a licence relevant to the type of work in which you are engaging.

PLUMBING CODE OF AUSTRALIA PART B6 AND AS/NZS 3500.1 'WATER SUPPLY'

Rebate schemes

There are many cost rebate schemes in place around Australia that property owners can access to offset the cost of system installation. In order that your customer is eligible for any available financial support, installations must satisfy certain tank sizes and applications. It is your responsibility to ensure you are conversant with what is on offer in your own region and confirm this with the customer prior to job commencement.

Regulatory requirements and rebate options vary widely around the country. The questions in Table 18.2 will prompt you to think about your own situation. For current and future use, get on the internet or contact the authorities listed above and write down the details in the space provided.

LEARNING TASK 18.3

1 Prior to excavation for an underground tank, list at least 4 property services that will need to be located.
2 What are 3 sources of water contamination that you need to be looking for when evaluating a roof catchment area?

Water storage requirements

For new construction work, the specification documents will detail the required storage volume of any particular site. However, for many smaller jobs that entail the installation of water storage systems

TABLE 18.2 What are your local authority requirements for water tank installation?

Details of local authority requirements	
Which local authority issues guidelines for water tank installation in your area?	Authority name.. Website ....
If a water tank rebate scheme exists in your area, who administers it?	Authority name.. Website ....
Which codes or standards apply to water tank installation in your area?	
What licensing requirements exist in your area relating to: • the installation of a water tank? • the connection of a water tank to a roof drainage system? • the connection of a water tank to the point of use?	

to existing premises, you may need to calculate the requirements yourself. The HB 230–2008 provides some additional sizing considerations relating to water storage capability.

To determine the system's water storage volume, you need to follow a series of basic calculations. Each of these steps is represented in the chart in Figure 18.3, and detailed below.

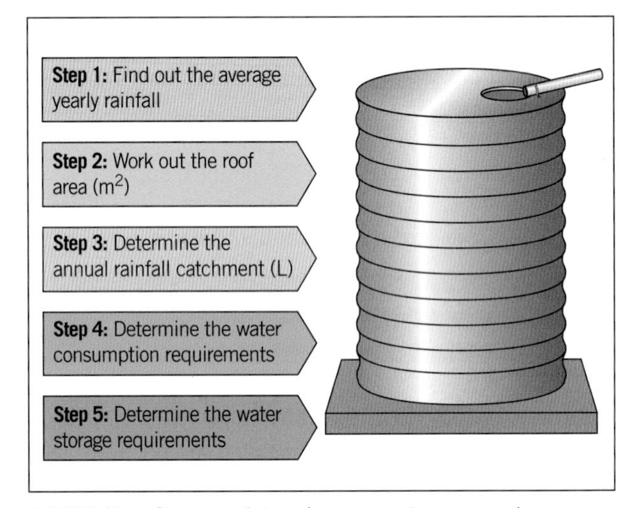

FIGURE 18.3 Steps to determine water storage needs

Step 1: Determine yearly rainfall

In a country the size of Australia, rainfall patterns and averages are regional and subject to considerable variation. Average rainfall figures for your particular area will act as a rough guide to the expected rainfall over a period of time, but they will not directly account for periods of extended drought or higher than normal rainfall. Figures can be sourced as both averages and median results. Median data is regarded as being more accurate for this application.

To find out what the average or median rainfall is in your particular area, there are at least two primary sources of information:

- local council
- Australian Bureau of Meteorology – go to www.bom.gov.au

Data is available for monthly and yearly rainfall averages over a range of years. The map of rainfall averages shown in Figure 18.4 has been compiled from data collected over a 30-year period. Similar information can be found for your own particular region. The HB 230–2008 also includes median monthly and yearly rainfall data for a selection of major centres and these will be referred to in this chapter.

Account for regional variability

The average yearly rainfall for major population centres is shown in Table 18.3. In certain areas an allowance for rainfall intensity will need to be taken into account when sizing your water tank.

For example, if you refer to the monthly rainfall data in the HB 230–2008 you will see that the rainfall in Hobart is historically uniform, with little variation in rainfall averages from month to month. Under normal circumstances regular rainfall patterns such as this will be a reliable guide to gross potential water collection capability during most times of the year.

Compare this with Darwin, where most rainfall is concentrated during the monsoon 'wet season', but very little during the 'dry'. In this case, referral only to the annual figure may be misleading as the distribution of this rain is not even across the year. During the wet season, tanks sized only on the yearly average may overflow and much of the yearly water supply may be lost in a short period of time. Therefore, an extra sizing allowance based on local experience may need to be considered to ensure that the household has the capacity to collect sufficient water during peak rainfall periods so that they will have enough to get through the dry times.

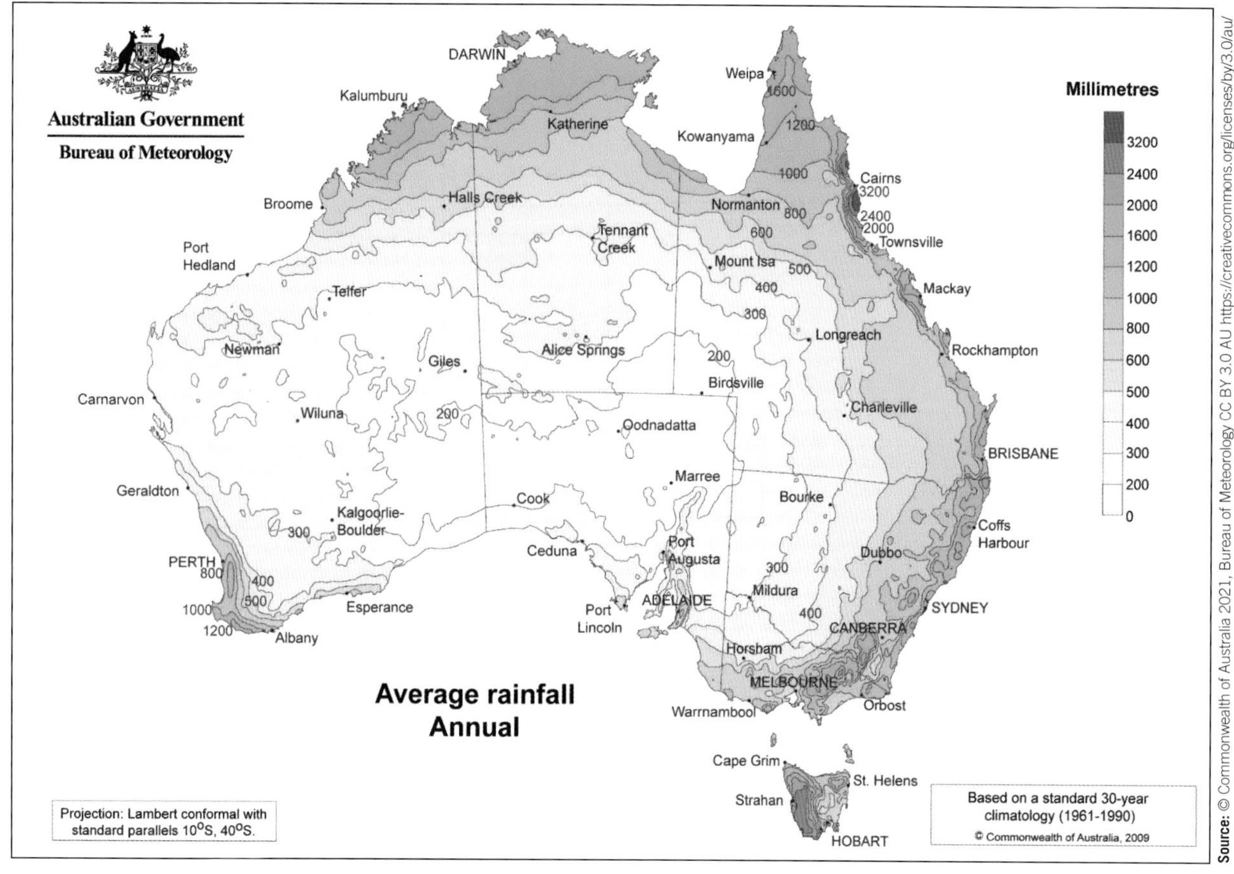

FIGURE 18.4 Annual rainfall averages around Australia

TABLE 18.3 Major city average rainfall

City	Average yearly rainfall (mm)
Adelaide	553
Alice Springs	274
Brisbane	1189
Canberra	631
Darwin	1666
Hobart	624
Melbourne	661
Perth	869
Sydney	1220

Source: © Bureau of Meteorology.

Step 2: Work out the roof area (m^2)

For the purposes of working out the area of roof that acts as a rainwater catchment, it is general practice to treat this as equivalent to the area of ground covered by a roof in square metres. You do not need to account for roof pitch, shape or profile.

Note that not all installations will drain the entire roof to the water tank. Only calculate the area of roof that drains towards the downpipes that supply the tank. Refer to Figure 18.5 to see how this works. Add all catchment areas of the roof together to arrive at a final figure in square metres.

Step 3: Determine the annual rainfall catchment (L)

Under perfect conditions the relationship between the catchment area and the amount of rainfall recorded is as follows:

1 mm of rain over a catchment area of 1 m^2 is equal in volume to 1 litre of water

This is shown pictorially in Figure 18.6.

Of course, rainfall does not collect on roof catchment areas in a perfect way. The following losses must be accounted for.

Actual run-off collected

For any amount of water that rains upon a roof area a certain percentage will be lost to wind gusts and splashing. Furthermore, the actual roof material has an impact. New steel roof sheeting and well-drained gutters are quite impervious and will allow a high percentage of total water that falls on the catchment to flow to the tank. On the other hand, roof materials such as unglazed concrete and terracotta tiles, old painted sheets and fibre cement slates will absorb a considerable amount of water before creating significant run-off. This effect is more pronounced in lower rainfall areas that characteristically experience light rainfall events.

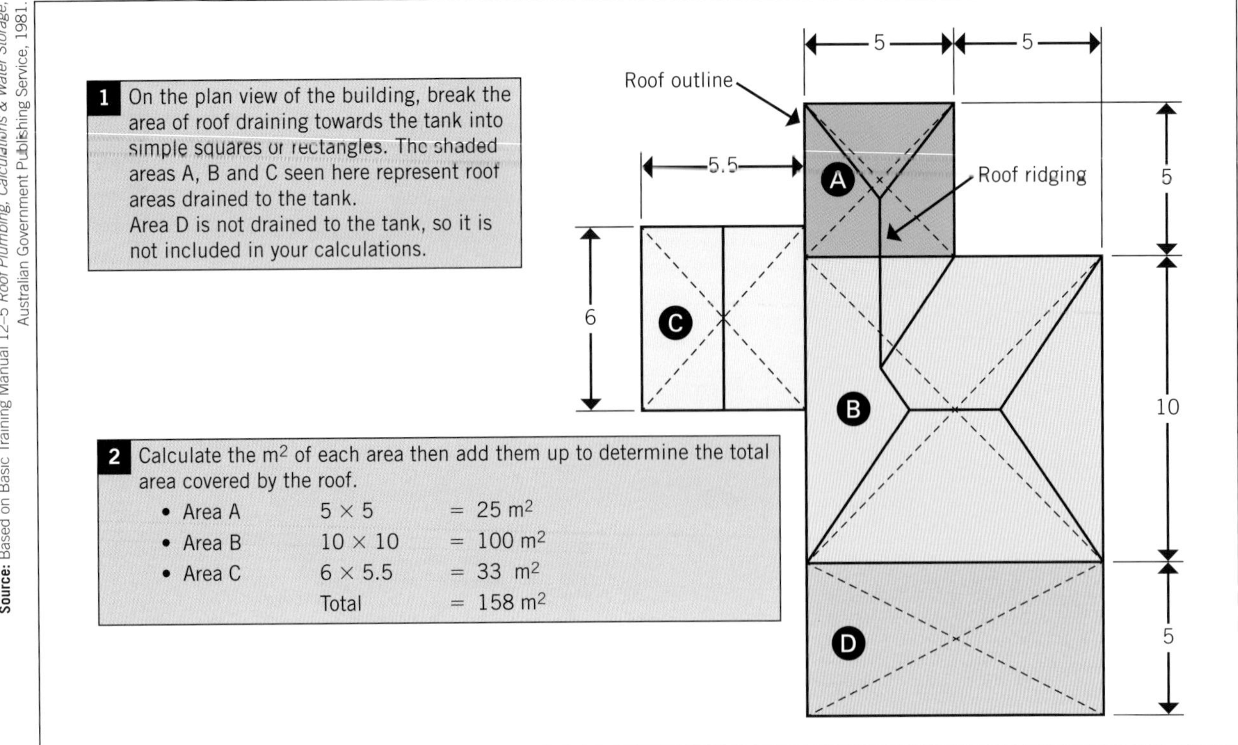

Source: Based on Basic Training Manual 12-5 *Roof Plumbing, Calculations & Water Storage*, Australian Government Publishing Service, 1981.

FIGURE 18.5 Calculating how much area the roof covers in square metres

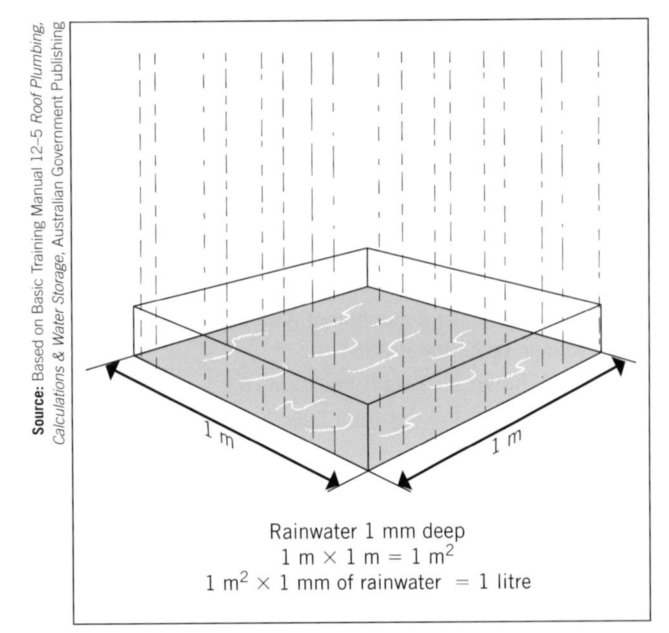

Source: Based on Basic Training Manual 12-5 *Roof Plumbing, Calculations & Water Storage*, Australian Government Publishing

FIGURE 18.6 The relationship between roof catchment area and rainfall

Run-off losses can be accounted for by using a coefficient for run-off that reduces the amount of collected water by a certain percentage. The actual coefficient is somewhat arbitrary and will always be subject to local experience. A combination of the figures in Example 18.1 may be useful to try out.

EXAMPLE 18.1

A coefficient of 90% may be appropriate for:

- new steel roof and gutter
- glazed tiles
- higher intensity regular rainfall events
- steep pitched roofs

A coefficient of 60–80% may be appropriate for:

- old painted roofs and flatter gutter installations
- unglazed concrete or terracotta tiles
- fibre cement roofs and gutters
- lower intensity regular rainfall events
- lower pitched roofs

Pre-treatment losses

To reduce the volume of roof and gutter contaminants reaching the tank, most modern tank systems include a pre-treatment 'first-flush' device that will divert the first 20 L or so of a rainfall event away from the tank.

Therefore, in calculating the actual rainfall catchment area, a correction for these losses is built into the following formula:

Maximum litres rainfall per year (L) = $(r \times a \times c) - d$

Where:

r = annual rainfall (mm)

a = catchment area (m^2)

c = coefficient of run-off (%)

d = diverted water total (L)

Using the annual median rainfall for Brisbane and the roof size example shown in Figure 18.5, follow Example 18.2 to see how this works.

EXAMPLE 18.2

Calculate the annual rainfall catchment (step 3)

- Median yearly rainfall for Brisbane: 886 mm
- Roof type – new steel sheet at 22.5° pitch. Assume 90% coefficient.
- Pre-treatment – 30 rainfall events at 20 L loss per event

Max rainfall (L) = (r × a × c) – d

Where:

r – annual rainfall (mm)

a – catchment area (m^2)

c – coefficient of run-off (%)

d – diverted water total (L)

Max rainfall (L) = (886 × 158 × 0.9) – (30 × 20)

= 125 989 – 600

= 125 389 L

Try this example again, but now insert the median or average rainfall for your own area into the formula.

This figure now represents an approximate volume of water you might expect a particular roof to collect under average conditions and basically tells you what you are working with. There is little point in catering to consumption and storage expectations that exceed the actual potential catchment. However, it should again be emphasised that this is only an approximate calculation, and it is subject to a number of variables. Additional examples are included in the HB 230–2008.

Step 4: Determine the water consumption requirements

Many factors influence the amount of water that a household will use. These include:

- the number of users
- what the water is used for:
  - drinking water only
  - washing only
  - garden use only
  - all fixtures
  - combined uses
- whether the stored water is the only source or whether it is supplementary to town water
- the behavioural habits of the user/s.

To determine the water use per year you may need to conduct a water use audit, or find data that will be sufficient for your calculations, as in Example 18.3. Such data may be sourced from household water bills, water authorities, councils and manufacturers.

Indicative water consumption data is included in the HB 230–2008 and will enable you to work out basic figures for your calculations. This includes:

- garden watering – drought tolerant
- garden watering – high water requirements
- toilet water use
- top- and front-loader washing machine consumption.

Step 5: Determine the water storage requirements

In these examples so far, you have done the following:

- worked out the average/median annual rainfall (mm)
- determined the roof area (m^2)
- calculated how much rainwater such a roof can collect (L) less the loss coefficient
- estimated the yearly consumption of tank water and confirmed that it is less than the potential catchment volume.

With this data, you can now size a tank or tanks to hold the required volume. However, this volume of water is not necessarily equivalent to the size of the tank required.

In fact, the volume of the tank/s to be installed on a property is determined by how long water needs to be stored, based on the probable period without rain. While the consumption per year or per day will not change, the frequency of rain and intervals of no rain will affect the volume of water that must be kept on hand during the dry period.

To determine the capacity of the tank/s, you need to carry out the following calculation, as evidenced in Example 18.4:

Tank/s volume (L) = Daily consumption (L) × Minimum storage days with no rain

Purchasing a tank of the right size

Having worked out your required storage volume, the best option is to take your requirements to a tank supplier and ask for some solution options.

They will need to know the following:

- available base area
- available height
- required stored volume
- access dimensions
- material and colour
- outlet and inlet size options.

Some manufacturers will be able to make tanks that are specific to your needs, while others will have a range of 'off-the-shelf' solutions from which you can choose.

EXAMPLE 18.3

Determine water consumption requirements (step 4)

You have been asked to connect a water tank to the house described in Example 18.1. The rainwater will only be used to wash clothes in an 8-kg front-loading washing machine. There are four people living at the residence. Reference to water use data reveals a probable total consumption of 25 584 L per year (see Figure 18.7).

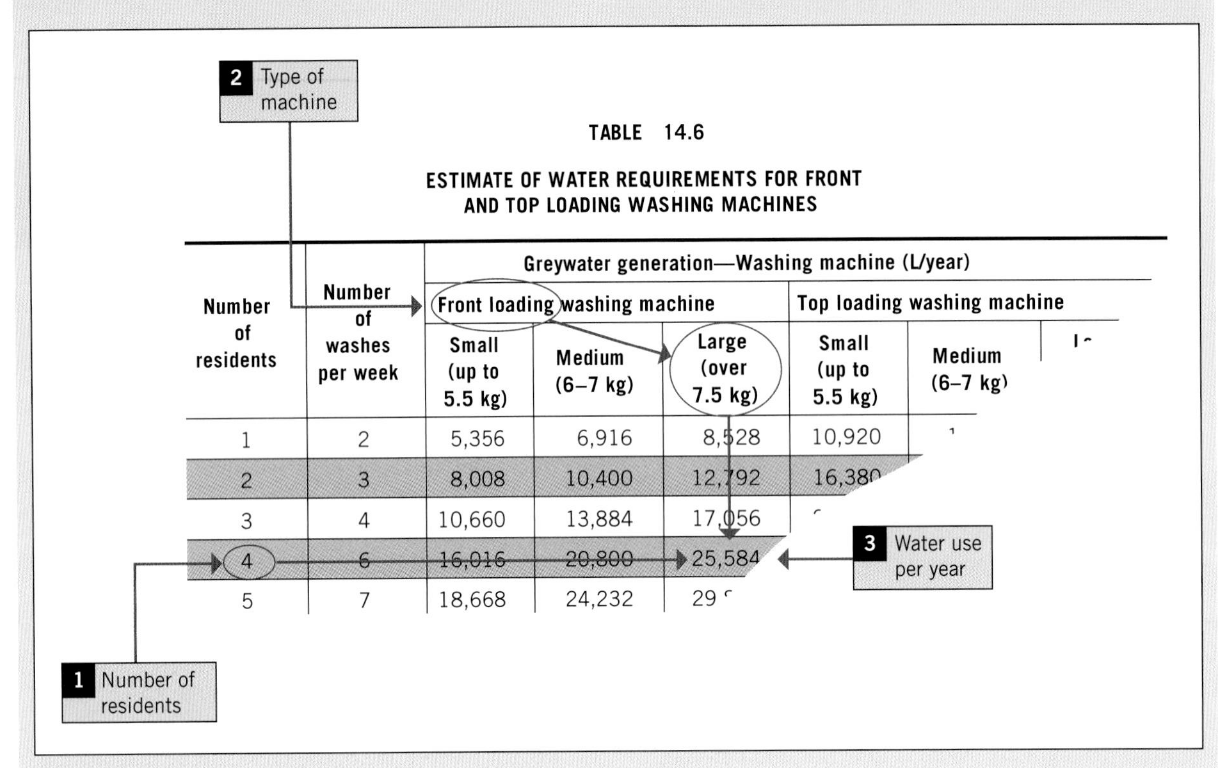

TABLE 14.6

ESTIMATE OF WATER REQUIREMENTS FOR FRONT AND TOP LOADING WASHING MACHINES

Number of residents	Number of washes per week	Greywater generation—Washing machine (L/year) Front loading washing machine: Small (up to 5.5 kg)	Front loading: Medium (6–7 kg)	Front loading: Large (over 7.5 kg)	Top loading washing machine: Small (up to 5.5 kg)	Top loading: Medium (6–7 kg)	L[illegible]
1	2	5,356	6,916	8,528	10,920		
2	3	8,008	10,400	12,792	16,380		
3	4	10,660	13,884	17,056			
4	6	16,016	20,800	25,584			
5	7	18,668	24,232	29[illegible]			

FIGURE 18.7 Working out water consumption requirements

Source: Based on information contained in the HB 203-2008 *Rainwater Tank Design and Installation Handbook* (2008), National Water Commission.

Comparing this result against the water catchment volume for Brisbane of 125389 L, as seen in step 3 – Example 18.1, you can see that this is well above the estimated use of 25584 L per year.

EXAMPLE 18.4

Determine water storage requirements (step 5)

- Estimated yearly use for clothes washing: 25584 L per year
- Estimated daily use for clothes washing: 25584 L ÷ 365 days = 70 L
- In this example the owners have asked for at least 6 months' supply.

Tank/s volume (L) = Daily consumption (L) × Minimum storage days with no rain

= 70 × (6 months × 30 days)

= 70 × 180

= 12 600 L

Depending on available space and system requirements, you may opt for a single tank of this volume, or multiple tanks to the equivalent capacity.

Safety factor

With the variable nature of the climate, you or the owners may wish to apply a safety factor to the calculation. A safety factor of 1.5 or 2 would be advisable. In this case, simply multiply the safety factor by the tank volume in litres:

Tank/s volume with safety factor of 2 = 12 600 L × 2

= 25 200 L

Where you need to work out the volume of a tank that might occupy a particular space, you need to apply the following formula, as shown in Example 18.5:

Volume of a cylinder (L) = $\pi r^2 h \times 1000$

EXAMPLE 18.5

Calculate the volume of a tank

For example, the set-back distance between a house and boundary means that you can only fit a tank having a maximum diameter of 1.8 m and a maximum height of 2.3 m. To work out the volume of such a tank, you apply the formula:

Volume of a cylinder (L) = $\pi r^2 h \times 1000$

$= (3.142 \times 0.9^2 \times 2.3) \times 1000$

$= (3.142 \times 0.81 \times 2.3) \times 1000$

$= 5.853\ 546\text{m} \times 1000$

$= 5853$ L

LEARNING TASK 18.4

1. What are the five steps required to determine water storage needs?
2. Which government agency website can you get the average and median yearly rain falls for your area from?
3. If we had a 1 m^2 impervious catchment area and received 10 mm of rain, how many litres would we have?

Installation considerations

This section will review some of the differences between water tank systems and cover some of the job requirements related to standard water tank installations.

Safety

As with all jobs, when installing water storage systems you need to apply appropriate hazard identification, risk control and general safety procedures. Refer to 'Basic roof safety' in Chapter 3 for revision of this subject.

Additional considerations particular to water tank installation include some or all of the following:

- *Manual handling.* Tanks are often very large and heavy. Trying to move something that is too heavy can result in serious strains and injury. The use of levers and rollers and/or the need to find more labour may need to be considered. Some tanks will require the use of machinery and cranes. Loads must be secured and manoeuvred by suitably qualified workers with the appropriate industrial licensing. Keep clear and be aware of what is going on at all times.
- *Working at heights.* You may find yourself working at heights when connecting the downpipe system to the tank inlet or where a tank is situated on an elevated tank stand. All relevant fall prevention processes and equipment must be used.
- *Confined spaces.* Where you are engaged in the maintenance or construction of a large concrete tank or any other tank that is fabricated onsite, you may end up working in what is classed as a confined space. Confined spaces present considerable hazards relating to air supply, dangerous gases and emergency access and exit. Where a hazard assessment identifies a confined space, only suitably trained staff may enter this area. Check with your supervisor for any clarification.
- *Trench work.* Unfortunately, deaths from trench collapse occur with tragic regularity. Where holes are excavated for underground tanks, the sides of the hole must be shored up or battened to a slope less than the angle of repose for the particular soil type.

  In addition, before any excavation, always ensure that the area has been checked for underground services such as power, gas, water and sewer lines.
- *Wind.* Wind can be a real problem with tank handling and installation. Apart from concrete, most tank materials are light in weight relative to their size. An unfilled tank can be easily dislodged by gusts of wind and may cause damage to the product and/or injury to those around it. Caution during the unloading procedure must be exercised. During storage and installation, ensure that tanks are securely lashed down to prevent dislodgement by wind. Subject to current water use restrictions, fill the tank as soon as possible.

Never enter a trench or hole where no shoring is in place! Trench sides can collapse without any warning.

Installation components

In addition to the usual plumbing fittings, modern water tank installations make use of two common components that work together to keep the water in a storage tank as clean as possible. These are *screened rainheads* and *first-flush diversion valves*.

Screened rainhead

This component is designed to collect water from a gutter outlet or downpipe while preventing leaves and foreign objects from entering the tank. The screened mesh that stops leaves from entering the system is angled so that leaves or objects do not build up and create a blockage, but simply fall to the ground (see **Figure 18.8**).

Screened rainheads are also specified as a form of vermin-proof overflow/air-break at the tank overflow outlet. Water can overflow freely from the tank without animals such as frogs and rodents crawling up the stormwater drain into the tank. Water from a blocked stormwater drain can also easily escape via this air gap so that the tank water is not contaminated.

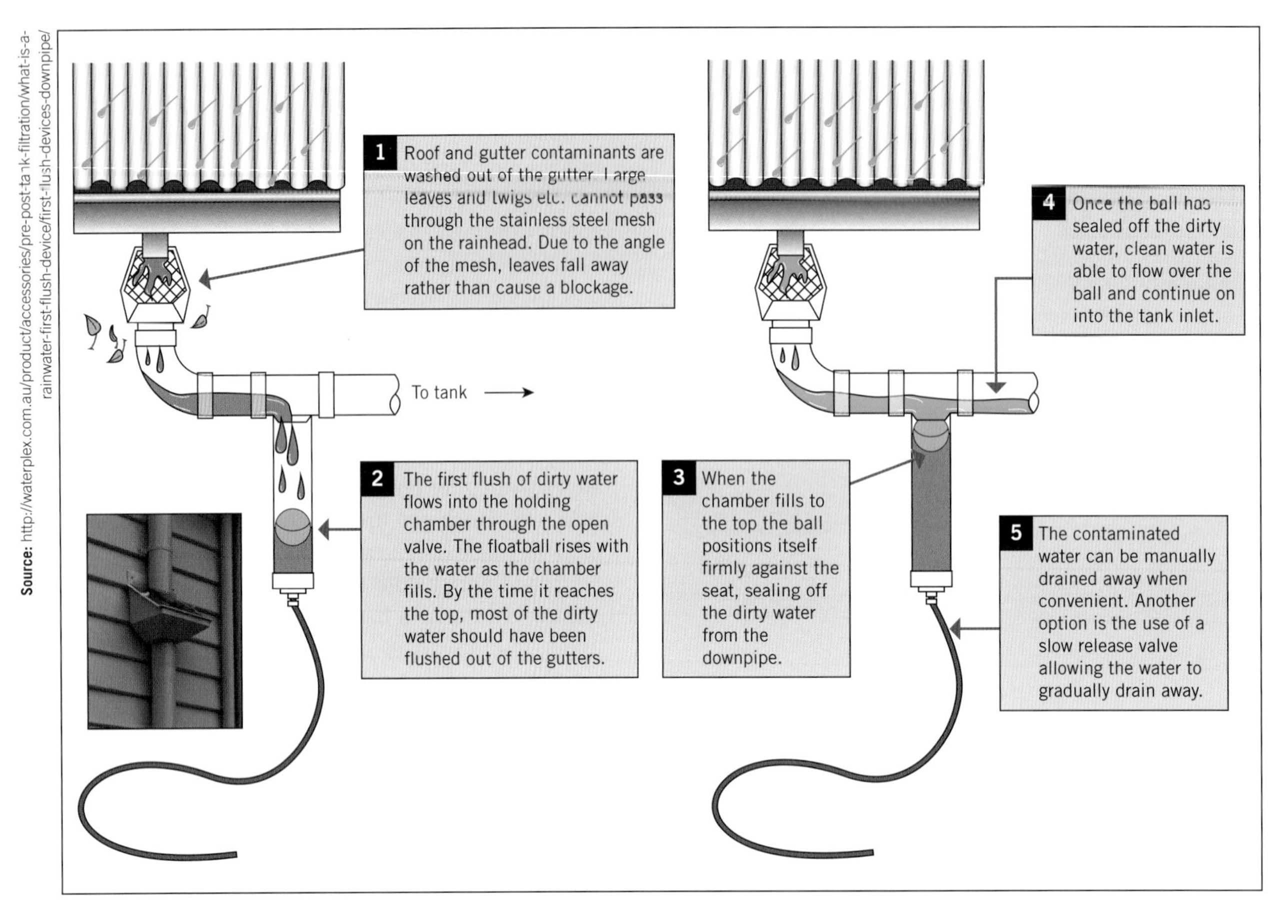

Source: http://waterplex.com.au/product/accessories/pre-post-tank-filtration/what-is-a-rainwater-first-flush-device/first-flush-devices-downpipe/

FIGURE 18.8 Operation of a screened rainhead and first-flush diversion valve

First-flush diversion valve

During extended periods between significant rainfall, the roof and gutters can build up with contaminants that would affect the quality of stored rainfall should they enter the tank. Such contaminants include:

- leaves
- litter
- dust
- industrial fallout
- droppings from cats, possums, birds and other animals.

The first-flush diversion valve is designed to capture the first 20–25 L of rainwater that has flushed most of the contaminants from the roof and gutters and divert it away from the tank.

Once the valve is full, it is designed to restore normal flow to the storage tank. The valve can be mounted on the wall below the gutter outlet or it may be installed just before the inlet of the tank itself. Underground first-flush devices are also available for 'wet' system connections to tanks.

Water captured within the valve can be manually released or it may be slowly drained to the garden through a slow-release valve fitted at the bottom of the chamber. Underground devices will divert the water straight to the stormwater connection.

A number of designs have been developed, but one of the most common and simple forms of this valve incorporates a floating ball and seat valve operation. This can be seen in Figure 18.8.

Inlet and outlet requirements

In order to prevent access from vermin and insects, in particular mosquitoes, specific requirements relate to all inlets, outlets and any access points for the tank. These include:

- use of removable and cleanable metallic screens to keep insects out
- vermin-proof flap valves on all openings.

Installation types

In relation to the siting of a rainwater tank, installations can be classified in three separate groups:

- on ground
- above ground
- below ground.

Some generic illustrations of such installations are included in this chapter. More detail and variations are also available in your HB 230–2008. Always read these in reference to local regulatory requirements.

In addition to these descriptions, which are based on the position of the tank, the installation can also be described by the form of connection used to connect the tank to the roof drainage system. These are:

- 'dry' connection – graded and open downpipes
- 'wet' connection – charged and sealed downpipes.

Examples of these installation types are given below.

On-ground installations

On-ground installations are those where a tank sits upon a manufacturer-approved base at ground level. The tank base may be one of the following:

- Concrete slab – A slab must be strong enough to handle the weight of the full tank and should be higher than the surrounding natural ground surface to prevent the pooling of water. Some tank manufacturers may recommend that their particular product also sit on hardwood slats to provide airflow between the slab and the base of the tank. Others may require a plastic or bituminous membrane to be placed under the tank. You must check the specifications relevant to the tank material type and the application.
- Sand/aggregate base – Tanks made from polyethylene, fibreglass and concrete can be placed on a sand or fine aggregate base. It is important that this base is deep enough to prevent any sharp rocks from protruding through and possibly puncturing the base of the tank.

'Dry' connection – graded and open downpipes

A water storage tank supplied with a 'dry' downpipe connection is one where the downpipe is run directly to the inlet of the tank on-grade and does not hold any water (see Figure 18.9). A key advantage of a dry

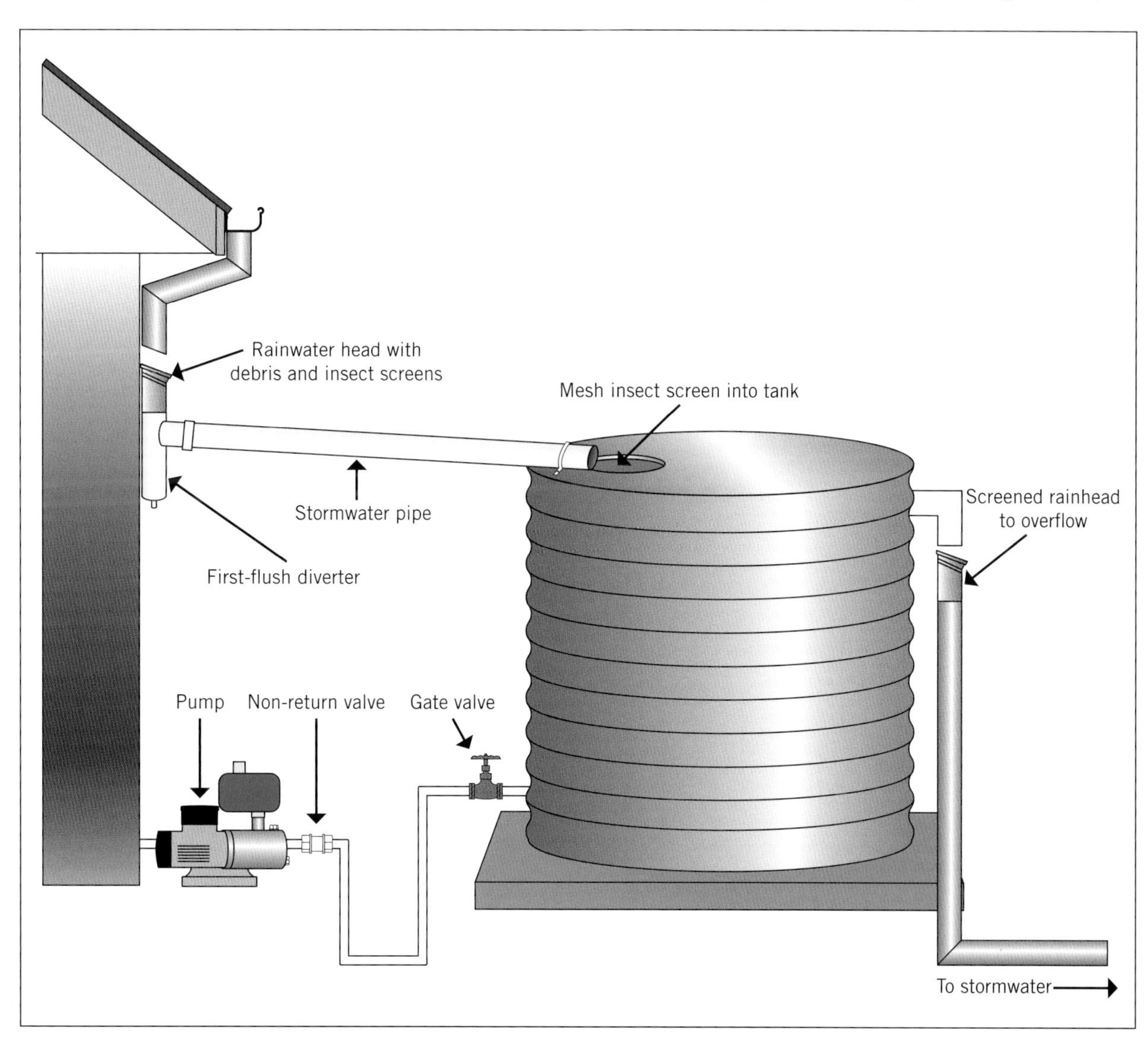

FIGURE 18.9 An on-ground tank with a 'dry' connection

connection is that there is no stagnant water held in the pipes that could become contaminated with pathogens or breed mosquitoes.

Dry connection systems require that the inlet to the tank be situated below the gutter or first-flush diversion valve outlet.

'Wet' connection – charged and sealed downpipes

Sometimes the storage tank is too far from the main building to allow for a direct graded connection. Downpipes from the other side of the house may also be connected into the one tank, making it impossible to use a 'dry' system. In this instance, a 'wet' downpipe connection is installed (see Figure 18.10).

The downpipes are connected to an underground pipe/s that then makes its way back up to the tank inlet. Water will always find its own level and therefore will simply fill the entire stormwater and downpipe system up to the level of the tank inlet. After rain, the pipes will remain charged full of water or 'wet'.

HB 230-2008 SECTION 5.9.4 'ROOF DRAINAGE'

All joints in the downpipes and stormwater lines must be fully sealed in accordance with the manufacturer's specifications. There must be at least 100 mm between the outlet of the screened rainhead and the tank inlet. There must also be a flush point installed at the lowest part of the system so that silt and stagnant water can be drained periodically.

Note that if the rainwater in the tank is going to be used for drinking, then all downpipes and underground stormwater lines must be made from pipe that is compliant with AS/NZS 4020 'Testing of products for use in contact with drinking water'. The use of some

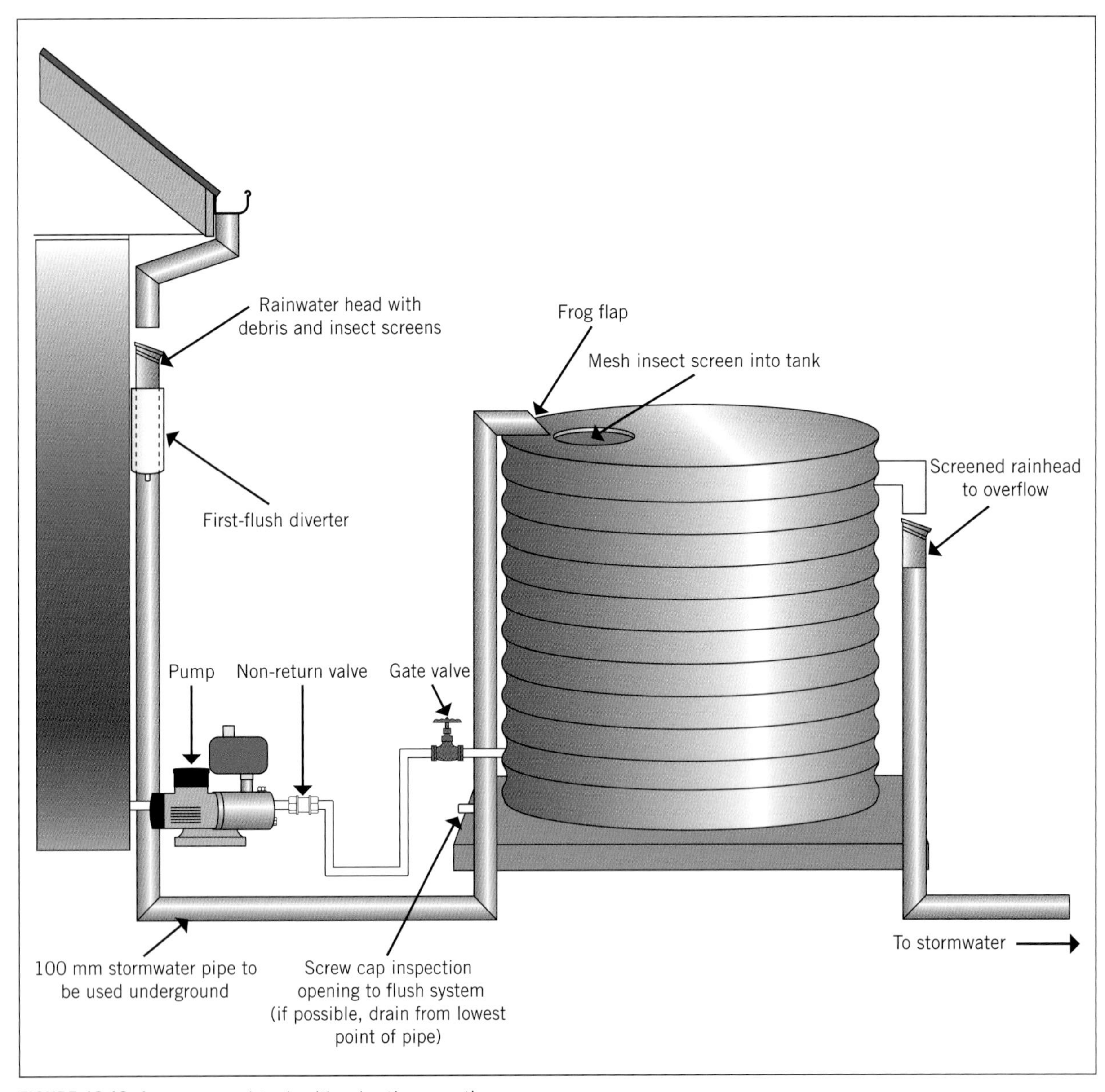

FIGURE 18.10 An on-ground tank with a 'wet' connection

non-compliant plastics may result in lead and other contaminants leaching into the drinking water.

Above-ground installations

Some high-set and split-level buildings may have roofs that are high enough to allow the water storage tank to be elevated on a stand, so that a gravity water supply may be drawn off to a point of use.

Particularly in rural areas, another system features a large-volume storage tank, sitting at ground level, supplying a smaller header tank on a stand with a float-switch activated pump. The header tank provides a gravity supply to all building draw-off points and is topped up when the water level is low enough to activate the float switch. Such a system can be seen in Figure 18.11.

Pressure head calculations

All fixture outlets and many appliances require a certain minimum pressure in order to operate effectively. Where you are installing any form of gravity supply system you must be able to calculate available pressure.

The higher the water tank, the greater the pressure or head at the outlet. The head of pressure is measured between the height of the water level and the outlet point. For every metre of height that water is stored above an outlet, the pressure increases 9.81 kPa (see Figure 18.12).

EXAMPLE 18.6

Calculate pressure head

a If the water level in an above-ground tank was 6 m above the laundry tap, you would calculate the outlet pressure as follows:

6 m × 9.81 kPa = 58.86 kPa

b A client asks you to install a header tank on the hill slope behind the main house. The tank must be high enough to provide at least 200 kPa to the house taps. How high up the slope must the tank be situated?

200 kPa ÷ 9.81 kPa = 20 m (rounded down)

Tank support

A fabricated and elevated tank stand is a Class 10b structure that must comply with the National Construction Code and will need to be constructed according to local authority permit requirements. This will normally be done by an accredited builder or steel fabricator. A litre of water weighs 1 kg, therefore a tank stand supporting 2000 L of stored volume will be actually holding over 2 tonnes of water weight alone.

The stands must be made of approved durable material and metal stands must be galvanised or appropriately coated. All stands must be designed to withstand wind action.

Below-ground installations

Some jobs require the installation of below-ground tanks (see Figure 18.13). Concrete tanks that are either pre-cast or fabricated onsite are commonly used. Polyethylene tanks are also available in various sizes and designs. Below-ground tanks offer the advantage of saving space and will keep the water at a cool and constant temperature.

Below-ground tank installations must have full building approvals in place, and the approved plans and specifications documents should detail all installation requirements. Some key points need to be considered in relation to below-ground installations:

- Where the ground surrounding a below-ground tank becomes saturated with water, hydrostatic pressure from the water is exerted against the sides and base of the tank. Unless the tank is constructed and anchored correctly, these forces will cause the tank to 'float' and pop out of the ground, a process known as hydrostatic uplift. To mitigate against hydrostatic uplift, tanks must be anchored as per the manufacturer's instructions prior to backfill around the tank, and all inlet and outlet pipes must be connected immediately to the tank so that it can be filled. The combined weight of the tank and water will prevent the tank from floating should it rain.
- Installations in areas with reactive clay soils should include pipe expansion joints to allow for movement between the tank and the pipes.
- All joints and openings into the tank must be made watertight to ensure that ground water does not seep in.
- Overflow connections must incorporate a reflux valve to prevent any build-up of water downstream of the overflow from making its way back to the tank.

REFER TO HB 230-2008 SECTION 5.15 'UNDERGROUND RAINWATER TANKS'

LEARNING TASK 18.5

1 What are two common components that work together to keep the water in a storage tank as clean as possible?
2 Rainwater tank installations can be classified in three separate groups. What are they?
3 A water storage tank where the downpipe is run directly to the inlet of the tank on-grade and does not hold any water is known as what type of connection?
4 A water storage tank where the downpipe is run underground to the inlet of the tank and holds water is known as what type of connection?

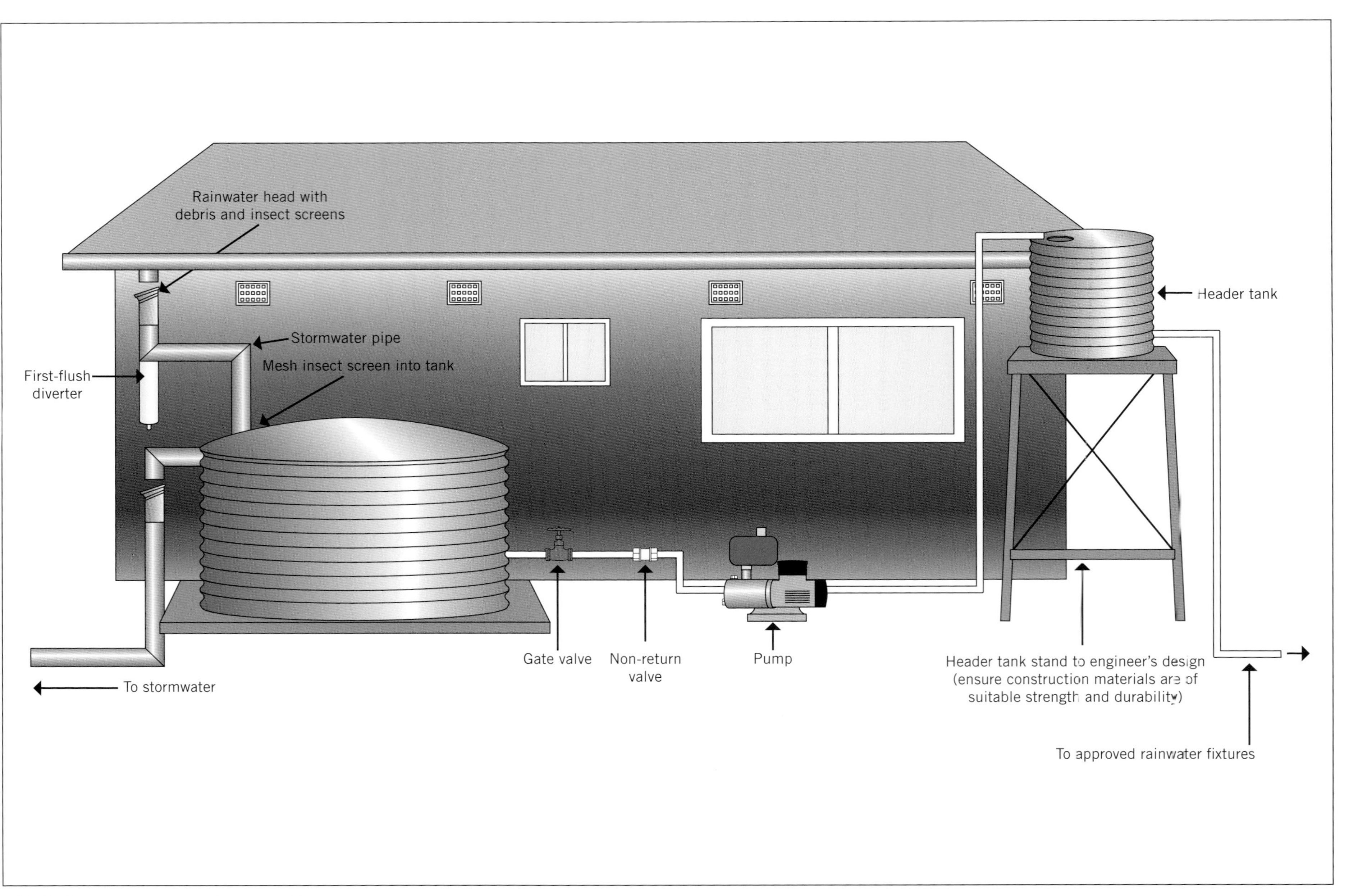

FIGURE 18.11 An above-ground system with a header tank

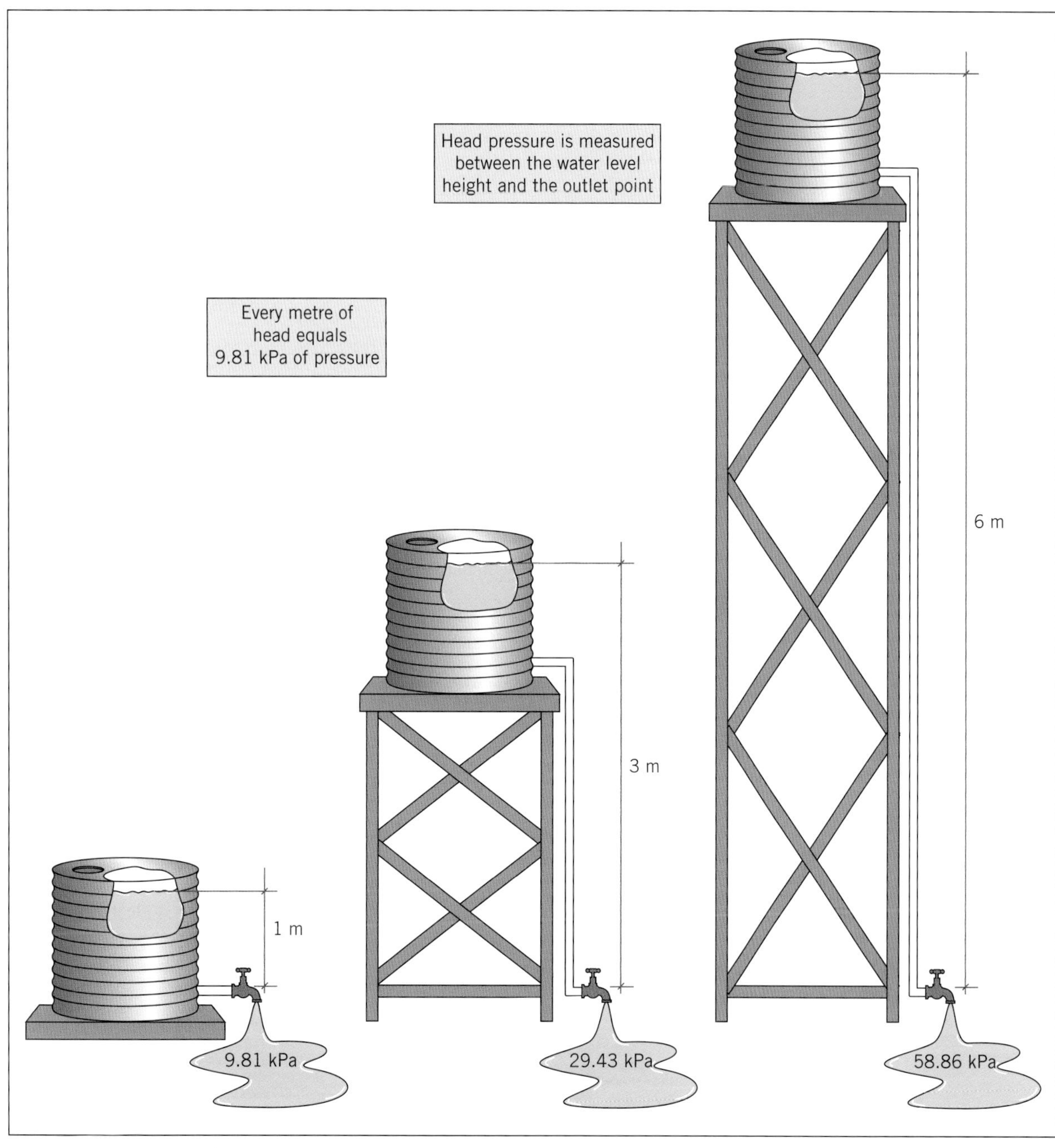

FIGURE 18.12 Relationship between height and pressure

Tank maintenance

A well-designed and correctly installed rainwater tank system should require minimal maintenance. The cleaner the catchment system is kept, the less major work is required in the longer term. You should recommend to the owner that regular attention and cleaning of the following should be observed:

- Check the roof for any build-up of dust and debris.
- Clean the gutters when necessary.
- Check that all gauze filters on inlets, rainheads and first-flush devices are in good condition and clear of debris.
- Clean first-flush diverters on a regular basis.
- Drain charged downpipe systems from the lowest flush point on a regular basis or as often as required to remove silt and stagnant water.
- Conduct a regular internal inspection for algae, debris, sediment and mosquitoes, and rectify the cause of these when necessary.

It is recommended that the tank be drained every three years to remove any sediment build-up. Draining of the tank should be done in accordance with local council requirements in respect to the prevention of water wastage.

Where tanks need to be treated to prevent a re-occurrence of algae or pathogens, the process should

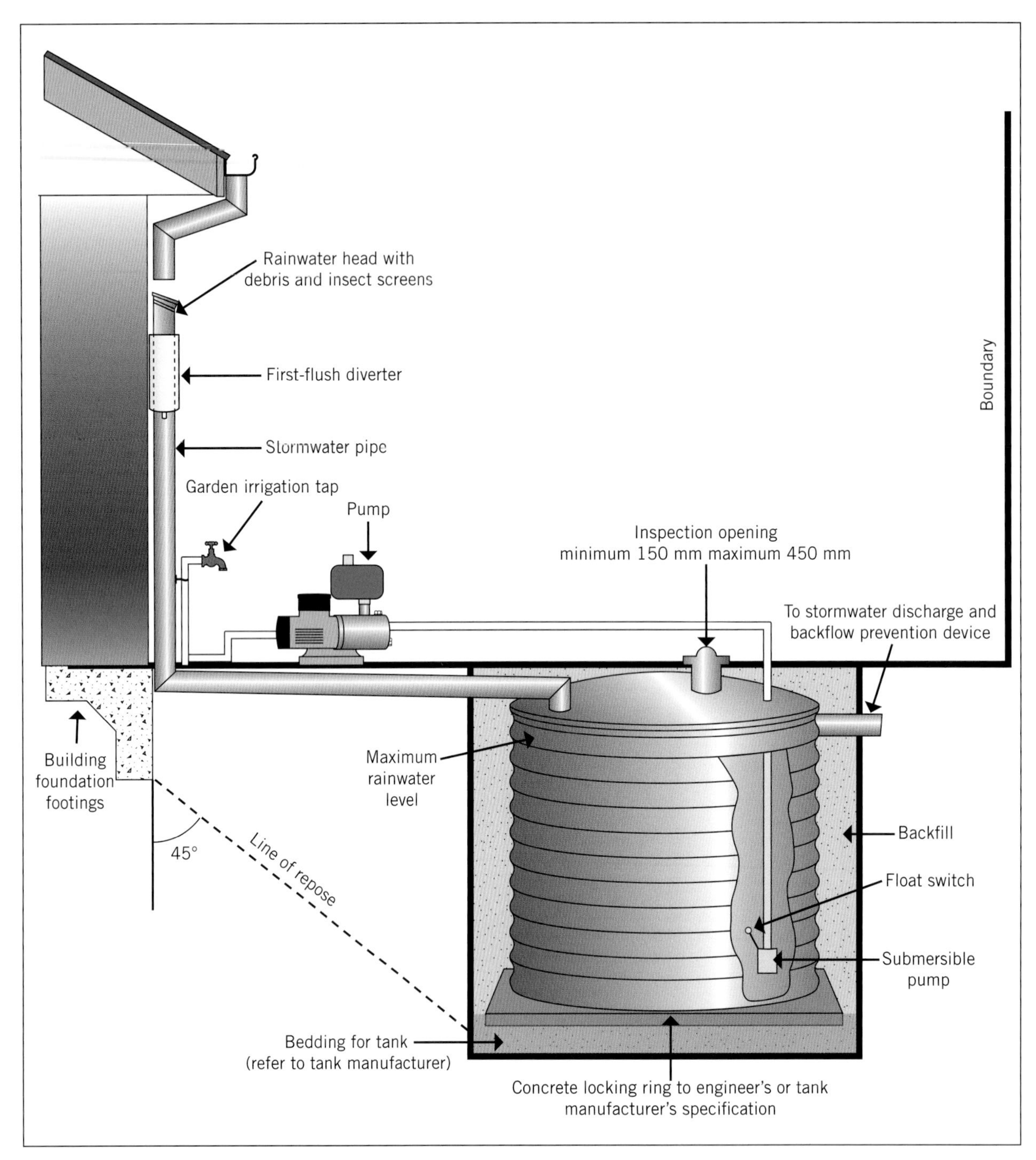

FIGURE 18.13 A below-ground tank installation

be done in reference to AS/NZS 3500.1 Appendix G and any local health department requirements.

Clean up

Remember that the job is not over until your work area is cleaned up. Cleaning up as you go creates a safer work environment and leaves less work for you to do at the end of the job.

The installation of rainwater tanks will generate a range of waste materials that need to be dealt with appropriately. Importantly, all waste must be cleaned from the roof, gutters and downpipes to prevent contamination of the water supply. Waste materials include:

- metal and plastic offcuts
- metal swarf
- trimmings
- rivet shanks.

Any offcuts that are large enough to be re-used or recycled should be placed in the appropriate place in accordance with company directives. Packaging debris is usually recyclable, so ensure that all waste is disposed of accordingly.

Ensure that all tools and equipment are checked for serviceability, and where faults are detected, report the details to your supervisor.

SUMMARY

In this chapter you have been provided with a basic introduction to rainwater tank installation requirements. The use of rainwater tanks will only increase into the future and if pursued may form a significant part of your work. You have seen throughout this chapter the importance of compliance with the Plumbing Code of Australia and relevant codes and standards in reference to:

- material selection and use
- calculation of rainwater catchment and the process in determining stored volume based upon consumption patterns and projected periods of no rainfall
- connection system types
- tank mounting options and compliance
- basic maintenance requirements.

COMPLETE WORKSHEET 1

WORKSHEET 1

To be completed by teachers	
Satisfactory	☐
Not satisfactory	☐

Student name: ____________________

Enrolment year: ____________________

Class code: ____________________

Competency name/Number: ____________________

Task: Working with your teacher/supervisor, refer to this text, HB 230–2008 and any other relevant resource to answer the following questions.

1 What are at least eight uses for stored rainwater?

____
____
____
____

2 In relation to new housing construction in your area, if applicable, what are the mandatory rainwater storage requirements?

____
____
____

3 What type of rainwater supply connection system can you identify from this photograph?

____
____
____
____
____
____
____
____
____
____
____
____

4 You need to purchase a polyethylene water tank for a domestic, non-drinking water, on-ground installation. When deciding what brand to buy, with what Australian Standard should you check that the tank is compliant?

5 A customer wants you to install a replacement galvanised corrugated water tank to a farm shed. Before ordering a new galvanised tank, what should you check for during a site visit and why?

6 What is the purpose of a first-flush diversion valve?

7 Who is the only person permitted to enter a large underground water tank?

8 What is the roof area of A and B in this plan view?

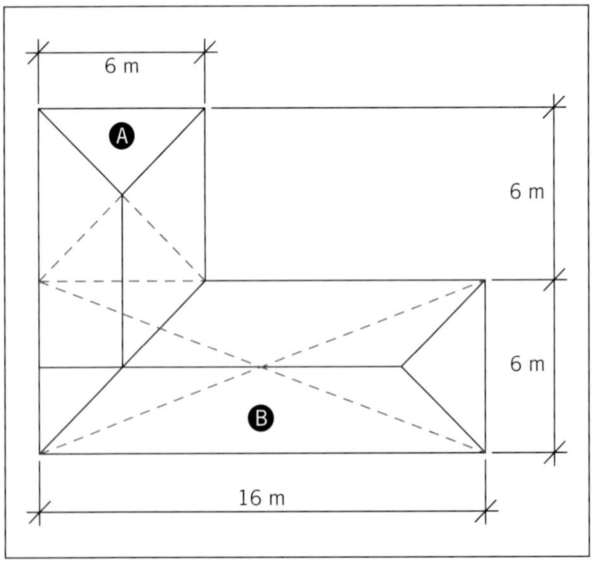

9 Using the answer from Question 8, determine the annual rainfall catchment of such a roof in your own area.

10 What should the stored volume of rainwater be for the following installation?

- yearly use of rainwater: 50 000 L
- 8 months' supply required with no rain
- safety factor of 1.5 applied

11 What would be the weight of stored water in a tank holding 5000 L?

12 Why should an air-gap be maintained at the tank overflow outlet?

13 You are installing a 'wet' system connection to an on-ground water tank that will supply drinking water to the household. With what Australian Standard should your pipes be compliant?

14 Calculate the water pressure at an outlet where the water level in an elevated header tank is 8 m above the outlet.

15 Having manoeuvred a polyethylene tank into position on a slab, what safety precaution should you immediately observe?

16 What is the name of the phenomenon that would cause an empty and poorly anchored below-ground tank to pop out of the ground?

____

____

____

17 What should be cleaned from a rainwater tank at least every three years?

____

____

____

18 What would be two advantages of a below-ground tank installation?

____

____

____

19 What should be done as part of the clean up or immediately before filling the tank with water?

____

____

20 Calculate the following rainwater catchment potential:

You have been asked to install an above-ground tank for a new dwelling in Bendigo (Vic). The following details apply:

- roof catchment area to be drained to tank – 180 m^2
- run-off coefficient – 90%
- pre-treatment diversion – 20 L per rainfall event (assume 20 rainfall events)

____

____

____

____

____

____

21 You are planning to install a rainwater tank/s for an existing 4-person household in Mount Gambier (SA). The property is connected to mains water supply so rainwater will only be used for toilet flushing, washing clothes and to water a drought-tolerant garden. Using your HB 230–2008, what is the probable water use estimate for the following:

Water demand	Water requirement per year (L)
6/3 dual flush toilet	
6-kg front-loader washing machine	
100-m^2 drought-tolerant garden	
Total	L

GLOSSARY

180* barrel vault roof the sheets of this style form a complete 180* arc from one side of the roof to the other

3:4:5 rule a rule based on the relationship between each side of a right-angle triangle: if each side of a triangle is equal to multiples of the units 3:4:5, then the angle must be square

A

ACM (asbestos-containing materials) products that have been identified as using asbestos in their manufacture

Act a written and approved law that has been passed through a state or federal parliament

Alloy a combination of two or more metals, usually to create desirable characteristics in a new matrix

Anti-capillary break a fold made in sheet metal with the purpose of breaking the force of molecular adhesion that permits water to rise up between close fitting surfaces

Asbestos a naturally occurring mineral, primarily known for its very effective resistance to heat and chemical breakdown, which led to its widespread use as insulation in many domestic, commercial and industrial applications

Australian Standards detailed technical documents that set the minimum design and performance standard for equipment and systems

B

Barge capping covers the area of a barge board, which is normally found at the intersection of a wall top with gable or skillion roof sheeting; it not only keeps water out but protects the edge of the roof sheets from wind lift and damage

Bonded ACM products manufactured with asbestos fibres that are mixed and firmly bound within the material itself; e.g. asbestos within cement sheets (fibro)

Box gutter a rectangular-shaped graded channel located within the building line that drains water to sumps and rainwater heads

Box louvre ventilators mounted on roofs to provide a large but protected opening for natural draught ventilation

Bullnose roof a pre-curved roof profile that has been popular in Australia since the colonial era. Most often used on verandas, the radius curve of the sheet into the gutter protects the building walls and windows against the sun's rays while still maintaining adequate roof height under the veranda

Butterfly roof two skillion style roofs that fall towards each other meeting above a central box gutter

C

Capillary action when the gap between two surfaces is small enough, the combination of water surface tension and the forces of molecular adhesion between the water and the close physical surfaces acts to lift the liquid upwards

Capping any flashing that sits on top of a lower flashing, wall or roof sheet intersection; e.g. parapet flashing, ridge capping, barge capping

Cladding lightweight wall sheeting material, often used in locations or circumstances where traditional construction methods and materials would be too expensive or heavy

Code of Practice a practical guide for achieving the requirements and standards stipulated in the Act and Regulations

Composite roof combination of a hip and gable roof and in some instances other roof structural variations

Concealed-fixed roof and wall cladding system a roofing system secured to the structure with the use of special clips that do not pass through the sheet; the clips are located beneath the sheet itself and are screwed directly to the purlin

Continuous ventilator; *see* ridge/slope ventilator

Corrosion the breakdown of a metal due to chemical reactions with its surroundings

Cover width *see* effective cover

Cupola effectively the same as a box louvre, though often more decorative in appearance and commonly seen on larger colonial-and Federation-era buildings

Curved roof any roof incorporating curved roof sheets as part of its structure

D

Decking a name often used for the metallic profile sheeting covering a roof

Deemed-to-satisfy solution a method of prescriptive compliance to which a plumbing installation must be designed. Use of the AS/NZS 3500 is a 'deemed-to-satisfy' solution

Downpipe carries the water from the roof to an approved point of discharge or storage

Dry downpipe connection downpipes that are run directly to a water storage tank on grade and do not hold water on a permanent basis

Dry-pan flashing (flat-tray flashing) a type of over-flashing where the sheet is inserted underneath the wall flashing or ridge capping and extends down to and around the penetration in one piece

Ductile the ability of a metal to be stretched under tensile stress

Durable in relation to roof plumbing materials, a product's ability to withstand corrosion and the effects of the weather and environment

Dutch gable a hip roof where the ridge has been extended past the normal intersection with the hips to form a small triangular gable

E

Eaves gutter generally installed external to the building line at the end of eaves or overhangs; some are installed internally and concealed behind a fascia. Roof water flows into the eaves gutter and is drained towards the downpipes

Effective cover (cover width) the difference between the actual width of the sheet, which is the measurement from one side of the sheet to the other, and the measurement of the amount of cover provided on the roof less the lap on one side

F

Fall calculating the overall angle of decline expressed as a ratio from the highest to lowest point

Fasteners the products used to secure or fasten another roofing product to the building structure

Fixings a form of bracket, clip or strap used to secure a roof sheet or rainwater product to the building structure

Flashings used to form a watertight seal at the intersections of roof sheets to walls, ridgelines, penetrations and valleys

Flat-tray flashing *see* dry-pan flashing

Friable ACM asbestos fibres only loosely bound within the parent material that, when dry, can be crumbled, pulverised or reduced to powder by hand pressure; e.g. brake linings, insulation, degraded roof sheets

G

Gable louvre a louvre installed vertically in the gable wall of a building

Gable roof a roof laid in two planes between the ridgeline and the eaves on either side

Galvanic series the ranking of metals in descending order according to their level of electrolytic activity

Gambrel roof a roof with two different pitches on either side of a ridgeline falling to the eaves gutter

Gravity intake ventilators non-powered, passive devices that operate through both thermal convection and the pressure differential that exists between the inside and outside of buildings

Gutter types include eaves gutters (also known as 'spouting' in some areas), concealed gutters, valley gutters and box gutters

Gutter fall is almost always expressed as a ratio. This is a ratio that is read as fall:run (fall is to run)

H

Hip roof a roof laid in all four planes between the ridgeline and the eaves of each four walls of the building

Hyperbolic paraboloid roof a double curved roof that is designed to enable the use of straight sheets to create a compound curved shape

I, J

Jerkin head roof a roof where the upper third of the gable end is replaced by a small hip section

K, L

Lean-to roof the roof of a lower building that falls in one plane towards the gutter and abuts the wall of a higher structure

Low-pitch roof generally a roof that has a fall less than 10°, often between 1–3°

M

Malleable the ability of a metal to be compressed or deformed under compressive stress

Mansard roof a roof with two different pitches on all four sides of the building

Metal fascia supports the eaves lining and the gutter itself

N

National Construction Code (NCC) an initiative of the Council of Australian Governments (COAG) developed to incorporate all onsite construction requirements into a single code. The NCC is comprised of the Building Code of Australia (BCA), Volume I and II; and the Plumbing Code of Australia (PCA), Volume III

Neutral cure sealant a silicone rubber sealant that is not corrosive and will not damage roofing materials

Noble less-active metals at the bottom of the galvanic series listing

O

Old Gothic (Ogee/cyma recta) a pre-curved profile featuring a convex curve above a concave curve; this creates a flowing S-curve that is used by designers to provide more visual interest and shape to a roof design

Overflows used to prevent water overflowing the gutter and into the building due to component blockage or during unusually high rainfall intensity or duration

P

Parapet capping designed to protect the top of the parapet wall (a wall that extends above and past the roofline)

Patina the passive and protective oxidation layer that forms on the surface of metals such as bronze, copper and zinc

Penetrations of a roof include vent pipes, flues, chimneys, skylights, roof ventilators and ducting

Performance solution a method of compliance that adopts a 'performance-based approach', allowing more flexibility for plumbing installation design as long as the solution meets or exceeds the performance requirements detailed in the Plumbing Code of Australia

pH scale runs from 0 to 14: solutions less than 7 are acidic and solutions greater than 7 are alkaline. A liquid solution with a pH of 7 is known as neutral (e.g. pure water)

Pierce-fixed roof and cladding system a roofing system secured to the building structure with fasteners that pass through the sheet

Pitch of a roof can be expressed and measured as either degrees or a ratio from the horizontal

Ponding occurs when a section of roof or gutter retains water for extended periods, which can cause premature deterioration of the material

Profiles the various shapes of eaves gutters, represented and named in relation to the cross-sectional shape of the gutter

R

Rainwater head (rainhead) a fabricated component connected to the top of a downpipe installed external to a building at the end of a box or within an eaves gutter run. A rainwater head increases the head of water at the entry to the downpipe, increasing its drainage capacity

Regulations typically are used to provide more detail and context in the application of the broader Act that they sit beneath

Relief ventilators *see* gravity intake ventilators

Rib The longitudinal fold or shape formed in a profile roof sheet

Ridge capping a capping required to prevent water ingress at the intersection where two roof planes meet at a ridge or hip; it extends an equal distance on either side of the ridge centreline

Ridge/slope ventilator ('continuous' ventilator) a prefabricated unit installed along the ridge of a building or along the slope of a roof; this system relies on the natural thermal convection movement of warm air rising to the highest point of the structure and then being vented to atmosphere

Roll-formed roof sheets custom roof sheets that take the exact dimensions of the roof from the approved plans and are created from flat sheets run through a special roll-forming machine; used when curves are too acute and the metal would buckle if you tried to spring-curve a standard high-tensile steel sheet around such a shape

Rotary turbine ventilators commonly found on domestic, commercial and industrial buildings, a unit consists of a bearing-mounted rotating turbine made from overlapping vanes that sits on top of a tube or turret and flashing

R-value a measure of thermal resistance, related to the thickness of the insulation material and its inherent thermal characteristics; the higher the figure, the better its potential performance

S

Sacrificial protection describes the chemical phenomenon where one metal will 'sacrifice' itself to protect another metal from corrosion

Saw-tooth roof commonly an industrial-style roof constructed in a saw-tooth configuration of multiple parallel roof slopes falling to box gutters

Scribing in tropical areas, a method of turning-down flashings above corrugated roof sheets which are cut into the profile

Skillion roof a stand-alone roof with sheets that fall in one plane towards the gutter

Spreader a discharge pipe that spreads water flow, used from a higher roof onto a lower roof to avoid a high volume of water discharging from the higher roof onto one point of the lower roof

Spring-arched (convex) roof while the spring-curved ridge roof is only curved over the ridge area itself, this roof is laid in a single radius curve from one eave to the other

Spring-arched roof a sprung-roof covering created when standard flat roofing sheets are gradually fixed to a purpose-designed roof featuring a gentle, free-form curve from eave to eave

Spring-curved (concave) roof the opposite of a convex roof: the roof sheets are spring-curved inwards onto specially located purlins. As with a convex roof, there is a minimum roof radius onto which the sheets are fixed

Spring-curved ridge roof created when flat sheets are fixed and curved over the ridge of a gable roof; the ridge is the only part of the sheet that is curved, while the rest of the sheet length remains flat

Spring-curved roof a sprung-roof covering which can be made across the ridge of the roof with the remainder of the sheets remaining flat

Sump installed across the full sole (base) of a box gutter, which creates a concentrated water collection point that increases the efficiency of water flow into the downpipe

Swarf the name given to the small particles of metal that are left behind from drilling and cutting processes

T, U, V

Valley gutter an inclined channel used to drain water away from the intersection of two sloping roof planes and direct it to eaves or box gutters

W, X, Y, Z

Wet downpipe connection a downpipe system that is deliberately designed to be permanently full of water

INDEX

* Page numbers in **bold** represent defined terms
* Page numbers in *italics* represent figures

D

E

F